Joseph Wolstenholme

Mathematical Problems

On the First and Second Divisions of the Schedule of Subjects. Second Edition

Joseph Wolstenholme

Mathematical Problems
On the First and Second Divisions of the Schedule of Subjects. Second Edition

ISBN/EAN: 9783337141615

Printed in Europe, USA, Canada, Australia, Japan

Cover: Foto ©berggeist007 / pixelio.de

MATHEMATICAL PROBLEMS

ON

THE FIRST AND SECOND DIVISIONS OF THE SCHEDULE OF SUBJECTS

FOR THE

CAMBRIDGE MATHEMATICAL TRIPOS EXAMINATION.

DEVISED AND ARRANGED

BY

JOSEPH WOLSTENHOLME, M.A.

LATE FELLOW AND TUTOR OF CHRIST'S COLLEGE;
SOMETIME FELLOW OF ST JOHN'S COLLEGE;
PROFESSOR OF MATHEMATICS IN THE ROYAL INDIAN ENGINEERING COLLEGE.

"Tricks to shew the stretch of human brain,
Mere curious pleasure, or ingenious pain."
POPE, *Essay on Man*.

SECOND EDITION, GREATLY ENLARGED.

London:
MACMILLAN
1878

PREFACE TO THE FIRST EDITION.

This "Book of Mathematical Problems" consists, mainly, of questions either proposed by myself at various University and College Examinations during the past fourteen years, or communicated to my friends for that purpose. It contains also a certain number, (between three and four hundred), which, as I have been in the habit of devoting considerable time to the manufacture of problems, have accumulated on my hands in that period. In each subject I have followed the order of the Text-books in general use in the University of Cambridge; and I have endeavoured also, to some extent, to arrange the questions in order of difficulty.

I had not sufficient boldness to seek to impose on any of my friends the task of verifying my results, and have had therefore to trust to my own resources. I have however done my best, by solving anew every question from the proof sheets, to ensure that few serious errors shall be discovered. I shall be much obliged to any one who will give me information as to those which still remain.

I have, in some cases, where I thought I had anything serviceable to communicate, prefixed to certain classes of problems fragmentary notes on the mathematical subjects to which they relate. These are few in number, and I hope will be found not altogether superfluous.

This collection will be found to be unusually copious in problems in the earlier subjects, by which I designed to make it useful to mathematical students, not only in the Universities, but in the higher classes of public schools.

I have to express my best thanks to Mr R. Morton, Fellow of Christ's College, for his great kindness in reading over the proof sheets of this work, and correcting such errors as were thereby discoverable.

NOTICE TO THE SECOND EDITION.

THE present edition has been enlarged by the addition of such other problems from my accumulated store as seemed to myself worthy of preservation. About one hundred of these, and probably the most interesting, have appeared in the mathematical columns of the "Educational Times," and many of the others have already been used for Examination purposes. The "Fragmentary Notes" have been increased, and I hope improved. Answers are given in the great majority of cases and sometimes hints for the solution. I have taken much pains to avoid mistakes, and although, from the nature of the case, I dare not venture to expect the errors to be few in number, I hope they will not often be found of much importance. The greater number of the proof sheets have been read over by my colleague, Professor Minchin, and many improvements are due to his suggestion. I am deeply grateful for his kind and efficient help.

I shall be thankful for information of misprints or other mistakes which are not in the list of Errata.

R. I. E. COLLEGE, *Nov.* 3, 1878.

CONTENTS.

CONIC SECTIONS, GEOMETRICAL.

CONIC SECTIONS, ANALYTICAL.

SOLID GEOMETRY.

STATICS.

DYNAMICS, ELEMENTARY.

DYNAMICS OF A PARTICLE.

DYNAMICS OF A RIGID BODY.

ERRATA.

Page 73, question 415 is wrong.

—— 107, line 6 from the bottom, for "all values of θ" read "values of θ between $-\frac{\pi}{2}$ and $\frac{\pi}{2}$."

—— 108, line 10, for $+336n+162$, read $-336n+164$.

—— 109, line 5, dele the second "that."

—— 211, question 1267, insert "Prove that."

—— 278, line 9, for $x+a_1$, $x+a_2$ read $x+b_1$, $x+b_2$.

—— 289, question 1686, for u^{1-2n} read u^{2n-1}; and for x^{1-2n}, read x^{2n-1}.

—— 327, line 6 from the bottom, for $m(m-1)$, read $m(n-1)$.

—— ,, line 3 from the bottom, for $+\ldots+dx_n$, read $\ldots\ldots dx_{2n}$.

—— 430, question 2519, for $\pi-\phi$, read $\pi+\phi$.

—— 444, line 3 from the bottom, for 23 read 32.

—— 446, line 2 from the bottom, for M read N.

GEOMETRY.

1. A point O is taken within a polygon $ABC...KL$; prove that OA, OB,...OL are together greater than half the perimeter of the polygon.

2. Two triangles are on the same base and between the same parallels; through the point of intersection of their sides is drawn a straight line parallel to the base and terminated by the sides which do not intersect: prove that the segments of this straight line are equal.

3. The sides AB, AC of a triangle are bisected in D, E, and CD, BE intersect in F: prove that the triangle BFC is equal in area to the quadrangle $ADFE$.

4. AB, CD are two parallel straight lines, E the middle point of CD, and F, G the respective points of intersection of AC, BE, and of AE, BD: prove that FG is parallel to AB.

5. Through the angular points of a triangle are drawn three parallel straight lines terminated by the opposite sides[1]: prove that the triangle formed by joining the ends of these lines will be double of the original triangle.

6. If a, b, c be the middle points of the sides of a triangle ABC, and if through A, B, C be drawn three parallels to meet bc, ca, ab respectively in A', B', C', the sides of the triangle $A'B'C'$ will pass through A, B, C respectively, and the triangle ABC will be double of the triangle $A'B'C'$.

7. In a right-angled triangle the straight line joining the right angle to the centre of the square on the hypotenuse will bisect the right angle.

8. Through the vertex of an equilateral triangle is drawn a straight line terminated by the two straight lines drawn through the ends of the base at right angles to the base, and on this straight line as base is described another equilateral triangle: prove that the vertex will lie either on the base of the former or on a fixed straight line parallel to that base.

[1] All straight lines are supposed to be produced if necessary.

9. Through the angle C of a parallelogram $ABCD$ is drawn a straight line meeting the two sides AB, AD in P, Q: prove that the rectangle under BP, DQ is of constant area.

10. In any quadrangle the squares on the sides together exceed the squares on the diagonals by the square on twice the line joining the middle points of the diagonals.

11. If a straight line be divided in extreme and mean ratio and produced so that the part produced is equal to the smaller of the segments, the rectangle contained by the whole line thus produced, and the part produced together with the square on the given line will be equal to four times the square on the larger segment.

12. Two equal circles touch at A, a circle of double the radius is drawn having internal contact with one of them at B and cutting the other in two points: prove that the straight line AB will pass through one of the points of section.

13. Two straight lines inclined at a given angle are drawn touching respectively two given concentric circles: their point of intersection will lie on one of two fixed circles concentric with the given circles.

14. A chord CD is drawn at right angles to a fixed diameter AB of a given circle, and DP is any other chord meeting AB in Q: prove that the angle PCQ is bisected by either CA or CB.

15. AB is the diameter of a circle, P a point on the circle, PM perpendicular on AB; on AM, MB as diameters are described two circles meeting AP, BP in Q, R respectively: prove that QR will touch both circles.

16. Given two straight lines in position and a point equidistant from them, prove that any circle through the given point and the point of intersection of the two given lines will intercept on the lines segments whose sum or whose difference will be equal to a given length.

17. A triangle circumscribes a circle and from each point of contact is drawn a perpendicular to the straight line joining the other two: prove that the straight lines joining the feet of these perpendiculars will be parallel to the sides of the original triangle.

18. From a fixed point O of a given circle are drawn two chords OP, OQ equally inclined to a fixed chord: prove that PQ will be fixed in direction.

19. Through the ends of a fixed chord of a given circle are drawn two other chords parallel to each other: prove that the straight line joining the other ends of these chords will touch a fixed circle.

20. Two circles with centres A, B cut each other at right angles and their common chord meets AB in C; DE is a chord of the first circle passing through B: prove that A, D, E, C lie on a circle.

21. Four fixed points lie on a circle, and two other circles are drawn touching each other, one passing through two fixed points of the four and the other through the other two: prove that their point of contact lies on a fixed circle.

22. A circle A passes through the centre of a circle B: prove that their common tangents will touch A in points lying on a tangent to B; and conversely.

23. On the same side of a straight line AB are described two segments of circles, AP, AQ are chords of the two segments including an angle equal to that between the tangents to the two circles at A: prove that P, Q, B are in one straight line.

24. The centre A of a circle lies on another circle which cuts the former in B, C; AD is a chord of the latter circle meeting BC in E and from D are drawn DF, DG to touch the former circle: prove that G, E, F lie on one straight line.

25. If the opposite sides of a quadrangle inscribed in a circle be produced to meet in P, Q, and if about two of the triangles so formed circles be described meeting again in R: P, R, Q will be in one straight line.

26. Two circles intersect in A and through A any two straight lines BAC, $B'AC'$ are drawn terminated by the circles: prove that the chords BB', CC' of the two circles are inclined at a constant angle.

27. If two circles touch at A and PQ be any chord of one circle touching the other, the sum or the difference of the chords AP, AQ will bear to the chord PQ a constant ratio.

28. Four points A, B, C, P are taken on a circle and chords PA', PB', PC' drawn parallel respectively to BC, CA, AB: prove that the angles APA', BPB', CPC' have common internal and external bisectors.

29. Two circles are drawn such that their two common points and the centre of either are corners of an equilateral triangle, P is one common point and PQ, PQ' tangents at P terminated each by the other circle: prove that QQ' will be a common tangent.

30. On a fixed diameter AB of a given circle is taken a fixed point C from which perpendiculars are let fall on the straight lines joining A, B to any point of the circle: prove that the straight line joining the feet of these perpendiculars will pass through a fixed point.

[If D be this fixed point and O the centre, the rectangle under OC, OD will be half the sum of the squares on OC, OA.]

31. Four points are taken on a circle and the three pairs of straight lines which can be drawn through the four points intersect respectively in E, F, G: prove that the three pairs of straight lines which bisect the angles at E, F, G respectively will be in the same directions.

32. Through one point of intersection of two circles is drawn a straight line at right angles to their common chord and terminated by the circles, and through the other point is drawn a straight line equally inclined to the straight lines joining that point to the extremities of the former straight line: prove that the tangents to the two circles at the points on this latter straight line will intersect in a point on the common chord.

33. Two circles cut each other at A and a straight line BAC is drawn terminated by the circles; with B, C as centre are described two circles each cutting at right angles one of the former circles: prove that these two circles and the circle of which BC is a diameter will have a common chord.

34. Circles are described on the sides of a triangle as diameters and each meets the perpendicular from the opposite angular point on its diameter in two points: prove that these six points lie on a circle.

35. The tangents from a point O to a circle are bisected by a straight line which meets a chord PQ of the circle in R: prove that the angles ROP, OQR are equal.

36. A straight line PQ of given length is intercepted between two straight lines OP, OQ given in position; through P, Q are drawn straight lines in given directions intersecting in a point R, and the angles POQ, PRQ are equal and on the same side of PQ (or supplementary and on opposite sides): prove that R lies on a fixed circle.

37. From the point of intersection of the diagonals of a quadrangle inscribed in a circle perpendiculars are let fall on the sides: prove that the sum of two opposite angles formed by the straight lines joining the feet of these perpendiculars is double of one of the angles between the two diagonals.

38. If OP, OQ be tangents to a circle, PR any chord through P, then will QR bisect the chord drawn through O parallel to PR.

39. Two chords AB, AC of a circle are drawn and the perpendicular from the centre on AB meets AC in D: prove that the straight line joining D to the pole of BC will be parallel to AB.

40. A circle is drawn subtending given angles at two given points: prove that its centre lies on a fixed circle with respect to which the two given points are reciprocal; and conversely that if a circle be drawn with its centre on a given circle and subtending a given angle at a fixed point it will also subtend a fixed angle at the reciprocal point.

41. Prove the following construction for finding a point P in the base BC of a triangle ABC such that the ratio of the square on AP to the rectangle under the segments BP, PC may be equal to a given ratio:—Take O the centre of the circle ABC and divide AO in O' so that the ratio of AO' to $O'O$ may be equal to the given ratio, the circle whose centre is O' and radius $O'A$ will meet BC in two points each satisfying the required condition. If P, Q be the two points AP, AQ will be equally inclined to the bisector of the angle A and will coincide with this bisector when the given ratio has its least possible value, which is when $O'O$ is to AO in the duplicate ratio of BC to the sum of the other two sides. Also the construction holds if O' lie in OA produced, AP, AQ being then equally inclined to the external bisector of A and coinciding with it when the given ratio has its least possible value outside the triangle, which is when OO' has to OA the duplicate ratio of BC to the difference of the other two sides.

42. If a circle touch each of two other circles the straight line passing through the points of contact will cut off similar segments from the two circles.

43. Two circles have internal contact at A, a straight line touches one circle at P and cuts the other in Q, Q': prove that QP, PQ' subtend equal angles at A.

If the contact be external, PA bisects the external angle between QA, $Q'A$.

44. A straight line touches one of two fixed circles which do not intersect in P and cuts the other in Q, Q': prove that there are two fixed points at either of which PQ, PQ' subtend angles equal or supplementary.

45. Any straight line is drawn through one corner A of a parallelogram to meet the diagonal and the two sides which do not pass through A in P, Q, R: prove that AP will be a mean proportional between PQ, PR.

46. In a triangle ABC are given the centres of the escribed circles opposite B, C and the length of the side BC: prove that (1) A lies on a fixed straight line; (2) AB, AC are fixed in direction; (3) the circle ABC is given in magnitude; and (4) the centre of the circle ABC lies on a fixed equal circle.

47. Any three points are taken on a given circle and from the middle point of the arc intercepted between two of the points perpendiculars are let fall on the straight lines joining them to the third point: prove that the sum of the squares on the distances of the feet of these perpendiculars from the centre is double the square on the radius.

48. At two fixed points A, B are drawn AC, BD at right angles to AB and on the same side of it and of such magnitude that the rectangle AC, BD is equal to the square on AB: prove that the circles whose diameters are AC, BD will touch each other and that their point of contact will lie on a fixed circle.

49. ABC is an isosceles triangle right angled at C and the parallelogram $ABCD$ is completed; with centre D and radius DC a circle is described: prove that if P be any point on this circle the squares on PA, PC will be together equal to the square on PB.

50. A circle is described about a triangle ABC and the tangents to the circle at B, C meet in A'; through A' is drawn a straight line meeting AC, AB in the points B', C': prove that BB', CC' will intersect on the circle.

51. If D be the middle point of the side BC of a triangle ABC and the tangents at B, C to the circumscribed circle meet in A', the angles BAA', DAC will be equal.

52. The side BC of a triangle ABC is bisected in D, and on DA is taken a point P such that the rectangle DP, DA is equal to the rectangle BD, DC: prove that the angles BPC, BAC are together equal to two right angles.

53. If the circle inscribed in a circle ABC touch BC in D, the circles inscribed in the triangles ABD, DAC will touch each other. Also a similar property holds for the escribed circles.

54. Given the base and the vertical angle of a triangle: prove that the centres of the four circles which touch the sides of the triangle will lie on two fixed circles passing through the extremities of the base.

55. A circle is drawn through B, C and the centre of perpendiculars of a triangle ABC; D is the middle point of BC and AD is produced to meet the circle in E: prove that AE is bisected in D.

56. The straight lines joining the centres of the four circles which touch the sides of a triangle are bisected by the circumscribed circle; also the middle point of the line joining any two of the centres and that of the line joining the other two are extremities of a diameter of the circumscribed circle.

57. With three given points not lying in one straight line as centres describe three circles which shall have three common tangents.

58. From the angular points of a triangle straight lines are drawn perpendicular to the opposite sides and terminated by the circumscribed circle: prove that the parts of these lines intercepted between their point of concourse and the circle are bisected by the corresponding sides respectively.

59. The radii from the centre of the circumscribed circle of a triangle to the angular points are respectively perpendicular to the straight lines joining the feet of the perpendiculars.

60. Three circles are described each passing through the centre of perpendiculars of a given triangle and through two of the angular points: prove that their centres are the angular points of a triangle equal in all respects to the given triangle and similarly situated: and that the relation between the two triangles is reciprocal.

61. If the centres of two of the circles which touch the sides of a triangle be joined, and also the centres of the other two, the squares on the joining lines are together equal to the square on a diameter of the circumscribed circle.

62. The centre of perpendiculars of a triangle is joined to the middle point of a side and the joining line produced to meet the circumscribed circle: prove that it will meet it in the same point as the diameter through the angular point opposite to the bisected side.

63. From any point of a given circle two chords are drawn touching another given circle whose centre is on the circumference of the former: prove that the straight line joining the ends of these chords is fixed in direction.

64. ABC is a triangle and O the centre of its circumscribed circle; $A'B'C'$ another triangle whose sides are parallel to OA, OB, OC; and through A', B', C' are drawn straight lines respectively parallel to the corresponding sides of the former triangle: prove that they will meet in a point which is the centre of one of the circles touching the sides of the triangle $A'B'C'$.

65. A triangle is drawn having its sides parallel to the straight lines joining the angular points of a given triangle to the middle points of the opposite sides: prove that the relation between the two triangles is reciprocal.

66. Two triangles are so related that straight lines drawn through the angular points of one parallel respectively to the sides of the other meet in a point: prove that straight lines drawn through the corresponding angular points of the second parallel to the sides of the first will also meet in a point; and that each triangle will be divided into three triangles which are each to each in the same ratio.

67. The diameter AB of a circle is produced to C so that $BC = AB$, the tangent at A and a parallel to it through C are drawn and any point P being taken on the latter the two tangents from P are drawn forming a triangle with the tangent at A: prove that this triangle will have a fixed centroid.

68. A common tangent AB is drawn to two circles, CD is their common chord and tangents are drawn from A to any other circle through C, D: prove that the chord of contact will pass through B.

69. Four straight lines in a plane form four finite triangles: prove that the centres of the four circumscribed circles lie on a circle which also passes through the common point of the four circumscribed circles.

70. A triangle ABC is inscribed in a circle and AA', BB', CC' are chords of the circle bisecting the angles of the triangle (or one internal and two external angles) and meeting in E: prove that $B'C'$, $C'A'$, $A'B'$ respectively bisect EA, EB, EC at right angles: also the circles EBC', $EB'C$ will touch each other at E, EA being the common tangent.

71. Two of the sides of a triangle are given in position and the area is given; through the middle point of the third side is drawn a straight line in a given direction and terminated by the two sides: prove that the rectangle under the segments of this straight line is constant.

72. In the hexagon $AB'CA'BC'$ the three sides AB', CA', BC' are parallel, as are also the three BA', CB', AC': prove that AA', BB', CC' will meet in a point.

73. Two parallelograms $ABCD$, $A'BC'D'$ have a common angle B: prove that AC', $A'C$, DD' will meet in a point; or, if the parallelograms be equal, will be parallel.

74. On two straight lines not in the same plane are taken points A, B, C; A', B', C' respectively: prove that the three straight lines each of which bisects two corresponding segments on the two straight lines will meet in a point.

75. Four planes can be drawn each of which cuts six edges of a given cube in the corners of a regular hexagon, and the other six produced in the corners of another regular hexagon, whose area is three times that of the first, and whose sides are respectively perpendicular to the central radii drawn to the corners of the first.

76. Given the circumscribed circle and the centre of perpendiculars of a triangle, prove that the feet of the perpendiculars lie on a fixed circle, and the straight lines joining the feet of the perpendiculars touch another fixed circle.

77. Given the circumscribed circle of a triangle and one of the circles which touch the sides, prove that the centres of the other three circles which touch the sides will lie on a fixed circle.

78. If O, K be the centres of the circumscribed and inscribed circles of a triangle, L the centre of perpendiculars, and OK be produced to H so that OH is bisected in K, then will $HL = R - 2r$, where R, r are the radii of the two circles.

79. In any triangle ABC, O, O' are the centres of the inscribed circle and of the escribed circle opposite A; OO' meets BC in D, any straight line through D meets AB, AC respectively in b, c, Ob, $O'c$ intersect in P, $O'b$, Oc in Q: prove that P, A, Q lie in one straight line perpendicular to OO'.

80. The centre of the circumscribed circle of a triangle and the centre of perpendiculars are joined: prove that the joining line is divided into segments in the ratio of 1 : 2 by each of the straight lines joining an angular point to the middle point of the opposite side.

81. The side BC of a triangle ABC is bisected in D, a straight line parallel to BC meeting AB, AC produced in P, P' respectively is divided in Q, so that PQ, BD, QP' are in continued proportion, and through Q is drawn a straight line RQR' terminated by AB, AC and bisected in Q: prove that the triangles ABC, ARR' are equal.

82. On AB, AC two sides of a triangle are taken two points D, E; AB, AC are produced to F, G so that BF is equal to AD and CG to AE; BG, CF, FG are joined, the two former meeting in H: prove that the triangle FHG is equal to the two triangles BHC, ADE together.

83. If two sides of a triangle be given in position, and their sum be also given, and if the third side be divided in a given ratio, the point of division will lie on one of two fixed straight lines.

84. Two circles intersect in A, B, PQ is a straight line through A terminated by the two circles: prove that BP has to BQ a constant ratio.

85. Through the centre of perpendiculars of a triangle is drawn a straight line at right angles to the plane of the triangle: prove that any tetrahedron of which the triangle is one face and whose opposite vertex lies on this straight line will be such that each edge is perpendicular to the direction of the opposite edge.

86. A, B, C, D are four points not in one plane, and AB, AC respectively lie in planes perpendicular to CD, BD: prove that AD lies in a plane perpendicular to BC; and that the middle points of these six edges lie on one sphere which also passes through the feet of the shortest distances between the opposite edges.

87. In a certain tetrahedron each edge is perpendicular to the direction of the opposite edge: prove that the straight line joining the centre of the circumscribed sphere to the middle point of any edge will be equal and parallel to the straight line joining the centre of perpendiculars of the tetrahedron to the middle point of the opposite edge.

88. Each edge of a tetrahedron is equal to the opposite edge: prove that the straight line joining the middle points of two opposite edges is at right angles to both: also in such a tetrahedron the centres of the inscribed and circumscribed spheres and the centres of gravity of the volume and of the surface of the tetrahedron coincide.

89. If from any point O be let fall perpendiculars Oa, Ob, Oc, Od on the faces of a tetrahedron $ABCD$, the perpendiculars from A, B, C, D on the corresponding faces of the tetrahedron $abcd$ will meet in a point O', and the relation between O and O' is reciprocal.

90. The greatest possible number of tetrahedrons which can be constructed having their six edges of lengths equal to six given straight lines all unequal is thirty; and when they are all possible the one of greatest volume is that in which the three shortest edges meet in a point, and to them are opposite the other three in opposite order of magnitude.

91. Two tetrahedrons $ABCD$, $abcd$ are so related that straight lines drawn from a, b, c, d perpendicular to the corresponding faces of $ABCD$ meet in a point O: prove that straight lines drawn from A, B, C, D perpendicular to the corresponding faces of $abcd$ will meet in a point o, and that $\text{vol. } OBCD : \text{vol. } ABCD :: \text{vol. } obcd : \text{vol. } abcd.$

92. A solid angle is contained by three plane angles: prove that any straight line through the vertex makes with the edges angles whose sum is greater than half the sum of the containing angles, and extend the proposition to any number of containing angles.

93. Two circles are drawn, one lying altogether within the other; O, O' are the two points which are reciprocals with respect to either circle, and PQ is a chord of the outer circle touching the inner: prove that if PP', QQ' be chords of the outer circle passing through O or O', $P'Q'$ will also touch the inner circle.

94. The circles described on the diagonals of a complete quadrilateral as diameters cut orthogonally the circle circumscribing the triangle formed by the diagonals.

95. Four points are taken on the circumference of a circle, and through them are drawn three pairs of straight lines, each intersecting in a point: prove that the straight line joining any one of these points to the centre will be perpendicular to the straight line joining the other two.

96. A sphere is described touching three given spheres: prove that the plane passing through the points of contact contains one of four fixed straight lines.

97. Four straight lines are given in position: prove that an infinite number of systems of three circles can be found such that the points of

intersection of the four straight lines shall be the centres of similarity of the circles taken two and two.

98. In two fixed circles are drawn two parallel chords PP', QQ'; PQ, $P'Q'$ are joined meeting the circles again in R, S; R', S', respectively: prove that the points of intersection of QQ', RR' and of PP', SS' lie on a fixed straight line, the radical axis of the two circles.

99. The six radical axes of the four circles taken two and two which touch the sides of a triangle are the straight lines bisecting internally and externally the angles of a triangle formed by joining the middle points of the sides of the former triangle.

100. If two circles have four common tangents the circles described on these tangents as diameters will have a common radical axis.

101. Four points are taken on a circle and from the middle point of the chord joining any two a straight line is drawn perpendicular to the chord joining the other two: prove that the six lines so drawn will meet in a point, which is also common to the four nine points' circles of the triangles each having three of the points for its angular points.

102. Given in position two sides of a triangle including an angle equal to that of an equilateral triangle; prove that the centre of the nine points' circle of the triangle lies on a fixed straight line.

103. Given in position two sides of a triangle and given the sum of those sides, prove that the centre of the nine points' circle lies on a fixed straight line.

104. The perpendiculars let fall from the centres of the escribed circles of a triangle on the corresponding sides meet in a point.

105. The straight lines bisecting each a pair of opposite edges of a tetrahedron $ABCD$ meet in O and through A, B, C, D respectively are drawn planes at right angles to OA, OB, OC, OD: prove that the faces of the tetrahedron bounded by these planes will be to one another as $OA : OB : OC : OD$.

106. A straight line meets the produced sides of a triangle ABC in A', B', C' respectively: prove that the triangles ABB', ACC', $A'CC'$, $A'BB'$ will be proportionals.

107. A point O is taken within a triangle ABC, and through A, B, C are drawn straight lines parallel to those bisecting the angles BOC, COA, AOB: prove that these lines will meet in a point.

108. Straight lines AA', BB', CC' are drawn through a point to meet the opposite sides of a triangle ABC: prove that the straight lines drawn from A, B, C to bisect $B'C'$, $C'A'$, $A'B'$ will meet in one point; and that straight lines drawn from A, B, C parallel to $B'C'$, $C'A'$, $A'B'$ will meet the respectively opposite sides in three points lying on one straight line.

109. If two circles lie entirely without each other and any straight line meet them in P, P'; Q, Q' respectively, there are two points O such that the straight lines bisecting the angles POP', QOQ' shall be always at right angles to each other.

110. Given two circles which do not intersect, a tangent to one at any point P meets the polar of P with respect to the other in P': prove that the circle whose diameter is PP' will pass through two fixed points.

111. A point has the same polar with respect to each of two circles: prove that any common tangent will subtend a right angle at that point.

112. Given two points A, B, a straight line PAQ is drawn through A so that the angle PBQ is equal to a given angle and that BP has to BQ a given ratio: prove that P, Q will lie on two fixed circles which pass through A and B.

113. If O be a fixed point, P any point on a fixed circle and the rectangle be constructed of which OP is a side and the tangent at P a diagonal, the angular point opposite O will lie on the polar of O.

114. If OA', OB', OC' be perpendiculars from a point O on the sides of a triangle ABC, then will

$$AB' . BC' . CA' + B'C . C'A . A'B = 2 \triangle A'B'C' \times \text{diameter of the circle } ABC.$$

115. From a fixed point O are let fall perpendiculars on two conjugate rays of a pencil in involution: prove that the straight line joining the feet of these perpendiculars passes through a fixed point.

116. If O be a fixed point, P, P' conjugate points of a range in involution and PQ, $P'Q$ be drawn at right angles to OP, OP'; Q will lie on a fixed straight line.

117. In any complete quadrilateral the common radical axis of the three circles whose diameters are the three diagonals will pass through the centres of perpendiculars of the four triangles formed by the four straight lines.

ALGEBRA.

I. *Highest Common Divisor.*

118. Reduce to their lowest terms the fractions

(1) $\dfrac{11x^4+24x^3+125}{x^4+24x+55}$, (2) $\dfrac{55x^4+24x^3+1}{125x^4+24x+1}$,

(3) $\dfrac{9x^5+5x-2}{27x^5-45x^4-16}$, (4) $\dfrac{9x^5+11x^2-2}{81x^5+11x+4}$,

(5) $\dfrac{2x^5-11x^3-9}{4x^5+11x^4+81}$, (6) $\dfrac{x^5+11x^3-54}{x^5+11x+12}$,

(7) $\dfrac{x^6-209x+56}{56x^6-209x^5+1}$, (8) $\dfrac{8x^7-377x^3+21}{21x^7-377x^4+8}$,

(9) $\dfrac{16x^9-x^5+16x^4+32}{32x^9+16x^5-x^4+16}$, (10) $\dfrac{x^8+2x^6+3x^4-2x^2+1}{6x^8+x^7+17x^5-7x^3-2}$,

(11) $\dfrac{1+x^2}{a+2hx+bx^2}+\dfrac{(a+hx)^2+(h+bx)^2}{a(h+bx)^2-2h(a+hx)(h+bx)+b(a+hx)^2}$.

119. Simplify the expressions

(1) $\dfrac{x(1-y^2)(1-z^2)+y(1-z^2)(1-x^2)+z(1-x^2)(1-y^2)-4xyz}{x+y+z-xyz}$,

(2) $\dfrac{a(b+c-a)^2+b(c+a-b)^2+c(a+b-c)^2+(b+c-a)(c+a-b)(a+b-c)}{a^2(b+c-a)+b^2(c+a-b)+c^2(a+b-c)-(b+c-a)(c+a-b)(a+b-c)}$,

(3) $\dfrac{a^3(b^2-c^2)+b^3(c^2-a^2)+c^3(a^2-b^2)}{a^2(b-c)+b^2(c-a)+c^2(a-b)}$,

(4) $\dfrac{1}{(a-b)(a-c)(a-d)}+\dfrac{1}{(b-c)(b-d)(b-a)}$

$+\dfrac{1}{(c-d)(c-a)(c-b)}+\dfrac{1}{(d-a)(d-b)(d-c)}$,

(5) $\dfrac{bcd}{(a-b)(a-c)(a-d)} + \dfrac{cda}{(b-c)(b-d)(b-a)}$ + two similar terms,

(6) $a^2\dfrac{(a+b)(a+c)}{(a-b)(a-c)} + b^2\dfrac{(b+c)(b+a)}{(b-c)(b-a)} + c^2\dfrac{(c+a)(c+b)}{(c-a)(c-b)}$,

(7) $a^3\dfrac{(a+b)(a+c)}{(a-b)(a-c)} + b^3\dfrac{(b+c)(b+a)}{(b-c)(b-a)} + c^3\dfrac{(c+a)(c+b)}{(c-a)(c-b)}$,

(8) $a^4\dfrac{(a+b)(a+c)}{(a-b)(a-c)} + b^4\dfrac{(b+c)(b+a)}{(b-c)(b-a)} + c^4\dfrac{(c+a)(c+b)}{(c-a)(c-b)}$,

(9) $\dfrac{8a^2b^2c^2 + (b^2+c^2-a^2)(c^2+a^2-b^2)(a^2+b^2-c^2)}{(a+b+c)(-a+b+c)(a-b+c)(a+b-c)}$.

120. Prove that

$$\frac{(ab-cd)(a^2-b^2+c^2-d^2)+(ac-bd)(a^2+b^2-c^2-d^2)}{(a^2-b^2+c^2-d^2)(a^2+b^2-c^2-d^2)+4(ab-cd)(ac-bd)} \equiv \frac{(b+c)(a+d)}{(b+c)^2+(a+d)^2}.$$

121. Prove that

$$\frac{(b-c)(1+b^2)(1+c^2)+(c-a)(1+c^2)(1+a^2)+(a-b)(1+a^2)(1+b^2)}{a(b-c)(1+b^2)(1+c^2)+b(c-a)(1+c^2)(1+a^2)+c(a-b)(1+a^2)(1+b^2)} \equiv \frac{1-bc-ca-ab}{a+b+c-abc}.$$

122. Prove that

$$\frac{\{(a+b)(a+c)+2a(b+c)\}^2-(a-b)^2(a-c)^2}{a}$$
$$\equiv \frac{\{(b+c)(b+a)+2b(c+a)\}^2-(b-c)^2(b-a)^2}{b}$$
$$\equiv \frac{\{(c+a)(c+b)+2c(a+b)\}^2-(c-a)^2(c-b)^2}{c}$$
$$\equiv 8(b+c)(c+a)(a+b).$$

123. Prove that

$$\{(b-c)^2+(c-a)^2+(a-b)^2\}\{a^2(b-c)^2+b^2(c-a)^2+c^2(a-b)^2\}$$
$$\equiv 3(b-c)^2(c-a)^2(a-b)^2+\{a(b-c)^2+b(c-a)^2+c(a-b)^2\}^2.$$

II. *Equations.*

124. Solve the equations

(1) $(x+1)(x+2)(x+3)=(x-3)(x+4)(x+5)$,

(2) $(x+1)(x+2)(x+3)=(x-1)(x-2)(x-3)+3(4x-1)(x+1)$,

(3) $(x+a)(x+a+b)=(x+b)(x+3a)$,

(4) $\dfrac{1}{x+1}+\dfrac{7}{x+5}=\dfrac{5}{x+3}+\dfrac{3}{x+7}$,

(5) $\dfrac{20}{x+10}-\dfrac{10}{x+20}=\dfrac{15}{x+5}-\dfrac{5}{x+15}$,

(6) $\dfrac{1}{x+6a}+\dfrac{2}{x-3a}+\dfrac{3}{x+2a}=\dfrac{6}{x+a}$,

(7) $\dfrac{x}{x+b-a}+\dfrac{b}{x+b-c}=1$,

(8) $\dfrac{a}{x+b-c}+\dfrac{b}{x+a-c}=\dfrac{a-c}{x+b}+\dfrac{b+c}{x+a}$,

(9) $\dfrac{(a-b)^2}{(a-b)x+\alpha-\beta}+\dfrac{(b-c)^2}{(b-c)x+\beta-\gamma}+\dfrac{(c-d)^2}{(c-d)x+\gamma-\delta}$

$$+\dfrac{(d-a)^2}{(d-a)x+\delta-\alpha}=0.$$

(10) $4(x-a)^2=9(x-b)(a-b)$,

(11) $2(x-2a)^2=(3x-2b)(3a-b)$,

(12) $x(x-5)(x-9)=(x-6)(x^2-27)$,

(13) $(x+7)(x^2-4)=(x+1)(x^2+14x+22)$,

(14) $\dfrac{x^2+3}{x-1}+\dfrac{x^2-x+1}{x-2}=2\dfrac{x^2-2x+1}{x-3}$,

(15) $\dfrac{x^2-x+1}{x-1}+\dfrac{x^2-3x+1}{x-3}=2x-\dfrac{1}{4x-8}$,

(16) $\dfrac{x+1}{x-1}+\dfrac{x+2}{x-2}=2\dfrac{11x+18}{11x-18}$,

(17) $\dfrac{1}{a}+\dfrac{1}{b}+\dfrac{1}{x}=\dfrac{1}{a+b+x}$,

(18) $a^2\dfrac{x-b}{a-b}+b^2\dfrac{x-a}{b-a}=x^2$,

(19) $(x-9)(x-7)(x-5)(x-1)=(x-2)(x-4)(x-6)(x-10)$,

(20) $(a+x)^{\frac{1}{2}}+(b+x)^{\frac{1}{2}}=(a-b)^{\frac{1}{2}}$,

(21) $(a+x)^{\frac{1}{3}}+(b+x)^{\frac{1}{3}}=(a-b)^{\frac{1}{3}}$,

(22) $\dfrac{(b-c)^2}{(b-c)^2-(x-a)^2}+\dfrac{(c-a)^2}{(c-a)^2-(x-b)^2}+\dfrac{(a-b)^2}{(a-b)^2-(x-c)^2}=2$,

(23) $\{x(a+b-x)\}^{\frac{1}{2}}+\{a(b+x-a)\}^{\frac{1}{2}}+\{b(a+x-b)\}^{\frac{1}{2}}=0$,

(24) $\dfrac{(x+a)(x+b)}{(x-a)(x-b)}+\dfrac{(x-a)(x-b)}{(x+a)(x+b)}=\dfrac{(x+c)(x+d)}{(x-c)(x-d)}+\dfrac{(x-c)(x-d)}{(x+c)(x+d)}$,

(25) $\dfrac{5}{x+2}+\dfrac{5}{x+4}=\dfrac{1}{x+1}+\dfrac{8}{x+3}+\dfrac{1}{x+5}$,

(26) $(x^2-18x-27)^2=(x+1)(x+9)^3$,

(27) $(x^2-27)^2=(x-5)(x-9)^3$,

(28) $\left(\dfrac{x^2-11x+19}{x^2+x-11}\right)^2+3\,\dfrac{x-2}{x+2}=0$,

(29) $(x+3)^2(x^2-9x+9)(2x^2-6x+9)+(x^2+3x-9)^3=0$,

(30) $x(x+4)+\dfrac{1}{x}\left(\dfrac{1}{x}+4\right)=10$,

(31) $x^6+1+(x+1)^6=2(x^2+x+1)^3$,

(32) $\dfrac{1}{10x-50}-\dfrac{2}{x-6}+\dfrac{9}{x-7}-\dfrac{14}{x-8}+\dfrac{7}{x-9}=0$,

(33) $13x^2=10\,\dfrac{8x-15}{x^2-4x+5}-12\,\dfrac{11x-15}{x^2-6x+6}$,

(34) $\dfrac{1}{x-1}-\dfrac{4}{x-2}+\dfrac{4}{x-3}-\dfrac{1}{x-4}=\dfrac{1}{30}$,

(35) $\dfrac{3}{x^2-7x+3}-\dfrac{2}{x^2+7x+2}=5$,

(36) $\dfrac{2}{x^2+2x-2}+\dfrac{3}{x^2-2x+3}=\dfrac{x}{2}$,

(37) $\dfrac{45}{x^2+3x+3}-\dfrac{22x+15}{x^2+x+1}=2x$,

(38) $\dfrac{7x+10}{x^2-4x+5}-\dfrac{2x+4}{x^2-2x+2}=x^2$,

(39) $\dfrac{7x-4}{x^2+1}-\dfrac{72x-32}{x^2-4x+8}+\dfrac{65x^2}{7}=0$,

(40) $3(x^2+2)+\dfrac{6}{x-1}+\dfrac{2}{(x-1)^2}=\dfrac{2}{x^2+x+1}$,

(41) $(2x^3-7x^2+9x-6)^2=4(x^2-x+1)(x^2-3x+3)^2$,

(42) $(x-2)(x+1)^2(x^2+2x+4)(x^2-x+1)^2+15x^3+8=0$,

(43) $x^8+1+(x+1)^8=2(x^2+x+1)^4$,

(44) $x^{10}+1+(x+1)^{10}=2(x^2+x+1)^5+15x^2(x^2+x+1)^2$,

(45) $16x(x+1)(x+2)(x+3)=9$,

(46) $x^4+2x^3-11x^2+4x+4=0$,

(47) $\frac{4}{x^2-2x} - \frac{2}{x^2-x} = x^2 - x,$

(48) $\frac{5}{x^2-7x+10} + \frac{5}{x^2-13x+40} = x^2 - 10x + 19,$

(49) $\frac{40}{x^2+2x-48} - \frac{20}{x^2+9x+8} + \frac{8}{x^2+10x} - \frac{12}{x^2+5x-50} + 1 = 0,$

(50) $\frac{1}{x(x-1)} + \frac{2}{(x-1)(x-3)} - \frac{6}{(x-1)(x+2)} + \frac{8}{x^2-4} + 1 = 0,$

(51) $(x^2+1)^2 = 4(2x-1),$

(52) $\frac{2x}{3} = \frac{7}{x^2+3x-7} - \frac{3}{x^2+x-3},$

(53) $8x^2 = \frac{4x+69}{x^2+2x+3} - \frac{9x+23}{x^2+x+1},$

(54) $x^3 = 6x + 6,$

(55) $4x^3 = 6x + 3,$

(56) $x^3 + 6x^2 = 36,$

(57) $xy(x+y) = 12x + 3y,\ xy(4x+y-xy) = 12(x+y-3);$

(58) $x\sqrt{1-y^2} - y\sqrt{1-x^2} = xy - \sqrt{1-x^2}\sqrt{1-y^2} = \frac{1}{2};$

(59) $x^3 + 6xy^2 + 2y^3 = 90,\ y(x^2+xy+y^2) = 21;$

(60) $x(y+z-x) = a^2,\ y(z+x-y) = b^2,\ z(x+y-z) = c^2;$

(61) $x^2 + 2xy^2 + 2y^2 = 1 = \frac{1}{x^2} - \frac{2}{xy^2} + \frac{2}{y^2};$

(62) $x^2 + 2yz = a,\ y^2 + 2zx = b,\ z^2 + 2xy = c;$

(63) $u^2 + v^2 + 2xy = a,\ x^2 + y^2 + 2uv = b,\ ux + vy = c,\ vx + uy = d;$

(64) $xyz = a(y^2+z^2) = b(z^2+x^2) = c(x^2+y^2);$

(65) $cy + bz = \frac{1}{a} - \frac{1}{x},\ az + cx = \frac{1}{b} - \frac{1}{y},\ bx + ay = \frac{1}{c} - \frac{1}{z};$

(66) $x + \frac{b^3-c^3}{yz} = y + \frac{c^3-a^3}{zx} = z + \frac{a^3-b^3}{xy} = \sqrt[3]{xyz};$

(67) $x + y + z = mxyz,\ yz + zx + xy = n,$

$$(1+x^2)(1+y^2)(1+z^2) = (1-n)^2;$$

(68) $10y^2 + 13z^2 - 6yz = 242,$

$5z^2 + 10x^2 - 2zx = 98,$

$13x^2 + 5y^2 - 16xy = 2;$

proving that an infinite number of solutions exist;

(69) $x_1^2 + 2x_2x_5 + 2x_3x_4 = a_1,$
$x_2^2 + 2x_3x_1 + 2x_4x_5 = a_2,$
$x_3^2 + 2x_4x_2 + 2x_5x_1 = a_3,$
$x_4^2 + 2x_5x_3 + 2x_1x_2 = a_4,$
$x_5^2 + 2x_1x_4 + 2x_2x_3 = a_5;$

(70) $x_1^2 + x_4^2 + 2x_2x_6 + 2x_3x_5 = a_1,\ x_1x_6 + x_2x_5 + x_3x_4 = b_1,$
$x_2^2 + x_5^2 + 2x_3x_1 + 2x_4x_6 = a_2,\ x_2x_1 + x_3x_6 + x_4x_5 = b_2,$
$x_3^2 + x_6^2 + 2x_4x_2 + 2x_5x_1 = a_3,\ x_3x_2 + x_4x_1 + x_5x_6 = b_3;$

(71) $x^3 + y^3 + z^3 + 6xyz = -4$ or a,
$xy^2 + yz^2 + zx^2 = 5$ or b,
$xz^2 + yx^2 + zy^2 = -1$ or c;

(72) $x_1^3 + 3x_1x_3^2 + 3x_3(x_2^2 + x_4^2) + 6x_1x_2x_4 = 17$ or a_1,
$x_2^3 + 3x_2x_4^2 + 3x_4(x_3^2 + x_1^2) + 6x_2x_3x_1 = 13$ or a_2,
$x_3^3 + 3x_3x_1^2 + 3x_1(x_4^2 + x_2^2) + 6x_3x_4x_2 = 15$ or a_3,
$x_4^3 + 3x_4x_2^2 + 3x_2(x_1^2 + x_3^2) + 6x_4x_1x_3 = 19$ or a_4;

and shew how to solve systems of equations like (69) and (71) with any odd number of unknown quantities, and systems like (70) and (72) with any even number.

[From (69) may be obtained

$$(x_1 + \omega x_2 + \omega^2 x_3 + \omega^3 x_4 + \omega^4 x_5)^2 = a_1 + \omega a_4 + \omega^2 a_2 + \omega^3 a_5 + \omega^4 a_3,$$

where ω is any fifth root of unity, and a like method applies to all such systems.]

(73) $x^3 - 3xy^2 - y^3 = a,\ xy(x + y) = b$;

$$[(x - \omega y)^3 = a - 3\omega b, \text{ where } \omega^2 + \omega + 1 = 0;]$$

(74) $x^4 - 6x^2y^2 - 4xy^3 = a,\ y^4 - 6x^2y^2 - 4x^3y = b$;

$$[(x - \omega y)^4 = a + \omega b.]$$

III.

125. In the equation

$$\frac{a}{x - md} + \frac{b}{x - mc} + \frac{c}{x + mb} + \frac{d}{x + ma} = 0,$$

prove that, if $a + b + c + d = 0$, the only finite value of x will be

$$\frac{m(ac + bd)}{a + b}.$$

126. In the equation

$$\frac{a_1}{x+b_1}+\frac{a_2}{x+b_2}+\frac{a_3}{x+b_3}+\frac{a_4}{x+b_4}=0,$$

prove that, if

$$a_1+a_2+a_3+a_4=0, \text{ and } a_1b_1+a_2b_2+a_3b_3+a_4b_4=0,$$

the only finite value of x will be

$$\frac{a_1b_1^3+a_2b_2^3+a_3b_3^3+a_4b_4^3}{a_1b_1^2+a_2b_2^2+a_3b_3^2+a_4b_4^2}-(b_1+b_2+b_3+b_4).$$

127. The equation

$$(x+\sqrt{x^2-bc})(y+\sqrt{y^2-ca})(z+\sqrt{z^2-ab})=abc$$

is equivalent to

$$ax^2+by^2+cz^2=abc+2xyz.$$

128. Find limits to the real values of x and y which can satisfy the equation

$$x^2+12xy+4y^2+4x+8y+20=0.$$

[x cannot lie between -2 and 1, nor y between -1 and $\frac{1}{2}$.]

129. If the roots of the equation

$$ax^2+2hx+b=0$$

be possible and different, the roots of the equation

$$(a+b)(ax^2+2hx+b)=2(ab-h^2)(x^2+1)$$

will be impossible: and *vice versâ*.

130. Prove that the equations

$$x+y+z=a+b+c,$$

$$\frac{x}{a}+\frac{y}{b}+\frac{z}{c}=1,$$

$$\frac{x}{a^3}+\frac{y}{b^3}+\frac{z}{c^3}=0,$$

are equivalent to only two independent equations, if $bc+ca+ab=0$.

131. Obtain the several equations for determining α, β, γ so that the equations

$$x^4+4px^3+6qx^2+4rx+s=0,\ (x^2+2px+\alpha)^2=(\beta x+\gamma)^2,$$

may coincide: and in this manner solve the equation

$$(x^2+3x-6)^2+3x^2=72.$$

132. Shew how to solve any biquadratic of the form

$$x^4 + 2ax^2 + \frac{(b^2 - a^2)^2}{4b^2} = 0,$$

[by putting it in the form

$$\left(x^2 + ax + \frac{b^2 - a^2}{2}\right)^2 = \left\{bx + a\frac{(b^2 - a^2)}{2b}\right\}^2;]$$

and hence solve the equations

(1) $x^4 - 8x^3 - 108 = 0,$

(2) $x^4 - 10x^3 - 3456 = 0.$

133. Prove that the equation

$$x^3 + 3ax^2 + 3bx + \frac{b^2}{a} = 0$$

can be solved directly, and that the complete cubic $x^3 + 3px^2 + 3qx + r = 0$ can be reduced to this form by the substitution $x \equiv y + h$.

Prove that the roots of the auxiliary quadratic are

$$\frac{\alpha(\beta-\gamma)^2 + \beta(\gamma-\alpha)^2 + \gamma(\alpha-\beta)^2 \pm \sqrt{-3}\,(\beta-\gamma)(\gamma-\alpha)(\alpha-\beta)}{(\beta-\gamma)^2 + (\gamma-\alpha)^2 + (\alpha-\beta)^2},$$

α, β, γ being the roots of the original cubic.

134. The roots of the equation

$$(x+a-c)(x+b+c)(x+a-d)(x+b+d) = e$$

will all be real if

$$16e < (a-b-2c)^2(a-b-2d)^2 \text{ and } > -4(c-d)^2(b+c+d-a)^2.$$

135. Determine λ so that the equation in x

$$\frac{2A}{x+a} + \frac{\lambda}{x} + \frac{2B}{x-a} = 0$$

may have equal roots; and if λ_1, λ_2 be the two values of λ, x_1, x_2 the corresponding values of x, prove that

$$x_1x_2 = a^2, \; \lambda_1\lambda_2 = (A-B)^4.$$

136. Prove that, if two relations be satisfied, the expression

$$(x^2+ax+m)(x^2+bx+m)(x^2+cx+m)$$

will contain no power of x except those whose index is a multiple of 3. Resolve $x^6 - 20x^3 + 343$ and $x^6 + 36x^3 + 1000$ into their real quadratic factors and identify the roots found from the expression in x^3 with those of the several quadratic factors.

137. The equations

$$\frac{x - 2\dfrac{xy - z^2}{x + y}}{a} = \frac{y - 2\dfrac{xy - z^2}{x + y}}{b} = \frac{z}{h} = \frac{xy - z^2}{ab - h^2}$$

have the unique solution

$$\frac{x}{a - 2\dfrac{ab - h^2}{a + b}} = \frac{y}{b - 2\dfrac{ab - h^2}{a + b}} = \frac{z}{h} = \frac{-(a + b)^2}{(a - b)^2 + 4h^2}.$$

138. The four equations

$$\frac{x + y - 2z}{a + b} = \frac{x^2 + y^2 - 2z^2}{a^2 + b^2} = \frac{xy - z^2}{ab} = \frac{xy(x + y) - 2z^3}{ab(a + b)} = \frac{x^2y^2 - z^4}{a^2b^2}$$

are consistent and equivalent to the three

$$\frac{x + y}{a + b} = \frac{xy}{ab} = \left(\frac{a - b}{a + b}\right)^2, \quad z = \frac{2ab}{a + b}.$$

139. The system of equations

$$x_1 - x_2 = a(x_3 - x_5),$$
$$x_2 - x_3 = a(x_4 - x_1),$$
$$x_3 - x_4 = a(x_5 - x_2),$$
$$x_4 - x_5 = a(x_1 - x_3),$$
$$x_5 - x_1 = a(x_2 - x_4),$$

will be equivalent to only two independent equations if $a(a - 1) = 1$.

[This may also be proved from Statical considerations.]

140. The six equations

$$a^2 = \frac{(cy + bz)(by + cz)}{bc + yz}, \quad x^2 = \frac{(by + cz)(bc + yz)}{cy + bz},$$

$$b^2 = \frac{(az + cx)(cz + ax)}{ca + zx}, \quad y^2 = \frac{(cz + ax)(ca + zx)}{az + cx},$$

$$c^2 = \frac{(bx + ay)(ax + by)}{ab + xy}, \quad z^2 = \frac{(ax + by)(ab + xy)}{bx + ay},$$

are equivalent only to the two independent equations

$$ax + by + cz = 0, \quad ayz + bzx + cxy + abc = 0.$$

[Geometrically these equations express relations between the six joining lines of a quadrangle inscribed in a circle.]

141. Having given the system of equations

$$ax+by+cz=0,\quad \frac{c'y+b'z}{a}=\frac{a'z+c'x}{b}=\frac{b'x+a'y}{c}=x+y+z,$$

prove that

$$a'(bb'+cc'-aa')+b'(cc'+aa'-bb')+c'(aa'+bb'-cc')=2a'b'c',$$
$$a'(bb'+cc'-aa')+b\,(cc'+aa'-bb')+c\,(aa'+bb'-cc')=2a'bc,$$
$$a\,(bb'+cc'-aa')+b'(cc'+aa'-bb')+c\,(aa'+bb'-cc')=2ab'c,$$
$$a\,(bb'+cc'-aa')+b\,(cc'+aa'-bb')+c'(aa'+bb'-cc')=2abc',$$

which are equivalent to the two-fold relation

$$(aa')^{\frac{1}{2}}+(bb')^{\frac{1}{2}}+(cc')^{\frac{1}{2}}=0,\quad \left(\frac{bc}{b'c'}\right)^{\frac{1}{2}}+\left(\frac{ca}{c'a'}\right)^{\frac{1}{2}}+\left(\frac{ab}{a'b'}\right)^{\frac{1}{2}}=1,$$

corresponding terms in the two being taken with the same sign.

142. From the equations

$$x^2+2yz=a,\quad y^2+2zx=b,\quad z^2+2xy=c,$$

obtain the result

$$3(yz+zx+xy)=a+b+c-\sqrt{a^2+b^2+c^2-bc-ca-ab}.$$

IV. *Theory of Divisors.*

143. Determine the condition necessary in order that

$$x^2+px+q \text{ and } x^3+p'x+q'$$

may have a common divisor $x+c$, and prove that such a divisor will also be a divisor of $px^2+(q-p')x-q'$.

144. The expression

$$x^6+3ax^5+3bx^4+cx^3+3dx^2+3ex+f$$

will be a complete cube if

$$\sqrt[3]{f}=\sqrt{\frac{e}{a}}=\frac{d}{b}=\frac{c-a^3}{6a}=b-a^2.$$

145. The expression $x^5-bx^3+cx^2+dx-e$ will be the product of a complete square and a complete cube if

$$\frac{12b}{5}=\frac{9d}{b}=\frac{5e}{c}=\frac{d^2}{c^2}.$$

146. Prove that ax^3+bx+c and $a+bx^4+cx^5$ will have a common quadratic factor if

$$b^2c^2=(c^2-a^2+b^2)(c^2-a^2+ab);$$

and that ax^3+bx^2+c and $a+bx^3+cx^5$ will have a common quadratic factor if

$$(a^2-c^2)(a^2-c^2+bc)=a^2b^2.$$

147. Prove that

$$a_0x^4+a_1x^3+a_2x^2+a_3x+a_4 \text{ and } a_0+a_1x+a_2x^2+a_3x^3+a_4x^4$$

will have a common quadratic factor if

$$(a_2-a_4-a_0)(a_4-a_0)^2+(a_3-a_1)(a_3a_0-a_4a_1)=0.$$

148. Prove that ax^7+bx^3+c and $a+bx^4+cx^7$ will have a common quadratic factor if

$$(a^2+ab-c^2)(a^2-ab-c^2)^2=b^2c^2(a^2-c^2).$$

149. The expression x^3+px^2+qx+r will be divisible by x^2+ax+b if $a^3-2pa^2+(p^2+q)a+r-pq=0$, and $b^3-qb^2+rpb-r^2=0$.

150. The expression x^4+px+q will be divisible by x^2+ax+b if

$$a^6-4qa^2=p^2 \text{ and } (b^2+q)(b^2-q)^2=p^2b^3.$$

151. The highest common divisor of $p(x^q-1)-q(x^p-1)$ and $(q-p)x^q-qx^{q-p}+p$ is $(x-1)^2$, p, q being numbers whose greatest common measure is 1 and q being greater than p.

152. If n be any positive whole number not a multiple of 3, the expression $x^{2n}+1+(x+1)^{2n}$ will be divisible by x^2+x+1; and, if n be of the form $3r-1$, by $(x^2+x+1)^2$.

153. Prove that

$$(ab-h^2)\{x(x-X)+y(y-Y)\}^2-b(x-X)^2+2h(x-X)(y-Y)-a(y-Y)^2$$

will be divisible by $(x-X)^2+(y-Y)^2$ if

$$\frac{X^2-Y^2}{a-b}=\frac{XY}{h}=\frac{1}{h^2-ab}.$$

V. *Identities and Equalities.*

154. Prove that

(1) $(a+b+c)^3\equiv a^3+b^3+c^3+3(b+c)(c+a)(a+b)$,

(2) $\dfrac{2}{b-c}+\dfrac{2}{c-a}+\dfrac{2}{a-b}+\dfrac{(b-c)^2+(c-a)^2+(a-b)^2}{(b-c)(c-a)(a-b)}\neq 0$,

(3) $(S-b^2)(S-c^2)+(S-c^2)(S-a^2)+(S-a^2)(S-b^2)$

$$\equiv 4s(s-a)(s-b)(s-c),$$

where $2S\equiv a^2+b^2+c^2$, and $2s\equiv a+b+c$;

(4) $(b-c)(1+ab)(1+ac)+(c-a)(1+bc)(1+ba)$

$+(a-b)(1+ca)(1+cb)\equiv(b-c)(c-a)(a-b)$,

(5) $a(b-c)(1+ab)(1+ac)$ + the two similar terms

$\equiv -abc(b-c)(c-a)(a-b)$,

(6) $(b-c)(1+a^2b)(1+a^2c)$ + the two similar terms

$\equiv -abc(a+b+c)(b-c)(c-a)(a-b)$,

(7) $(b^2-c^2)(1-a^2b)(1-a^2c)$ + the two similar terms

$\equiv(1+abc)(a^2+b^2+c^2+bc+ca+ab)(b-c)(c-a)(a-b)$,

(8) $2b^2c^2(c+a)^2(a+b)^2$ + the two similar terms

$\equiv a^4(b+c)^4+b^4(c+a)^4+c^4(a+b)^4+16a^2b^2c^2(bc+ca+ab)$,

(9) $(a^2+2bc)^3+(b^2+2ca)^3+(c^2+2ab)^3-3(a^2+2bc)(b^2+2ca)(c^2+2ab)$

$\equiv(a^3+b^3+c^3-3abc)^2$,

(10) $8a^2b^2c^2+(b^2+c^2-a^2)(c^2+a^2-b^2)(a^2+b^2-c^2)$

$\equiv(a^2+b^2+c^2)(a+b+c)(-a+b+c)(a-b+c)(a+b+c)$,

(11) $(a-b)^2(a-c)^2+(b-c)^2(b-a)^2+(c-a)^2(c-b)^2$

$\equiv(a^2+b^2+c^2-bc-ca-ab)^2$

$\equiv\frac{1}{2}\{(b-c)^4+(c-a)^4+(a-b)^4\}$,

(12) $a(b-c)(b+c-a)^4$ + the two similar terms

$\equiv 16abc(b-c)(a-b)(a-c)$,

(13) $a(b-c)(b+c-a)^6$ + the two similar terms

$\equiv 16abc(b-c)(a-b)(a-c)\{(a+b+c)^2-4(a^2+b^2+c^2)\}$,

(14) $(bcd+cda+dab+abc)^2-abcd(a+b+c+d)^2$

$\equiv(bc-ad)(ca-bd)(ab-cd)$,

(15) $(a+b+c+d)^3-4(a+b+c+d)(bc+ad+ca+bd+ab+cd)$

$+8(bcd+cda+dab+abc)\equiv-(b+c-a-d)(c+a-b-d)(a+b-c-d)$,

(16) $$\frac{(b+c)^3+(c+a)^3+(a+b)^3-3(b+c)(c+a)(a+b)}{a^3+b^3+c^3-3abc}\equiv 2,$$

(17) $(x^2-x+1)(x^4-x^2+1)\dots(x^{2^n}-x^{2^{n-1}}+1)$

$$\equiv\frac{x^{2^{n+1}}+x^{2^n}+1}{x^2+x+1},$$

(18) $\{(b-c)^2+(c-a)^2+(a-b)^2\}\{a^2(b-c)^2+b^2(c-a)^2+c^2(a-b)^2\}$

$\equiv 3(b-c)^2(c-a)^2(a-b)^2$.

155. Prove the following identities, where a, b, c, d are the roots of the equation

$$x^4 - Lx^3 + Mx^2 - Nx + P = 0,$$

and the product of their differences

$$(b-c)(c-a)(a-b)(a-d)(b-d)(c-d)$$

is denoted by Δ,

(1) $(b^2c^2\,\overline{a+d} + a^2d^2\,\overline{b+c})(b-c)(a-d)$
$+ (c^2a^2\,\overline{b+d} + b^2d^2\,\overline{c+a})(c-a)(b-d)$
$+ (a^2b^2\,\overline{c+d} + c^2d^2\,\overline{a+b})(a-b)(c-d) \equiv 0,$

(2) $\{(b+c)^2 + (a+d)^2\}(b-c)(a-d) +$ the two similar terms $\equiv 0$,

(3) $(b^2c^2 + a^2d^2)(b-c)(a-d) + \ldots\ldots \equiv -\Delta$,

(4) $(b^2-c^2)(a^2-d^2)(bc+ad) + \ldots\ldots \equiv \Delta$,

(5) $(b^2-c^2)(a^2-d^2)(b+c)(a+d) + \ldots\ldots \equiv -\Delta$,

(6) $\{bc\,(b+c)^2 + ad\,(a+d)^2\}(b-c)(a-d) + \ldots\ldots \equiv -\Delta$,

(7) $(b^2+c^2)(a^2+d^2)(b-c)(a-d) + \ldots\ldots \equiv \Delta$,

(8) $(bc\,\overline{b^3+c^3} + ad\,\overline{a^3+d^3})(b-c)(a-d) + \ldots\ldots \equiv L\Delta$,

(9) $(bc\,\overline{b+c} + ad\,\overline{a+d})(b^2-c^2)(a^2-d^2) + \ldots\ldots \equiv L\Delta$,

(10) $\{(b+c)^5 + (a+d)^5\}(b-c)(a-d) + \ldots\ldots \equiv -5L\Delta$,

(11) $(b^3c^3 + a^3d^3)(b-c)(a-d) + \ldots\ldots \equiv -M\Delta$,

(12) $(bc+ad)(b^3-c^3)(a^3-d^3) + \ldots\ldots \equiv M\Delta$,

(13) $(bc+ad)(b-c)^3(a-d)^3 + \ldots\ldots \equiv M\Delta$,

(14) $(b+c)^3(a+d)^3(b-c)(a-d) + \ldots\ldots \equiv -2M\Delta$,

(15) $(b^2c^2 + a^2d^2)(b^3-c^3)(a^3-d^3) + \ldots\ldots \equiv (LN-P)\Delta$,

(16) $(b^2+c^2)(a^2+d^2)(b^3-c^3)(a^3-d^3) + \ldots\ldots \equiv (P-LN)\Delta$,

(17) $(bc+ad)^2(b^3-c^3)(a^3-d^3) + \ldots\ldots \equiv (4P-LN)\Delta$,

(18) $(b+c-a-d)^4(b-c)(a-d) + \ldots\ldots \equiv \Delta$.

156. The expression

$$(ax^2 + by^2 + cz^2) \div \{bc\,(y-z)^2 + ca\,(z-x)^2 + ab\,(x-y)^2\}$$

will have a constant value for all values of x, y, z which satisfy the equation $ax + by + cz = 0$.

157. If $xy + x + y = 2$, then will

$$\frac{x^4 - 8x}{1 + x^3} = \frac{y^4 - 8y}{1 + y^3} = \frac{xy\,(4 - xy)}{xy - 1}.$$

158. If $l_2l_3 + m_2m_3 = l_3l_1 + m_3m_1 = l_1l_2 + m_1m_2$,

then will $$\frac{(m_2-m_3)(m_3-m_1)(m_1-m_2)}{(l_2-l_3)(l_3-l_1)(l_1-l_2)} + \frac{l_1l_2l_3}{m_1m_2m_3} = 0.$$

159. If $$\frac{x}{a} + \frac{y}{b} = 1 \text{ and } \frac{x^2}{a} + \frac{y^2}{b} = \frac{ab}{a+b},$$

then will $$\frac{x^{n+1}}{a} + \frac{y^{n+1}}{b} = \left(\frac{ab}{a+b}\right)^n.$$

160. Having given $\frac{x}{y+z} = a$, $\frac{y}{z+x} = b$, $\frac{z}{x+y} = c$, find the relation between a, b, c; and prove that

$$\frac{x^2}{a-abc} = \frac{y^2}{b-abc} = \frac{z^2}{c-abc}.$$

161. Having given the equations

$$\left.\begin{aligned} x + y + z &= 1, \\ ax + by + cz &= d, \\ a^2x + b^2y + c^2z &= d^2, \end{aligned}\right\}$$

prove that $$a^3x + b^3y + c^3z = d^3 - (d-a)(d-b)(d-c),$$

and $$a^4x + b^4y + c^4z = d^4 - (d-a)(d-b)(d-c)(a+b+c+d).$$

162. If $\frac{1}{a} + \frac{1}{b} + \frac{1}{c} = \frac{1}{a+b+c}$, then, for all integral values of n,

$$\frac{1}{a^{2n+1}} + \frac{1}{b^{2n+1}} + \frac{1}{c^{2n+1}} = \frac{1}{(a+b+c)^{2n+1}}.$$

163. If $x + y + z = xyz$, or if $yz + zx + xy = 1$,

$$\frac{2x}{1-x^2} + \frac{2y}{1-y^2} + \frac{2z}{1-z^2} = \frac{2x}{1-x^2}\,\frac{2y}{1-y^2}\,\frac{2z}{1-z^2},$$

and $$\frac{y+z}{1-yz} + \frac{z+x}{1-zx} + \frac{x+y}{1-xy} = \frac{y+z}{1-yz}\,\frac{z+x}{1-zx}\,\frac{x+y}{1-xy}.$$

164. If $yz + zx + xy = (yz)^{-1} + (zx)^{-1} + (xy)^{-1} = m$, then will

$$\frac{(1+yz)(1+zx)(1+xy)}{(1+x^2)(1+y^2)(1+z^2)} = \frac{1+m}{(1-m)^2}.$$

165. Having given the system of equations

$$\frac{bz+cy}{b-c} = \frac{cx+az}{c-a} = \frac{ay+bx}{a-b} = \frac{ax+by+cz}{a+b+c},$$

prove that $(b+c)x + (c+a)y + (a+b)z = 0$, $bcx + cay + abz = 0$, and that either $a+b+c=0$ or $abc = (b-c)(c-a)(a-b)$.

166. If a, b, c be real quantities satisfying the equation

$$a^2 + b^2 + c^2 + 2abc = 1,$$

then will a^2, b^2, c^2 be all less than 1, or all greater than 1.

167. If x, y, z be finite quantities satisfying the equations

$$ax(y + z - x) + by(z + x - y) + cz(x + y - z) = 0,$$

$$a^2x^2 + b^2y^2 + c^2z^2 = 2bcyz + 2cazx + 2abxy,$$

then will

$$\frac{x}{a(b-c)^2} = \frac{y}{b(c-a)^2} = \frac{z}{c(a-b)^2} = \frac{xyz}{abc(x^2 + y^2 + z^2 - yz - zx - xy)}.$$

168. If

$$yz + zx + xy = 0 \text{ and } (b-c)^2 x + (c-a)^2 y + (a-b)^2 z = 0,$$

then will $$x(b-c) = y(c-a) = z(a-b).$$

169. If $x(b-c) + y(c-a) + z(a-b) = 0$, then will

$$\frac{bz - cy}{b - c} = \frac{cx - az}{c - a} = \frac{ay - bx}{a - b}.$$

170. If x, y, z, u be all finite and satisfy the equations

$$x = by + cz + du,$$

$$y = ax + cz + du,$$

$$z = ax + by + du,$$

$$u = ax + by + cz,$$

then will $$\frac{a}{1+a} + \frac{b}{1+b} + \frac{c}{1+c} + \frac{d}{1+d} = 1.$$

171. If $$\frac{y^2 - z^2}{b - c} = \frac{yz}{x}, \text{ and } \frac{z^2 - x^2}{c - a} = \frac{zx}{y},$$

then will $$\frac{x^2 - y^2}{a - b} = \frac{xy}{z};$$

and if $$a + \frac{y^2 - z^2}{b - c} = b + \frac{z^2 - x^2}{c - a},$$

then will each member of the equation be equal to $c + \dfrac{x^2 - y^2}{a - b}$.

172. Having given the equations

$$\frac{yz}{c^4} = \frac{a^2 + x^2}{c^4 + a^2x^2}, \quad \frac{zx}{c^4} = \frac{a^2 + y^2}{c^4 + a^2y^2},$$

prove that $$\frac{xy}{c^4} = \frac{a^2 + z^2}{c^4 + a^2z^2}, \quad yz + zx + xy = a^2,$$

and $$a^2 xyz = c^4 (x + y + z).$$

173. Having given

$$\frac{x}{1-x^2} = \frac{y+z}{m+nyz}, \quad \frac{y}{1-y^2} = \frac{z+x}{m+nzx},$$

prove that, if x, y be unequal,

$$\frac{z}{1-z^2} = \frac{x+y}{m+nxy}, \quad yz+zx+xy+m+1=0,$$

and

$$(yz)^{-1}+(zx)^{-1}+(xy)^{-1}=n-1.$$

174. If $\dfrac{x-\dfrac{yz}{x}}{1-yz} = \dfrac{y-\dfrac{zx}{y}}{1-zx}$ and x, y be unequal, then will each member

of this equation be equal to $\dfrac{z-\dfrac{xy}{z}}{1-xy}$, to $x+y+z$, and to $\dfrac{1}{x}+\dfrac{1}{y}+\dfrac{1}{z}$.

175. If $\dfrac{ayz}{(y-z)^2} = \dfrac{bzx}{(z-x)^2} = \dfrac{cxy}{(x-y)^2}$, each member will be equal to

$$abc \div (a^2+b^2+c^2-2bc-2ca-2ab).$$

176. If x, y be unequal and if

$$\frac{(2x-y-z)^3}{x} = \frac{(2y-z-x)^3}{y},$$

each member will be equal to

$$\frac{(2z-x-y)^3}{z}, \text{ to } 9\left(x^2+y^2+z^2-yz-zx-xy\right),$$

and to $\quad -27\left\{x(y-z)^2+y(z-x)^2+z(x-y)^2\right\} \div (x+y+z).$

177. Having given the equations

$$alx+bmy+cnz = al'x+bm'y+cn'z = ax^2+by^2+cz^2 = 0,$$

prove that $\quad x(mn'-m'n)+y(nl'-n'l)+z(lm'-l'm)=0,$

and that

$$\frac{(\overline{m-m'}z-\overline{n-n'}y)^2}{a} + \frac{(\overline{n-n'}x-\overline{l-l'}z)^2}{b} + \frac{(\overline{l-l'}y-\overline{m-m'}x)^2}{c} = 0.$$

178. Having given the equations

$$lx+my+nz = 0, \quad (b-c)\frac{x}{l}+(c-a)\frac{y}{m}+(a-b)\frac{z}{n} = 0,$$

$$x^2+y^2+z^2 = \frac{(b-c)^2}{l^2}+\frac{(c-a)^2}{m^2}+\frac{(a-b)^2}{n^2},$$

prove that $\quad l^2yz(mz-ny)+m^2zx(nx-lz)+n^2xy(ly-mx)$

$$= \frac{(l^2+m^2+n^2)^{\frac{3}{2}}}{lmn}(b-c)(c-a)(a-b).$$

179. Having given

$$a+b+c+d=a'-b'-c'+d'=0=aa'+bb'+cc'+dd',$$

prove that $$\frac{aa'^2+bb'^2+cc'^2+dd'^2}{\frac{a}{a'}+\frac{b}{b'}+\frac{c}{c'}+\frac{d}{d'}}=b'c'\frac{a+d}{\frac{a}{a'}+\frac{d}{d'}}=a'd'\frac{b+c}{\frac{b}{b'}+\frac{c}{c'}}.$$

180. The equation

$$\frac{a_1}{x+b_1}+\frac{a_2}{x+b_2}+\ldots\ldots+\frac{a_n}{x+b_n}=0$$

will reduce to a simple equation if

$$a_1+a_2+\ldots+a_n=0,$$
$$a_1b_1+a_2b_2+\ldots+a_nb_n=0,$$
$$\ldots\ldots\ldots\ldots\ldots\ldots\ldots\ldots=0,$$
$$a_1b_1^{n-3}+a_2b_2^{n-3}+\ldots+a_nb_n^{n-3}=0,$$

and the single value of x will then be equal to

$$b_1b_2\ldots b_n\frac{\frac{a_1}{b_1}+\frac{a_2}{b_2}+\ldots+\frac{a_n}{b_n}}{a_1b_1^{n-2}+\ldots+a_nb_n^{n-2}}.$$

181. Having given the equations

$$\frac{x}{l(mb+nc-la)}=\frac{y}{m(nc+la-mb)}=\frac{z}{n(la+mb-nc)},$$

prove that $$\frac{mz+ny}{a}=\frac{nx+lz}{b}=\frac{ly+mx}{c};$$

and that $$\frac{l}{x(by+cz-ax)}=\frac{m}{y(cz+ax-by)}=\frac{n}{z(ax+by-cz)}.$$

182. If a, b, c, x, y, z be any six quantities, and

$$a_1=bc-x^2,\quad b_1=ca-y^2,\quad c_1=ab-z^2,$$
$$x_1=yz-ax,\quad y_1=zx-by,\quad z_1=xy-cz;$$

and a_2, b_2, c_2, x_2, y_2, z_2 be similarly formed from a_1, b_1, c_1, x_1, y_1, z_1, and so on; then will

$$\frac{a_{2n}}{a}=\frac{b_{2n}}{b}=\frac{c_{2n}}{c}=\frac{x_{2n}}{x}=\frac{y_{2n}}{y}=\frac{z_{2n}}{z}$$
$$=(ax^2+by^2+cz^2-abc-2xyz)^{\frac{2^{2n}-1}{3}}.$$

183. Prove the following equalities, having given that $a+b+c=0$,

$$\frac{a^5+b^5+c^5}{5}=\frac{a^3+b^3+c^3}{3}\cdot\frac{a^2+b^2+c^2}{2},$$

$$\frac{a^7+b^7+c^7}{7}=\frac{a^5+b^5+c^5}{5}\cdot\frac{a^2+b^2+c^2}{2}=\frac{a^3+b^3+c^3}{3}\cdot\frac{a^4+b^4+c^4}{2},$$

$$\frac{a^{11}+b^{11}+c^{11}}{11}=\frac{a^3+b^3+c^3}{3}\cdot\frac{a^8+b^8+c^8}{2}-\frac{(a^3+b^3+c^3)^3}{9}\cdot\frac{a^2+b^2+c^2}{2}.$$

184. Prove that

$$\frac{(a+b+c)^5-a^5-b^5-c^5}{(a+b+c)^3-a^3-b^3-c^3}\equiv\frac{5}{3}(a^2+b^2+c^2+bc+ca+ab).$$

185. If $a+b+c+d=0$, prove that

$$\frac{a^5+b^5+c^5+d^5}{5}=\frac{a^3+b^3+c^3+d^3}{3}\cdot\frac{a^2+b^2+c^2+d^2}{2}.$$

186. Having given

$$a+b+c+a'+b'+c'=0,\quad a^3+b^3+c^3+a'^3+b'^3+c'^3=0,$$

prove that

$$\frac{a^7+b^7+c^7+a'^7+b'^7+c'^7}{7}=\frac{a^2+b^2+c^2+a'^2+b'^2+c'^2}{2}\cdot\frac{a^5+b^5+c^5+a'^5+b'^5+c'^5}{5}.$$

187. Prove that

$$(2a-b-c+\overline{b-c}\sqrt{-3})^3\equiv(2b-c-a+\overline{c-a}\sqrt{-3})^3.$$

188. If $X=ax+cy+bz,\quad Y=cx+by+az,\quad Z=bx+ay+cz,$ then will

$$X^3+Y^3+Z^3-3XYZ=(a^3+b^3+c^3-3abc)(x^3+y^3+z^3-3xyz).$$

Hence shew how to express the product of any number of factors of this form in a similar form.

[By means of the identity

$$a^3+b^3+c^3-3abc\equiv(a+b+c)(a+\omega b+\omega^2c)(a+\omega^2b+\omega c),$$

where $\omega^2+\omega+1=0$.]

The same equation will be true if

$$X=ax+by+cz,\quad Y=ay+bz+cx,\quad Z=az+bx+cy,$$

and these two are the only essentially different arrangements.

189. If $x+y+z=xyz$ and $x^2=yz$, then will y and z be capable of all values, but x^2 cannot be less than 3.

190. If $x+y+z=x^2+y^2+z^2=2$, then will

$$x(1-x)^2=y(1-y)^2=z(1-z)^2;$$

also the greatest of the three x, y, z lies between $\frac{4}{3}$ and 1, the next between 1 and $\frac{1}{3}$, and the least between $\frac{1}{3}$ and 0; and the difference between the greatest and least cannot be less than 1 nor greater than $\frac{2}{\sqrt{3}}$.

191. Having given the equations

$$(y+z)^2=4a^2yz, \quad (z+x)^2=4b^2zx, \quad (x+y)^2=4c^2xy;$$

prove that $$a^2+b^2+c^2\pm 2abc=1.$$

192. Having given the equations

$$\frac{y}{z}-\frac{z}{y}=a, \quad \frac{z}{x}-\frac{x}{z}=b, \quad \frac{x}{y}-\frac{y}{x}=c;$$

prove that $$a^4+b^4+c^4=2b^2c^2+2c^2a^2+2a^2b^2+a^2b^2c^2.$$

193. Having given the equations

$$\frac{x}{a}(y^2-z^2)+\frac{y}{b}(z^2-x^2)+\frac{z}{c}(x^2-y^2)=0,$$

$$\frac{x}{x'(by'+cz'-ax')}=\frac{y}{y'(cz'+ax'-by')}=\frac{z}{z'(ax'+by'-cz')};$$

prove that, if x, y, z be all finite,

$$\frac{x'}{a}(y'^2-z'^2)+\frac{y'}{b}(z'^2-x'^2)+\frac{z'}{c}(x'^2-y'^2)=0.$$

194. Having given the equations

$$x^3+y^3+z^3=(y+z)(z+x)(x+y),$$
$$a(y^2+z^2-x^2)=b(z^2+x^2-y^2)=c(x^2+y^2-z^2);$$

prove that $$a^3+b^3+c^3=(b+c)(c+a)(a+b).$$

195. If $a(b+c-a)(b^2+c^2-a^2)=b(c+a-b)(c^2+a^2-b^2)$ and a, b be unequal, then will each member be equal to $c(a+b-c)(a^2+b^2-c^2)$ and to $2abc(a+b+c)$; also $4abc+(b+c-a)(c+a-b)(a+b-c)=0$. [This relation is equivalent to $a^3+b^3+c^3=(b+c)(c+a)(a+b)$.]

196. If $x=b^2+c^2-a^2$, $y=c^2+a^2-b^2$, and $z=a^2+b^2-c^2$, prove that

$$y^3z^3+z^3x^3+x^3y^3-xyz(y+z)(z+x)(x+y)$$

is the product of four factors, one of which is

$$4abc+(b+c-a)(c+a-b)(a+b-c),$$

and the other three are formed from this by changing the signs of a, b, c respectively.

197. If $$\frac{x}{a}+\frac{y}{b}+\frac{z}{c}=\frac{x}{a'}+\frac{y}{b'}+\frac{z}{c'}=0;$$
then will
$$\frac{x}{a}\left(\frac{b'}{b}+\frac{c'}{c}-\frac{a'}{a}\right)^2+\frac{y}{b}\left(\frac{c'}{c}+\frac{a'}{a}-\frac{b'}{b}\right)^2+\frac{z}{c}\left(\frac{a'}{a}+\frac{b'}{b}-\frac{c'}{c}\right)^2=0.$$

198. Simplify the fraction
$$\frac{a(b^2-c^2)+b(c^2-a^2)+c(a^2-b^2)}{b(c-a)^2+c(a-b)^2-a(b-c)^2};$$
and thence the fraction whose numerator is
$$a^2(b-c)^4-2bc(a-b)^2(a-c)^2+\text{the two similar expressions},$$
and denominator
$$a^2(b^2-c^2)^2-2bc(a^2-b^2)(a^2-c^2)+\text{the two similar expressions}.$$
[The numerator and denominator in the last case are each equivalent to
$$(b-c)^2(c-a)^2(a-b)^2.]$$

199. If $b^2+bc+c^2=3y^2+2yz+3z^2$, $c^2+ca+a^2=3z^2+2zx+3x^2$, and $a^2+ab+b^2=3x^2+2xy+3y^2$, then will
$$3(bc+ca+ab)^2=32\{y^2z^2+z^2x^2+x^2y^2+xyz(x+y+z)\}.$$

VI. *Inequalities.*

[The symbols employed in the following questions are always supposed to denote real quantities.

The fundamental proposition on which the solution generally depends is $a^2+b^2>2ab$.

Limiting values of certain expressions involving an unknown quantity in the second degree only may be found from the condition that a quadratic equation shall have real roots :—*e.g.* "To find the greatest and least values of $\frac{x^2-4x+7}{x^2-2x+4}$." Assuming the expression $\equiv y$, we obtain the quadratic in x,
$$x^2(1-y)-2(2-y)x+7-4y=0,$$
and if x be a real quantity satisfying this equation we must have
$$(2-y)^2>(1-y)(7-4y),$$
or
$$3y^2-7y+3<0,$$
so that y must lie between $\frac{7-\sqrt{13}}{6}$ and $\frac{7+\sqrt{13}}{6}$, which are accordingly the least and greatest possible values of the expression.]

200. If x, y, z be three positive quantities whose sum is unity, then will

$$(1-x)(1-y)(1-z) > 8xyz.$$

201. Prove that

$$4(a^4+b^4+c^4+d^4) > (a+b+c+d)(a^3+b^3+c^3+d^3) > (a^2+b^2+c^2+d^2)^2 > 16abcd.$$

202. Prove that

$$\{8a^2b^2c^2 + (b^2+c^2-a^2)(c^2+a^2-b^2)(a^2+b^2-c^2)\}^2$$
$$> 3\{2b^2c^2 + 2c^2a^2 + 2a^2b^2 - a^4 - b^4 - c^4\}^3$$

except when $a = b = c$.

203. If a, b, c be positive and not all equal

$$a^3 + b^3 + c^3 + 3abc > a^2(b+c) + b^2(c+a) + c^2(a+b).$$

204. If $(a+b+c)^2 < 4(b+c)(c+a)(a+b)$, then will

$$a^2 + b^2 + c^2 < 2bc + 2ca + 2ab.$$

205. If a, b, c be positive and not all equal, the expression

$$a^n(a-b)(a-c) + b^n(b-c)(b-a) + c^n(c-a)(c-b)$$

will be positive for all integral values of n, and for the values 0 and -1.

206. Prove that, if n be a positive whole number,

$$\left(\frac{n+1}{2}\right)^n > |\underline{n} > n^{\frac{n}{2}};$$

and that $$\left(2-\frac{1}{n}\right)\left(2-\frac{3}{n}\right)\ldots\left(2-\frac{2n-1}{n}\right) > \frac{1}{|\underline{n}}.$$

207. If a, b, c be the sides of a triangle, then will

$$\frac{1}{b+c-a} + \frac{1}{c+a-b} + \frac{1}{a+b-c} > \frac{1}{a} + \frac{1}{b} + \frac{1}{c} > \frac{9}{a+b+c};$$

and $(b+c-a)^2(c+a-b)^2(a+b-c)^2 > (b^2+c^2-a^2)(c^2+a^2-b^2)(a^2+b^2-c^2)$;

also x, y, z being any real quantities,

$$a^2(x-y)(x-z) + b^2(y-z)(y-x) + c^2(z-x)(z-y)$$

cannot be negative. If $x+y+z=0$, $a^2yz + b^2zx + c^2xy$ cannot be positive.

208. If $$xyz = (1-x)(1-y)(1-z)$$

the greatest value of either of these equals is $\frac{1}{8}$, x, y, z being each positive and less than 1.

209. Prove that

$$(ax\overline{b+c} + by\overline{c+a} + cz\overline{a+b})^2 > 4abc(x+y+z)(ax+by+cz);$$

a, b, c, x, y, z being all positive and a, b, c unequal.

210. Prove that, for real values of x,

$$2(a-x)(x+\sqrt{x^2+b^2})<a^2+b^2.$$

211. Find the greatest numerical values without regard to sign which the expression

$$(x-8)(x-14)(x-16)(x-22)$$

can have for values of x between 8 and 22.

[When x lies between 8 and 14 the expression is negative and has the greatest numerical value 576 when $x=10$; when x lies between 14 and 16 the expression is positive and has its greatest value 49 when $x=15$; and when x lies between 16 and 22 the expression is again negative and has again the greatest numerical value 576 when $x=20$.]

212. If $a>b$, and c be positive, the greatest value which the expression

$$16(x-a)(x-b)(x-a-c)(x-b+c)$$

can have for values of x between $b-c$ and $a+c$ is $(a-b)^2(a-b+2c)^2$.

213. If $p>m$,

$$\frac{x^2-2mx+p^2}{x^2+2mx+p^2}>\frac{p-m}{p+m} \text{ and } <\frac{p+m}{p-m}.$$

214. The expression $\dfrac{ax^2+bx+c}{cx^2+bx+a}$

will be capable of all values whatever if

$$b^2>(a+c)^2;$$

there will be two values between which it cannot lie if

$$b^2<(a+c)^2 \text{ and } >4ac;$$

and two values between which it must lie if

$$b^2<4ac.$$

215. The expression $\dfrac{(x-a)(x-b)}{(x-c)(x-d)}$

can have any real value whatever if one and only one of the two a, b lie between c and d: otherwise there will be two values between which it cannot lie.

216. The expression $\dfrac{x+a}{x^2+bx+c^2}$

will always lie between two fixed limits if $b^2<4c^2$; there will be two limits between which it cannot lie if $a^2+c^2>ab$ and $b^2>4c^2$; and the expression will be capable of all values if $a^2+c^2<ab$.

217. The expression $\dfrac{ax^2 + 2hx + b}{a'x^2 + 2h'x + b'}$

will be capable of all values, provided that

$$(ab' - a'b)^2 < 4(a'h - ah')(h'b - hb')$$

{or, which is equivalent, $(2hh' - ab' - a'b)^2 < 4(h^2 - ab)(h'^2 - a'b')$}.

Prove that this inequality involves the two

$$h^2 > ab,\ h'^2 > a'b';$$

and investigate the condition (1) that two limits exist between which the value of the expression cannot lie, (2) that two limits exist between which the value of the expression must lie.

[(1) $(ab' - a'b)^2 > 4(a'h - ah')(bh' - b'h),\ h^2 > ab,$

(2) $(ab' - a'b)^2 > 4(a'h - a'h)(bh' - b'h),\ h^2 < ab.$]

218. If $x_1, x_2, x_3, \ldots x_n$ be real quantities such that

$$x_1^2 + x_2^2 + \ldots + x_n^2 - x_1x_2 - x_2x_3 - \ldots - x_{n-1}x_n - x_n + \frac{n-1}{2n} = 0,$$

then will $0, x_1, x_2, x_3, \ldots x_n, 1$ be in ascending order of magnitude.

219. If $x_1^2 + x_2^2 + \ldots + x_n^2 + x_2x_3 + x_3x_1 + x_1x_2 + \ldots = 1$, then none of the quantities $x_1^2, x_2^2, \ldots x_n^2$ can be greater than $\dfrac{2n}{n+1}$; and their sum must lie between 2 and $\dfrac{2}{n+1}$.

220. If $x_1^2 + x_2^2 + \ldots x_n^2 + 2m(x_2x_3 + x_3x_1 + x_1x_2 + \ldots) = 1$, m being positive and < 1, then none of the quantities $x_1^2, x_2^2, \ldots x_n^2$ can be greater than $\dfrac{1 + m\overline{n-2}}{(1-m)(1 + m\overline{n-1})}$: and their sum must lie between $\dfrac{1}{1-m}$ and $\dfrac{1}{1 + m\overline{n-1}}$.

221. If $x_1^2 + x_2^2 + \ldots + x_n^2 - x_1x_2 - x_2x_3 - \ldots - x_{n-1}x_n = \dfrac{n+1}{2}$, then will $x_r^2 < r(n+1-r)$, and the greatest and least values of $x_1^2 + x_2^2 + \ldots + x_n^2$ will be $\dfrac{n+1}{a_1}, \dfrac{n+1}{a_2}, \dfrac{n+1}{a_3}, \ldots \dfrac{n+1}{a_n}$, where $a_1, a_2, \ldots a_n$ are the roots of the equation in z,

$$(z-2)^n - (n-1)(z-2)^{n-2} + \frac{(n-2)(n-3)}{\lfloor 2}(z-2)^{n-4}$$

$$- \frac{(n-3)(n-4)(n-5)}{\lfloor 3}(z-2)^{n-6} + \ldots = 0.$$

222. If $2x_2 - x_1 = \frac{2x_3 - x_2}{2} = \frac{2x_4 - x_3}{2^2} = \ldots = \frac{2x_{n+1} - x_n}{2^{n-1}} = a$, then will

$$\frac{x_{p+1} - x_{q+1}}{x_{p+r+1} - x_{q-r+1}} = \frac{2^p - 2^q}{2^{p+r} - 2^{q-r}},$$

so that if $x_{n+1} = x_1$, $x_{r+2} = x_{n-r}$: and in this case $x_{r+1} = \frac{a}{3}(2^r + 2^{n-r})$, and has its least value when $r = \frac{n}{2}$, or $\frac{n+1}{2}$.

223. If $2x_2 - x_1 = \frac{2x_3 - x_2}{2} = \frac{2x_4 - x_3}{3} = \ldots = \frac{2x_{n+1} - x_n}{n} = a$, then will $2^n x_{n+1} - x_1 = a(1 + \overline{n-1}\,2^n)$: and if $x_{n+1} = x_1$ each will be equal to $a\left(n - 1 + \frac{n}{2^n - 1}\right)$, and $x_{r+1} = a\left(r - 1 + \frac{n2^{n-r}}{2^n - 1}\right)$, and will have its least value when r is the integer next below $\log_2 n$.

VII. *Proportion, Variation, Scales of Notation.*

224. If $b+c+d$, $c+d+a$, $d+a+b$, $a+b+c$ be proportionals, then will

$$\frac{a^3 - d^3}{a - d} = \frac{b^3 - c^3}{b - c}.$$

225. If y vary as the sum of three quantities of which the first is constant, the second varies as x, and the third as x^2: and if $(a, 0)$, $(2a, a)$, $(3a, 4a)$ be three pairs of simultaneous values of x and y, then when $x = na$, $y = (n-1)^2 a$.

226. A triangle has two sides given in position and a given perimeter $2s$: if c be the length of the side opposite to the given angle, the area of the triangle will vary as $s - c$.

227. The radix of the scale in which 49 denotes a square number must be of the form $(r+1)(r+4)$, where r is some whole number.

228. The radix of a scale being $4r+2$, prove that if the digit in the units' place of any number N be either $2r+1$ or $2r+2$, N^2 will have the same digit in the units' place.

229. Find a number (1) of three digits, (2) of four digits, in the denary scale such that if the first and last digits be interchanged the result represents the same number in the nonary scale: and prove that there is only one solution in each case.

[The numbers are 445, 5567 respectively.]

230. If the radix of any scale have more than one prime factor there will exist two and only two digits different from unity such that if any number N have one of these digits in the units' place, N^x will have the same digit in the units' place.

231. Prove that the product of the numbers denoted by 10, 11, 12, 13, increased by 1, will be the square of the number denoted by 131, whatever be the scale of notation.

232. Prove that $\lfloor 2m-1 \div \lfloor m \lfloor m-1$ is always an even number except when m is a power of 2, and the index of the power of 2 contained in it $= q - p$, where q is the sum of the digits of $2m-1$ when expressed in the binary scale and 2^p is the highest power of 2 which is a divisor of m.

233. The index of the highest power of q which is a divisor of $\lfloor pq \div (\lfloor p)^q$ is the sum of the digits of p when expressed in the scale whose radix is q.

VIII. *Arithmetical, Geometrical, and Harmonical Progressions.*

234. If the sum of m terms of an A. P. be to the sum of n terms as $m^2 : n^2$; prove that the mth term will be to the nth term as

$$2m-1 : 2n-1.$$

235. The series of natural numbers is divided into groups 1; 2, 3, 4; 5, 6, 7, 8, 9; and so on: prove that the sum of the numbers in the nth group is $n^3 + (n-1)^3$.

236. The sum of the products of every two of n terms of an A. P., whose first term is a and last term l, is

$$\frac{n(n-2)(3n-1)(a+l)^2 + 4n(n+1)al}{24(n-1)}.$$

237. The sum of the products of every three of n terms of an A. P., whose first term is a and last term l, is

$$\frac{n(n-2)(a+l)}{48(n-1)}\{n(n-3)(a+l)^2 + 4(n+1)al\},$$

and lies between $\dfrac{n(n-1)(n-2)}{12} al(a+l)$ and $\dfrac{n(n-1)(n-2)}{48}(a+l)^3$; and the sum of the products of every three of n consecutive whole numbers beginning with r is

$$\frac{n(n-1)(n-2)}{48}\{(n+2r-1)^3 - (n+1)(n+2r-1)\}.$$

238. Having given that $\dfrac{a}{b-c}$, $\dfrac{b}{c-a}$, and $\dfrac{c}{a-b}$ are in A. P.; prove that

$$\frac{a^3+c^3-2b^3}{a^2+c^2-2b^2} = \frac{a+b+c}{2}.$$

239. If a, b, c; b, c, a; or c, a, b be in A. P., then will

$$\tfrac{2}{9}(a+b+c)^3 = a^2(b+c) + b^2(c+a) + c^2(a+b);$$

and if in G. P.,

$$b^3c^3 + c^3a^3 + a^3b^3 = abc(a^3 + b^3 + c^3).$$

240. If a, l be the first and n^{th} terms of an A. P. the continued product of all the n terms will be

$$> (al)^{\frac{n}{2}} \text{ and } < \left(\frac{a+l}{2}\right)^n.$$

241. The first term of a G. P. is a and the nth term l; prove that the rth term is

$$(a^{n-r} l^{r-1})^{\frac{1}{n-1}}.$$

242. If a, b, c be in A. P., α, β, γ in H. P., and $a\alpha, b\beta, c\gamma$ in G. P., then will

$$a : b : c :: \frac{1}{\gamma} : \frac{1}{\beta} : \frac{1}{\alpha}.$$

243. The first term of an H. P. is a and the nth term l, prove that the rth term is

$$\frac{(n-1)\,al}{(n-r)\,l + (r-1)\,a}.$$

Prove that the sum of these n terms is $< (a+l)\frac{n}{2}$; and their continued product $< (al)^{\frac{n}{2}}$.

244. If a, b, c be in H. P., then will

$$\frac{1}{b-c} + \frac{4}{c-a} + \frac{1}{a-b} = \frac{1}{c} - \frac{1}{a}.$$

245. If a, b, c, d be four positive quantities in H. P.,

$$a + d > b + c.$$

246. Prove that $b+c$, $c+a$, $a+b$ will be in H. P., if a^2, b^2, c^2 be in A. P.

247. If three numbers be in G. P. and the mean be added to each of the three, the three sums will be in H. P.

248. Prove that, for all values of x except -1,

$$\frac{x + x^2 + x^3 + \dots + x^{2n}}{1 + x^{2n+1}} < n.$$

[For $\frac{x^{r+1} + x^{2n-r}}{1 + x^{2n+1}} < 1$ if $\frac{(1 - x^{r+1})(1 - x^{2n-r})}{1 + x^{2n+1}}$ be positive.]

249. An A. P., a G. P., and an H. P. have each the same first and last terms and the same number of terms (n), and the rth terms are a_r, b_r, c_r; prove that

$$a_{r+1} : b_{r+1} = b_{n-r} : c_{n-r};$$

and thence that if A, B, C be the respective continued products of the n terms

$$AC = B^2.$$

250. If n harmonic means be inserted between two positive quantities a and b, the difference between the first and last of these means bears to the difference between a and b a ratio less than

$$n - 1 : n + 1.$$

251. If a_0, a_1, a_2 ... be an A. P., b_0, b_1, b_2 ... a G. P., and A, B, C any three consecutive terms of the series a_0b_0, a_1b_1, a_2b_2, ..., then will

$$b_0^2C - 2b_0b_1B + b_1^2A = 0;$$

and if A, B, C, D be any four consecutive terms of the series $a_0 + b_0$, $a_1 + b_1$, $a_2 + b_2$, ..., then will

$$Ab_1 - B(b_0 + 2b_1) + C(b_1 + 2b_0) - Db_0 = 0.$$

IX. *Permutations and Combinations.*

[The number of permutations of n different things taken r together is denoted by ${}_nP_r$, and the corresponding number of combinations by ${}_nC_r$.]

252. Prove *a priori* that

$$\begin{aligned}{}_nP_r &\equiv {}_{n-2}P_r + 2r\,{}_{n-2}P_{r-1} + r(r-1)\,{}_{n-2}P_{r-2},\\ &\equiv {}_{n-3}P_r + 3r\,{}_{n-3}P_{r-1} + 3r(r-1)\,{}_{n-3}P_{r-2} + r(r-1)(r-2)\,{}_{n-3}P_{r-3},\\ &\equiv {}_{n-p}P_r + pr\,{}_{n-p}P_{r-1} + \frac{p(p-1)}{2}r(r-1)\,{}_{n-p}P_{r-2} + \dots\\ &\qquad + r(r-1)\dots(r-p+1)\,{}_{n-p}P_{r-p},\end{aligned}$$

p being a whole number $< r$.

253. In the expansion of $(a_1 + a_2 + \dots + a_p)^n$, where n is any whole number not greater than p, prove that the coefficient of any term in which none of the quantities a_1, a_2 ... a_p appears more than once is $\lfloor n$.

254. The number of permutations of n different letters taken all together in which no letter occupies the same place as in a certain given permutation is

$$\lfloor n \left\{ \frac{1}{\lfloor 2} - \frac{1}{\lfloor 3} + \frac{1}{\lfloor 4} - \dots + \frac{(-1)^n}{\lfloor n} \right\}.$$

255. Prove that

$$ {}_nC_r \equiv {}_{n+2}C_r - 2\,{}_{n+2}C_{r-1} + 3\,{}_{n+2}C_{r-2} - \dots + (-1)^r (r+1), $$

$$ \equiv {}_{n+p}C_r - p\,{}_{n+p}C_{r-1} + \frac{p(p+1)}{\underline{|2}}\,{}_{n+p}C_{r-2} - \dots + (-1)^r \frac{p(p+1)\dots(p+r-1)}{\underline{|r}}. $$

256. The number of combinations of $2n$ things taken n together when n of the things and no more are alike is 2^n; and the number of combinations of $3n$ things, n together, when n of the things and no more are alike is

$$ 2^{2n-1} + \frac{\underline{|2n}}{2\,(\underline{|n})^2}. $$

257. The number of ways in which mn different things can be distributed among m persons so that each person shall have n of them is

$$ \frac{\underline{|mn}}{(\underline{|n})^m}. $$

258. There are p suits of cards, each suit consisting of q cards numbered from 1 to q; prove that the number of sets of q cards numbered from 1 to q which can be made from all the suits is p^q.

259. If there be n straight lines lying in one plane the number of different n-sided polygons formed by them is $\frac{1}{2}\underline{|n-1}$.

260. The number of ways in which p things may be distributed among q persons so that everybody may have one at least is

$$ q^p - q(q-1)^p + \frac{q(q-1)}{\underline{|2}}(q-2)^p - \dots $$

261. The number of ways in which r things may be distributed among $n+p$ persons so that certain n of those persons may each have one at least is (S_r)

$$ (n+p)^r - n(n+p-1)^r + \frac{n(n-1)}{\underline{|2}}(n+p-2)^r \dots $$

Hence prove that

$$ S_1 = S_2 = \dots = S_{n-1} = 0,\ S_n = \underline{|n},\ S_{n+1} = \left(\frac{n}{2} + p\right)\underline{|n+1}. $$

X. *Binomial Theorem.*

262. Prove that

(1) $$ 1 - n\frac{1+x}{1+nx} + \frac{n(n-1)}{\underline{|2}}\frac{1+2x}{(1+nx)^2} - \frac{n(n-1)(n-2)}{\underline{|3}}\frac{1+3x}{(1+nx)^3} + \dots = 0; $$

(2) $1+3\dfrac{2n+1}{2n-1}+5\left(\dfrac{2n+1}{2n-1}\right)^2+\dots$

$$+(2n-1)\left(\frac{2n+1}{2n-1}\right)^{n-1}\equiv n(2n-1),$$

n being a whole number.

263. Determine a, b, c, d, e in order that the nth term in the expansion of

$$\frac{a+bx+cx^2+dx^3+ex^4}{(1-x)^5}$$

may be n^4x^{n-1}.

[The numerator is $1+11x+11x^2+x^3$.]

264. Prove that the series

$$1^n+2^nx+3^nx^2+\dots+r^nx^{r-1}+\dots$$

is the expansion of a function of x of the form

$$\frac{a_0+a_1x+a_2x^2+\dots+a_nx^n}{(1-x)^{n+1}};$$

also prove that

$$a_n=0,\ a_{n-1}=a_0=1,\ a_{n-2}=a_1=2^n-(n+1);\ a_{n-r}=a_{r-1}.$$

265. The sum of the first $r+1$ **coefficients** of the expansion of $(1-x)^{-m}$ **is equal to** $\dfrac{\lfloor m+r}{\lfloor m \lfloor r}$.

266. If from the sum of n different quantities be severally subtracted r times each one of the quantities and the n remainders be multiplied together, the coefficient in this product of the term which involves all the n quantities **is**

$$\lfloor n\left\{1-r+\frac{r^2}{\lfloor 2}-\frac{r^3}{\lfloor 3}+\dots+\frac{(-r)^n}{\lfloor n}\right\}.$$

267. Prove that

(1) $2^n-(n-1)2^{n-2}+\dfrac{(n-2)(n-3)}{\lfloor 2}2^{n-4}-\dfrac{(n-3)(n-4)(n-5)}{\lfloor 3}2^{n-6}$

$$+\dots\equiv n+1,$$

(2) $(m+1)^n-(n-1)m(m+1)^{n-2}+\dfrac{(n-2)(n-3)}{\lfloor 2}m^2(m+1)^{n-4}$

$$-\frac{(n-3)(n-4)(n-5)}{\lfloor 3}m^3(m+1)^{n-6}+\dots\equiv\frac{m^{n+1}-1}{m-1},$$

(3) $(p+q)^n-(n-1)pq(p+q)^{n-2}+\dfrac{(n-2)(n-3)}{\lfloor 2}p^2q^2(p+q)^{n-4}$

$$-\frac{(n-3)(n-4)(n-5)}{\lfloor 3}p^3q^3(p+q)^{n-6}+\dots\equiv\frac{p^{n+1}-q^{n+1}}{p-q},$$

n being a positive integer.

268. **If p be nearly equal to** q, then will $\frac{q+2p}{p+2q}$ be nearly equal to $\sqrt[3]{\frac{p}{q}}$.

269. **If p** be nearly equal to **q**, $\frac{(n+1)p+(n-1)q}{(n-1)p+(n+1)q}$ is **a** close approximation to $\left(\frac{p}{q}\right)^{\frac{1}{n}}$; and if $\frac{p}{q}$ differ from 1 only **in the $\overline{r+1}^{th}$** decimal place, this approximation will be correct to $2r$ places.

270. If a_r denote the coefficient of x^r in the expansion of $\left(\frac{1+x}{1-x}\right)^n$ in **a series** of ascending powers of x, the following relation will hold among any three consecutive coefficients,

$$(r+1)a_{r+1} - 2na_r - (r-1)a_{r-1} = 0.$$

271. If $\frac{(1+x)^n}{(1-x)^2}$ be expanded in ascending **powers of x, the coefficient** of x^{n+r-1} is $(n+2r)2^{n-1}$, **n, r being positive** integers (including zero).

272. If $(1+x)^n \equiv a_0 + a_1x + \dots + a_rx^r + \dots$, **then will**

$$\frac{a_1^2 + 2a_2^2 + 3a_3^2 + \dots + na_n^2}{a_0^2 + a_1^2 + a_2^2 + \dots + a_n^2} = \frac{n}{2},$$

n being a positive **integer.**

273. Prove that

$$2^n + \frac{n(n-1)}{1^2}2^{n-2} + \frac{n(n-1)(n-2)(n-3)}{1^2 2^2}2^{n-4} + \dots = {}_{2n}C_n.$$

274. The sum of the first **n coefficients of the expansion in ascending** powers of x of $\frac{(1+x)^n}{(1-x)^3}$ is $\frac{n(n+2)(n+7)}{3}2^{n-4}$, **$n$ being a positive** integer.

275. Prove that, if n be a positive integer,

$$1 + 3n + \frac{3\,.\,4}{1\,.\,2}\frac{n(n-1)}{\lfloor 2} + \frac{4\,.\,5}{1\,.\,2}\frac{n(n-1)(n-2)}{\lfloor 3} + \dots$$
$$+ \frac{(n+1)(n+2)}{2} \equiv (n^2 + 7n + 8)2^{n-3}.$$

276. The coefficient of x^{n+r-2} in the expansion of $(1+x)^n(1-x)^{-3}$ is $2^{n-3}\{(n+2r)(n+2r-2)+n\}$; and that of x^{n+r-3} in the expansion of $(1+x)^n(1-x)^{-4}$ is $\frac{1}{3}(n+2r-2)2^{n-4}\{(n+2r)(n+2r-4)+3n\}$.

277. If the expansion of $(1+x)^n(1+x^2)^n(1+x^4)^n$ be $1 + a_1x + a_2x^2 + \dots + a_rx^r + \dots$ and $S_1 \equiv a_1 + a_8 + a_{15} + \dots$, $S_2 = a_2 + a_9 + a^{16} + \dots$, and so **on to S_7, then will** $S_1 = S_2 = S_3 = \dots = S_7 = \frac{1}{7}(2^{7n} - 1)$.

278. If the expansion of $(1 + x + x^2 + \ldots + x^p)^n$ be $1 + a_1x + a_2x^2 + \ldots + a_rx^r + \ldots$ and $S_1 = a_1 + a_{p+1} + a_{2p+1} + \ldots$, $S_2 = a_2 + a_{p+2} + a_{2p+2} + \ldots$, and so on to S_p, then will $S_1 = S_2 = \ldots = S_p$.

279. Prove that $\dfrac{a^{n+2}(b-c) + b^{n+2}(c-a) + c^{n+2}(a-b)}{a^2(b-c) + b^2(c-a) + c^2(a-b)}$ is equal to the sum of the homogeneous products of n dimensions of a, b, c.

280. Prove that the coefficient of x^r in the expansion of

$$(1-ax)^{-2}(1-bx)^{-2}$$

in ascending powers of x is

$$\frac{(r+1)(a^{r+3}-b^{r+3}) - (r+3)ab(a^{r+1}-b^{r+1})}{(a-b)^3}.$$

281. If $x = \dfrac{a + n(a-b)}{b + n(a-b)}$, $\dfrac{a - bx}{(1-x)^2}$ will be equal to the sum of the first n terms of its expansion in ascending powers of x; a, b being any unequal quantities.

XI. *Exponential and Logarithmic Series.*

[In the questions under this head, n always denotes a positive whole number.]

282. Prove the following identities:—

(1) $n^n - (n+1)(n-1)^n + \dfrac{(n+1)n}{\underline{|2}}(n-2)^n - \ldots$ to n terms $\equiv 1$,

(2) $(n-1)^n - n(n-2)^n + \dfrac{n(n-1)}{\underline{|2}}(n-3)^n - \ldots$ to $\overline{n-1}$ terms $\equiv \underline{|n} - 1$,

(3) $(n-2)^n - n(n-3)^n + \dfrac{n(n-1)}{\underline{|2}}(n-4)^n - \ldots$ to $\overline{n-2}$ terms $\equiv \underline{|n} + n - 2^n$,

(4) $1^n - n2^n + \dfrac{n(n-1)}{\underline{|2}}3^n - \ldots$ to $\overline{n+1}$ terms $\equiv (-1)^n \underline{|n}$.

[(1) is obtained by means of the expansion of $\epsilon^{-x}(\epsilon^x - 1)^{n+1}$, (2) from that of $\epsilon^{-x}(\epsilon^x - 1)^n$, (3) from that of $\epsilon^{-2x}(\epsilon^x - 1)^{n+1}$, and (4) from that of $\epsilon^{\overline{n+1}x}(\epsilon^{-x} - 1)^n$.]

283. Prove the following identities by the consideration of the coefficients of powers of x in the expansions of $(\epsilon^x - \epsilon^{-x})^n$ and of its equivalent $\left\{2x + \dfrac{2x^3}{\underline{|3}} + \ldots\right\}^n$:—

(1) $n^{n-2r} - n(n-2)^{n-2r} + \frac{n(n-1)}{\underline{|2}}(n-4)^{n-2r} - \ldots \equiv 0,$

(2) $n^n - n(n-2)^n + \frac{n(n-1)}{\underline{|2}}(n-4)^n - \ldots \equiv \underline{|n}\, 2^{n-1},$

(3) $n^{n+2} - n(n-2)^{n+2} + \frac{n(n-1)}{\underline{|2}}(n-4)^{n+2} - \ldots \equiv \frac{n}{\underline{|3}}\underline{|n+2}\, 2^{n-1};$

the number of terms in each series being $\frac{n}{2}$ or $\frac{n+1}{2}$, and r a whole number $< \frac{n}{2}$.

284. If S_r denote the series

$$(2n+1)^r - (2n+1)(2n-1)^r + \frac{(2n+1)\,2n}{\underline{|2}}(2n-3)^r - \ldots \text{ to } n+1 \text{ terms,}$$

then will $S_1 = S_3 = S_5 = \ldots = S_{2n-1} \equiv 0,\ S_{2n+1} \equiv 2^{2n}\,\underline{|2n+1},$

and $$S_{2n+3} \equiv \frac{2^{2n}(2n+1)\,\underline{|2n+3}}{\underline{|3}}.$$

285. Prove that the sums of the infinite series

(1) $\frac{1}{1.2.3} + \frac{1}{3.4.5} + \frac{1}{5.6.7} + \ldots$

(2) $\frac{1}{1.2.3.4} + \frac{1}{3.4.5.6} + \frac{1}{5.6.7.8} + \ldots$

(3) $\frac{1}{1.2.3.4} + \frac{4}{3.4.5.6} + \frac{9}{5.6.7.8} + \ldots$

are respectively $\log 2 - \frac{1}{2}$, $\frac{2}{3}\log 2 - \frac{5}{12}$, and $\frac{1}{6}\log 2 - \frac{1}{24}$; and that, if S_n denote the sum of the infinite series

$$\frac{1}{1.2.3\ldots(n+2)} + \frac{1}{3.4.5\ldots(n+4)} + \frac{1}{5.6.7\ldots(n+6)} + \ldots,$$

$$(n+1)\,S_n = 2S_{n-1} - \frac{1}{n\,\underline{|n}}.$$

286. The coefficient of x^r in the expansion of $(1+x)^n$ being denoted by a_r, prove that

$$a_0 p^{n-1} - a_1(p-1)^{n-1} + a_2(p-2)^{n-1} - \ldots + (-1)^{p-1}a_{p-1}$$
$$\equiv a_0(n-p)^{n-1} - a_1(n-p-1)^{n-1} + a_2(n-p-2)^{n-1} - \ldots + (-1)^{n-p-1}a_{n-p-1},$$

p being a whole number $< n$.

[From the expansion of $\epsilon^{-px}(\epsilon^x - 1)^n$ containing no lower power of x than x^n, so that the coefficient of x^{n-1} is zero. This result might be used to prove (264).]

287. By means of the identity

$$\log(1-x^3) \equiv \log(1-x) + \log(1+x+x^2),$$

prove that the sum of n terms of the series

$$1 - \frac{n(n+1)}{\lfloor 3} + \frac{(n-1)n(n+1)(n+2)}{\lfloor 5} - \ldots$$

is 0 if n be of the forms $3r$ or $3r-1$; and $\frac{3(-1)^{n-1}}{2n+1}$ if n be of the form $3r+1$; also that the sum of n terms of the series

$$1 - \frac{n^2}{\lfloor 2} + \frac{n^2(n^2-1^2)}{\lfloor 4} - \ldots$$

is $(-1)^{n-1}$ if n be of the forms $3r \pm 1$ and $\frac{(-1)^n}{2}$ if n be of the form $3r$.

288. By means of the identity

$$\log(1-x+x^2) + \log(1+x+x^2) \equiv \log(1+x^2+x^4),$$

prove that, if $f(n)$ denote the sum of $n+1$ terms of the series

$$1 - \frac{n^2}{\lfloor 2} + \frac{n^2(n^2-1^2)}{\lfloor 4} - \frac{n^2(n^2-1^2)(n^2-2^2)}{\lfloor 6} + \ldots,$$

$f(2n) = (-1)^n f(n)$; and that, if $F(n)$ denote the sum of $n+1$ terms of the series

$$1 - \frac{n(n+1)}{\lfloor 3} + \frac{(n-1)n(n+1)(n+2)}{\lfloor 5} - \ldots,$$

$$(2n+1)F(n) = (-1)^n \text{ or } 2(-1)^{n-1}.$$

289. By means of the identity

$$\log\frac{x^3-3x+2}{x^3-3x-2} \equiv \log\frac{x+2}{x-2} - 2\log\frac{x+1}{x-1},$$

prove that

$$3^{n-1} + 2^2 3^{n-4}\frac{(n-3)(n-2)}{\lfloor 3} + 2^4 3^{n-7}\frac{(n-6)(n-5)(n-4)(n-3)}{\lfloor 5}$$
$$+ 2^6 3^{n-10}\frac{(n-9)(n-8)\ldots(n-4)}{\lfloor 7} + \ldots \equiv \frac{2^{2n}-1}{2n+1};$$

the series being continued so long as the indices of the powers of 3 are positive.

290. Denoting by u_n the series

$$1^n + 2^n + \frac{3^n}{\lfloor 2} + \ldots + \frac{(r+1)^n}{\lfloor r} + \ldots \text{ to infinity,}$$

prove that

$$u_{n-1}+u_n\equiv u_{n+1}-nu_n+\frac{n(n-1)}{\lfloor 2}u_{n-1}-\frac{n(n-1)(n-2)}{\lfloor 3}u_{n-2}+\ldots+(-1)^n u_1$$
$$\equiv \Delta^n u_1,$$

and that $u_{n+1}-u_n\equiv u_n+nu_{n-1}+\frac{n(n-1)}{\lfloor 2}u_{n-2}+\ldots+nu_1+u_0;$

and by means of either of these prove that $u_7=4140e$.

291. If u_n denote the infinite series

$$1^{n-1}-2^{n-1}+\frac{3^{n-1}}{\lfloor 2}-\frac{4^{n-1}}{\lfloor 3}+\ldots$$

then will $1=u_{n+1}+u_n+nu_{n-1}+\frac{n(n-1)}{\lfloor 2}u_{n-2}+\ldots+nu_1+u_0$.

292. If $$v_n\equiv\frac{1}{\lfloor 2}-\frac{1}{\lfloor 3}+\frac{1}{\lfloor 4}-\ldots+\frac{(-1)^{n+1}}{\lfloor n+1};$$

then will $$1=v_n+v_{n-1}+\frac{v_{n-2}}{\lfloor 2}+\ldots+\frac{v_1}{\lfloor n-1}+\frac{1}{\lfloor n+1}.$$

293. Having given

$$u_n=nv_{n-1},\ v_n=u_n+u_{n-1},$$

prove that the limit of $\frac{u_n}{\lfloor n+1}$ when n is indefinitely increased is

$$u_0+\frac{u_1-2u_0}{e}.$$

294. If there be a series of terms $u_0, u_1, u_2, \ldots u_n \ldots$, of which any one is obtained from the preceding by the formula

$$u_n=nu_{n-1}+(-1)^n$$

and if $u_0=1$, then will

$$\lfloor n\equiv u_n+nu_{n-1}+\frac{n(n-1)}{\lfloor 2}u_{n-2}+\ldots+\frac{n(n-1)}{\lfloor 2}u_2+nu_1+u_0.$$

Prove also that $\frac{u_n}{\lfloor n}$ tends to become equal to $\frac{1}{e}$ as n increases indefinitely.

295. Prove that

$$\frac{2^{n+2}-2}{n+2}\equiv 2^n-\frac{n-1}{\lfloor 2}2^{n-2}+\frac{(n-2)(n-3)}{\lfloor 4}2^{n-4}-\ldots$$

and that

$$p^n + q^n \equiv (p+q)^n - npq\,(p+q)^{n-2} + \frac{n\,(n-3)}{\underline{|2}}\,p^2q^2\,(p+q)^{n-4}$$
$$- \frac{n\,(n-4)\,(n-5)}{\underline{|3}}\,p^3q^3\,(p+q)^{n-6} + \ldots\ldots$$

[By means of the expansions of the identicals

$$\log(1-px) + \log(1-qx),\ \log\{1 - x\,(p+q-pqx)\};$$

or by expanding in ascending powers of x both members of the identity

$$\frac{1}{1-px} + \frac{1}{1-qx} \equiv \frac{2-(p+q)\,x}{1-x\,(p+q-pqx)}.]$$

296. **If** there be n **quantities** $a, b, c, \ldots$ and s_n denote their sum, s_{n-1} **the sum** of *any* $n-1$ **of them, and so on,** and if

$$S_r \equiv (s_n)^r - \Sigma\,(s_{n-1})^r + \Sigma\,(s_{n-2})^r - \ldots + (-1)^{n-1}\,\Sigma\,(s_1)^r;$$

prove that $$S_1 = S_2 = S_3 = \ldots = S_{n-1} \equiv 0,$$

$$S_n \equiv \underline{|n}\,abc\ldots,\quad 2S_{n+1} \equiv \underline{|n+1}\,abc\ldots(a+b+c+\ldots),$$

and $$12S_{n+2} \equiv \underline{|n+2}\,abc\ldots\{2\Sigma\,(a^2) + 3\Sigma\,(ab)\}.$$

Also if α be any other quantity and if S_r now denote

$$(\alpha+s_n)^r - \Sigma\,(\alpha+s_{n-1})^r + \Sigma\,(\alpha+s_{n-2})^r - \ldots + (-1)^{n-1}\,\Sigma\,(\alpha+s_1)^r,$$

then will $$S_1 = S_2 = S_3 = \ldots = S_{n-1} \equiv 0,$$

$$S_n \equiv \underline{|n}\,abc\ldots,\quad 2S_{n+1} \equiv \underline{|n+1}\,abc\ldots(2\alpha+a+b+c+\ldots),$$

and $$12S_{n+1} \equiv \underline{|n+2}\,abc\ldots\{2\Sigma\,(a^2) + 3\Sigma\,(ab) + 6\alpha\Sigma\,(a) + 6\alpha^2\}.$$

[These results are deduced from the identities

(1) $(\epsilon^{ax}-1)(\epsilon^{bx}-1)\ldots \equiv \epsilon^{s_nx} - \Sigma\,(\epsilon^{s_{n-1}x}) + \Sigma\,(\epsilon^{s_{n-2}x}) - \ldots$

(2) $\epsilon^{\alpha x}(\epsilon^{ax}-1)(\epsilon^{bx}-1)(\ldots) \equiv \epsilon^{(\alpha+s_n)x} - \Sigma\,\epsilon^{(\alpha+s_{n-1})x} + \Sigma\,\{\epsilon^{(\alpha+s_{n-2})x}\} - \ldots$

by taking the expansions of every term of the form ϵ^{mx} and equating the coefficients of like powers of x up to x^{n+2}.]

XII. *Summation of Series.*

[If u_n denote a certain function of n and

$$S_n \equiv u_1 + u_2 + \ldots + u_n,$$

the summation of the series means expressing S_n as a function of n involving only a fixed number (independent of n) of terms. The usual artifice by which this is effected consists in expressing u_n as the difference of two quantities, one of which is the same function of n as the other is of $n-1$, $(U_n - U_{n-1})$. This being effected we have at once

$$S_n \equiv (U_1 - U_0) + (U_2 - U_1) + \dots + (U_n - U_{n-1}) \equiv U_n - U_0.$$

Thus if u_n be the product of r consecutive terms of a given A. P., beginning with the nth, we have

$$u_n \equiv \{a + \overline{n-1}\, b\}\,(a+nb)\dots\{a+(n+r-2)\,b\}$$

$$\equiv \frac{(a+\overline{n-1}\,b)\,(a+n\,b)\dots(a+\overline{n+r-1}\,b) - (a+\overline{n-2}\,b)\dots(a+\overline{n+r-2}\,b)}{(r+1)\,b}$$

whence $$U_n \equiv \frac{(a+\overline{n-1}\,b)(a+nb)\dots(a+\overline{n+r-1}\,b)}{(r+1)\,b}$$

and

$$S_n \equiv \frac{1}{(r+1)\,b}\{(a+\overline{n-1}\,b)\,(a+nb)\dots(a+\overline{n+r-1}\,b)$$
$$-(a-b)\,a\,(a+b)\dots(a+\overline{r-1}\,b)\}.$$

The sums of many series can also be expressed in a finite form by equating the coefficients of x^n in the expansions of the same function of x effected by two different methods of which examples have been already given in the Binomial, Exponential, and Logarithmic Series.

In the examples under this head, n always means a positive whole number.]

297. Sum the series :—

(1) $\frac{1}{1\,.\,2} + \frac{3}{2\,.\,5} + \frac{5}{5\,.\,10} + \dots + \frac{2n-1}{(1+\overline{n-1}^2)(1+n^2)}$,

(2) $\frac{1}{2\,.\,4} + \frac{1\,.\,3}{2\,.\,4\,.\,6} + \frac{1\,.\,3\,.\,5}{2\,.\,4\,.\,6\,.\,8} + \dots + \frac{1\,.\,3\,.\,5\dots(2n-1)}{2\,.\,4\dots 2n\,(2n+2)}$,

(3) $\frac{1\,.\,2}{\lfloor 3} + \frac{2\,.\,2^2}{\lfloor 4} + \dots + \frac{n2^n}{\lfloor n+2}$,

(4) $\frac{3}{\lfloor 4} + \frac{2\,.\,3^2}{\lfloor 5} + \dots + \frac{n3^n}{\lfloor n+3}$,

(5) $\frac{r}{\lfloor r+1} + \frac{2r^2}{\lfloor r+2} + \dots + \frac{nr^n}{\lfloor n+r}$,

(6) $\frac{1}{(1+x)(1+2x)} + \frac{1}{(1+2x)(1+3x)} + \dots + \frac{1}{(1+nx)(1+\overline{n-1}\,x)}$,

(7) $\frac{1}{1\,.\,3} + \frac{2}{1\,.\,3\,.\,5} + \dots + \frac{n}{1\,.\,3\,.\,5\dots(2n+1)}$,

(8) $\frac{1}{3} + \frac{3}{3\,.\,7} + \frac{5}{3\,.\,7\,.\,11} + \dots + \frac{2n-1}{3\,.\,7\,.\,11\dots(4n-1)}$,

(9) $\frac{1}{\lfloor 3} + \frac{5}{\lfloor 4} + \ldots + \frac{n^2+n-1}{\lfloor n+2}$,

(10) $\frac{4}{\lfloor 3} + \frac{9}{\lfloor 4} + \ldots + \frac{2n^2+3n-1}{\lfloor n+2}$,

(11) $\frac{1}{2 \cdot 5} + \frac{5}{5 \cdot 10} + \frac{11}{10 \cdot 17} + \ldots + \frac{n^2+n-1}{(1+n^2)(1+\overline{n-1}^2)}$,

(12) $\frac{1}{(1+x)(1+x^2)} + \frac{x}{(1+x^2)(1+x^3)} + \ldots + \frac{x^{n-1}}{(1+x^n)(1+x^{n+1})}$,

(13) $\frac{x}{1+x^2} + \frac{x}{1+x^2}\frac{x^2}{1+x^4} + \frac{x}{1+x^2}\frac{x^2}{1+x^4}\frac{x^4}{1+x^8} + \ldots$

$+ \frac{x}{1+x^2}\frac{x^2}{1+x^4} \ldots \frac{x^{2^{n-1}}}{1+x^{2^n}}$,

(14) $\frac{x(1-ax)}{(1+x)(1+ax)(1+a^2x)} + \frac{ax(1-a^2x)}{(1+ax)(1+a^2x)(1+a^3x)} + \ldots$

$+ \frac{a^{n-1}x(1-a^nx)}{(1+a^{n-1}x)(1+a^nx)(1+a^{n+1}x)}$,

(15) $\frac{1}{p+r} + \frac{r+1}{(p+r)(p+2r)} + \frac{(r+1)(2r+1)}{(p+r)(p+2r)(p+3r)} + \ldots$

$+ \frac{(r+1)(2r+1)\ldots(\overline{n-1}\,r+1)}{(p+r)(p+2r)\ldots(p+nr)}$.

298. **Prove that**

$$1 \equiv \frac{n(n-1)}{\lfloor 2} - 2\frac{n(n-1)(n-2)}{\lfloor 3} + 3\frac{n(n-1)(n-2)(n-3)}{\lfloor 4} - \ldots$$
$$+ (n-1)(-1)^n.$$

299. **Prove that**

$$(1+r+r^2+r^3)(1+r^2+r^4+r^6)\ldots(1+r^{2^{n-1}}+r^{2^n}+r^{3.2^{n-1}})$$
$$\equiv \frac{1-r^{2^n}-r^{2^{n+1}}+r^{3.2^n}}{1-r-r^2+r^3}.$$

300. **Prove that**

(1) $\frac{1}{4} + \frac{1 \cdot 3}{4 \cdot 6} + \frac{1 \cdot 3 \cdot 5}{4 \cdot 6 \cdot 8} + \ldots$ to $\infty = 1$,

(2) $1 + \frac{3}{8} + \frac{3 \cdot 5}{8 \cdot 10} + \frac{3 \cdot 5 \cdot 7}{8 \cdot 10 \cdot 12} + \ldots$ to $\infty = 2$,

(3) $1 + \frac{11}{14} + \frac{11 \cdot 13}{14 \cdot 16} + \frac{11 \cdot 13 \cdot 15}{14 \cdot 16 \cdot 18} + \ldots$ to $\infty = 12$;

and generally that, if p, q, r and $q-p-r$ be positive, the sum of the infinite series

$$1+\frac{p+r}{q+r}+\frac{(p+r)(p+2r)}{(q+r)(q+2r)}+\frac{(p+r)(p+2r)(p+3r)}{(q+r)(q+2r)(q+3r)}+\dots$$

will be $\dfrac{q}{q-p-r}$.

[The sum of n terms of the last series is

$$\frac{q}{q-p-r}\left\{1-\frac{(p+r)(p+2r)\dots(p+nr)}{q(q+r)\dots(q+\overline{n-1}\,r)}\right\}.]$$

301. Prove that

$$1+\frac{2}{3}+\frac{2.4}{3.5}+\dots+\frac{2.4.6\dots 2n}{3.5.7\dots(2n+1)}\equiv\frac{2.4.6\dots(2n+2)}{3.5.7\dots(2n+1)}-1.$$

302. Prove that, m being not less than n,

(1) $$1+\frac{n}{m}+\frac{n(n-1)}{m(m-1)}+\frac{n(n-1)(n-2)}{m(m-1)(m-2)}+\dots \text{ to } n+1 \text{ terms}$$

$$\equiv\frac{m+1}{m-n+1},$$

(2) $$1+2\frac{n}{m}+3\frac{n(n-1)}{m(m-1)}+4\frac{n(n-1)(n-2)}{m(m-1)(m-2)}+\dots\dots\dots$$

$$\equiv\frac{(m+1)(m+2)}{(m-n+1)(m-n+2)},$$

(3) $$1+3\frac{n}{m}+6\frac{n(n-1)}{m(m-1)}+10\frac{n(n-1)(n-2)}{m(m-1)(m-2)}+\dots\dots$$

$$\equiv\frac{(m+1)(m+2)(m+3)}{(m-n+1)(m-n+2)(m-n+3)}.$$

303. Prove that

$$1-n\frac{a}{b}+\frac{n(n-1)}{\lfloor 2}\frac{a(a-1)}{b(b-1)}-\frac{n(n-1)(n-2)}{\lfloor 3}\frac{a(a-1)(a-2)}{b(b-1)(b-2)}$$

$$+\dots \text{ to } n+1 \text{ terms}\equiv\left(1-\frac{a}{b}\right)\left(1-\frac{a}{b-1}\right)\dots\left(1-\frac{a}{b-n+1}\right).$$

304. Prove that, if x be less than 1,

$$\frac{1-\dfrac{(r-1)(r-2)}{\lfloor 3}x+\dfrac{(r-1)(r-2)(r-3)(r-4)}{\lfloor 5}x^2-\dots}{1-\dfrac{(r+1)(r+2)}{\lfloor 3}x+\dfrac{(r+1)(r+2)(r+3)(r+4)}{\lfloor 5}x^2-\dots}$$

$$=(1+x)^r.$$

[Obtained by expanding numerator and denominator of the fraction

$$\frac{(1+\sqrt{-x})^r-(1-\sqrt{-x})^r}{(1-\sqrt{-x})^{-r}-(1+\sqrt{-x})^{-r}}.]$$

305. If $$u_n \equiv 1-\frac{n(n-1)}{|\underline{2}}+\frac{n(n-1)(n-2)(n-3)}{|\underline{4}}-\dots,$$

and $$v_n \equiv n-\frac{n(n-1)(n-2)}{|\underline{3}}+\frac{n(n-1)(n-2)(n-3)(n-4)}{|\underline{5}}-\dots,$$

then will $$u_n^2+v_n^2=2(u_n u_{n-1}+v_n v_{n-1}).$$

306. If $$u_n \equiv 1-\frac{n(n-1)}{|\underline{2}}x^2+\frac{n(n-1)(n-2)(n-3)}{|\underline{4}}x^4+\dots,$$

and $$v_n \equiv nx-\frac{n(n-1)(n-2)}{|\underline{3}}x^3+\dots,$$

then will $$u_n^2+v_n^2=(1+x^2)^n=(1+x^2)(u_n u_{n-1}+v_n v_{n-1}).$$

307. Prove the identity

$$n+1-\frac{\overline{n+1}\,n\,\overline{n-1}}{|\underline{3}}(2u+1)+\frac{\overline{n+1}\,n\,\overline{n-1}\,\overline{n-2}\,\overline{n-3}}{|\underline{5}}(2u+1)^2-\dots$$

$$\equiv 2^n-\overline{n-1}\,2^{n-1}(u+1)+\frac{\overline{n-2}\,\overline{n-3}}{|\underline{2}}2^{n-2}(u+1)^2$$

$$-\frac{\overline{n-3}\,\overline{n-4}\,\overline{n-5}}{|\underline{3}}2^{n-3}(u+1)^3+\dots$$

XIII. *Recurring Series.*

[The series $u_0+u_1+u_2+\dots+u_n$ is a recurring series if a fixed number (r) of consecutive terms are connected by a relation of the form

$$a_n+p_1 a_{n-1}+p_2 a_{n-2}+\dots+p_{r-1}a_{n-r+1}=0,\quad (A)$$

in which n may have any integral value, but $p_1, p_2, \dots p_{r-1}$ are independent of n. It follows that the series $a_0+a_1x+a_2x^2+\dots+a_nx^n+\dots$ is the expansion in ascending powers of x of a function of x of the form $\dfrac{A_0+A_1x+\dots+A_{r-2}x^{r-2}}{1+p_1x+p_2x^2+\dots+p_{r-1}x^{r-1}}$ (the *generating function* of the series); and if the *scale of relation* (A) and the first $r-1$ terms of the series be given this function will be completely determined; when by separating this function into its partial fractions $\frac{B_1}{1-a_1x}+\frac{B_2}{1-a_2x}+\ldots$ and expanding each we obtain the nth term of the series and the sum of n terms. Thus the nth term of such a series is $B_1a_1^{n-1}+B_2a_2^{n-1}+\ldots$ where $a_1, a_2, \ldots$ are the roots of the *auxiliary equation* $x^{r-1}+p_1x^{r-2}+p_2x^{r-3}+\ldots+p_{r-1}=0$, and $B_1, B_2, \ldots$ constants which can be determined from the first $r-1$ terms of the series. If however two roots of the equation be equal (say $a_2=a_1$) we must write B_2n instead of B_2, if three be equal ($a_1=a_2=a_3$) the corresponding terms will be $(B_1+nB_2+n^2B_3)\,a_1^{n-1}$ and so on.

If the scale of relation is not given we shall require $2(r-1)$ terms of the series to be known to determine completely the generating function; thus if four terms are given we can find a recurring series with a scale of relation between any three consecutive terms and whose first four terms are the given terms.]

308. Prove that every A. P. is a recurring series and that its generating function is $\frac{a+(b-a)x}{(1-x)^2}$, a being the first term and b the common difference.

309. Find the generating functions of the following series:—

(1) $1+3x+5x^2+7x^3+\ldots$

(2) $2+5x+13x^2+35x^3+\ldots$

(3) $2+4x+14x^2+52x^3+\ldots$

(4) $4+5x+7x^2+11x^3+\ldots$

(5) $2+2x+8x^2+20x^3+\ldots$

(6) $1+3x+12x^2+54x^3+\ldots$

and employ the last to prove that the integer next greater than $(\sqrt{3}+1)^{2n}$ is divisible by 2^{n+1}, n being any integer.

310. The generating function of the recurring series whose first four terms are a, b, c, d, is

$$\frac{ab^2-ca^2+x(a^2d-2abc+b^3)}{b^2-ac+x(ad-bc)+x^2(c^2-bd)}.$$

311. If the scale of relation of a recurring series be

$$a_n-7a_{n-1}+12a_{n-2}=0,$$

and if $u_0=2$, $u_1=7$, find u_n and the sum of the series $u_0+u_1+\ldots+u_{n-1}$.

312. Prove that, if $a_0, a_1, a_2 \ldots a_n$ be an A. P. and $b_0, b_1, \ldots b_n$ a G. P., the series

$$a_0+b_0,\; a_1+b_1, \ldots a_n+b_n, \ldots$$
$$a_0b_0,\quad a_1b_1, \ldots\quad a_nb_n, \ldots$$

will be recurring series.

313. The series $u_0, u_1, u_2, \ldots$, and $v_0, v_1, v_2, \ldots$ are both recurring series, the scales of relation being

$$u_{n+1} + p_1 u_n + p_2 u_{n-1} = 0, \quad v_{n+1} + q_1 v_n + q_2 v_{n-1} = 0;$$

prove that the series $u_0 v_0, u_1 v_1, u_2 v_2, \ldots$ is a recurring series whose scale of relation is

$$u_{n+2} - p_1 q_1 u_{n+1} + (p_1^2 q_2 + q_1^2 p_2 - 2p_2 q_2) u_n - p_1 p_2 q_1 q_2 u_{n-1} + p_2^2 q_2^2 u_{n-2} = 0.$$

[It is obvious that the series $u_0 + v_0, u_1 + v_1, u_2 + v_2, \ldots$ is a recurring series whose generating function is the sum of the generating functions of the two series.]

314. Prove that the series

$$1^2 + 2^2 + 3^2 + \ldots + n^2,$$
$$1^3 + 2^3 + 3^3 + \ldots + n^3,$$
$$\ldots\ldots\ldots\ldots\ldots\ldots\ldots\ldots$$
$$1^r + 2^r + 3^r + \ldots + n^r$$

are recurring series, the scales of relation being between 4, 5, ... $r + 2$ terms respectively.

315. Find the generating functions of the recurring series

(1) $1 + 2x + 5x^2 + 10x^3 + 17x^4 + 26x^5 + \ldots$

(2) $1 + 3x + 4x^2 + 8x^3 + 12x^4 + 20x^5 + \ldots$

(3) $3 + 6x + 14x^2 + 36x^3 + 98x^4 + 276x^5 + \ldots$

(4) $3 - x + 13x^2 - 9x^3 + 41x^4 - 53x^5 + \ldots$

and prove that the nth terms of the series are respectively (1) $1 + \overline{n-1}^2$, (2) $\frac{1}{9}\{4(-1)^n + (29 - 3n)2^{n-2}\}$, (3) $1^{n-1} + 2^{n-1} + 3^{n-1}$, and (4) $2n - 1 - (-2)^n$.

316. Find the generating function of the recurring series

$$2 + 9x + 6x^2 + 45x^3 + 99x^4 + 189x^5 + \ldots$$

and prove that the coefficient of x^{n-2} is one third the sum of the nth powers of the roots of the equation $z^3 - 3z - 9 = 0$, and that the coefficients of x^{3n-1} and of x^{3n} are each divisible by 3^n.

317. If the terms of the series $a_0, a_1, a_2, \ldots$ be derived each from the preceding by the formula

$$a_{n+1} = \frac{pq}{p + q - a_n},$$

prove that

$$a_n = pq \frac{(a_0 - p) p^{n-1} - (a_0 - q) q^{n-1}}{(a_0 - p) p^n - (a_0 - q) q^n}.$$

[If we assume $a_n = \dfrac{u_n}{u_{n+1}}$, we get at once a scale of relation for $u_0, u_1, u_2 \ldots$

$$pq u_{n+2} - (p + q) u_{n+1} + u_n = 0, \text{ so that } u_n = \frac{A}{p^{n-1}} + \frac{B}{q^{n-1}}.]$$

XIV. *Convergent Fractions.*

[If $\frac{p_n}{q_n}$ be the nth convergent to the continued fraction

$$\frac{a_1}{b_1+}\frac{a_2}{b_2+}\frac{a_3}{b_3+}\dots$$

we have the equations

$$p_n = b_n p_{n-1} + a_n p_{n-2},\ q_n = b_n q_{n-1} + a_n q_{n-2};$$

and for the fraction

$$\frac{a_1}{b_1-}\frac{a_2}{b_2-}\frac{a_3}{b_3-}\dots$$

the equations

$$p_n = b_n p_{n-1} - a_n p_{n-2},\ q_n = b_n q_{n-1} - a_n q_{n-2}.$$

The solution of each equation, a_n, b_n being functions of n, must involve two constants, since it is necessary that two terms be known in order to determine the remaining terms by this formula. These constants may conveniently be taken to be p_1, p_2, q_1, q_2 respectively. The fraction $\frac{p_n}{q_n}$ thus determined will not generally be in its lowest terms.

We will take as an example the question, "To find the nth convergent to the continued fraction

$$\frac{1}{1-}\frac{1}{3-}\frac{4}{6-}\frac{12}{9-}\dots\frac{2n(n-1)}{3n}-\dots"$$

Take u_n to represent *either* p_n or q_n (since the same law holds for both), then $$u_{n+1} = 3nu_n - 2n(n-1)u_{n-1};$$

or $$u_{n+1} - 2nu_n = n\{u_n - 2(n-1)u_{n-1}\}.$$

So $$u_n - 2(n-1)u_{n-1} = (n-1)\{u_{n-1} - 2(n-2)u_{n-2}\},$$

$$\dots\dots\dots\dots = \dots\dots\dots\dots\dots\dots$$

$$u_3 - 4u_2 = 2(u_2 - 2u_1),\ (= 2 \text{ or } 0 \text{ as } u = p \text{ or } q).$$

Hence, $$u_{n+1} - 2nu_n = \underline{|n} \text{ or } 0,$$

or $$\frac{u_{n+1}}{\underline{|n}} - \frac{2u_n}{\underline{|n-1}} = 1 \text{ or } 0.$$

So $$\frac{2u_n}{\underline{|n-1}} - \frac{2^2u_{n-1}}{\underline{|n-2}} = 2 \text{ or } 0,$$

$$\dots\dots\dots\dots = \dots\dots$$

$$\frac{2^{n-1}u_2}{\underline{|1}} - 2^n u_1 = 2^{n-1} \text{ or } 0,$$

and
$$2^n u_1 = 2^n;$$

whence
$$\frac{u_{n+1}}{\underline{|n}} = 2^{n+1} - 1 \text{ or } 2^n,$$

and the nth convergent is $\dfrac{2^n - 1}{2^{n-1}}$.]

318. The nth convergent to $\frac{1}{2+}\,\frac{1}{2+}\,\frac{1}{2+}\dots$ is

$$\frac{(1+\sqrt{2})^n - (1-\sqrt{2})^n}{(1+\sqrt{2})^{n+1} - (1-\sqrt{2})^{n+1}}.$$

319. The nth convergent to the continued fraction

$$\frac{x}{x+1-}\,\frac{x}{x+1-}\,\frac{x}{x+1-}\dots$$

is equal to $\dfrac{x^{n+1}-x}{x^{n+1}-1}$; and that to

$$\frac{x}{x-1+}\,\frac{x}{x-1+}\,\frac{x}{x-1+}\dots$$

is equal to $\dfrac{x^{n+1}-(-1)^n x}{x^{n+1}+(-1)^n}$; and the numerator and denominator of any convergent to either fraction differ by unity.

320. Prove that the continued fraction

$$\frac{a_1}{a_1+}\,\frac{a_2}{a_2+}\,\frac{a_3}{a_3+}\dots+\frac{a_n}{a_n}$$

is equal to the continued fraction

$$\frac{1}{1+}\,\frac{1}{a_1+}\,\frac{a_1}{a_2+}\,\frac{a_2}{a_3+}\dots+\frac{a_{n-2}}{a_{n-1}}.$$

321. If $\dfrac{p_n}{q_n}$ be the nth convergent to the infinite continued fraction $\frac{1}{a+}\,\frac{1}{a+}\,\frac{1}{a+}\dots$, p_n, q_n will be the coefficients of x^{n-1} and x^n respectively in the expansion of $\dfrac{1}{1-ax-x^2}$.

322. Prove that, $\dfrac{p_n}{q_n}$ being the nth convergent to the continued fraction

$$\frac{1}{a+}\,\frac{1}{b+}\,\frac{1}{a+}\,\frac{1}{b+}\dots$$

$$p_{n+2} - (ab+2)\,p_n + p_{n-2} = 0, \quad q_{n+2} - (ab+2)\,q_n + q_{n-2} = 0.$$

323. Prove that the products of the infinite continued fractions

$$(1)\quad \frac{1}{a+}\frac{1}{b+}\frac{1}{c+}\frac{1}{a+}\dots,\quad c+\frac{1}{b+}\frac{1}{a+}\frac{1}{c+}\dots,$$

$$(2)\quad \frac{1}{a+}\frac{1}{b+}\frac{1}{c+}\frac{1}{d+}\frac{1}{a+}\dots,\quad d+\frac{1}{c+}\frac{1}{b+}\frac{1}{a+}\frac{1}{d+}\dots$$

are, (1) $\dfrac{1+bc}{1+ab}$, (2) $\dfrac{b+d+bcd}{a+c+abc}$.

324. Prove that the differences of the infinite continued fractions

$$(1)\quad \frac{1}{a+}\frac{1}{b+}\frac{1}{c+}\frac{1}{a+}\dots,\quad \frac{1}{b+}\frac{1}{a+}\frac{1}{c+}\frac{1}{b+}\frac{1}{a+}\dots,$$

$$(2)\quad \frac{1}{a+}\frac{1}{b+}\frac{1}{c+}\frac{1}{d+}\frac{1}{a+}\dots,\quad \frac{1}{c+}\frac{1}{b+}\frac{1}{a+}\frac{1}{d+}\frac{1}{c+}\dots$$

are, (1) $\dfrac{a-b}{1+ab}$, (2) $\dfrac{b(a-c)}{a+c+abc}$.

325. If $x=\dfrac{a}{1+}\dfrac{b}{1+}\dfrac{a}{1+}\dfrac{b}{1+}\dots$, and $y=\dfrac{b}{1+}\dfrac{a}{1+}\dfrac{b}{1+}\dfrac{a}{1+}\dots$,

then will $x-y=a-b$, and $xy+\dfrac{ab}{xy}=a+b+1$.

326. If $x=\dfrac{a}{a'+}\dfrac{b}{b'+}\dfrac{a}{a'+}\dfrac{b}{b'+}\dots$, and $y=\dfrac{b}{b'+}\dfrac{a}{a'+}\dfrac{b}{b'+}\dfrac{a}{a'+}\dots$,

then will $a'x-b'y=a-b$, and $xy+\dfrac{ab}{xy}=a+b+a'b'$.

327. If $x=\dfrac{a}{1+}\dfrac{b}{1+}\dfrac{c}{1+}\dfrac{a}{1+}\dots$ and $y=\dfrac{c}{1+}\dfrac{b}{1+}\dfrac{a}{1+}\dfrac{c}{1}\dots$,

then will $x-y=\dfrac{a-c}{1+b}$, and $x(1+y)=\dfrac{a(1+c)}{1+b}$.

328. If $x=\dfrac{a}{1+}\dfrac{b}{1+}\dfrac{c}{1+}\dfrac{d}{1+}\dfrac{a}{1+}\dots$ and $y=\dfrac{d}{1+}\dfrac{c}{1+}\dfrac{b}{1+}\dfrac{a}{1+}\dfrac{d}{1+}\dots$,

prove that $x(1+y)=a\dfrac{1+c+d}{1+b+c}$, and $y(1+x)=d\dfrac{1+a+b}{1+b+c}$.

329. The convergents to the infinite continued fraction

$$1-\frac{1}{5-}\frac{2}{1-}\frac{1}{5-}\frac{2}{1-}\frac{1}{5-}\dots$$

recur after eight.

330. The continued fractions

$$4+\frac{4}{8+}\frac{4}{8+}\frac{4}{8+}\dots,\quad 2+\frac{1}{4+}\frac{1}{4+}\frac{1}{4+}\dots,$$

each to n quotients are in the ratio $2:1$.

[This can be readily proved without calculating either.]

331. Prove that

$$\frac{n}{n+}\frac{n-1}{n-1+}\frac{n-2}{n-2+}\dots+\frac{2}{2+}\frac{1}{1+}\frac{1}{2}=\frac{n+1}{n+2}.$$

332. Prove that, if $a>1$, the infinite continued fraction

$$\frac{1}{a-}\frac{a}{a+1-}\frac{a+1}{a+2-}\dots=\frac{1}{a-1},$$

that

$$\frac{1}{a+1+}\frac{a}{a+}\frac{a}{a+1+}\frac{a+1}{a+2+}\dots+\frac{a+n-1}{a+n}$$

$$=\frac{1}{a+1}-\frac{1}{(a+1)(a+2)}+\frac{1}{(a+1)(a+2)(a+3)}-\dots \text{ to } n+2 \text{ terms},$$

and that on reducing this to a single fraction the factor $a+n+1$ divides out.

333. The nth convergent to $1-\frac{1}{4-}\frac{1}{4-}\dots$ is equal to the $(2n-1)^{\text{th}}$ convergent to

$$\frac{1}{1+}\frac{1}{2+}\frac{1}{1+}\frac{1}{2+}\dots.$$

334. If $\frac{p_n}{q_n}$ be the nth convergent to $\frac{r}{r-1+}\frac{r}{r+1+}\frac{r}{r-1+}\dots$, then will

$$(r-1)p_{2n+1}=r(q_{2n+1}-rq_{2n-1}),$$

$$(r-1)p_{2n}=r(r+1)q_{2n-1},$$

$$r(r+1)q_{2n}=(r^2+r-1)p_{2n}-r^2p_{2n-2}.$$

335. If $x=\frac{n}{n+}\frac{n+1}{n+1+}\frac{n+2}{n+2+}\dots$ to ∞, prove that

$$\frac{1}{n+x}=\frac{1}{n}-\frac{1}{n(n+1)}+\frac{1}{n(n+1)(n+2)}-\dots \text{ to } \infty.$$

336. Prove that the value of the infinite continued fraction

$$\frac{1}{1+}\frac{2}{2+}\frac{3}{3+}\dots$$

is $\frac{1}{\epsilon-1}$; and that $\epsilon=2+\frac{1}{1+}\frac{1}{2+}\frac{2}{3+}\frac{3}{4+}\dots$ to ∞.

337. **Prove that**

$$\frac{1}{1+}\frac{n}{1+}\frac{n(n+1)}{2}+\frac{n(n+2)}{3}+\ldots+\frac{n(n+r-1)}{r}+$$

is equal to

$$1-\frac{n}{n+1}+\frac{n^2}{(n+1)(n+2)}-\frac{n^3}{(n+1)(n+2)(n+3)}+\ldots;$$

and thence that the value of the fraction continued to infinity is for $n=1, 2, 3, 4, 5$ respectively

$$\frac{\epsilon-1}{\epsilon},\ \frac{\epsilon^2+1}{2\epsilon^2},\ \frac{5\epsilon^3-2}{9\epsilon^3},\ \frac{17\epsilon^4+3}{32\epsilon^4},\ \text{and}\ \frac{329\epsilon^5-24}{625\epsilon^5}.$$

338. Any two consecutive terms of the series $a_1, a_2, \ldots a_n, \ldots$ satisfy the equation

$$a_{n+1}=\frac{n(n+1)}{2n-a_n},$$

find a_n in terms of a_1; and prove that when n is indefinitely increased the limit of

$$\frac{a_1a_2\ldots a_{n+1}}{\lfloor n}\ \text{is}\ \frac{a_1}{1-a_1}.$$

339. Having expressed $\sqrt{(n^2+a)}$ as a continued fraction in the form $n+\frac{a}{2n+}\frac{a}{2n+}\ldots$, $\frac{p_r}{q_r}$ is the rth convergent; prove that

$$p_r^2-(n^2+a)\,q_r^2=(-a)^r, \qquad q_{r+1}+aq_{r-1}=2p_r,$$

$$p_rp_{r+1}-(n^2+a)\,q_rq_{r+1}=n(-a)^r, \qquad p_{r+1}+ap_{r-1}=2(n^2+a)\,q_r.$$

340. Prove that $\sqrt{(n^2+a)}$ can be expressed as a continued fraction in the form $n+\frac{a}{n+}\frac{n^2+a}{n}\frac{a}{+n+}\frac{n^2+a}{n}+\ldots$; and that if $\frac{p_r}{q_r}$ be the rth convergent,

$$p_{2r-1}=q_{2r}=n\frac{\alpha^r-\beta^r}{\alpha-\beta},$$

$$p_{2r}=(n^2+a)\,q_{2r-1}=\frac{1}{2}(\alpha^r+\beta^r);$$

where α, β are the roots of the equation in x

$$x^2-2(n^2+a)\,x+a(n^2+a)=0.$$

341. If $$\frac{1+x}{1-2x-x^2}\equiv 1+a_1x+a_2x^2+\ldots+a_nx^n+\ldots,$$

and $$\frac{1}{1-2x-x^2}\equiv 1+b_1x+b_2x^2+\ldots+b_nx^n+\ldots,$$

prove that $$a_n^2-2b_n^2=(-1)^{n-1}.$$

342. If $$\frac{2-x}{1-4x+x^2} \equiv 2 + a_1x + a_2x^2 + \dots + a_nx^n + \dots,$$
and $$\frac{1}{1-4x+x^2} \equiv 1 + b_1x + b_2x^2 + \dots + b_nx^n + \dots,$$
prove that $$a_n^2 - 3b_n^2 = 1.$$

343. If $$\frac{r-x}{1-2rx+x^2} \equiv r + a_1x + a_2x^2 + \dots + a_nx^n + \dots,$$
and $$\frac{1}{1-2rx+x^2} \equiv 1 + b_1x + b_2x^2 + \dots + b_nx^n + \dots,$$
prove that $$a_n^2 - (r^2-1)\,b_n^2 = 1.$$

344. In the equation
$$x^4 - 2nx^2 - x + (n-1)\,n = 0,$$
prove that $$x = \pm\sqrt{n \pm \sqrt{n+x}},$$
and find all the roots of the equation. Prove that
$$\sqrt{7-\sqrt{7+\sqrt{7-\sqrt{\dots}}}} \text{ to } \infty = 2,$$
and express the other roots of the biquadratic in the same form.

345. Prove that
$$\sqrt{p+\sqrt{p+\sqrt{p+\dots}}} = \frac{m^2+mn+n^2}{2mn},$$
$$\sqrt{p-\sqrt{p-\sqrt{p-\dots}}} = \frac{m^2-mn+n^2}{2mn},$$
$$\sqrt{p-\sqrt{p+\sqrt{p-\dots}}} = \frac{m^2-mn-n^2}{2mn},$$
$$-\sqrt{p+\sqrt{p-\sqrt{p+\sqrt{\dots}}}} = \frac{n^2-mn-m^2}{2mn};$$
where $p \equiv \dfrac{m^4+m^2n^2+n^4}{4m^2n^2}$, and $m > n$.

346. Prove that

(1) $$\frac{r(r+1)}{1}\,+\,\frac{r(r+1)}{1}\,+\dots \text{ to } n \text{ quotients}$$
$$\equiv \frac{r(r+1)^{n+1} + (r+1)(-r)^{n+1}}{(r+1)^{n+1} - (-r)^{n+1}},$$

(2) $$\frac{1}{1}\,+\,\frac{1}{2}\,+\,\frac{3^2}{2}\,+\,\frac{5^2}{2}\,+\dots+\,\frac{(2n-1)^2}{2}$$
$$\equiv 1 - \frac{1}{3} + \frac{1}{5} - \dots + \frac{(-1)^n}{2n+1},$$

(3) $\frac{1}{2+}\frac{2}{2+}\frac{6}{2+}\frac{12}{2+}\dots+\frac{n(n+1)}{2}$

$$\equiv \frac{1}{1.2}-\frac{1}{2.3}+\frac{1}{3.4}\dots+\frac{(-1)^n}{(n+1)(n+2)},$$

(4) $\frac{1}{1+}\frac{1}{r+}\frac{r+1}{r+1+}\frac{r+2}{r+2+}\dots$ to n quotients

$$\equiv 1-\frac{1}{r+1}+\frac{1}{(r+1)(r+2)}-\frac{1}{(r+1)(r+2)(r+3)}+\dots \text{ to } n \text{ terms},$$

(5) $\frac{x}{1+}\frac{x}{2-x+}\frac{4x}{3-2x-}\dots+\frac{n^2x}{n+1-nx}$

$$\equiv x-\frac{x^2}{2}+\frac{x^3}{3}-\dots+\frac{(-1)^n x^{n+1}}{n+1},$$

(6) $\frac{x}{1+}\frac{x^2}{3-x^2+}\frac{(3x)^2}{5-3x^2+}\dots+\frac{(\overline{2n-1}\,x)^2}{2n+1-\overline{2n-1}\,x^2}$

$$\equiv x-\frac{x^3}{3}+\frac{x^5}{5}-\dots+\frac{(-1)^n x^{2n+1}}{2n+1},$$

(7) $\frac{1}{1+}\frac{r}{1+}\frac{r(r+1)}{2}+\dots+\frac{r(r+n-1)}{n}$

$$\equiv 1-\frac{r}{r+1}+\frac{r^2}{(r+1)(r+2)}-\dots+\frac{(-r)^n}{(r+1)(r+2)\dots(r+n)},$$

(8) $\frac{1}{1+}\frac{1}{3+}\frac{16}{5+}\frac{81}{7+}\dots+\frac{n^4}{2n+1+}\dots$ to $\infty=\frac{\pi^2}{12}$,

(9) $\frac{1}{1+}\frac{1}{1+}\frac{4}{1+}\frac{9}{1+}\dots+\frac{n^2}{1+}\dots$ to ∞ $=\log 2$,

(10) $\frac{4}{1+}\frac{6}{2+}\frac{8}{3+}\dots+\frac{2n}{n-1+}\dots$ to ∞ $=2\frac{\epsilon^2-1}{\epsilon^2+1}$,

(11) $\frac{3^2}{1+}\frac{3.4}{2+}\frac{3.5}{3+}\dots+\frac{3n}{n-2+}\dots$ to $\infty=6\frac{2\epsilon^3+1}{5\epsilon^3-2}$,

(12) $\frac{4^2}{1+}\frac{4.5}{2+}\frac{4.6}{3+}\dots+\frac{4(n+3)}{n+}\dots$ to $\infty=12\frac{5\epsilon^4-1}{17\epsilon^4+3}$.

347. Prove that

(1) $\frac{1}{1-}\frac{1}{4-}\frac{1}{1-}\frac{1}{4-}\dots$ to n quotients $\equiv\frac{2n}{n+1}$,

(2) $\frac{1}{4-}\frac{1}{1-}\frac{1}{4-}\frac{1}{1-}\dots$ to n quotients $\equiv\frac{n}{2(n+1)}$,

(3) $\frac{1}{1-}\frac{1}{3-}\frac{4}{5-}\frac{9}{7-}\dots-\frac{n^2}{2n+1}\equiv 1+\frac{1}{2}+\frac{1}{3}+\dots+\frac{1}{n+1}$,

(4) $\frac{1}{1} - \frac{1}{4} - \frac{9}{8} - \frac{25}{12} - \dots - \frac{(2n-1)^2}{4n} \equiv 1 + \frac{1}{3} + \frac{1}{5} + \dots + \frac{1}{2n+1}$,

(5) $\frac{1}{r} - \frac{r^2}{2r+1} - \frac{(r+1)^2}{2r+3} - \dots - \frac{(r+n-1)^2}{2(r+n)-1}$

$$\equiv \frac{1}{r} + \frac{1}{r+1} + \dots + \frac{1}{r+n},$$

(6) $\frac{1}{a} - \frac{a^2}{2a+b} - \frac{(a+b)^2}{2a+3b} - \dots - \frac{(a+\overline{n-1}b)^2}{2a+(2n-1)b}$

$$\equiv \frac{1}{a} + \frac{1}{a+b} + \dots + \frac{1}{a+nb},$$

(7) $\frac{a^2}{1} - \frac{(a-1)^2}{1} - \frac{a^2}{1} - \frac{(a-1)^2}{1} - \dots$ to n quotients

$$\equiv \frac{a(n+2a-1)}{n+1} \text{ or } \frac{na}{n-2a+2}, \text{ as } n \text{ is odd or even,}$$

(8) $\frac{a(a+1)}{1} - \frac{a(a-1)}{1} - \frac{a(a+1)}{1} - \dots$ to n quotients

$$\equiv a - \frac{a}{(a+1)(1-a^{-2})^{\frac{n}{2}} - a}, \text{ or } a - \frac{1}{(1-a^{-2})^{\frac{n+1}{2}} - 1};$$

(9) $\frac{x}{1} - \frac{x}{2+x} - \frac{4x}{3+2x} - \dots - \frac{n^2x}{n+1+nx}$

$$\equiv x + \frac{x^2}{2} + \frac{x^3}{3} + \dots + \frac{x^{n+1}}{n+1},$$

(10) $\frac{2}{2} - \frac{3}{3} - \frac{4}{4} - \dots - \frac{n+1}{n} \equiv 1 + 1 + |\underline{2} + |\underline{3} + \dots + |\underline{n}$,

(11) $\frac{1}{1} - \frac{4}{5} - \frac{9}{13} - \dots - \frac{(n^2-1)^2}{n^2+(n+1)^2} \equiv \frac{n(n+1)(2n+1)}{6}$,

(12) $\frac{2}{1} - \frac{3}{5} - \frac{8}{7} - \dots - \frac{n^2-1}{2n+1} \equiv \frac{n(n+3)}{2}$,

(13) $\frac{1}{1} - \frac{1^4}{1^2+2^2} - \frac{2^4}{2^2+3^2} - \dots - \frac{n^4}{n^2+\overline{n+1}^2}$

$$\equiv \frac{1}{1^2} + \frac{1}{2^2} + \dots + \frac{1}{(n+1)^2},$$

(14) $\frac{1}{1} - \frac{a}{a+1} - \frac{a+1}{a+2} - \dots$ to $n+1$ quotients

$$\equiv 1 + a + a(a+1) + a(a+1)(a+2) + \dots + a(a+1)\dots(a+n-1),$$

(15) $\frac{1}{1} - \frac{1}{2} - \frac{3}{4} - \frac{5}{6} - \dots - \frac{2n-1}{2n}$

$$\equiv 2 + 1 \,.\, 3 + 1 \,.\, 3 \,.\, 5 + \dots + 1 \,.\, 3 \,.\, 5 \,..\, 2n-1,$$

(16) $\dfrac{2n}{2n-1-}\,\dfrac{2n-2}{2n-3-}\dots\dfrac{4}{3-}\,\dfrac{2}{1} \equiv 2n,$

(17) $\dfrac{1}{1-}\,\dfrac{1}{3-}\,\dfrac{2}{4-}\,\dfrac{3}{5-}\dots\dfrac{n}{n+2-}\dots$ to $\infty = \epsilon - 1,$

(18) $1 + \dfrac{r}{1-}\,\dfrac{r}{r+2-}\,\dfrac{2r}{r+3-}\,\dfrac{3r}{r+4-}\dots$ to $\infty = \epsilon^r,$

(19) $\dfrac{r^2}{2r+1-}\,\dfrac{(r+1)^2}{2r+3-}\,\dfrac{(r+2)^2}{2r+5-}\dots$ to $\infty = r,$

(20) $\dfrac{r}{r-}\,\dfrac{r+1}{r+1-}\,\dfrac{r+2}{r+2-}\dots$ to $\infty = \dfrac{r-1}{r-2}.$

XV. *Poristic Systems of Equations.*

[Any system of algebraical equations

$$\frac{a}{x_1x_2} + bx_1x_2 + c + f(x_1+x_2) + g\left(\frac{1}{x_1}+\frac{1}{x_2}\right) + h\left(\frac{x_1}{x_2}+\frac{x_2}{x_1}\right) = 0,$$

$$\frac{a}{x_2x_3} + bx_2x_3 + c + f(x_2+x_3) + g\left(\frac{1}{x_2}+\frac{1}{x_3}\right) + h\left(\frac{x_2}{x_3}+\frac{x_3}{x_2}\right) = 0,$$

....

$$\frac{a}{x_{n-1}x_n} + bx_{n-1}x_n + c + f(x_{n-1}+x_n) + g\left(\frac{1}{x_{n-1}}+\frac{1}{x_n}\right) + h\left(\frac{x_{n-1}}{x_n}+\frac{x_n}{x_{n-1}}\right) = 0,$$

$$\frac{a}{x_nx_1} + bx_nx_1 + c + f(x_n+x_1) + g\left(\frac{1}{x_n}+\frac{1}{x_1}\right) + h\left(\frac{x_n}{x_1}+\frac{x_1}{x_n}\right) = 0,$$

is poristic if a certain relation holds between the coefficients a, b, c, f, g, h: when $n=3$, this relation is $M \equiv h^2 - ab - ch + fg = 0$; when $n=4$, it is $-N \equiv abc + 2fgh - af^2 - bg^2 - ch^2 = 0$, and when $n=5$, it is

$$M^3 - LMN + N^2 = 0,$$

where $L \equiv c - 4h$. For any number of such equations, if there is one solution for which $x_1, x_2, \dots, x_n$ are all unequal there is an infinite number of such solutions, but this cannot be the case unless a certain relation hold, which relation involves L, M, N only. See *Proceedings of London Math. Soc.* Vol. IV. page 312.]

348. If x_1, x_2 be the roots of the equation

$$\frac{a}{(1-m)(1-x)} + b + \frac{c}{mx} = 0,$$

then will
$$\frac{a}{(1-x_1)(1-x_2)} + b + \frac{c}{x_1x_2} = 0.$$

349. If x_1, x_2 be the two roots of the equation

$$a^2 + a\left(mx + \frac{1}{mx}\right) = \frac{x}{m} + \frac{m}{x} + 1,$$

then will $$a^2 + a\left(x_1x_2 + \frac{1}{x_1x_2}\right) = \frac{x_1}{x_2} + \frac{x_2}{x_1} + 1,$$

and $$x_1x_2 + m(x_1 + x_2) = \frac{1}{x_1x_2} + \frac{1}{mx_1} + \frac{1}{mx_2} = -a.$$

350. If x_1, x_2 be the two roots of the equation

$$a^2(1 + m^2)(1 + x^2) + a(m + x)(mx - 1) = mx,$$

then will $$a^2(1 + x_1^2)(1 + x_2^2) + a(x_1 + x_2)(x_1x_2 - 1) = x_1x_2,$$

and $$m + x_1 + x_2 + \frac{1}{m} + \frac{1}{x_1} + \frac{1}{x_2} = mx_1x_2 + \frac{1}{mx_1x_2}.$$

351. If x_1, x_2 be the roots of the equation

$$\frac{(1 - m^2)(1 - x^2)}{b - c} + \frac{4mx}{c - a} = \frac{(1 + m^2)(1 + x^2)}{a - b},$$

then will $$\frac{(1 - x_1^2)(1 - x_2^2)}{b - c} + \frac{4x_1x_2}{c - a} = \frac{(1 - x_1^2)(1 + x_2^2)}{a - b}.$$

352. If the quantities x, y, z be all unequal and satisfy the equations

$$\frac{a(y^2z^2 + 1) + y^2 + z^2}{yz} = \frac{a(z^2x^2 + 1) + z^2 + x^2}{zx} = \frac{a(x^2y^2 + 1) + x^2y^2}{xy};$$

each member of the equations $= a^2 - 1$, and $xyz(yz + zx + xy) = xyz$.

353. Having given the equations

$$yz + \frac{1}{yz} - ax - \frac{b}{x} = zx + \frac{1}{zx} - ay - \frac{b}{y} = xy + \frac{1}{xy} - az - \frac{b}{z};$$

prove that, if x, y, z be all unequal, $ab = 1$, and each member of these equations $= 0$.

354. Having given the equations

$$y^2 + z^2 + ayz = z^2 + x^2 + azx = x^2 + y^2 + axy,$$

prove that, if x, y, z be all unequal,

$$a = 1, \quad \text{and} \quad x + y + z = 0.$$

355. The system

$$(a^2 - x^2)(b^2 + yz) = (a^2 - y^2)(b^2 + zx) = (a^2 - z^2)(b^2 + xy)$$

is poristic if $b^2 = a^2$; in which case each member will be equal to

$$-\frac{xyz(y + z)(z + x)(x + y)}{(x + y + z)^2}.$$

356. Prove that the system of equations

$$x(a-y)=y(a-z)=z(a-x)=b^2$$

can only be satisfied if $x=y=z$; unless $b^2=a^2$, in which case the equations are not independent.

357. Prove that the system of equations

$$u(2a-x)=x(2a-y)=y(2a-z)=z(2a-u)=b^2$$

can only be satisfied if $u=x=y=z$; unless $a^2=2b^2$, in which case the equations are not independent.

358. Prove that the system of equations

$$x_1(1-x_2)=x_2(1-x_3)=\ldots=x_n(1-x_{n+1})=x_{n+1}(1-x_1)=u$$

can only be satisfied if $x_1=x_2=x_3=\ldots=x_{n+1}$, unless u be a root of the equation $(1+\sqrt{1-4u})^{n+1}=(1-\sqrt{1-4u})^{n+1}$ different from $\frac{1}{4}$, in which case the equations are not independent.

[By putting $1-4u=-\tan^2\theta$, it will appear that the roots of the auxiliary equation are

$$\frac{1}{4}\sec^2\frac{\pi}{n+1},\ \frac{1}{4}\sec^2\frac{2\pi}{n+1},\ \ldots,\ \text{to } \frac{n}{2} \text{ or } \frac{n-1}{2} \text{ terms.}$$

359. Prove that the system of equations

$$\frac{a}{1-x_1}+\frac{b}{x_2}=\frac{a}{1-x_2}+\frac{b}{x_1}=c$$

can only be satisfied if $x_1=x_2$, unless $c=a+b$, in which case the equations are not independent.

360. Prove that the system of equations

$$\frac{a}{1-x_1}+\frac{b}{x_2}=\frac{a}{1-x_2}+\frac{b}{x_3}=\frac{a}{1-x_3}+\frac{b}{x_1}=c$$

is poristic if $(a+b-c)^2=ab$, and the system

$$\frac{a}{1-x_1}+\frac{b}{x_2}=\frac{a}{1-x_2}+\frac{b}{x_3}=\frac{a}{1-x_3}+\frac{b}{x_4}=\frac{a}{1-x_4}+\frac{b}{x_1}=c$$

if $(a+b-c)^2=2ab$.

361. In general the system of equations

$$\frac{a}{1-x_1}+\frac{b}{x_2}=\frac{a}{1-x_2}+\frac{b}{x_3}=\ldots=\frac{a}{1-x_n}+\frac{b}{x_{n+1}}=\frac{a}{1-x_{n+1}}+\frac{b}{x_1}=c$$

is poristic if $\dfrac{\alpha^{n+1}-\beta^{n+1}}{\alpha-\beta}=0$, where α, β are the roots of the quadratic

$$x^2+(a+b-c)x+ab=0.$$

XVI. *Properties of Numbers.*

362. **If** n be a positive whole number, prove **that**

(1)	$2^{2n}+15n+1$	is divisible by 9,
(2)	$(2n+1)^3-2n-1$	 240,
(3)	$3^{2n+2}-8n-9$	 64,
(4)	$3^{2n+3}+40n-27$	 64,
(5)	$3^{2n+5}+160n^2-56n-243$	 512,
(6)	$3^{2n+1}+2^{n+2}$	 7,
(7)	$3^{2n+2}+2^{6n+1}$	 11,
(8)	$3^{4n+2}+4^{3n+3}$	 17,
(9)	$3\,.\,5^{2n+1}+2^{3n+1}$	.... 17.

363. If $2p+1$ be a prime **number,** $(\underline{|p})^2+(-1)^p$ will be divisible by $2p+1$.

364. If p be a prime number, ${}_{p-1}C_n+(-1)^{n-1}$ will be divisible by p.

365. **If** p be a prime **number** $>$ **3,** ${}_{2p-1}C_p-1$ will be divisible by p^3.

366. **If** $n-1$ and $n+1$ be both prime numbers >5, n must be of one of **the forms** $30t$, or $30t\pm 12$, and $n^2(n^2+16)$ will be divisible by 720.

367. If $n-2$ and $n+2$ be both prime numbers >5, n must be of one of the forms $30t+15$ or $30t\pm 9$.

368. Prove that there are never more than two proper solutions of the question "to find a number which exceeds p times the integral part of its square root by q"; that if q be any number between $rp+r^2$ and $(r+1)p+r^2$ the two numbers $p(p+r)+q$ and $p(p+r-1)+q$ are solutions; but if q be any number from $rp+(r-1)^2$ to $rp+r^2$ there is only the single solution $p(p+r-1)+q$.

369. If n be **a** whole number, $n+1$ and n^2-n+1 cannot both be square numbers.

370. The whole number next greater **than** $(3+\sqrt{5})^n$ is divisible by 2^n.

371. The integral part of $\dfrac{1}{\sqrt{3}}(\sqrt{3}+\sqrt{5})^{2n+1}$, and the integer next greater than $(\sqrt{3}+\sqrt{5})^{2n}$, are each **divisible by** $\mathbf{2^{n+1}}$.

372. If n, r be whole numbers, the integer next greater than $(\sqrt{n+1}+\sqrt{n-1})^{2r}$ is divisible by 2^{r+1}; as is also the integer next greater than $\dfrac{1}{\sqrt{n+1}}(\sqrt{n+1}+\sqrt{n-1})^{2r+1}$ and the integer next less than $\dfrac{1}{\sqrt{n-1}}(\sqrt{n+1}+\sqrt{n-1})^{2r+1}$.

373. The equation $x^3-2y^3=\pm 1$ cannot be satisfied by any integral values of x and y different from unity.

374. The sum of the squares of all the numbers less than a given number N and prime to it is

$$\frac{N^3}{3}\left(1-\frac{1}{a}\right)\left(1-\frac{1}{b}\right)\left(1-\frac{1}{c}\right)\ldots+\frac{N}{6}(1-a)(1-b)(1-c)\ldots;$$

the sum of the cubes is

$$\frac{N^4}{4}\left(1-\frac{1}{a}\right)\left(1-\frac{1}{b}\right)\left(1-\frac{1}{c}\right)\ldots+\frac{N^2}{4}(1-a)(1-b)(1-c)\ldots;$$

and the sum of the fourth powers is

$$\frac{N^5}{5}\left(1-\frac{1}{a}\right)\left(1-\frac{1}{b}\right)\left(1-\frac{1}{c}\right)\ldots+\frac{N^3}{3}(1-a)(1-b)(1-c)\ldots;$$

$$-\frac{N}{30}(1-a^3)(1-b^3)(1-c^3)\ldots;$$

where $a, b, c, \ldots$ are the different prime factors of N.

375. The product of any r consecutive terms of the series

$$1-c,\ 1-c^2,\ 1-c^3,\ldots$$

is completely divisible by the product of the first r terms.

XVII. *Probabilities.*

376. *A* and *B* throw for a certain stake, each one throw with one die; *A*'s die is marked 2, 3, 4, 5, 6, 7 and *B*'s 1, 2, 3, 4, 5, 6; and equal throws divide the stake: prove that *A*'s expectation is $\frac{47}{72}$ of the stake. What will *A*'s expectation be if equal throws go for nothing?

[$\frac{21}{31}$ of the stake.]

377. A certain sum of money is to be given to the one of three persons *A*, *B*, *C* who first throws 10 with three dice; supposing them to throw in the order named until the event happen, prove that *A*'s chance of winning is $\left(\frac{8}{13}\right)^2$, *B*'s $\frac{8 \cdot 7}{13^2}$, and *C*'s $\left(\frac{7}{13}\right)^2$.

378. Ten persons each write down one of the digits 0, 1, 2, ... 9 at random; find the probability of all ten digits being written.

379. A throws a pair of dice each of which is a cube; B throws a pair one of which is a regular tetrahedron and the other a regular octahedron whose faces are marked from 1 to 4 and from 1 to 8 respectively; which throw is likely to be the higher? (The number on the lowest face is taken in the case of the tetrahedron.) If A throws 6, what is the chance that B will throw higher?

[The chances are even in the first case; in the second B's chance is $\frac{9}{16}$.]

380. A, B, C throw three dice for a prize, the highest throw winning and equal highest throws continuing the trial: at the first throw A throws 13, prove that his chance of the prize is ·623864 nearly.

381. The sum of two positive quantities is known, prove that it is an even chance that their product will be not less than three-fourths of their greatest possible product.

382. Two points are taken at random on a given straight line of length a: prove that the probability of their distance exceeding a given length c $(< a)$ is $\left(\frac{a-c}{a}\right)^2$.

383. Three points are taken at random on the circumference of a circle: the probability of their lying on the same semicircle is $\frac{3}{4}$.

384. If q things be distributed among p persons, the chance that every one of the persons will have at least one is the coefficient of x^q in the expansion of $\lfloor q \,(e^{\frac{x}{p}} - 1)^p$.

385. If a rod be marked at random in n points and divided at those points, the chance that none of the parts shall be greater than $\frac{1}{n}$ th of the rod is $\frac{1}{n^n}$.

386. If a rod be marked at random in $p-1$ points and divided at those points, prove that (1) the chance that none of the parts shall be $< \frac{1}{m}$ th the whole is $\left(1 - \frac{p}{m}\right)^{p-1}$, $(m > p)$: (2) the chance that none of the parts shall be $> \frac{1}{n}$ th the whole is

$$\left(\frac{p}{n} - 1\right)^{p-1} - p\left(\frac{p-1}{n} - 1\right)^{p-1} + \frac{p(p-1)}{\lfloor 2}\left(\frac{p-2}{n} - 1\right)^{p-1} - \dots$$

to r terms where r is the integer next greater than $p-n$, $(n < p)$; or the equivalent

$$1 - p\left(1 - \frac{1}{n}\right)^{p-1} + \frac{p(p-1)}{\lfloor 2}\left(1 - \frac{2}{n}\right)^{p-1} - $$

to r terms where r is the integer next greater than n. Also (3) the chance that none of the parts shall be $< \frac{1}{m}$ th and none greater than $\frac{1}{n}$ th of the whole is

$$\left(1-\frac{p}{m}\right)^{r-1} - p\left(1-\frac{p-1}{m}-\frac{1}{n}\right)^{r-1} + \frac{p(p-1)}{\lfloor 2}\left(1-\frac{p-2}{m}-\frac{2}{n}\right)^{r-1} - \ldots$$

to r terms where r is the integer next greater than $n\,\frac{m-p}{m-n}$; provided that $\frac{1}{m}+\frac{p-1}{n} > 1$. If $\frac{1}{m}+\frac{p-1}{n} < 1$, and none of the parts be $> \frac{1}{n}$ th the whole, it follows that none can be $< \frac{1}{m}$ th the whole, so that the case is then reduced to (2).

387. At an examination each candidate is distinguished by an index number; there are n successful candidates, and the highest index number is $m+n$: prove that the chance that the number of candidates exceeded $m+n+r-1$ is

$$\frac{\lfloor m+r \;\lfloor m+n-1}{\lfloor m \;\lfloor m+n+r-1}.$$

[It is assumed that all numbers are *a priori* equally likely.

388. There are $2m$ black balls and m white balls, from which six balls are drawn at random; prove that when m is very large the chance of drawing four white and two black is $\frac{20}{243}$, and the chance of drawing two white and four black is $\frac{80}{243}$.

389. If n whole numbers taken at random be multiplied together, the chance of the digit in the units' place of the product being 1, 3, 7, or 9 is $(\frac{2}{5})^n$, and the chances of the several digits are equal; the chance of its being 2, 4, 6, or 8 is $\frac{4^n-2^n}{5^n}$, and the chances are equal; the chance of its being 5 is $\frac{5^n-4^n}{10^n}$; and of its being 0 is

$$\frac{10^n-8^n-5^n+4^n}{10^n}.$$

390. If ten things be distributed among three persons the chance of a particular person having more than five of them is $\frac{1507}{19683}$, and of his having five at least is $\frac{4195}{19683}$.

391. If on a straight line of length $a+b$ be measured at random two lengths a, b, the probability that the common part of these lengths shall not exceed c is $\frac{c^2}{ab}$, ($c < a$ or b): and the probability of the smaller b lying entirely within the larger a is $\frac{a-b}{a}$.

392. If on a straight line of length $a+b+c$ be measured at random two lengths a, b, the chance of their having a common part not greater than d is $\frac{(c+d)^2-c^2}{(c+a)(c+b)}$, ($d < a$ or b); the chance of their not having a common part greater than d is $\frac{(c+d)^2}{(c+a)(c+b)}$; and the chance of the smaller b lying altogether within the larger a is $\frac{a-b}{a+c}$.

393. There are $m+p+q$ coins in a bag each of which is equally likely to be a shilling or a sovereign; $p+q$ being drawn p are shillings and q sovereigns: prove that the value of the expectation of the remaining sovereigns in the bag is $\frac{m(q+1)}{p+q+2}$ £. If $m=5$, $p=2$, $q=1$, find the chance that if two more coins be drawn they will be a shilling and a sovereign, (1) when the coins previously drawn are not replaced, (2) when they are replaced.

[In case (1) $\frac{2}{5}$, in case (2) $\frac{13}{28}$.]

394. From an unknown number of balls each equally likely to be white or black three are drawn of which two are white and one black: if five more balls be drawn the chances of drawing five white, four white and one black, three white and two black, and so on, are as 7 : 10 : 10 : 8 : 5 : 2.

395. A bag contains ten balls each equally likely to be white or black; three balls being drawn turn out two white and one black; these are replaced and five balls are then drawn, two white and three black: prove that the chance of a draw from the remaining five giving a white ball is $\frac{71}{128}$.

396. From a very large number of balls each equally likely to be white and black a ball is drawn and replaced p times, and each drawing gives a white ball: prove that the chance of drawing a white ball at the next draw is $\frac{p+1}{p+2}$.

397. A bag contains four white and four black balls; from these four are drawn at random and placed in another bag; three draws are made from the latter, the ball being replaced after each draw, and each draw gives a white ball: prove that the chance of the next draw giving a black ball is ·33.

398. From an unknown number of balls each equally likely to be white or black a ball is drawn and turns out to be white; this is not replaced and $2n$ more balls are drawn: prove that the chance that in the $2n+1$ balls there are more white than black is $\frac{3n+2}{4n+2}$. If the first draw be of three balls and they turn out two white and one black and $2n$ more balls be then drawn from the remainder, the chance that the

majority of the $2n+3$ balls are white is $\dfrac{11n^2+25n+12}{4(2n+1)(2n+3)}$; and the chance that in the $2n$ balls there are more white than black is

$$\frac{11n^2+13n}{4(2n+1)(2n+3)}.$$

399. From a large number of balls each equally likely to be white or black $p+q$ being drawn turn out to be p white and q black: prove that if it is an even chance that on three more balls being drawn two will be white and one black

$$\frac{p}{q}=1+\sqrt[3]{2}$$

nearly, p and q being both large.

400. A bag contains m white balls and n black balls and from it balls are drawn one by one until a white ball is drawn; A bets B at each draw $x : y$ that a black ball is drawn; prove that the value of A's expectation at the beginning of the drawing is $\dfrac{ny}{m+1}-x$. If balls be drawn one by one so long as all drawn are of the same colour, and if for a sequence of r white balls A is to pay B rx£, but for a sequence of r black balls B is to pay A ry£, the value of A's expectation will be $\dfrac{ny}{m+1}-\dfrac{mx}{n+1}$: and if A pay B x for the first white ball drawn, rx for the second, $\dfrac{r(r+1)}{2}x$ for the third, and so on, and B pay A y for the first black ball drawn, ry for the second, $\dfrac{r(r+1)}{2}y$ for the third, and so on, the value of A's expectation at the beginning of a drawing will be

$$\frac{\lfloor m+n+r-1 \; \lfloor m \; \lfloor n}{\lfloor m+n \; \lfloor m+r \; \lfloor n+r}\{n(n+1)\ldots(n+r)\,y-m(m+1)\ldots(m+r)\,x\}.$$

401. From an unknown number of balls each equally likely to be red, white, or blue, ten are drawn and turn out to be five red, three white, and two blue; prove that if three more balls be drawn the chance of their being one red, one white, and one blue is $\frac{72}{455}$; the chance of three red is $\frac{8}{65}$; of three white is $\frac{4}{91}$; and of three blue is $\frac{2}{91}$; and the chance of there being no white ball in the three is $\frac{33}{91}$.

PLANE TRIGONOMETRY.

I. *Equations.*

[In the solution of Trigonometrical Equations, it must be remembered that when an equation has been reduced to the forms (1) $\sin x = \sin \alpha$, (2) $\cos x = \cos \alpha$, (3) $\tan x = \tan \alpha$, the solutions are respectively (1) $x = n\pi + (-1)^n \alpha$, (2) $x = 2n\pi \pm \alpha$, (3) $x = n\pi + \alpha$, where n denotes a positive or negative integer.

The formulæ most useful in Trigonometrical reductions are

$$2 \sin A \cos B \equiv \sin (A + B) + \sin (A - B),$$
$$2 \cos A \cos B \equiv \cos (A - B) + \cos (A + B),$$
$$2 \sin A \sin B \equiv \cos (A - B) - \cos (A + B),$$

and (which are really the same with a different notation)

$$\sin A + \sin B \equiv 2 \sin \frac{A+B}{2} \cos \frac{A-B}{2},$$
$$\cos A + \cos B \equiv 2 \cos \frac{A-B}{2} \cos \frac{A+B}{2},$$
$$\cos A - \cos B \equiv 2 \sin \frac{B-A}{2} \sin \frac{A+B}{2};$$

which enable us to transform products of Trigonometrical functions (sines or cosines) into sums of such functions or conversely sums into products. Thus to transform

$$\sin 2 (B - C) + \sin 2 (C - A) + \sin 2 (A - B).$$

We have

$$\sin 2 (C - A) + \sin 2 (A - B) \equiv 2 \sin (C - B) \cos (B + C - 2A)$$

and

$$\sin 2 (B - C) \equiv 2 \sin (B - C) \cos (B - C),$$

whence the sum of the three

$$\equiv 2 \sin (B - C) \{\cos (B - C) - \cos (B + C - 2A)\},$$
$$\equiv -4 \sin (B - C) \sin (C - A) \sin (A - B).$$

Again, to transform

$$\cos(B-C)\cos(C-A)\cos(A-B),$$

we have

$$2\cos(C-A)\cos(A-B) \equiv \cos(C-B)+\cos(B+C-2A),$$

whence

$$4\cos(B-C)\cos(C-A)\cos(A-B) \equiv 1+\cos 2(B-C)+\cos 2(C-A) + \cos 2(A-B).]$$

402. Solve the equations

$$2\sin x\sin 3x = 1,$$
$$\cos x\cos 3x = \cos 2x\cos 6x,$$
$$\sin 5x\cos 3x = \sin 9x\cos 7x,$$
$$\sin 9x + \sin 5x + 2\sin^2 x = 1,$$
$$\cos mx\cos nx = \cos(m+p)x\cos(n-p)x,$$
$$\sin mx\sin nx = \cos(m+p)x\cos(n+p)x,$$
$$\tan^2 2x + \tan^2 x = 10,$$
$$\cos x + \cos(x-a) = \cos(x-\beta) + \cos(x+\beta-a),$$
$$2\sin^2 2x\cos 2x = \sin^2 3x,$$
$$2\cot 2x - \tan 2x = 3\cot 3x,$$
$$8\cos x = \frac{\sqrt{3}}{\sin x} + \frac{1}{\cos x},$$
$$\sin 2x + \cos 2x + \sin x - \cos x = 0,$$
$$(1+\sin x)(1-2\sin x)^2 = (1-\cos a)(1+2\cos a)^2,$$
$$\frac{\sin a\cos(\beta+x)}{\sin\beta\cos(a+x)} = \frac{\tan\beta}{\tan a},$$
$$\cos 2x + 2\cos x\cos a - 2\cos 2a = 1,$$
$$\sin a\cos 3x - 3\sin 3a\cos x + \sin 4a + 2\sin 2a = 0,$$
$$\frac{\cos^3 a}{\cos x} + \frac{\sin^3 a}{\sin x} = 1,$$
$$(\cos 2x - 4\cos x - 6)^2 = 3(\sin 2x + 4\sin x)^2,$$
$$(\cos 5x - 10\cos 3x + 10\cos x)^2 = 3(\sin 5x - 10\sin x)^2.$$

403. If $\cos(x+3y) = \sin(2x+2y)$,

and $\sin(3x+y) = \cos(2x+2y)$;

then will

$$\left.\begin{aligned} x &= (5m-3n)\frac{\pi}{8}+\frac{\pi}{16} \\ y &= (5n-3m)\frac{\pi}{8}+\frac{\pi}{16} \end{aligned}\right\}; \text{ or } x-y = 2r\pi+\frac{\pi}{2},$$

m, n, r being integers.

404. The real roots of the equation $\tan^2 x \tan \frac{x}{2} = 1$ satisfy the equation
$$\cos 2x = 2 - \sqrt{5}.$$

405. Given $\cos 3x = -\frac{3\sqrt{3}}{4\sqrt{2}}$, prove that the three values of $\cos x$ are
$$\sqrt{\frac{3}{2}} \sin \frac{\pi}{10},\quad \sqrt{\frac{3}{2}} \sin \frac{\pi}{6},\quad -\sqrt{\frac{3}{2}} \sin \frac{3\pi}{10}.$$

406. If the equation $\tan \frac{x}{2} = \frac{\tan x + a - 1}{\tan x + a + 1}$ have real roots, $a^2 > 1$.

407. Find the limits of $\frac{\tan (x + a)}{\tan (x - a)}$ for possible values of x.

$$\left[\text{It cannot lie between } \frac{1 - \sin 2a}{1 + \sin 2a} \text{ and } \frac{1 + \sin 2a}{1 - \sin 2a}\right].$$

408. The ambiguities in the equations
$$\cos \frac{A}{2} + \sin \frac{A}{2} = \pm \sqrt{1 + \sin A},\quad \cos \frac{A}{2} - \sin \frac{A}{2} = \pm \sqrt{1 - \sin A},$$
may be replaced by $(-1)^m$, $(-1)^n$, where m, n denote the greatest integers in $\frac{A + 90°}{360°}$, $\frac{A + 270°}{360°}$ respectively.

409. The solutions of the equation $\sec^2 x + \sec^2 2x = 12$ are $x = n\pi \pm \frac{\pi}{5}$, $x = n\pi \pm \frac{2\pi}{5}$, $x = \frac{1}{2} \cos^{-1} (-\frac{1}{3})$.

410. The roots of the equation in θ, $\tan \frac{a + \theta}{2} \tan \theta = m$, are all roots of the equation
$$\sin a \sin \theta \{1 - \cos (a + \theta)\} + m^2 \cos a \cos \theta \{1 + \cos (a + \theta)\} = m \sin^2 (a + \theta).$$

411. The equations
$$\tan (\theta + \beta) \tan (\theta + \gamma) + \tan (\theta + \gamma) \tan (\theta + a) + \tan (\theta + a) \tan (\theta + \beta) = -3,$$
$$\cot (\theta + \beta) \cot (\theta + \gamma) + \cot (\theta + \gamma) \cot (\theta + a) + \cot (\theta + a) \cot (\theta + \beta) = -3,$$
will be satisfied for all values of θ if they are satisfied by $\theta = 0$.

412. If $\frac{\cos (a + \theta)}{\sin^2 a} = \frac{\cos (\beta + \theta)}{\sin^2 \beta} = \frac{\cos (\gamma + \theta)}{\sin^2 \gamma}$, a, β, γ being unequal and less than π, then will $a + \beta + \gamma = \pi$, and
$$\tan \theta = \frac{3 - \cos 2a - \cos 2\beta - \cos 2\gamma}{\sin 2a + \sin 2\beta + \sin 2\gamma} = \frac{1 + \cos a \cos \beta \cos \gamma}{\sin a \sin \beta \sin \gamma}.$$

413. If $a + \beta + \gamma = \pi$ and θ be an angle determined by the equation
$$\sin (a - \theta) \sin (\beta - \theta) \sin (\gamma - \theta) = \sin^3 \theta,$$
then will
$$\frac{\sin (a - \theta)}{\sin^2 a} = \frac{\sin (\beta - \theta)}{\sin^2 \beta} = \frac{\sin (\gamma - \theta)}{\sin^2 \gamma} = \frac{\sin \theta}{\sin a \sin \beta \sin \gamma} = \frac{\cos \theta}{1 + \cos a \cos \beta \cos \gamma}.$$

[These equations occur when a, β, γ are the angles of a triangle ABC and O is a point such that $\angle OBC = \angle OCA = \angle OAB = \theta$.]

414. If a, β, γ, δ be the four roots of the equation

$$\sin 2\theta - m\cos\theta - n\sin\theta + r = 0,$$

then will

$$a + \beta + \gamma + \delta = (2p+1)\pi \ (p \text{ integral});$$

also

$\sin a + \sin\beta + \sin\gamma + \sin\delta = m$, $\cos a + \cos\beta + \cos\gamma + \cos\delta = n$,
$\sin 2a + \sin 2\beta + \sin 2\gamma + \sin 2\delta = 2mn - 4r$, $\cos 2a + \cos 2\beta + \cos 2\gamma + \cos 2\delta = n^2 - m^2$.

415. If two roots θ_1, θ_2 of the equation

$$\cos(\theta - a) - e\cos(2\theta - a) = m(1 - e\cos\theta)^2$$

satisfy the equation $\tan\frac{\theta_1}{2}\tan\frac{\theta_2}{2} = \sqrt{\frac{1-e}{1+e}}$, then will $me\sin^2 a = \cos a$.

416. Find x and y from the equations

$$x(1 + \sin^2\theta - \cos\theta) - y\sin\theta(1 + \cos\theta) = c(1 + \cos\theta),$$
$$y(1 + \cos^2\theta) - x\sin\theta\cos\theta = c\sin\theta;$$

also eliminate θ from the two equations.

$$\left[\text{The results are } x = c\cot^2\frac{\theta}{2},\ y = c\cot\frac{\theta}{2},\ y^2 = cx.\right]$$

417. The equation

$$\frac{2\sin(a+\beta+\gamma-\theta) + \sin 2\theta - \sin(\beta+\gamma) - \sin(\gamma+a) - \sin(a+\beta)}{2\cos(a+\beta+\gamma-\theta) + \cos 2\theta - \cos(\beta+\gamma) - \cos(\gamma+a) - \cos(a+\beta)}$$
$$= \frac{2\sin 2\theta + \sin(a+\beta+\gamma-\theta) - \sin(a+\theta) - \sin(\beta+\theta) - \sin(\gamma+\theta)}{2\cos 2\theta + \cos(a+\beta+\gamma-\theta) - \cos(a+\theta) - \cos(\beta+\theta) - \cos(\gamma+\theta)}$$

is satisfied if $\theta = a$, β, or γ, or if

$$\cot\frac{\theta - a}{2} + \cot\frac{\theta - \beta}{2} + \cot\frac{\theta - \gamma}{2} = 0.$$

418. The equation

$$\frac{2\sin(a+\beta+\gamma+\theta) + \sin\theta - \sin(\beta+\gamma+\theta) - \sin(\gamma+a+\theta) - \sin(a+\beta+\theta)}{2\cos(a+\beta+\gamma+\theta) + \cos\theta - \cos(\beta+\gamma+\theta) - \cos(\gamma+a+\theta) - \cos(a+\beta+\theta)}$$
$$= \frac{2\sin\theta + \sin(a+\beta+\gamma+\theta) - \sin(a+\theta) - \sin(\beta+\theta) - \sin(\gamma+\theta)}{2\cos\theta + \cos(a+\beta+\gamma+\theta) - \cos(a+\theta) - \cos(\beta+\theta) - \cos(\gamma+\theta)}$$

is independent of θ; and equivalent to

$$\sin\frac{a}{2}\sin\frac{\beta}{2}\sin\frac{\gamma}{2}\left(\cot\frac{a}{2} + \cot\frac{\beta}{2} + \cot\frac{\gamma}{2}\right) = 0.$$

419. Having given the equations

$$\frac{x + y\cos c + z\cos b}{\cos(s-a)} = \frac{y + z\cos a + x\cos c}{\cos(s-b)} = \frac{z + x\cos b + y\cos a}{\cos(s-c)} = 2m,$$

where $2s \equiv a + b + c$; prove that

$$\frac{x}{\sin a} = \frac{y}{\sin b} = \frac{z}{\sin c} = \frac{m}{\sin s}.$$

420. If $\alpha, \beta, \gamma, \delta$ be the four roots of the equation

$$a\cos 2\theta + b\sin 2\theta - c\cos\theta - d\sin\theta + e = 0,$$

and $2s \equiv \alpha + \beta + \gamma + \delta$, then will

$$\frac{a}{\cos s} = \frac{b}{\sin s} = \frac{c}{\cos(s-\alpha) + \cos(s-\beta) + \cos(s-\gamma) + \cos(s-\delta)}$$

$$= \frac{d}{\sin(s-\alpha) + \sin(s-\beta) + \sin(s-\gamma) + \sin(s-\delta)}$$

$$= \frac{e}{\cos(s-\alpha-\delta) + \cos(s-\beta-\delta) + \cos(s-\gamma-\delta)}.$$

421. Reduce to the simplest forms

(1) $(x\cos 2\alpha + y\sin 2\alpha - 1)(x\cos 2\beta + y\sin 2\beta - 1)$

$$- \{x\cos\overline{\alpha+\beta} + y\sin\overline{\alpha+\beta} - \cos\overline{\alpha-\beta}\}^2,$$

(2) $(x\cos\overline{\alpha+\beta} + y\sin\overline{\alpha+\beta} - \cos\overline{\alpha-\beta})(x\cos\overline{\gamma+\delta} + y\sin\overline{\gamma+\delta} - \cos\overline{\gamma-\delta})$
$- (x\cos\overline{\alpha+\gamma} + y\sin\overline{\alpha+\gamma} - \cos\overline{\alpha-\gamma})(x\cos\overline{\beta+\delta} + y\sin\overline{\beta+\delta} - \cos\overline{\beta-\delta})$.

[(1) $(x^2 + y^2 - 1)\sin^2(\alpha-\beta)$, (2) $(x^2 + y^2 - 1)\sin(\beta-\gamma)\sin(\alpha-\delta)$.]

422. If β, γ be different values of x given by the equation $\sin(\alpha + x) = m\sin 2\alpha$

$$\cos\frac{\beta-\gamma}{2} \pm m\sin(\beta+\gamma) = 0.$$

423. The real values of x which satisfy the equation

$\sin\left(\frac{\pi}{2}\cos x\right) = \cos\left(\frac{\pi}{2}\sin x\right)$ are $2n\pi$ or $2n\pi \pm \frac{\pi}{2}$; n being integral.

424. If x, y be real and if

$$\sin^2 x\sin^2 y + \sin^2(x+y) = (\sin x + \sin y)^2,$$

x or y must be a multiple of π.

[The equation is satisfied if $\sin x = 0$, $\sin y = 0$, or

$$\cos(x+y) + \cos x\cos y = 2:$$

and this last can only be satisfied if

$$\cos x = \cos y = \pm 1, \text{ and } \cos(x+y) = 1.]$$

425. If α, β, γ be three angles unequal and less than 2π which satisfy the equation

$$\frac{a}{\cos x} + \frac{b}{\sin x} + c = 0,$$

then will $$\sin(\beta+\gamma) + \sin(\gamma+\alpha) + \sin(\alpha+\beta) = 0.$$

426. If β, γ be angles unequal and less than π which satisfy the equation

$$\frac{\cos\alpha\cos x}{a} + \frac{\sin\alpha\sin x}{b} = \frac{1}{c},$$

then will

$$(b^2+c^2-a^2)\cos\beta\cos\gamma + (c^2+a^2-b^2)\sin\beta\sin\gamma = a^2+b^2-c^2.$$

427. If α, β be angles unequal and less than π which satisfy the equation

$$a\cos 2x + b\sin 2x = 1,$$

and if $$(l\cos^2 2\alpha + m\sin^2 2\alpha)(l\cos^2 2\beta + m\sin^2 2\beta) = \{l\cos^2(\alpha+\beta) + m\sin^2(\alpha+\beta)\}^2,$$

then will either $$l = m, \text{ or } a^2 - b^2 = \frac{m-l}{m+l}.$$

428. If α, β, γ, δ be angles unequal and less than π, and if β, γ be roots of the equation

$$a\cos 2x + b\sin 2x = 1,$$

and α, δ roots of the equation

$$a'\cos 2x + b'\cos 2x = 1;$$

and if $$\{l\cos^2(\alpha+\beta) + m\sin^2(\alpha+\beta)\}\{l\cos^2(\gamma+\delta) + m\sin^2(\gamma+\delta)\} = \{l\cos^2(\alpha+\gamma) + m\sin^2(\alpha+\gamma)\}\{l\cos^2(\beta+\delta) + m\sin^2(\beta+\delta)\},$$

then will either $$l = m, \text{ or } aa' - bb' = \frac{m-l}{m+l}.$$

II. *Identities and Equalities.*

429. If $\tan^2 A = 1 + 2\tan^2 B$, then will $\cos 2B = 1 + 2\cos 2A$.

430. Having given that $\sin(B+C-A)$, $\sin(C+A-B)$, $\sin(A+B-C)$ are in A. P.; prove that $\tan A$, $\tan B$, $\tan C$ are in A. P.

431. Having given that

$$1 + \cos(\beta-\gamma) + \cos(\gamma-\alpha) + \cos(\alpha-\beta) = 0;$$

prove that $\beta-\gamma$, $\gamma-\alpha$, or $\alpha-\beta$ is an odd multiple of π.

432. If $\cot\alpha$, $\cot\beta$, $\cot\gamma$ be in A. P. so also will $\cot(\beta-\alpha)$, $\cot\beta$, $\cot(\beta-\gamma)$; and

$$\frac{\sin(\beta+\gamma)}{\sin\alpha}, \quad \frac{\sin(\gamma+\alpha)}{\sin\beta}, \quad \frac{\sin(\alpha+\beta)}{\sin\gamma};$$ be respectively in A. P.

433. Having given

$$\cos\theta = \cos\alpha\cos\beta,\ \cos\theta' = \cos\alpha'\cos\beta,\ \tan\frac{\theta}{2}\tan\frac{\theta'}{2} = \tan\frac{\beta}{2};$$

prove that $\sin^2\beta = (\sec\alpha - 1)(\sec\alpha' - 1)$.

434. If $\tan 2\alpha = 2\dfrac{ab+cd}{a^2-b^2+c^2-d^2}$, and $\tan 2\beta = 2\dfrac{ac+bd}{a^2-c^2+b^2-d^2}$, then will $\tan(\alpha-\beta)$ be equal to $\dfrac{a+d}{c-b}$ or to $\dfrac{b-c}{a+d}$.

435. If α, β, γ be all unequal and less than 2π, and if

$$\cos\alpha + \cos\beta + \cos\gamma = \sin\alpha + \sin\beta + \sin\gamma = 0;$$

then will $\cos^2\alpha + \cos^2\beta + \cos^2\gamma = \sin^2\alpha + \sin^2\beta + \sin^2\gamma = \frac{3}{2}$;

$$\cos(\beta+\gamma) + \cos(\gamma+\alpha) + \cos(\alpha+\beta) = \sin(\beta+\gamma) + \ldots = 0;$$

$$\cos 2\alpha + \cos 2\beta + \cos 2\gamma = \sin 2\alpha + \sin 2\beta + \sin 2\gamma = 0;$$

and generally if n be a whole number not divisible by 3,

$$\cos n\alpha + \cos n\beta + \cos n\gamma = \sin n\alpha + \sin n\beta + \sin n\gamma = 0.$$

436. Prove that

$$\left(1-\tan^2\frac{\alpha}{2}\right)\left(1-\tan^2\frac{\alpha}{2^2}\right)\left(1-\tan^2\frac{\alpha}{2^3}\right)\ldots \text{ to } \infty = \frac{\alpha}{\tan\alpha}.$$

437. If $C \equiv 2\cos\theta - 5\cos^3\theta + 4\cos^5\theta$, $S \equiv 2\sin\theta - 5\sin^3\theta + 4\sin^5\theta$; then will

$$C\cos 3\theta + S\sin 3\theta \equiv \cos 2\theta, \text{ and } C\sin 3\theta - S\cos 3\theta \equiv \sin\theta\cos\theta.$$

438. Having given

$$x\cos\phi + y\sin\phi = x\cos\phi' + y\sin\phi' = 2a,\quad 2\cos\frac{\phi}{2}\cos\frac{\phi'}{2} = 1;$$

prove that $y^2 = 4a(a+x)$; ϕ, ϕ' being unequal and less than 2π.

439. Prove that

$$\frac{1}{a\cos^2\theta + 2h\sin\theta\cos\theta + b\sin^2\theta} + \frac{1}{a\cos^2\phi + 2h\sin\phi\cos\phi + b\sin^2\phi} = \frac{a+b}{ab-h^2}$$

if $$a + h(\tan\theta + \tan\phi) + b\tan\theta\tan\phi = 0,$$

or $$= 2\frac{(ab-h^2)}{a+b}(1+\tan\theta\tan\phi).$$

440. Having given the equations

$$\frac{e^2-1}{1+2e\cos\alpha+e^2} = \frac{1+2e\cos\beta+e^2}{e^2-1};$$

prove that each member $= \dfrac{e+\cos\beta}{e+\cos\alpha} = \pm\dfrac{\sin\beta}{\sin\alpha}$; and that

$$\tan\frac{\alpha}{2}\tan\frac{\beta}{2} = \pm\frac{1+e}{1-e}.$$

441. Having given the equations

$$\cos\alpha+\cos\beta+\cos\gamma+\cos(\beta+\gamma-\alpha)+\cos(\gamma+\alpha-\beta)+\cos(\alpha+\beta-\gamma)=0,$$
$$\sin\alpha+\sin\beta+\sin\gamma+\sin(\beta+\gamma-\alpha)+\sin(\gamma+\alpha-\beta)+\sin(\alpha+\beta-\gamma)=0;$$

prove that

$$\cos\alpha+\cos\beta+\cos\gamma=\sin\alpha+\sin\beta+\sin\gamma=0.$$

442. Having given the equations

$$\frac{\cos(\beta'+\gamma')+\cos(\gamma'+\alpha')+\cos(\alpha'+\beta')-\cos(\beta+\gamma)-\cos(\gamma+\alpha)-\cos(\alpha+\beta)}{\sin(\beta'+\gamma')+\sin(\gamma'+\alpha')+\sin(\alpha'+\beta')-\sin(\beta+\gamma)-\sin(\gamma+\alpha)-\sin(\alpha+\beta)}$$
$$=\frac{\cos\alpha'+\cos\beta'+\cos\gamma'-\cos\alpha-\cos\beta-\cos\gamma}{\sin\alpha'+\sin\beta'+\sin\gamma'-\sin\alpha-\sin\beta-\sin\gamma};$$

and $$\alpha'+\beta'+\gamma'=\alpha+\beta+\gamma;$$

prove that either

$$\cos\frac{\alpha'}{2}\cos\frac{\beta'}{2}\cos\frac{\gamma'}{2}=\cos\frac{\alpha}{2}\cos\frac{\beta}{2}\cos\frac{\gamma}{2},$$

or $$\sin\frac{\alpha'}{2}\sin\frac{\beta'}{2}\sin\frac{\gamma'}{2}=\sin\frac{\alpha}{2}\sin\frac{\beta}{2}\sin\frac{\gamma}{2}.$$

443. Eliminate θ from the equations

$$\frac{\cos(\alpha-3\theta)}{\cos^3\theta}=\frac{\sin(\alpha-3\theta)}{\cos^3\theta}=m.$$

[The resultant is $m^2+m\cos\alpha=2$.]

444. If x, y satisfy the equations

$$x^2+y^2=\frac{x\cos 3\theta+y\sin 3\theta}{\cos^3\theta}=\frac{y\cos 3\theta-x\sin 3\theta}{\sin^3\theta},$$

then will $$x^2+y^2+x=2.$$

445. Having given the equations

$$x\cos\frac{\theta-\phi}{2}=\cos\theta\cos\phi\cos\frac{\theta+\phi}{2},$$
$$y\cos\frac{\theta-\phi}{2}=\sin\theta\sin\phi\sin\frac{\theta+\phi}{2},$$
$$c=a\cos\theta\cos\phi+b\sin\theta\sin\phi,$$
$$0=bc+ca+ab;$$

prove that $$a^2x^2+b^2y^2=c^2.$$

Also eliminate θ, ϕ from the first three equations when the fourth equation does not hold.

[The resultant in general is $4x^2y^2=z^2(x^2+y^2+z^2-1)^2$, where

$$z^2\equiv\frac{(b+c)(c+a)(a+b)}{bc+ca+ab}\left\{\frac{1}{a+b}-\frac{x^2}{b+c}-\frac{y^2}{c+a}\right\};$$

or $$4x^2y^2=\left(\frac{c^2-a^2x^2-b^2y^2}{bc+ca+ab}\right)^2+\left(\frac{c^2-a^2x^2-b^2y^2}{bc+ca+ab}\right)^2(1-x^2-y^2).]$$

446. Having given

$$(x-a)\cos\theta + y\sin\theta = (x-a)\cos\theta' + y\sin\theta' = a,$$

$$\tan\frac{\theta}{2} - \tan\frac{\theta'}{2} = 2c;$$

prove that $$y^2 = 2ax - (1-c^2)x^2,$$

θ, θ' being unequal and less than 2π.

447. If $(1+\sin\theta)(1+\sin\phi)(1+\sin\psi) = \cos\theta\cos\phi\cos\psi$, then will each member $= (1-\sin\theta)(1-\sin\phi)(1-\sin\psi)$, and

$$\sec^2\theta + \sec^2\phi + \sec^2\psi - 2\sec\theta\sec\phi\sec\psi = 1.$$

448. Eliminate a from the equations

$$\frac{\cos\theta}{e+\cos a} = \frac{\sin\theta}{\sin a} = m.$$

[The resultant is $1 = 2em\cos\theta + m^2(1-e^2)$.]

449. If θ_1, θ_2 be the roots of the equation in θ,

$$(a^2-b^2)\sin(a+\beta)\{a^2\cos a(\cos a - \cos\theta) + b^2\sin a(\sin a - \sin\theta)\} = 2a^2b^2\sin(\beta-\theta),$$

then will

$$a^2\sin\frac{a+\theta_1}{2}\sin\frac{a+\theta_2}{2} + b^2\cos\frac{a+\theta_1}{2}\cos\frac{a+\theta_2}{2} = 0.$$

450. The rational equation equivalent to

$$(m^2 - 2mn\cos A + n^2)^{\frac{1}{2}} + (n^2 - 2nl\cos B + l^2)^{\frac{1}{2}} + (l^2 - 2lm\cos C + m^2)^{\frac{1}{2}} = 0,$$

where $A + B + C = 180°$, is

$$(mn\sin A + nl\sin B + lm\sin C)^2 = 4lmn(l\cos A + m\cos B + n\cos C).$$

451. Having given the equation

$$\cos A = \cos B\cos C \pm \sin B\sin C\cos A,$$

prove that

$$\cos B = \cos C\cos A \pm \sin C\sin A\cos B,$$
$$\cos C = \cos A\cos B \pm \sin A\sin B\cos C;$$

and that

$$\sec^2 A + \sec^2 B + \sec^2 C - 2\sec A\sec B\sec C = 1,$$

$$\tan\left(45° \pm \frac{A}{2}\right)\tan\left(45° \pm \frac{B}{2}\right)\tan\left(45° \pm \frac{C}{2}\right) = 1.$$

452. If

$$\tan^2 A\tan A' = \tan^2 B\tan B' = \tan^2 C\tan C' = \tan A\tan B\tan C,$$

and $$\operatorname{cosec} 2A + \operatorname{cosec} 2B + \operatorname{cosec} 2C = 0;$$

then will

$$\tan(A-A') = \tan(B-B') = \tan(C-C') = \tan A + \tan B + \tan C.$$

453. Having given

$$x \sin 3(\beta-\gamma) + y \sin 3(\gamma-\alpha) + z \sin 3(\alpha-\beta) = 0;$$

prove that

$$\frac{x \sin(\beta-\gamma) + y \sin(\gamma-\alpha) + z \sin(\alpha-\beta)}{x \cos(\beta-\gamma) + y \cos(\gamma-\alpha) + z \cos(\alpha-\beta)} + \frac{\sin 2(\beta-\gamma) + \sin 2(\gamma-\alpha) + \sin 2(\alpha-\beta)}{\cos 2(\beta-\gamma) + \cos 2(\gamma-\alpha) + \cos 2(\alpha-\beta)} = 0.$$

454. Having given the equations

$$\left.\begin{aligned} x^2 &= \beta^2 + \gamma^2 - 2\beta\gamma \cos\theta \\ y^2 &= \gamma^2 + \alpha^2 - 2\gamma\alpha \cos\phi \\ z^2 &= \alpha^2 + \beta^2 - 2\alpha\beta \cos\psi \end{aligned}\right\}, \qquad \left.\begin{aligned} x + y + z &= 0 \\ \theta + \phi + \psi &= 0 \end{aligned}\right\};$$

prove that $\beta\gamma \sin\theta + \gamma\alpha \sin\phi + \alpha\beta \sin\psi = 0.$

455. Reduce to its simplest form the equation

$$\{x \cos(\alpha+\beta) + y \sin(\alpha+\beta) - \cos(\alpha-\beta)\}\{x \cos(\gamma+\delta) + y \sin(\gamma+\delta) - \cos(\gamma-\delta)\}$$
$$= \{x \cos(\alpha+\gamma) + y \sin(\alpha+\gamma) - \cos(\alpha-\gamma)\}\{x \cos(\beta+\delta) + y \sin(\beta+\delta) - \cos(\beta-\delta)\}.$$

[The reduced equation is $\sin(\beta-\gamma)\sin(\alpha-\delta)(1 - x^2 - y^2) = 0.$]

456. Having given the equations

$$yz' - y'z + zx' - z'x + xy' - x'y = 0, \quad A + B + C = 180°,$$
$$xx' \sin^2 A + yy' \sin^2 B + zz' \sin^2 C = (yz' + y'z) \sin B \sin C \cos A + (zx' + z'x) \sin C \sin A \cos B + (xy' + x'y) \sin A \sin B \cos C;$$

prove that either $x = y = z$; or $x' = y' = z'$.

457. Having given the equations

$$(yz' - y'z) \sin A + (zx' - z'x) \sin B + (xy' - x'y) \sin C = 0,$$
$$xx' + yy' + zz' = (yz' + y'z) \cos A + (zx' + z'x) \cos B + (xy' + x'y) \cos C,$$
$$x \sin(S-A) + y \sin(S-B) + z \sin(S-C) = 0$$
$$= x' \sin(S-A) + y' \sin(S-B) + z' \sin(S-C),$$

where $2S = A + B + C;$

prove that

$$\frac{x}{\sin A} = \frac{y}{\sin B} = \frac{z}{\sin C}, \text{ and } \frac{x'}{\sin A} = \frac{y'}{\sin B} = \frac{z'}{\sin C}.$$

[It is assumed that $\cos S$ is not zero.]

458. Having given the equations

$$\frac{y^2 + z^2 - 2yz \cos\alpha}{\sin^2\alpha} = \frac{z^2 + x^2 - 2zx \cos\beta}{\sin^2\beta} = \frac{x^2 + y^2 - 2xy \cos\gamma}{\sin^2\gamma};$$

prove that, if $2s$ denote $\alpha+\beta+\gamma$, one of the following systems of equations will hold :—

$$\frac{x}{\cos(s-\alpha)}=\frac{y}{\cos(s-\beta)}=\frac{z}{\cos(s-\gamma)},$$

$$\frac{x}{\cos s}=\frac{y}{\cos(s-\gamma)}=\frac{z}{\cos(s-\beta)},$$

$$\frac{x}{\cos(s-\gamma)}=\frac{y}{\cos s}=\frac{z}{\cos(s-\alpha)},$$

$$\frac{x}{\cos(s-\beta)}=\frac{y}{\cos(s-\alpha)}=\frac{z}{\cos s}.$$

459. Having given the equation

$$\frac{\cos\alpha}{\cos\theta}+\frac{\sin\alpha}{\sin\theta}=-1;$$

prove that $$\frac{\cos^3\theta}{\cos\alpha}+\frac{\sin^3\theta}{\sin\alpha}=1.$$

460. Eliminate θ, ϕ from the equations

$$\frac{x}{a}\cos\theta+\frac{y}{b}\sin\theta=\frac{x}{a}\cos\phi+\frac{y}{b}\sin\phi=1,$$

$$4\cos\frac{\theta-\phi}{2}\cos\frac{\alpha-\theta}{2}\cos\frac{\alpha-\phi}{2}=1.$$

[The resultant equation is $\left(\frac{x}{a}-\cos\alpha\right)^2+\left(\frac{y}{b}-\sin\alpha\right)^2=3.$]

461. Having given the equations

$$a^2+b^2-2ab\cos\alpha=c^2+d^2-2cd\cos\gamma,$$
$$b^2+c^2-2bc\cos\beta=a^2+d^2-2ad\cos\delta,$$
$$ab\sin\alpha+cd\sin\gamma=bc\sin\beta+ad\sin\delta;$$

prove that $$\cos(\alpha+\gamma)=\cos(\beta+\delta).$$

[These are the equations connecting the sides and angles of a quadrangle.]

462. Having given the equations

$$\sin\theta+\sin\phi=a,\quad \cos\theta+\cos\phi=b;$$

prove that

(1) $$\tan\frac{\theta}{2}+\tan\frac{\phi}{2}=\frac{4a}{a^2+b^2+2b},$$

(2) $$\tan\theta+\tan\phi=\frac{8ab}{(a^2+b^2)^2-4a^2},$$

(3) $$\cos\theta\cos\phi=\frac{(a^2+b^2)^2-4a^2}{4(a^2+b^2)},$$

(4) $\sin\theta\sin\phi = \dfrac{(a^2+b^2)^2-4b^2}{4(a^2+b^2)}$,

(5) $\cos 2\theta + \cos 2\phi = \dfrac{(b^2-a^2)(a^2+b^2-2)}{a^2+b^2}$,

(6) $\cos 3\theta + \cos 3\phi = b^3 - 3a^2b - 3b + \dfrac{12a^2b}{a^2+b^2}$.

463. Eliminate θ from the equations

$$\frac{x\cos\theta}{a} + \frac{y\sin\theta}{b} = 1, \quad x\sin\theta - y\cos\theta = \sqrt{a^2\sin^2\theta + b^2\cos^2\theta}.$$

[The resultant equation is $\dfrac{x^2}{a} + \dfrac{y^2}{b} = a + b$, provided that $a\sin^2\theta + b\cos^2\theta$ does not vanish, in which case the two equations coincide.]

464. Prove that

$$\cos^2\theta + \cos^2(\alpha+\theta) - 2\cos\alpha\cos\theta\cos(\alpha+\theta)$$

is independent of θ.

[It is always equal to $\sin^2\alpha$.]

465. Prove that

$$\sin 2\alpha\cos\beta\cos\gamma\sin(\beta-\gamma) + \sin 2\beta\cos\gamma\cos\alpha\sin(\gamma-\alpha) + \sin 2\gamma\cos\alpha\cos\beta\sin(\alpha-\beta) \equiv 0,$$

$$\cos 2\alpha\cos\beta\cos\gamma\sin(\beta-\gamma) + \ldots + \ldots \equiv \sin(\beta-\gamma)\sin(\gamma-\alpha)\sin(\alpha-\beta).$$

466. Prove that

$$2\sin\frac{3\theta}{2}\left\{\sin^3\frac{5\theta}{2} - \sin^3\frac{3\theta}{2}\right\} \equiv \cos^3\theta + \cos^3 2\theta + \cos^3 3\theta - 3\cos\theta\cos 2\theta\cos 3\theta;$$

and

$$2\sin\frac{3\theta}{2}\left\{\cos^3\frac{3\theta}{2} - \cos^3\frac{5\theta}{2}\right\} \equiv \sin^3\theta + \sin^3 2\theta + \sin^3 3\theta - 3\sin\theta\sin 2\theta\sin 3\theta.$$

467. Prove that

(1) $$\frac{\sin 2\alpha\sin(\beta-\gamma) + \sin 2\beta\sin(\gamma-\alpha) + \sin 2\gamma\sin(\alpha-\beta)}{\sin(\gamma-\beta) + \sin(\alpha-\gamma) + \sin(\beta-\alpha)} \equiv \sin(\beta+\gamma) + \sin(\gamma+\alpha) + \sin(\alpha+\beta);$$

(2) $$\frac{\cos 2\alpha\sin(\beta-\gamma) + \cos 2\beta\sin(\gamma-\alpha) + \cos 2\gamma\sin(\alpha-\beta)}{\sin(\gamma-\beta) + \sin(\alpha-\gamma) + \sin(\beta-\alpha)} \equiv \cos(\beta+\gamma) + \cos(\gamma+\alpha) + \cos(\alpha+\beta);$$

(3) $$\frac{\sin 4\alpha\sin(\beta-\gamma) + \sin 4\beta\sin(\gamma-\alpha) + \sin 4\gamma\sin(\alpha-\beta)}{\sin(\gamma-\beta) + \sin(\alpha-\gamma) + \sin(\beta-\alpha)} \equiv 2\Sigma\sin(2\alpha+\beta+\gamma) + \Sigma\sin 2(\beta+\gamma) + \Sigma\sin(3\beta+\gamma);$$

(4) $$\frac{\cos 4\alpha \sin(\beta-\gamma) + \cos 4\beta \sin(\gamma-\alpha) + \cos 4\gamma \sin(\alpha-\beta)}{\sin(\gamma-\beta) + \sin(\alpha-\gamma) + \sin(\beta-\alpha)}$$
$$\equiv 2\Sigma \cos(2\alpha+\beta+\gamma) + \Sigma \cos 2(\beta+\gamma) + \Sigma \cos(3\beta+\gamma).$$

468. Prove that

(1) $$\frac{\sin 3\alpha \sin(\beta-\gamma) + \text{two similar terms}}{\cos 3\alpha \sin(\beta-\gamma) + \text{two similar terms}} \equiv \tan(\alpha+\beta+\gamma),$$

(2) $$\frac{\sin 5\alpha \sin(\beta-\gamma) + \ldots + \ldots}{\cos 5\alpha \sin(\beta-\gamma) + \ldots + \ldots} \equiv \frac{\sin(3\alpha+\beta+\gamma) + \ldots + \ldots}{\cos(3\alpha+\beta+\gamma) + \ldots + \ldots},$$

(3) $$\frac{\sin 7\alpha \sin(\beta-\gamma) + \ldots + \ldots}{\cos 7\alpha \sin(\beta-\gamma) + \ldots + \ldots} \equiv \frac{\sin(\alpha+3\beta+3\gamma) + \sin(5\alpha+\beta+\gamma) + \ldots}{\cos(\alpha+3\beta+3\gamma) + \cos(5\alpha+\beta+\gamma) + \ldots},$$

(4) $$\frac{\sin^3 \alpha \sin(\beta-\gamma) + \ldots + \ldots}{\cos^3 \alpha \sin(\beta-\gamma) + \ldots + \ldots} \equiv -\tan(\alpha+\beta+\gamma).$$

469. Prove that, if n be any positive whole number,

$$\frac{\sin n\alpha \sin(\beta-\gamma) + \sin n\beta \sin(\gamma-\alpha) + \sin n\gamma \sin(\alpha-\beta)}{\sin(\gamma-\beta) + \sin(\alpha-\gamma) + \sin(\beta-\alpha)}$$
$$= 2\Sigma \sin(p\alpha + q\beta + r\beta);$$

where p, q, r are three positive integers whose sum is n, no two vanishing together, and the coefficient when one vanishes being 1 instead of 2.

[If we write $\alpha + \frac{\pi}{2n}$, $\beta + \frac{\pi}{2n}$, $\gamma + \frac{\pi}{2n}$ for α, β, γ we get a similar equation with cosines instead of sines as the functions whose argument is the sum of n angles.]

470. Prove that, if n be any odd positive whole number,

$$\frac{\sin n\alpha \sin(\beta-\gamma) + \ldots + \ldots}{\sin 2(\gamma-\beta) + \ldots + \ldots} \equiv \Sigma \sin(p\alpha + q\beta + r\gamma);$$

where p, q, r are odd positive integers whose sum is n.

[The same remark as in the last.]

471. The resultant of the equations

$$x\cos\theta + y\sin\theta = x\cos\phi + y\sin\phi = 1,$$
$$a\cos\theta\cos\phi + b\sin\theta\sin\phi + c + f(\cos\theta + \cos\phi) + g(\sin\theta + \sin\phi) + h\sin(\theta+\phi)$$
is $(c-b)x^2 + (c-a)y^2 + a + b + 2fy + 2gx + 2hxy = 0$.

472. Prove that

$$\frac{1}{2\cos\alpha -}\,\frac{1}{2\cos\alpha -}\,\frac{1}{2\cos\alpha -}\ldots -\frac{1}{2\cos\alpha + p} \equiv \frac{\sin n\alpha + p\sin(n-1)\alpha}{\sin(n+1)\alpha + p\sin n\alpha};$$

there being n quotients in the left-hand member.

473. Prove that

$$\frac{\sec^2 a}{4-}\ \frac{\sec^2 a}{1-}\ \frac{\sec^2 a}{4-}\ \ldots\ldots \text{ to } n \text{ quotients} \equiv \frac{\sin na}{\sin(n+2)a+\sin na}.$$

[If we call the dexter u_n, $1-u_{n-1}=\dfrac{\sin(n+1)a}{2\sin na\cos a}$, whence the equation $u_n(1-u_{n-1})=\dfrac{1}{4\cos^2 a}=\dfrac{\sec^2 a}{4}$.]

474. Prove that, if $\sin\alpha+\sin\beta+\sin\gamma=0$,

$$\frac{\sin\frac{\beta+\gamma}{2}\sin(\beta-\gamma)+\sin\left(\frac{\beta+\gamma}{2}-\alpha\right)\sin\gamma}{\sin\frac{\beta+\gamma}{2}\sin(\gamma-\beta)+\sin\left(\frac{\beta+\gamma}{2}-\alpha\right)\sin\beta}=\frac{\sin(\alpha-\gamma)}{\sin(\alpha-\beta)}.$$

475. From the identity

$$a^2\frac{(x-b)(x-c)}{(a-b)(a-c)}+b^2\frac{(x-c)(x-a)}{(b-c)(b-a)}+c^2\frac{(x-a)(x-b)}{(c-a)(c-b)}\equiv x^2,$$

deduce the identities

$$\cos 2(\theta+\alpha)\frac{\sin(\theta-\beta)\sin(\theta-\gamma)}{\sin(\alpha-\beta)\sin(\alpha-\gamma)}+\ldots+\ldots\equiv\cos 4\theta,$$

$$\sin 2(\theta+\alpha)\frac{\sin(\theta-\beta)\sin(\theta-\gamma)}{\sin(\alpha-\beta)\sin(\alpha-\gamma)}+\ldots+\ldots\equiv\sin 4\theta.$$

476. Prove the identities

(1) $\cos 2(\beta+\gamma-\alpha-\delta)\sin(\beta-\gamma)\sin(\alpha-\delta)$
$+\cos 2(\gamma+\alpha-\beta-\delta)\sin(\gamma-\alpha)\sin(\beta-\delta)$
$+\cos 2(\alpha+\beta-\gamma-\delta)\sin(\alpha-\beta)\sin(\gamma-\delta) \equiv -8K,$

(2) $\cos 3(\beta+\gamma-\alpha-\delta)\sin(\beta-\gamma)\sin(\alpha-\delta)+$ two similar terms $\equiv -16KL,$

(3) $\cos(\beta+\gamma-\alpha-\delta)\sin^3(\beta-\gamma)\sin^3(\alpha-\delta)+\ldots+\ldots \equiv KL,$

(4) $\cos(\beta-\gamma)\cos(\alpha-\delta)\sin^3(\beta-\gamma)\sin^3(\alpha-\delta)+\ldots+\ldots \equiv KL,$

(5) $\cos(\beta+\gamma-\alpha-\delta)\sin^5(\beta-\gamma)\sin^5(\alpha-\delta)+\ldots+\ldots \equiv -16KL;$

where K denotes

$$\sin(\beta-\gamma)\sin(\gamma-\alpha)\sin(\alpha-\beta)\sin(\alpha-\delta)\sin(\beta-\delta)\sin(\gamma-\delta);$$

and L denotes

$$\cos(\beta+\gamma-\alpha-\delta)+\cos(\gamma+\alpha-\beta-\delta)+\cos(\alpha+\beta-\gamma-\delta).$$

[The first is deduced from the identity $(b^2c^2+a^2d^2)(b-c)(a-d)$ + two similar terms $\equiv(b-c)(c-a)(a-b)(d-a)(d-b)(d-c)$ by putting $\cos 2a+i\sin 2a$ for a and the like: the same substitutions in other identities of (155) give (2), (3), (4), (5).]

477. Prove the identities

(1) $$\frac{\sin 4\alpha}{\sin\overline{\alpha-\beta}\sin\overline{\alpha-\gamma}\sin\overline{\alpha-\delta}}+\frac{\sin 4\beta}{\sin\overline{\beta-\gamma}\sin\overline{\beta-\delta}\sin\overline{\beta-\alpha}}+\ldots+\ldots\equiv-8\cos(\alpha+\beta+\gamma+\delta),$$

(2) $$\frac{\cos 4\alpha}{\sin\overline{\alpha-\beta}\sin\overline{\alpha-\gamma}\sin\overline{\alpha-\delta}}+\frac{\cos 4\beta}{\sin\overline{\beta-\gamma}\sin\overline{\beta-\delta}\sin\overline{\beta-\alpha}}+\ldots+\ldots\equiv 8\sin(\alpha+\beta+\gamma+\delta).$$

[In the following questions, up to (484) inclusive, A, B, C are the angles of a finite triangle.]

478. Having given the equations

$$\frac{y^2+z^2+2yz\cos A}{\sin^2 A}=\frac{z^2+x^2+2zx\cos B}{\sin^2 B}=\frac{x^2+y^2+2xy\cos C}{\sin^2 C};$$

prove that either $x\sin A+y\sin B+z\sin C=0$, or

$$x\sec A=y\sec B=z\sec C.$$

479. Prove that

$$\frac{1+\cos B\cos C}{\sin B\sin C}\tan\frac{B-C}{2}+\text{two similar terms}$$

$$=-\tan\frac{B-C}{2}\tan\frac{C-A}{2}\tan\frac{A-B}{2},$$

$$\frac{1-\cos A}{\sin B\sin C}\tan\frac{B-C}{2}+\ldots+\ldots=-2\tan\frac{B-C}{2}\tan\frac{C-A}{2}\tan\frac{A-B}{2}.$$

480. Prove that

$$(m\sin C+n\sin B\cos A)(n\sin A+l\sin C\cos B)(l\sin B+m\sin A\cos C)$$
$$+(n\sin B+m\sin C\cos A)(l\sin C+n\sin A\cos B)(m\sin A+l\sin B\cos C)$$
$$=(l+m+n)(mn\sin^2 A+nl\sin^2 B+lm\sin^2 C)\sin A\sin B\sin C.$$

481. If x, y, z be real quantities such that

$$\frac{y\sin C-z\sin B}{x-y\cos C-z\cos B}=\frac{z\sin A-x\sin C}{y-z\cos A-x\cos C},$$

prove that each member $=\dfrac{x\sin B-y\sin A}{z-x\cos B-y\cos A}$, and that

$$\frac{x}{\sin A}=\frac{y}{\sin B}=\frac{z}{\sin C}.$$

[The given equation is also satisfied if $\sin C=0$, but this is excluded by the condition that A, B, C are angles of a finite triangle. It may be noticed that each term must be of the form $\frac{0}{0}$.]

482. Prove that

$$(3 + 2\cos 2A + 3\cos 2B + 3\cos 2C + \cos 2\overline{B - C})^2 + (3\sin 2B - 3\sin 2C + \sin 2\overline{B - C})^2$$
$$= 8(\cos 2A \cos 2B \cos 2C + 3\cos \overline{B - C}\cos\overline{C - A}\cos\overline{A - B} - 15\cos A\cos B\cos C - 1).$$

483. Prove that

$$\sin^4 A(\sin^2 B + \sin^2 C - \sin^2 A) + \ldots + \ldots = 2\sin^2 A \sin^2 B \sin^2 C\,(1 + 4\cos A\cos B\cos C);$$
$$(\sin A + \sin B + \sin C)(-\sin A + \sin B + \sin C)(\sin A - \sin B + \sin C)(\sin A + \sin B - \sin C) = 4\sin^2 A\sin^2 B\sin^2 C.$$

484. Prove that

$$\sin A(\sin A - \sin B)(\sin A - \sin C) + \ldots + \ldots = \sin A\sin B\sin C(3 - 2\cos A - 2\cos B - 2\cos C);$$
$$\sin^2 A(\sin A - \sin B)(\sin A - \sin C) + \ldots + \ldots = (\cos A + \cos B + \cos C - 1)^2(\cos^2 A + \ldots - \cos B\cos C - \ldots);$$
$$\sin^2 B\sin 2C - 2\sin B\sin C\sin(B - C) - \sin^2 C\sin 2B = 0;$$
$$\sin^2 B\cos 2C - 2\sin B\sin C\cos(B - C) + \sin^2 C\cos 2B = \sin^2 A.$$

III. *Poristic Systems of Equations.*

[In all the examples under this head, solutions arising from the equality of any two angles are excluded, and all angles are supposed to lie between 0 and 2π.

A system of n equations of which the type is

$$a\cos(\alpha_r - \alpha_{r+1}) + b\cos(\alpha_r + \alpha_{r+1}) + c + 2f(\sin\alpha_r + \sin\alpha_{r+1}) + 2g(\cos\alpha_r + \cos\alpha_{r+1}) + 2h\sin(\alpha_r + \alpha_{r+1}) = 0,$$

where r has successively integral values from 1 to n, and $\alpha_{n+1} = \alpha_1$, is a poristic system, when solutions in which angles are equal are excluded: that is, the equations cannot be satisfied unless a certain relation hold connecting the coefficients a, b, c, f, g, h, and if this relation be satisfied the number of solutions is infinite, the n equations being equivalent to $n - 1$ independent equations only so that there is one solution for each value of α_1. All the examples here given are reducible to this type.]

485. Having given

$$\tan\beta\cot\frac{\gamma + \alpha}{2} = \tan\gamma\cot\frac{\alpha + \beta}{2},$$

prove that each $= \tan\alpha\cot\dfrac{\beta + \gamma}{2}$; and that

$$\sin(\beta + \gamma) + \sin(\gamma + \alpha) + \sin(\alpha + \beta) = 0.$$

486. Having given

$$\tan\frac{\alpha+\theta}{2}\tan\beta = \tan\frac{\beta+\theta}{2}\tan\alpha,$$

prove that each $= -\cot\frac{\alpha+\beta}{2}\tan\theta$, and that

$$\sin(\alpha+\theta)+\sin(\beta+\theta)=\sin(\alpha+\beta).$$

487. Having given the equations

$$\tan\frac{\beta+\gamma+\theta}{2}\tan\alpha = \tan\frac{\gamma+\alpha+\theta}{2}\tan\beta = \tan\frac{\alpha+\beta+\theta}{2}\tan\gamma = -\frac{m}{n};$$

prove that

$$\sin(\beta+\gamma)+\sin(\gamma+\alpha)+\sin(\alpha+\beta)=0,$$

$$\frac{\cos\beta\cos\gamma}{n}+\frac{\sin\beta\sin\gamma}{m}+\frac{1}{m+n}=0,$$

&c.

and that $\theta=\pi$.

488. Having given the system of equations

$$a\cos\beta\cos\gamma + b\sin\beta\sin\gamma = a\cos\gamma\cos\alpha + b\sin\gamma\sin\alpha$$
$$= a\cos\alpha\cos\beta + b\sin\alpha\sin\beta = c,$$

prove that $bc+ca+ab=0$,

and that the given system is equivalent to any one of the following systems:

(1) $\tan\frac{\beta+\gamma}{2}\cot\alpha = \tan\frac{\gamma+\alpha}{2}\cot\beta = \tan\frac{\alpha+\beta}{2}\cot\gamma = \frac{b}{a}$;

(2) $\dfrac{\cos\frac{\beta+\gamma}{2}}{\cos\alpha\cos\frac{\beta-\gamma}{2}} = \ldots\ldots\ldots\ldots = \ldots\ldots\ldots\ldots = \frac{a}{c}$;

(3) $\dfrac{\sin\frac{\beta+\gamma}{2}}{\sin\alpha\cos\frac{\beta-\gamma}{2}} = \ldots\ldots\ldots\ldots = \ldots\ldots\ldots\ldots = \frac{b}{c}$;

(4) $\cot\frac{\beta}{2}\cot\frac{\gamma}{2}+\cot\frac{\gamma}{2}\cot\frac{\alpha}{2}+\cot\frac{\alpha}{2}\cot\frac{\beta}{2}$

$$=\tan\frac{\beta}{2}\tan\frac{\gamma}{2}+\ldots+\ldots=1+\frac{2b}{c};$$

(5) $\sin(\beta+\gamma)+\sin(\gamma+\alpha)+\sin(\alpha+\beta)=0,$

$$\cos(\beta+\gamma)+\cos(\gamma+\alpha)+\cos(\alpha+\beta)=\frac{a-b}{a+b},$$

$$1+\cos(\beta-\gamma)+\cos(\gamma-\alpha)+\cos(\alpha-\beta)=-\frac{2ab}{(a+b)^2};$$

(6) $\dfrac{\cos\alpha+\cos\beta+\cos\gamma}{\cos(\alpha+\beta+\gamma)}=\dfrac{\sin\alpha+\sin\beta+\sin\gamma}{\sin(\alpha+\beta+\gamma)}=\dfrac{a-b}{a+b}$;

(7) $\dfrac{\cos\alpha\cos\beta\cos\gamma}{b^2\cos(\alpha+\beta+\gamma)}=\dfrac{-\sin\alpha\sin\beta\sin\gamma}{a^2\sin(\alpha+\beta+\gamma)}=\dfrac{1}{(a+b)^2}$;

(8) $\dfrac{\tan\frac{\alpha}{2}+\tan\frac{\beta}{2}+\tan\frac{\gamma}{2}}{\tan\frac{\alpha+\beta+\gamma}{2}}=\dfrac{\cot\frac{\alpha}{2}+\cot\frac{\beta}{2}+\cot\frac{\gamma}{2}}{\cot\frac{\alpha+\beta+\gamma}{2}}=1+\dfrac{2b}{a}$;

(9) $\dfrac{b(\tan\alpha+\tan\beta+\tan\gamma)}{(2a+b)\tan(\alpha+\beta+\gamma)}=\dfrac{a(\cot\alpha+\cot\beta+\cot\gamma)}{(a+2b)\cot(\alpha+\beta+\gamma)}=1$;

(10) $a^2(\tan\beta\tan\gamma+\tan\gamma\tan\alpha+\tan\alpha\tan\beta-1)$
$=b^2(\cot\beta\cot\gamma+\cot\gamma\cot\alpha+\cot\alpha\cot\beta-1)=-(a+b)^2$;

and that α, β, γ are roots of the equation in θ,

$$\frac{b\cos(\alpha+\beta+\gamma)}{\cos\theta}-\frac{a\sin(\alpha+\beta+\gamma)}{\sin\theta}+a+b=0.$$

489. Having given the equations

$$e\cos(\beta+\gamma)+\cos(\beta-\gamma)=e\cos(\gamma+\alpha)+\cos(\gamma-\alpha)$$
$$=e\cos(\alpha+\beta)+\cos(\alpha-\beta);$$

prove that each member of the equations is equal to $\dfrac{e^2-1}{2}$, and that

$$\sin(\beta+\gamma)+\sin(\gamma+\alpha)+\sin(\alpha+\beta)=0,$$
$$\cos(\beta+\gamma)+\cos(\gamma+\alpha)+\cos(\alpha+\beta)=e.$$

490. Having given the equations

$$\frac{\sin(\alpha-\beta)+\sin(\alpha-\gamma)}{\sin\beta+\sin\gamma-2\sin\alpha}=\frac{\sin(\beta-\gamma)+\sin(\beta-\alpha)}{\sin\gamma+\sin\alpha-2\sin\beta}=e;$$

prove that

$$\sin\alpha+\sin\beta+\sin\gamma=0,\quad \cos\alpha+\cos\beta+\cos\gamma=-3e.$$

491. Having given the equations

$$\frac{\sin\left(\alpha-\frac{\beta+\gamma}{2}\right)}{\cos\left(\alpha+\frac{\beta+\gamma}{2}\right)}=\frac{\sin\left(\beta-\frac{\gamma+\alpha}{2}\right)}{\cos\left(\beta+\frac{\gamma+\alpha}{2}\right)}=e;$$

prove that each member is equal to

$$\frac{\sin\left(\gamma-\frac{\alpha+\beta}{2}\right)}{\cos\left(\gamma+\frac{\alpha+\beta}{2}\right)},$$

and that $$\cos(\beta+\gamma)+\cos(\gamma+\alpha)+\cos(\alpha+\beta)=0,$$
$$\sin(\beta+\gamma)+\sin(\gamma+\alpha)+\sin(\alpha+\beta)=-e.$$

492. If $$\frac{\cos(\alpha+\beta-\theta)}{\sin(\alpha+\beta)\cos^2\gamma}=\frac{\cos(\alpha+\gamma-\theta)}{\sin(\alpha+\gamma)\cos^2\beta},$$
each member will be equal to
$$\frac{\cos(\beta+\gamma-\theta)}{\sin(\beta+\gamma)\cos^2\alpha},$$
and $$\cot\theta=\frac{\sin(\beta+\gamma)\sin(\gamma+\alpha)\sin(\alpha+\beta)}{\cos(\beta+\gamma)\cos(\gamma+\alpha)\cos(\alpha+\beta)+\sin^2(\alpha+\beta+\gamma)}.$$

493. Having given the equations
$$a^2\cos\alpha\cos\beta+a(\sin\alpha+\sin\beta)=a^2\cos\alpha\cos\gamma+a(\sin\alpha+\sin\gamma)=-1,$$
prove that $$a^2\cos\beta\cos\gamma+a(\sin\beta+\sin\gamma)+1=0,$$
that $$\cos\alpha+\cos\beta+\cos\gamma=\cos(\alpha+\beta+\gamma),$$
$$\sin\alpha+\sin\beta+\sin\gamma=\sin(\alpha+\beta+\gamma)-\frac{2}{a}:$$
and that
$$\tan\frac{\beta+\gamma}{2}\cos\alpha=\tan\frac{\gamma+\alpha}{2}\cos\beta=\tan\frac{\alpha+\beta}{2}\cos\gamma=\frac{1}{a}.$$

494. Having given the equations
$$\sin\beta=m\tan\frac{\alpha+\gamma}{2},\qquad \sin\gamma=m\tan\frac{\alpha+\beta}{2};$$
prove that
$$\sin\alpha=m\tan\frac{\beta+\gamma}{2};\quad \sin\alpha+\sin\beta+\sin\gamma+\sin(\alpha+\beta+\gamma)=0,$$
$$\cos\alpha+\cos\beta+\cos\gamma+\cos(\alpha+\beta+\gamma)=-2m,$$
and $$\sin\alpha\sin\beta\sin\gamma=m^2\sin(\alpha+\beta+\gamma)$$
$$=-\frac{m}{2}\sin\{(\beta+\gamma)+\sin(\gamma+\alpha)+\sin(\alpha+\beta)\}.$$

495. The system of equations
$$\cos(\beta+\gamma)+m(\sin\beta+\sin\gamma)+n=0,$$
$$\cos(\gamma+\alpha)+m(\sin\gamma+\sin\alpha)+n=0,$$
$$\cos(\alpha+\beta)+m(\sin\alpha+\sin\beta)+n=0,$$
is equivalent to
$$m^2=1,\ m=\sin(\alpha+\beta+\gamma),\ mn+\sin\alpha+\sin\beta+\sin\gamma=0.$$

496. Having given the equations
$$\sin(\beta+\gamma)+k\sin(\alpha+\theta)=\sin(\gamma+\alpha)+k\sin(\beta+\theta)$$
$$=\sin(\alpha+\beta)+k\sin(\gamma+\theta);$$
prove that $k^2=1$, and that each member $=0$.

497. **Having** given the equations

$$a\cos(\beta-\gamma)+b(\cos\beta+\cos\gamma)+c(\sin\beta+\sin\gamma)+d=0,$$
$$a\cos(\gamma-\alpha)+b(\cos\gamma+\cos\alpha)+c(\sin\gamma+\sin\alpha)+d=0,$$
$$a\cos(\alpha-\beta)+b(\cos\alpha+\cos\beta)+c(\sin\alpha+\sin\beta)+d=0;$$

prove that

$$a^2+b^2+c^2=2ad,\quad a(\cos\alpha+\cos\beta+\cos\gamma)+b=0,$$
$$a(\sin\alpha+\sin\beta+\sin\gamma)+c=0.$$

498. Having given the equations

$$(m+\cos\beta)(m+\cos\gamma)+n(\sin\beta+\sin\gamma)=(m+\cos\gamma)(m+\cos\alpha)$$
$$+n(\sin\gamma+\sin\alpha)=(m+\cos\alpha)(m+\cos\beta)+n(\sin\alpha+\sin\beta);$$

prove that each $=-n^2$; and that

$$\cos(\alpha+\beta+\gamma)-\cos\alpha-\cos\beta-\cos\gamma=2m,$$
$$\sin(\alpha+\beta+\gamma)-\sin\alpha-\sin\beta-\sin\gamma=2n.$$

499. Having given the equations

$$\cos(\theta-\beta)+\cos(\theta-\gamma)+\cos(\beta-\gamma)=\cos(\theta-\gamma)+\cos(\theta-\alpha)+\cos(\gamma-\alpha)$$
$$=\cos(\theta-\alpha)+\cos(\theta-\beta)+\cos(\alpha-\beta);$$

prove that each $=\cos(\beta-\gamma)+\cos(\gamma-\alpha)+\cos(\alpha-\beta)=-1$; and that

$$\cos\alpha+\cos\beta+\cos\gamma+\cos\theta=0=\sin\alpha+\sin\beta+\sin\gamma+\sin\theta.$$

500. **Having given**

$$\frac{m\cos\alpha+n\sin\alpha-\sin(\beta+\gamma)}{\cos(\beta+\gamma)}=\frac{m\cos\beta+n\sin\beta-\sin(\gamma+\alpha)}{\cos(\gamma+\alpha)};$$

prove that each

$$=\frac{m\cos\gamma+n\sin\gamma-\sin(\alpha+\beta)}{\cos(\alpha+\beta)}=m\cos(\alpha+\beta+\gamma)+n\sin(\alpha+\beta+\gamma).$$

501. The three equations

$$\frac{m\cos\frac{\beta+\gamma}{2}}{\cos\alpha}-\frac{n\sin\frac{\beta+\gamma}{2}}{\sin\alpha}=(m-n)\cos\frac{\beta-\gamma}{2},$$

$$\frac{m\cos\frac{\gamma+\alpha}{2}}{\cos\beta}-\frac{n\sin\frac{\gamma+\alpha}{2}}{\sin\beta}=(m-n)\cos\frac{\gamma-\alpha}{2},$$

$$\frac{m\cos\frac{\alpha+\beta}{2}}{\cos\gamma}-\frac{n\sin\frac{\alpha+\beta}{2}}{\sin\gamma}=(m-n)\cos\frac{\alpha-\beta}{2},$$

are equivalent to only two independent equations.

502. Having given

$$\frac{\cos(\alpha+\theta)}{\sin(\beta+\gamma)} = \frac{\cos(\beta+\theta)}{\sin(\gamma+\alpha)};$$

prove that each $= \dfrac{\cos(\gamma+\theta)}{\sin(\alpha+\beta)} = \pm 1$.

503. Having given the equations

$$m(\cos\beta+\cos\gamma) - n(\sin\beta+\sin\gamma) + \sin(\beta+\gamma)$$
$$= m(\cos\gamma+\cos\alpha) - n(\sin\gamma+\sin\alpha) + \sin(\gamma+\alpha)$$
$$= m(\cos\alpha+\cos\beta) - n(\sin\alpha+\sin\beta) + \sin(\alpha+\beta);$$

prove that $\quad m = \sin(\alpha+\beta+\gamma), \quad n = \cos(\alpha+\beta+\gamma)$.

504. Having given the equations

$$\frac{e\cos\frac{\beta+\gamma}{2} + \cos\frac{\beta-\gamma}{2}}{e\sin\frac{\beta+\gamma}{2} + \sin\frac{\beta-\gamma}{2}} = \frac{e\cos\frac{\gamma+\alpha}{2} + \cos\frac{\gamma-\alpha}{2}}{e\sin\frac{\gamma+\alpha}{2} + \sin\frac{\gamma-\alpha}{2}} = \frac{e\cos\frac{\alpha+\beta}{2} + \cos\frac{\alpha-\beta}{2}}{e\sin\frac{\alpha+\beta}{2} + \sin\frac{\alpha-\beta}{2}};$$

prove that each $= \pm\sqrt{\dfrac{1-e^2}{3+e^2}}$.

505. Having given the equations

$$\frac{e^2\sin\beta\sin\gamma + e(\cos\beta+\cos\gamma)+1}{\cos^2\frac{\beta-\gamma}{2}} = \frac{e^2\sin\gamma\sin\alpha + e(\cos\gamma+\cos\alpha)+1}{\cos^2\frac{\gamma-\alpha}{2}}$$
$$= \frac{e^2\sin\alpha\sin\beta + e(\cos\alpha+\cos\beta)+1}{\cos^2\frac{\alpha-\beta}{2}};$$

prove that each $= 0$ or 4. If each $= 0$,

$$\sin\alpha + \sin\beta + \sin\gamma + \sin(\alpha+\beta+\gamma) = 0,$$

and $\quad e\{\cos\alpha + \cos\beta + \cos\gamma + \cos(\alpha+\beta+\gamma)\} = -2;$

and, if each $= 4$,

$$\frac{e^2}{\sin\alpha+\sin\beta+\sin\gamma} = \frac{2e}{\sin(\beta+\gamma)+\sin(\gamma+\alpha)+\sin(\alpha+\beta)}$$
$$= \frac{8}{\sin\alpha+\sin\beta+\sin\gamma+\sin(\alpha+\beta+\gamma)}.$$

Also the given system is equivalent to the system

$$e^2\sin\frac{\beta+\gamma}{2}\cos\alpha - e\cos\frac{\beta-\gamma}{2}\sin\alpha = \sin\left(\frac{\beta+\gamma}{2}-\alpha\right),$$

and the two corresponding equations.

506. Having given the system of three equations whose type is

$$a\cos\beta\cos\gamma + b\sin\beta\sin\gamma + c + f(\sin\beta + \sin\gamma) + g(\cos\beta + \cos\gamma) + h\sin(\beta+\gamma) = 0,$$

prove that $$ab - h^2 = bc - f^2 + ca - g^2.$$

507. Prove that any system of three poristic equations between α, β, γ is equivalent to two independent equations of the form

$$l(\cos\alpha + \cos\beta + \cos\gamma) + m\cos(\alpha+\beta+\gamma) + n(\cos\overline{\beta+\gamma} + \cos\overline{\gamma+\alpha} + \cos\overline{\alpha+\beta}) = p,$$
$$l(\sin\alpha + \sin\beta + \sin\gamma) + m\sin(\alpha+\beta+\gamma) + n(\sin\overline{\beta+\gamma} + \sin\overline{\gamma+\alpha} + \sin\overline{\alpha+\beta}) = q.$$

[The equation between β, γ will be

$$2(n^2 - l^2)\cos(\beta - \gamma) + 2(lm + np)\cos(\beta+\gamma) + m^2 + n^2 - p^2 - q^2 - l^2$$
$$+ 2lq(\sin\beta + \sin\gamma) + 2(mn + lp)(\cos\beta + \cos\gamma) + 2nq\sin(\beta+\gamma) = 0.]$$

508. The condition for the coexistence of four equations of the type in (506) between α, β; β, γ; γ, δ; and δ, α respectively is

$$\Delta \equiv abc + 2fgh - af^2 - bg^2 - ch^2 = 0.$$

509. Having given the system of five equations

$$\frac{1}{a}\cos\alpha\cos\beta + \frac{1}{b}\sin\alpha\sin\beta = \frac{1}{c},$$

and the like equations between β, γ; γ, δ; δ, ϵ; ϵ, α respectively; prove that

$$a^3 + b^3 + c^3 = (b+c)(c+a)(a+b).$$

[An equivalent form is $4abc + (b+c-a)(c+a-b)(a+b-c) = 0$.]

510. Prove that the system of n equations

$$\tan\frac{\alpha_2+\alpha_n}{2}\cot\alpha_1 = \tan\frac{\alpha_3+\alpha_1}{2}\cot\alpha_2 = \tan\frac{\alpha_4+\alpha_2}{2}\cot\alpha_3 = \dots$$
$$= \tan\frac{\alpha_1+\alpha_{n-1}}{2}\cot\alpha_n = \frac{b}{a},$$

is equivalent to only $n-1$ independent equations.

IV. *Inequalities.*

511. Prove that $\cot\frac{\theta}{2} > 1 + \cot\theta$, for values of θ between 0 and π: and that, for all values of θ, $\frac{3\sin\theta}{\theta} < 2 + \cos\theta$.

512. Prove that, for real values of x, $\dfrac{x^2 - 2x\cos\alpha + 1}{x^2 - 2x\cos\beta + 1}$ lies between $\dfrac{1-\cos\alpha}{1-\cos\beta}$ and $\dfrac{1+\cos\alpha}{1+\cos\beta}$.

513. If x, y, z be any real quantities and A, B, C the angles of a triangle, prove that

$$x^2 + y^2 + z^2 > 2yz\cos A + 2zx\cos B + 2xy\cos C,$$

unless $$x\operatorname{cosec} A = y\operatorname{cosec} B = z\operatorname{cosec} C.$$

514. Under the same conditions as the last, prove that

$$(x\sin^2 A + y\sin^2 B + z\sin^2 C)^2 > 4(yz + zx + xy)\sin^2 A\sin^2 B\sin^2 C,$$

unless $$x\tan A = y\tan B = z\tan C.$$

515. If A, B, C be the angles of a triangle, prove

$$\frac{\sin B\sin C + \sin C\sin A + \sin A\sin B}{\sin^2 A + \sin^2 B + \sin^2 C}$$

lies between the values $\frac{1}{2}$ and 1; and

$$\frac{\sin A\sin B\sin C}{1 + \cos A\cos B\cos C}$$

between 0 and $\dfrac{1}{\sqrt{3}}$.

516. Having given the equation

$$\sec\beta\sec\gamma + \tan\beta\tan\gamma = \tan\alpha,$$

prove that, for real values of β and γ, $\cos 2\alpha$ must be negative; and that

$$\frac{\tan\beta + \tan\alpha\tan\gamma}{\tan\gamma + \tan\alpha\tan\beta} = \frac{\cos\beta}{\cos\gamma}.$$

517. Prove that, A, B, C being the angles of a triangle,

$$\tfrac{1}{8} > \sin\frac{A}{2}\sin\frac{B}{2}\sin\frac{C}{2} > (1-\cos A)(1-\cos B)(1-\cos C) > \cos A\cos B\cos C;$$

and that

$$\cos\frac{A}{2}\cos\frac{B}{2}\cos\frac{C}{2} > \sin A\sin B\sin C > \sin 2A\sin 2B\sin C;$$

except when $A = B = C$.

518. Prove that, A, B, C being the angles of a triangle,

$$8\sin A\sin B\sin C < \left(\frac{1 + \cos\frac{A}{2}\cos\frac{B}{2}\cos\frac{C}{2}}{\sin\frac{A}{2}\sin\frac{B}{2}\sin\frac{C}{2}}\right)^2,$$

$$1 + \cos A\cos B\cos C > \sqrt{3}\sin A\sin B\sin C.$$

519. On a fixed straight line AB is taken a point C such that $AC = 2CB$ and any other point P between A and C; prove that, if $CP \equiv CA \sin A$, $AP \cdot BP^2$ will vary as $1 + \sin 3\theta$ and thence that $AP \cdot BP^2$ has its greatest value when P bisects AC.

520. If $\alpha + A$, $\beta + B$, $\gamma + C$ be the angles subtended at a point by the sides of the triangle ABC, then will

$$\left(\frac{\sin^2 \alpha}{\sin A} + \frac{\sin^2 \beta}{\sin B} + \frac{\sin^2 \gamma}{\sin C}\right)^2 > \frac{2 \sin^2 \alpha \sin^2 \beta \sin^2 \gamma}{\sin \frac{A}{2} \sin \frac{B}{2} \sin \frac{C}{2}},$$

except when the point is the centre of the inscribed circle.

521. Having given the equations

$$\cot \beta \cot \gamma + \cot \gamma \cot \alpha + \cot \alpha \cot \beta = \tan \beta \tan \gamma + \tan \gamma \tan \alpha + \tan \alpha \tan \beta;$$

prove that $\cos 2(\beta - \gamma) + \cos 2(\gamma - \alpha) + \cos 2(\alpha - \beta) > -\frac{3}{2}$.

522. If α, β, γ be angles between 0 and $\frac{\pi}{2}$, and if $\tan \alpha \tan \beta \tan \gamma = 1$, then will

$$\sin \alpha \sin \beta \sin \gamma < \frac{1}{2\sqrt{2}},$$

unless $\alpha = \beta = \gamma$.

523. Prove that

$$\left\{\frac{\cos^2(\alpha - \theta)}{a^2} + \frac{\sin^2(\alpha - \theta)}{b^2}\right\}\left\{\frac{\cos^2(\alpha + \theta)}{a^2} + \frac{\sin^2(\alpha + \theta)}{b^2}\right\}$$

cannot be less than $\frac{\sin^2 2\alpha}{a^2 b^2}$; and can never be equal to it unless $\tan^2 \alpha$ lie between $\frac{b^2}{a^2}$ and $\frac{a^2}{b^2}$.

524. If ω, x, y, z be any real quantities, and a, b, c, a', b', c' cosines of angles satisfying the condition

$$\begin{vmatrix} -1, & c', & b', & a \\ c', & -1, & a', & b \\ b', & a', & -1, & c \\ a, & b, & c, & -1 \end{vmatrix} = 0;$$

prove that

$$x^2 + y^2 + z^2 + \omega^2 > 2a'yz + 2b'zx + 2c'xy + 2a\omega x + 2b\omega y + 2c\omega z,$$

except when

$$\frac{x^2}{1 - a'^2 - b^2 - c^2 - 2a'bc} = \frac{y^2}{1 - a^2 - b'^2 - c^2 - 2ab'c}$$

$$= \frac{z^2}{1 - a^2 - b^2 - c'^2 - 2abc'} = \frac{\omega^2}{1 - a'^2 - b'^2 - c'^2 - 2a'b'c'}.$$

V. *Properties of Triangles.*

[In these questions a, b, c denote the sides and A, B, C the respectively opposite angles of a triangle, R is the radius of the circumscribed circle, and r, r_1, r_2, r_3 the radii of the inscribed circle and of the escribed circles respectively opposite A, B, C.]

525. Prove that if

$$1 + \cos A = \cos B + \cos C, \quad \sec A - 1 = \sec B + \sec C.$$

526. Prove that

$$a > b\cos B + c\cos C, \quad b > c\cos C + a\cos A, \text{ and } c > a\cos A + b\cos B.$$

527. If a triangle $A'B'C'$ be drawn whose sides are $b + c$, $c + a$, $a + b$ respectively, and if the angle $A' = A$, then will $2a$ lie between $b + c$ and $2(b + c)$, and

$$\cos\frac{B - C}{2} = 4\sin\frac{A}{2} - \sin\frac{3A}{2}.$$

528. If θ, ϕ, ψ be acute angles given by the equations

$$\cos\theta = \frac{a}{b + c}, \quad \cos\phi = \frac{b}{c + a}, \quad \cos\psi = \frac{c}{a + b},$$

then will

$$\tan^2\frac{\theta}{2} + \tan^2\frac{\phi}{2} + \tan^2\frac{\psi}{2} = 1;$$

and

$$\tan\frac{\theta}{2}\tan\frac{\phi}{2}\tan\frac{\psi}{2} = \tan\frac{A}{2}\tan\frac{B}{2}\tan\frac{C}{2}.$$

529. If $\sin A$, $\sin B$, $\sin C$ be in harmonical progression so also will be

$$1 - \cos A, \quad 1 - \cos B, \quad 1 - \cos C.$$

530. From the three relations between the sides and angles given in the forms

$$a^2 = b^2 + c^2 - 2bc\cos A, \text{ \&c.}$$

deduce the equations

$$\frac{\sin A}{a} = \frac{\sin B}{b} = \frac{\sin C}{c}, \quad A + B + C = 180°;$$

assuming that each angle lies between 0 and 180°.

531. In the side BC produced if necessary find a point P such that the square on PA may be equal to the sum of the squares on PB, PC; and prove that this is only possible when A, B, C are all acute and $\tan A < \tan B + \tan C$, or when B or C is obtuse. When possible, prove that there are in general two such points which lie both between B and C, one between and one beyond, or both beyond, according as A is the greatest, the mean, or the least angle of the triangle.

532. The sides of a triangle are $2pq+p^2$, p^2+pq+q^2, and p^2-q^2; prove that the angles are in A. P., the common difference being

$$2\tan^{-1}\left(\frac{q\sqrt{3}}{2p+q}\right).$$

533. The line joining the middle points of BC and of the perpendicular from A on BC makes with BC the angle $\cot^{-1}(\cot B-\cot C)$.

534. The line joining the centres of the inscribed and circumscribed circles makes with BC the angle

$$\tan^{-1}\left(\frac{\cos B+\cos C-1}{\sin B-\sin C}\right).$$

535. The line joining the centre of the circumscribed circle and the centre of perpendiculars makes with BC the angle

$$\tan^{-1}\left(\frac{\tan B\tan C-3}{\tan B-\tan C}\right).$$

536. The line joining the centre of the inscribed circle and the centre of perpendiculars makes with BC the angle

$$\tan^{-1}\left\{\cot\frac{C-B}{2}+\frac{\cos A}{2\sin\frac{B}{2}\sin\frac{C}{2}\sin\frac{B-C}{2}}\right\}.$$

537. In a triangle, right-angled at A, prove that

$$r_1=r_2+r_3+r.$$

538. If $BA=AC+\frac{1}{2}BC$ and BC be divided in O in the ratio $1:3$, then will the angle ACO be double of the angle AOC.

539. If the sides of a triangle be in A. P. and the greatest angle exceed the less by (1) 60°, (2) 90°, (3) 120°, the sides of the triangle will be as

(1) $\sqrt{13}-1:\sqrt{13}:\sqrt{13}+1$,
(2) $\sqrt{7}-1:\sqrt{7}:\sqrt{7}+1$,
(3) $\sqrt{5}-1:\sqrt{5}:\sqrt{5}+1$.

[In general if the sides be in A. P. and the greatest angle exceed the least by α, the sides will be as

$$\sqrt{7-\cos\alpha}+\sqrt{1-\cos\alpha}:\sqrt{7-\cos\alpha}:\sqrt{7-\cos\alpha}-\sqrt{1-\cos\alpha}.]$$

540. If O be the centre of the circumscribed circle and AO meet BC in D,

$$OD:AO=\cos A:\cos(B-C).$$

541. Three parallel straight lines are drawn through the angular points of the triangle ABC to meet the opposite sides in A', B', C': prove that

$$\frac{A'B\,.\,A'C}{AA'^2}+\frac{B'C\,.\,B'A}{BB'^2}+\frac{C'A\,.\,C'B}{CC'^2}=1;$$

the segments of a side being affected with opposite signs when they fall on opposite sides of the point of section.

[The convention stated in the last clause ought always to be attended to, but it is not yet so sufficiently recognised in our elementary books as to make the mention superfluous. $BC + CA + AB \equiv 0$ ought always to be an allowed identity.]

542. The perimeter of a triangle bears to the perimeter of an inscribed circle the same ratio as the area of the triangle to the area of the circle, which is

$$\cot\frac{A}{2}\cot\frac{B}{2}\cot\frac{C}{2} : \pi.$$

[The first part of this proposition is true for any polygon circumscribed to a circle, and a similar one for any polyhedron circumscribed to a sphere.]

543. A triangle is formed by joining the feet of the perpendiculars of the triangle ABC, and the circle inscribed in this triangle touches the sides in A', B', C': prove that

$$\frac{B'C'}{BC} = \frac{C'A'}{CA} = \frac{A'B'}{AB} = 2\cos A\cos B\cos C.$$

544. A circle is drawn to touch the circumscribed circle and the sides AB, AC; prove that its radius is $r\sec^2\frac{A}{2}$: and if it touch the circumscribed circle and the sides AB, AC produced its radius is $r_1\sec^2\frac{A}{2}$. If $B = C$ and the latter radius $= R$, $\cos A = \frac{7}{9}$.

545. Having given the equations

$$c^2y + b^2z = a^2z + c^2x = b^2x + a^2y;$$

prove that
$$\frac{x}{\sin 2A} = \frac{y}{\sin 2B} = \frac{z}{\sin 2C}.$$

546. Determine a triangle having a base c, an altitude h, and a given difference a of the base angles; and if θ_1, θ_2 be the two values obtainable for the vertical angle, prove that $\cot\theta_1 + \cot\theta_2 = \frac{4h}{c\sin^2 a}$. Prove that only one of these values corresponds to a proper solution; and if this be θ_1, that

$$\tan\frac{\theta_1}{2} = \frac{\sqrt{4h^2 + c^2\sin^2 a} - 2h}{c(1 - \cos a)}.$$

Account for the appearance of the other value.

547. Determine a triangle in which are given a side a, the opposite angle A, and the rectangle m^2 under the other two sides: and prove that no such triangle exists if $2m\sin\frac{A}{2} > a$.

548. Find the angles of a triangle in which the greatest side is twice the least, and the greatest angle twice the mean angle. Prove that a triangle whose sides are as 17156 : 13395 : 8578 is a very approximate solution.

549. A triangle $A'B'C'$ has its angles respectively complementary to the half angles of the triangle ABC and its side $B'C'$ equal to BC; prove that

$$\Delta A'B'C' : \Delta ABC = \sin\frac{A}{2} : 2\sin\frac{B}{2}\sin\frac{C}{2}.$$

550. Two triangles ABC, $A'B'C'$ are such that

$$\cot A + \cot A' = \cot B + \cot B' = \cot C + \cot C';$$

a point P is taken within ABC such that its distances from A, B, C are as $B'C' : C'A' : A'B'$: prove that the angles subtended at P by the sides of the triangle ABC are $B' + C'$, $C' + A'$, $A' + B'$; also that

$$\cot A' + \cot B' + \cot C' = \cot A + \cot B + \cot C.$$

551. If A', B', C' be the angles subtended at the centroid of a triangle ABC by the sides,

$$\cot A - \cot A' = \cot B - \cot B' = \cot C - \cot C'$$
$$= \tfrac{2}{3}(\cot A + \cot B + \cot C).$$

552. If $8\cos A\cos B\cos C = \cos^2\alpha$, each angle of the triangle ABC must lie between the acute angles $\cos^{-1}\left(\frac{1 \pm \sin\alpha}{2}\right)$, and the difference between the greatest and least angles cannot exceed α.

553. If a straight line can be drawn not intersecting the sides of a triangle ABC, and such that the perpendiculars on it from the angular points are respectively equal to the opposite sides, then will

$$\tan^2\frac{A}{2} + \tan^2\frac{B}{2} + \tan^2\frac{C}{2} = \tan\frac{A}{2} + \tan\frac{B}{2} + \tan\frac{C}{2} = 2,$$

$$\cos^2\frac{A}{2} + \cos^2\frac{B}{2} + \cos^2\frac{C}{2} = \sin A + \sin B + \sin C.$$

[Of course these equations are equivalents.]

554. With A, B, C as centres are described circles whose radii are $a\cos A$, $b\cos B$, $c\cos C$ respectively, and the internal common tangents are drawn to each pair: prove that three of these will pass through the centre of the circumscribed circle and the other three through the centre of perpendiculars.

555. The centroid of ABC is G, a triangle $A'B'C'$ is drawn whose sides are GA, GB, GC, and circles described with centres A, B, C and radii $a'\sin A'$, $b'\sin B'$, $c'\sin C'$ respectively: prove that three of the internal common tangents to a pair of circles intersect in G, and the other three in the point of concourse of the lines joining A, B, C to the corresponding intersections of the tangents to the circumscribed circle at A, B, C.

556. If a triangle ABC be inscribed in a circle and the tangent at A meet BC in A', then a straight line drawn through A', perpendicular to the internal bisector of the angle A, will meet the circle in two points (P, Q), whose distance from A is a mean proportional between the distances from B, C respectively; and $\angle PAB - \angle CAQ = \frac{1}{2}(B - C)$. Also if $a^2 > 4bc$ the straight line drawn through A' parallel to the bisector will meet the circle in two other points having the same property. If B, C be fixed points and A any point such that $a^2 = 4bc$, the two last points will coincide, and its locus will be a rectangular hyperbola whose foci are B, C.

557. If O be the point within the triangle ABC at which the sum of the distances of A, B, C is a minimum, straight lines drawn through A, B, C at right angles to OA, OB, OC respectively will form the maximum equilateral triangle which can be circumscribed to ABC; and if $A'B'C'$ be this maximum triangle, then will

$$OA' : OB' : OC' = BC : CA : AB.$$

Prove also that

$$OA \sin^2 A \sin (B - C) + OB \sin^2 B \sin (C - A) + OC \sin^2 C \sin (A - B) = 0.$$

558. If x, y, z be perpendiculars from the angular points on any straight line; prove that

$$a^2 (x - y)(x - z) + b^2 (y - z)(y - x) + c^2 (z - x)(z - y) = (2\Delta ABC)^2,$$

any perpendicular being reckoned negative which is drawn from its angular point in the opposite sense to the other two.

559. If perpendiculars OD, OE, OF be let fall from any point O on the sides of the triangle ABC, and x, y, z be the radii of the circles AEF, BFD, CDE, respectively; prove that

$$16a^2 (x^2 - y^2)(x^2 - z^2) + \ldots - 8abc\,(ax^2 \cos A + \ldots) + a^2b^2c^2 = 0.$$

560. If O be the centre of the circle inscribed in ABC, OD, OE, OF perpendiculars on the sides, and x, y, z radii of the circles inscribed in the quadrilaterals $OEAF$, $OFBD$, $ODCE$; prove that

$$(r - 2x)(r - 2y)(r - 2z) = r^3 - 4xyz.$$

561. A triangle $A'B'C'$ is circumscribed to the triangle ABC, prove that when its perimeter is the least possible $BC = B'C'\sqrt{1 - \sin B' \sin C'}$; and, if x, y, z be the sides of the triangle $A'B'C'$, that

$$\frac{x^2 - a^2}{x} = \frac{y^2 - b^2}{y} = \frac{z^2 - c^2}{z} = \frac{(x + y + z)(y + z - x)(z + x - y)(x + y - z)}{4xyz}.$$

562. If p_1, p_2, p_3 be the perpendiculars of the triangle

$$\frac{1}{p_1} + \frac{1}{p_2} + \frac{1}{p_3} = \frac{1}{r}, \quad \frac{\cos A}{p_1} + \frac{\cos B}{p_2} + \frac{\cos C}{p_3} = \frac{1}{R};$$

also p_1 is a harmonic mean between r_2, r_3.

563. The distances between the centres of the escribed circles being α, β, γ; prove that

$$4R = \frac{\alpha^2}{r_2+r_3} = \frac{\beta^2}{r_3+r_1} = \frac{\gamma^2}{r_1+r_2}$$
$$= \frac{(r_2+r_3)(r_3+r_1)(r_1+r_2)}{r_2r_3+r_3r_1+r_1r_2} = \frac{\alpha\beta\gamma}{2\sqrt{\sigma(\sigma-\alpha)(\sigma-\beta)(\sigma-\gamma)}},$$

where $2\sigma \equiv \alpha+\beta+\gamma$. Prove also that

$$a = \frac{r_1(r_2+r_3)}{\sqrt{r_2r_3+r_3r_1+r_1r_2}}, \quad \Delta ABC = \frac{r_1r_2r_3}{\sqrt{r_2r_3+r_3r_1+r_1r_2}}.$$

564. The distances of the centre of the inscribed circle from those of the escribed circles being α', β', γ'; prove that

$$4R = \frac{\alpha'^2}{r_1-r} = \frac{\beta'^2}{r_2-r} = \frac{\gamma'^2}{r_3-r}$$
$$= \frac{(r_2+r_3)(r_2-r)(r_3-r)}{r_2r_3-rr_2-rr_3},$$

that $32R^3 - 2R(\alpha'^2+\beta'^2+\gamma'^2) - \alpha'\beta'\gamma' = 0$.

565. Prove that the area of the triangle

$$= r_2r_3\sqrt{\frac{4R}{r_2+r_3}-1} = rr_1\sqrt{\frac{4R}{r_1-r}-1} = r_2r_3\sqrt{\frac{r_1-r}{r_2+r_3}}$$
$$= \sqrt{rr_1r_2r_3} = \frac{rr_2r_3}{\sqrt{r_2r_3-r(r_2+r_3)}}.$$

566. Taking ρ to be the radius of the polar circle of the triangle, prove that the area

$$= r\sqrt{(2R+r)^2-\rho^2} = r_1\sqrt{(2R-r_1)^2-\rho^2} = \ldots,$$

and that

$$4\rho^2(r_1+r_2+r_3-r) + (r_2+r_3-r_1+r)(r_3+r_1-r_2+r)(r_1+r_2-r_3+r) = 0.$$

567. The cosines of the angles of a triangle are the roots of the equation

$$4R^2x^3 - 4R(R+r)x^2 + (2r^2+4Rr-\rho^2)x + \rho^2 = 0;$$

and the radii r_1, r_2, r_3 are roots of the equation

$$x^2(4R+r-x) = (x-r)(\overline{2R+r}^2-\rho^2).$$

568. Prove that

$$\frac{8r_1r_2r_3}{\alpha\beta\gamma} = \sin A \sin B \sin C,$$

and
$$\frac{8r_1r_2r_3}{\alpha'\beta'\gamma'} = (1+\cos A)(1+\cos B)(1+\cos C).$$

569. Prove that

$$4R = \frac{a}{\cos\frac{A}{2}} = \frac{\beta}{\cos\frac{B}{2}} = \frac{\gamma}{\cos\frac{C}{2}} = \frac{a'}{\sin\frac{A}{2}} = \frac{\beta'}{\sin\frac{B}{2}} = \frac{\gamma'}{\sin\frac{C}{2}}.$$

570. The radii of two of the circles which touch the sides of a triangle are p, q, and the distance between their centres δ; prove that the area is $pq\sqrt{\frac{\delta^2}{(p \pm q)^2} - 1}$, the lower sign being taken when either circle is the inscribed circle.

571. The points O, O' are the centres of the circumscribed circle, and of a circle which touches the sides of the triangle in the points A', B', C'; L is the centre of perpendiculars of the triangle $A'B'C'$; prove that O, O', L are in one straight line, and that

$$O'L : OO' = r \text{ (or } r_1, r_2, r_3) : R.$$

572. If an isosceles triangle be constructed whose vertical angle is $\cos^{-1}(\frac{1}{9})$, the inscribed circle will pass through the centre of perpendiculars.

[In general the inscribed circle will pass through the centre of perpendiculars if $2\cos A\cos B\cos C = (1-\cos A)(1-\cos B)(1-\cos C)$.]

573. If O, o be the centres of the circumscribed and inscribed circles, and L the centre of perpendiculars

$$OL^2 - 2oL^2 = R^2 - 4r^2;$$

and if o_1 be the centre of the escribed circle opposite A,

$$OL^2 - 2o_1L^2 = R^2 - 4r_1^2.$$

574. If the centre of the inscribed circle be equidistant from the centre of the circumscribed circle and from the centre of perpendiculars, one angle of the triangle must be 60°; and with a similar property for an escribed circle, one angle must be 60° or 120°.

575. The cosine of the angle at which the circumscribed circle intersects the escribed circle opposite A is

$$\frac{1 + \cos A - \cos B - \cos C}{2};$$

and if a, β, γ be the three such cosines

$$(\beta + \gamma)(\gamma + a)(a + \beta) = 2(a + \beta + \gamma - 1)^2.$$

576. If P be any point on the circumscribed circle,

$$PA\sin A + PB\sin B + PC\sin C = 0;$$

a certain convention being made in respect to sign: also

$$PA^2\sin 2A + PB^2\sin 2B + PC^2\sin 2C = 4\Delta ABC.$$

577. If P be any point in the plane of the triangle, and O the centre of the circumscribed circle,

$$PA^2 \sin 2A + \ldots + \ldots - 4OP^2 \sin A \sin B \sin C = 2\Delta ABC.$$

578. If P be any point on the inscribed circle

$$PA^2 \sin A + PB^2 \sin B + PC^2 \sin C$$

will be constant; and if on the escribed circle opposite A,

$$-PA^2 \sin A + PB^2 \sin B + PC^2 \sin C$$

will be constant.

579. Prove that, if P be any point on the nine points' circle,

$$PA^2(\sin 2B + \sin 2C) + PB^2(\sin 2C + \sin 2A) + PC^2(\sin 2A + \sin 2B)$$
$$= 8R^2 \sin A \sin B \sin C\,(1 + 2\cos A \cos B \cos C).$$

580. If P be any point on the polar circle,

$$PA^2 \tan A + PB^2 \tan B + PB^2 \tan C$$

will be constant. If ρ be the radius of this circle and δ the distance of its centre from the centre of the circumscribed circle, then will

$$\delta^2 = R^2 + 2\rho^2;$$

and if δ' be the distance of its centre from the centre of the inscribed circle, then will

$$\delta'^2 = \rho^2 + 2r^2.$$

581. The straight line joining the centres of the circumscribed and inscribed circles will subtend a right angle at the centre of perpendiculars if

$$1 + (1 - 2\cos A)(1 - 2\cos B)(1 - 2\cos C) = 8\cos A \cos B \cos C.$$

582. If P be a point within a triangle at which the sides subtend angles $A + \alpha$, $B + \beta$, $C + \gamma$ respectively,

$$PA\frac{\sin A}{\sin \alpha} = PB\frac{\sin B}{\sin \beta} = PC\frac{\sin C}{\sin \gamma}.$$

583. Any point P is taken within the triangle ABC and the angles BPC, CPA, APC are A', B', C' respectively; prove that

$$\Delta BPC(\cot A - \cot A') = \Delta CPA(\cot B - \cot B') = \Delta APB(\cot C - \cot C').$$

584. Having given the equations

$$\cos^2 A\,(p\sin^2\theta + q\cos^2\theta) = \cos^2 B\,(p\sin^2\overline{C+\theta} + q\cos^2\overline{C+\theta})$$
$$= \cos^2 C\,(p\sin^2\overline{B-\theta} + q\cos^2\overline{C-\theta});$$

prove that each $= (p+q)\cos A \cos B \cos C$; and that

$$\frac{(p+q)^2}{pq} = \frac{4\cos^2 A \cos^2 B \cos^2 C}{(\cos A - \cos B\cos C)(\cos B - \cos C\cos A)(\cos C - \cos A\cos B)}.$$

VI. *Heights and Distances. Polygons.*

585. At a point A are measured the angle (α) subtended by two objects (points) P, Q in the same horizontal plane as A and the distances b, c at right angles to AP, AQ respectively to points at which PQ subtends the same angle (α); find the length of PQ.

586. An object is observed at three points A, B, C lying in a horizontal straight line which passes directly underneath the object; the angular elevations at A, B, C are θ, 2θ, 3θ, and $AB = a$, $BC = b$; prove that the height of the object is

$$\frac{a}{2b}\sqrt{(a+b)(3b-a)}.$$

If $\cot\theta = 3$, $a : b = 13 : 5$.

587. The sides of a rectangle are $2a$, $2b$, and the angles subtended by its diagonals at a point whose distance from its centre is c are α, β prove that

$$\frac{16a^2b^2c^2}{(a^2+b^2-c^2)^2} = a^2(\tan\alpha - \tan\beta)^2 + b^2(\tan\alpha + \tan\beta)^2;$$

$2a$ being that side which is cut by the distance c.

588. The diagonals $2a$, $2b$ of a rhombus subtend angles α, β at a point whose distance from the centre is c: prove that

$$b^2(a^2-c^2)^2\tan^2\alpha + a^2(b^2-c^2)^2\tan^2\beta = 4a^2b^2c^2.$$

589. Three circles A, B, C touch each other two and two and one common tangent to A and B is parallel to a common tangent of A and C: prove that if a, b, c be the radii, and p, q the distances of the centres of B and C from that diameter of A which is normal to the two parallel tangents

$$pq = 2a^2 = 8bc.$$

590. Three circles A, B, C touch each other two and two, prove that the distances from the centre of A of the common tangents to B and C are equal to

$$\frac{2bc(b+c) - a(b-c)^2 \pm 4bc\sqrt{a(a+b+c)}}{(b+c)^2},$$

and that one of these distances $= 0$ if $a(b + c \pm 2\sqrt{2bc}) = 2bc$.

591. Circles are described on the sides of the triangle ABC as diameters: prove that the rectangle under the radii of the two circles, which can be described touching the three, is

$$\frac{4R^2(1+\cos A)(1+\cos B)(1+\cos C)\cos A\cos B\cos C}{(1+\cos A)(1+\cos B)(1+\cos C)-(\cos B-\cos C)^2-(\cos C-\cos A)^2-(\cos A-\cos B)^2}.$$

592. Four points A, P, Q, B lie in a straight line and the distances AQ, BP, AB are $2a$, $2b$, $2c$ respectively; circles are described with diameters AQ, BP, AB: prove that the radius of the circle which touches the three is

$$\frac{c(c-a)(c-b)}{c^2-ab}.$$

593. A polygon of n sides inscribed in a circle is such that its sides subtend angles $2a, 4a, 6a, \ldots 2na$ at the centre; prove that its area is to the area of the inscribed regular n-gon in the ratio

$$\sin na : n \sin a.$$

594. A point P is taken within a parallelogram $ABCD$; prove that the value of

$$\Delta APC \cot APC - \Delta BPD \cot BPD$$

is independent of the position of P.

595. The distances of any point P on a circle from the angular points of an inscribed regular n-gon are the positive roots of the equation

$$(d^2 - x^2)^n - \frac{2n(2n-1)}{|\!\underline{2}} x^2 (d^2 - x^2)^{n-1}$$

$$+ \frac{2n(2n-1)(2n-2)(2n-3)}{|\!\underline{4}} x^4 (d^2 - x^2)^{n-2} - \ldots = d^{2n} \cos n\theta;$$

d being the diameter of the circle and θ the angle subtended at the centre by any one of the distances. Prove that if we take $d = 2$, the equation may also be written

$$x^{2n} - 2nx^{2n-2} + \frac{2n(2n-3)}{|\!\underline{2}} x^{2n-4} - \frac{2n(2n-4)(2n-5)}{|\!\underline{5}} x^{2n-6} - \ldots$$

$$+ (-1)^n 2(1 - \cos n\theta) = 0.$$

596. The sides of a convex quadrilateral are a, b, c, d and $2s$ is their sum: prove that

$$\sqrt{s(s-a-d)(s-b-d)(s-c-d)}$$

cannot be greater than the area.

597. The equation giving the length x of the diagonal joining the angles (a, d), (b, c) of a quadrilateral whose sides taken in order are a, b, c, d, is

$$\{x^2(ab+cd) - (ac+bd)(bc+ad)\}^2 \sin^2 \alpha$$

$$+ \{x^2(ab-cd) - (ac-bd)(bc-ad)\}^2 \cos^2 \alpha$$

$$= 4a^2b^2c^2d^2 \sin^2 2\alpha;$$

where 2α is the sum of two opposite angles.

[This equation, being a quadratic in $\cos 2\alpha$, leads to the equation giving the extreme possible values of x; which can be reduced to the form

$$3(x^2 - a^2 - b^2)^2 (x^2 - c^2 - d^2)^2$$

$$- 4\{x^2(ab+cd) - (ac+bd)(bc+ad)\}^2 + 16a^2b^2c^2d^2 = 0.]$$

598. In any quadrangle $ABCD$, the vertices are E, F, G, (the intersections of BC, AD; CA, BD; and AB, CD respectively); prove that

$$\frac{(EB \cdot EC - EA \cdot ED)^2}{EA \cdot EB \cdot EC \cdot ED \sin^2 E} = \frac{(FC \cdot FA - FB \cdot FD)^2}{FA \cdot FB \cdot FC \cdot FD \sin^2 F}$$
$$= \frac{(GA \cdot GB - GC \cdot GD)^2}{GA \cdot GB \cdot GC \cdot GD \sin^2 G}.$$

VII. *Expansions of Trigonometrical Functions. Inverse Functions.*

599. By means of the equivalence of the expansions of

$$2e^x \sin x \times e^x \cos x, \text{ and } e^{2x} \sin 2x;$$

prove that

$$\frac{\Sigma_{r=0}^{r=n} \sin (n-r)\frac{\pi}{4} \cos \frac{r\pi}{4}}{\lfloor n-r \, \lfloor r} = \frac{2^{n-1} \sin \frac{n\pi}{4}}{\lfloor n}.$$

600. Prove by comparing the coefficients of θ^{2n-1} that the expansions of $\sin\theta$ and $\cos\theta$ in terms of θ satisfy the identity

$$2 \sin\theta \cos\theta \equiv \sin 2\theta.$$

601. Prove that

$$2^n - (n-1)\,2^{n-1} + \frac{(n-2)(n-3)}{\lfloor 2} 2^{n-2} - \ldots \equiv \frac{\sin (n+1)\frac{\pi}{4}}{\left(\sin\frac{\pi}{4}\right)^{n+1}}, \text{ (n integral).}$$

602. Prove, from the identity

$$\frac{1}{1-xz} - \frac{1}{1-xz^{-1}} = \frac{2i \sin\theta}{1-2x\cos\theta+x^2},$$

that

$$\frac{\sin (n+1)\theta}{\sin\theta} = (2\cos\theta)^n - (n-1)(2\cos\theta)^{n-2}$$
$$+ \frac{(n-2)(n-3)}{\lfloor 2}(2\cos\theta)^{n-4} - \ldots,$$

and deduce the expansion of $\cos n\theta \left(\equiv \frac{\sin (n+1)\theta - \sin (n-1)\theta}{2\sin\theta}\right)$ in terms of $\cos\theta$ when n is a positive integer.

603. From the identity

$$\log (1 + xz^{-1}) + \log (1 - xz) \equiv \log (1 - 2ix\sin\theta - x^2),$$

or from the identity

$$\frac{1}{1-xz} + \frac{1}{1+xz^{-1}} \equiv \frac{2(1-ix\sin\theta)}{1-2ix\sin\theta - x^2},$$

deduce the equations

$$\cos n\theta = 1 - \frac{n^2}{2}\sin^2\theta + \frac{n^2(n^2-2^2)}{\lfloor 4}\sin^4\theta - \frac{n^2(n^2-2^2)(n^2-4^2)}{\lfloor 6}\sin^6\theta + \dots,$$

$$\sin n\theta = n\sin\theta - \frac{n(n^2-1^2)}{\lfloor 3}\sin^3\theta + \frac{n(n^2-1^2)(n^2-3^2)}{\lfloor 5}\sin^5\theta - \dots,$$

n being an integer, even for the first and odd for the second. Also prove that, if θ lie between $-\frac{\pi}{2}$ and $\frac{\pi}{2}$, both are true for all values of n; and thence deduce the true expansions of $\cos n\theta$ and $\sin n\theta$ in terms of $\sin\theta$ for any value of θ.

604. From the expansion of $(\sin\theta)^{2n+1}$ in terms of sines of multiples of θ, prove that

$$0 \equiv 1-(2n-1)+\frac{2n(2n-3)}{\lfloor 2}-\frac{2n(2n-1)(2n-5)}{\lfloor 3}-\dots \text{ to } n+1 \text{ terms.}$$

605. Prove that

$$1-\frac{n-1}{n}\cos\theta+\frac{n-1}{n}\,\frac{2n-1}{2n}\cos 2\theta-\frac{n-1}{n}\,\frac{2n-1}{2n}\,\frac{3n-1}{3n}\cos 3\theta+\dots \text{ to } \infty$$

$$=\frac{\cos\frac{n-1}{n}\frac{\theta}{2}}{\left(2\cos\frac{\theta}{2}\right)^{\frac{n-1}{n}}},$$

if θ lie between $-\pi$ and π.

606. Prove that

$$1-(n-1)+\frac{(n-2)(n-3)}{\lfloor 2}-\frac{(n-3)(n-4)(n-5)}{\lfloor 3}+\dots$$

$$\equiv (-1)^n \sin 2(n+1)\frac{\pi}{3} \div \sin\frac{2\pi}{3}.$$

607. Prove that, if $\tan\theta\ (\equiv t)$ be less than 1,

$$\sin n\theta\cos^n\theta = nt - \frac{n(n+1)(n+2)}{\lfloor 3}t^3 + \frac{n(n+1)\dots(n+4)}{\lfloor 5}t^5 - \dots$$

and $$\cos n\theta\cos^n\theta = 1 - \frac{n(n+1)}{\lfloor 2}t^2 + \frac{n(n+1)(n+2)(n+3)}{\lfloor 4}t^4 - \dots$$

[These results are obviously true when n is a negative whole number.]

608. The sum of the infinite series

$$1+\frac{1}{2}\cos 2\theta-\frac{1}{2\,.\,4}\cos 4\theta+\frac{1\,.\,3}{2\,.\,4\,.\,6}\cos 6\theta-\dots$$

is $$\sqrt{\cos\theta(1+\cos\theta)}, \text{ if } \theta \text{ lie between } -\frac{\pi}{2} \text{ and } \frac{\pi}{2}.$$

609. Prove that the identity

$$\cos n\theta \equiv 1 - \frac{n^2}{\lfloor 2} \sin^2\theta + \frac{n(n^2-2^2)}{\lfloor 4} \sin^4\theta - \dots$$

may be deduced from the identity

$$2\cos n\theta \equiv (2\cos\theta)^n - n(2\cos\theta)^{n-2} + \frac{n(n-3)}{\lfloor 2}(2\cos\theta)^{n-4} - \dots$$

when n is an even integer, by writing $\frac{\pi}{2} - \theta$ for θ and taking the terms in reverse order; and similarly for $\sin n\theta$ when n is odd.

610. If

$$F(n) = 1 - \frac{n^2}{\lfloor 2} + \frac{n^2(n^2-1^2)}{\lfloor 4} - \frac{n^2(n^2-1^2)(n^2-2^2)}{\lfloor 6} + \dots \text{ to } n+1 \text{ terms,}$$

prove that $F(2n) = (-1)^n F(n)$.

$$\left[F(n) = \cos\frac{n\pi}{3} \text{ for all values of } n.\right]$$

611. If the constants $a_1, a_2, a_3 \dots a_n$ be so determined in the expression

$$a_1 \sin x + a_2 \sin 2x + \dots + a_n \sin nx + \sin(n+1)x$$

that the coefficients of $x, x^3, x^5, \dots x^{2n-1}$ shall vanish, the value of the expression will be $2^n \sin x(\cos x - 1)^n$; and if, in

$$a_1\cos x + a_2 \cos 2x + \dots + a_n \cos nx + \cos(n+1)x$$

the coefficients of all the powers of x up to x^{2n-2} inclusive vanish, the value will be

$$2^n(\cos x - 1)^n\left(\cos x + \frac{n}{n+1}\right).$$

612. Prove that

$$\sum_{r=1}^{r=n} \frac{\sin^2 ra \cos^{2n-2} ra}{x^2 + \tan^2 ra} \equiv \frac{(2n+1)x}{(1+x)^{2n+1} - (1-x)^{2n+1}}$$

where $(2n+1)a = \pi$.

613. From the identity

$$\epsilon^x - 2\cos\theta + \epsilon^{-x} \equiv 4\sin\frac{\theta - ix}{2}\sin\frac{\theta+ix}{2},$$

resolve the former into its quadratic factors.

[The result is $4\sin^2\frac{\theta}{2}\left(1+\frac{x^2}{\theta^2}\right)\left(1+\frac{x^2}{(2\pi-\theta)^2}\right)\left(1+\frac{x^2}{(2\pi+\theta)^2}\right)\dots$ all factors of the form $1 + \frac{x^2}{(2r\pi+\theta)^2}$ being taken where r is a positive or negative integer.

Similarly $\dfrac{\epsilon^x + 2\cos\theta + \epsilon^{-x}}{4\cos^2\dfrac{\theta}{2}} \equiv$ the product of all factors of the form $1 + \dfrac{x^2}{(r\pi+\theta)^2}$ where r is an odd positive or negative integer.]

614. From the result in the last question deduce

(1) $\dfrac{1}{\sin^2\theta} = \Sigma_{r=-\infty}^{r=\infty} \dfrac{1}{(r\pi+\theta)^2}$,

(2) $\dfrac{3-2\sin^2\theta}{3\sin^4\theta} \equiv \Sigma_{r=-\infty}^{r=\infty} \dfrac{1}{(r\pi+\theta)^4}$;

zero being included among the values of r.

[By equating coefficients of x^{2n} in the results, it appears that $\dfrac{2^{2n-1}}{|\underline{2n}\,\sin^{2n}\theta} \equiv$ sum of the products n together of all expressions included in $\dfrac{1}{(r\pi+\theta)^2}$ for integral values of r from $-\infty$ to ∞ including zero.]

615. Prove that

$$\frac{1}{\sin\theta} = \frac{1}{\theta} + \frac{1}{\pi-\theta} - \frac{1}{\pi+\theta} - \frac{1}{2\pi-\theta} + \frac{1}{2\pi+\theta} + \frac{1}{3\pi-\theta} - \dots \text{ to } \infty,$$

$$\frac{\theta}{2\sin\theta} = \frac{\pi^2}{\pi^2-\theta^2} - \frac{(2\pi)^2}{(2\pi)^2-\theta^2} + \frac{(3\pi)^2}{(3\pi)^2-\theta^2} - \dots \text{ to } \infty.$$

616. Prove that

$$\frac{\tan\theta}{8\theta} = \frac{1}{\pi^2-4\theta^2} + \frac{1}{(3\pi)^2-4\theta^2} + \frac{1}{(5\pi)^2-4\theta^2} + \dots \text{ to } \infty.$$

617. Prove that, if θ be an angle between $-\dfrac{\pi}{4}$ and $\dfrac{\pi}{4}$,

$$\theta^2 = \sin^2\theta + \frac{2}{3}\,\frac{\sin^4\theta}{2} + \frac{2\,.\,4}{3\,.\,5}\,\frac{\sin^6\theta}{3} + \dots \text{ to } \infty$$

$$= \tan^2\theta - \left(1+\frac{1}{3}\right)\frac{\tan^4\theta}{2} + \left(1+\frac{1}{3}+\frac{1}{5}\right)\frac{\tan^6\theta}{3} - \dots \text{ to } \infty.$$

[The former is true for all values of θ.]

618. Prove that

(1) $\dfrac{1^2}{1^2+1}\,\dfrac{2^2}{2^2+1}\,\dfrac{3^2}{3^2+1} \dots \text{ to } \infty = \dfrac{2\pi}{\epsilon^\pi - \epsilon^{-\pi}}$,

(2) $\dfrac{1^4}{1^4+1}\,\dfrac{2^4}{2^4+1}\,\dfrac{3^4}{3^4+1} \dots \text{ to } \infty = \dfrac{4\pi^2}{\epsilon^{\pi\sqrt{2}} + \epsilon^{-\pi\sqrt{2}} - 2\cos(\pi\sqrt{2})}$.

619. In a triangle the sides a, b, and the angle $\pi - \theta$ opposite b are given, and θ is small: prove that, approximately,

$$\frac{c}{b-a} = 1 + \frac{a}{b}\frac{\theta^2}{2} + \frac{a(3a^2+3ab-b^2)}{b^2}\frac{\theta^4}{\lfloor 4}.$$

620. Prove that the expansion of tan tan ... tan x is

$$x + 2n\frac{x^3}{\lfloor 3} + 4n(5n-1)\frac{x^5}{\lfloor 5} + \frac{8n}{3}(175n^2 - 84n + 11)\frac{x^7}{\lfloor 7} + \dots;$$

when the tangent is taken n times.

621. Prove that the expansion of sin sin... sin x is

$$x - n\frac{x^3}{\lfloor 3} + n(5n-4)\frac{x^5}{\lfloor 5} - \frac{n}{3}(175n^2 + 336n + 162)\frac{x^7}{\lfloor 7} + \dots;$$

the sine being taken n times.

622. Prove that the expansion of $\tan^{-1}\tan^{-1}\dots\tan^{-1}x$ is

$$x - 2n\frac{x^3}{\lfloor 3} + 4n(5n+1)\frac{x^5}{\lfloor 5} - \frac{8n}{3}(175n^2 + 84n + 11)\frac{x^7}{\lfloor 7} + \dots$$

[That is, the expansion of $\tan^{-n}x$ might be deduced from the expansion of $\tan^n x$ by putting $-n$ for n, the index applied to the function denoting repetition of the functional operation.]

623. Prove that

(1) $\tan^{-1}\frac{1}{70} = \tan^{-1}\frac{1}{83} + \tan^{-1}\frac{1}{447}$,

(2) $\tan^{-1}\frac{1}{99} = \tan^{-1}\frac{1}{157} + \tan^{-1}\frac{1}{268}$,

(3) $0 = \tan^{-1}\frac{1}{19} - \tan^{-1}\frac{1}{27} - \tan^{-1}\frac{1}{46} + \tan^{-1}\frac{1}{162}$,

(4) $\frac{\pi}{4} = \tan^{-1}\frac{1}{4} + \tan^{-1}\frac{1}{5} + \tan^{-1}\frac{1}{7} + 2\tan^{-1}\frac{1}{13} + \tan^{-1}\frac{1}{21}$,

$= \tan^{-1}\frac{1}{3} + \tan^{-1}\frac{1}{5} + \tan^{-1}\frac{1}{6} + \tan^{-1}\frac{1}{8} - \tan^{-1}\frac{1}{43}$,

$= 3\tan^{-1}\frac{1}{4} + \tan^{-1}\frac{1}{20} + \tan^{-1}\frac{1}{1985}$.

624. The convergents to $\sqrt{2}$ are $1, \frac{3}{2}, \frac{7}{5}, \dots \frac{p_n}{q_n} \dots$; prove that

$$\tan^{-1}\frac{1}{q_{2n}} - \tan^{-1}\frac{1}{p_{2n}} = \tan^{-1}\frac{1}{p_{2n+1}},$$

and that $$\tan^{-1}\frac{1}{q_{2n+1}} - \tan^{-1}\frac{1}{p_{2n+1}} = \tan^{-1}\left(\frac{1}{p_{2n+2} + \frac{1}{p_n q_n}}\right).$$

625. Find x from the equation

$$\cot^{-1}x + \cot^{-1}(n^2 - x + 1) = \cot^{-1}(n-1);$$

and find the tangent of the angle

$$\tan^{-1}3 + 3\tan^{-1}7 + \tan^{-1}26 - \frac{\pi}{4}.$$

626. Prove that, if $\tan(\alpha + i\beta) = i$, α, β being real, that α will be indeterminate and β infinite.

627. Prove that if $\cos(\alpha + i\beta) = \cos\phi + i\sin\phi$, where α, β, ϕ are real, $\sin\phi = \pm\sin^2\alpha$, and that the relation between α and β is

$$\epsilon^{\beta} - \epsilon^{-\beta} = \pm 2\sin\alpha.$$

628. Prove that, if $\tan(\alpha + i\beta) = \cos\phi + i\sin\phi$, and α, β, ϕ be real,

$$\alpha = n\frac{\pi}{2} + \frac{\pi}{4}, \quad 4\beta = \log\left(\frac{1+\sin\phi}{1-\sin\phi}\right).$$

629. Prove that, if $\tan(\alpha + i\beta) = \tan\phi + i\sec\phi$, and α, β, ϕ be real,

$$2\alpha = n\pi + \frac{\pi}{2} + \phi, \quad 4\beta = \log\left(\frac{1+\cos\phi}{1-\cos\phi}\right).$$

VIII. *Series.*

[In the summation of many Trigonometric series in which the rth term is of the form $a_r\cos r\theta$, or $a_r\sin r\theta$, a_r being a function of r, it is convenient to sum the series in the manner exemplified by the following solution of the question :—

"To find the sum of the series $1 + 2\cos\theta + 3\cos 2\theta + \ldots + n\cos\overline{n-1}\,\theta$."

Let C denote the proposed series and S the corresponding series with sines in place of cosines, namely,

$$S \equiv 2\sin\theta + 3\sin 2\theta + \ldots + n\sin\overline{n-1}\,\theta,$$

the first term being $\sin 0\,.\,\theta$ or 0, then, if $\cos\theta + i\sin\theta \equiv z$,

$$C + iS \equiv 1 + 2z + 3z^2 + \ldots + nz^{n-1} \equiv \frac{1 - z^n(n + 1 - nz)}{(1-z)^2}$$

$$\equiv \frac{1 - (\cos n\theta + i\sin n\theta)(n+1) + n(\cos\overline{n+1}\,\theta + i\sin\overline{n+1}\,\theta)}{(1 - \cos\theta - i\sin\theta)^2}$$

$$\equiv \frac{\text{same numerator}}{\left(2\sin\frac{\theta}{2}\right)^2\left(\sin\frac{\theta}{2} - i\cos\frac{\theta}{2}\right)^2 \equiv -\left(2\sin\frac{\theta}{2}\right)^2(\cos\theta + i\sin\theta)}$$

$$\equiv \frac{\cos\theta - i\sin\theta - (n+1)(\cos\overline{n-1}\,\theta + i\sin\overline{n-1}\,\theta) + n(\cos n\theta + i\sin n\theta)}{-\left(2\sin\frac{\theta}{2}\right)^2};$$

whence, equating possibles and impossibles on the two sides,

$$C \equiv \frac{(n+1)\cos(n-1)\theta - \cos\theta - n\cos n\theta}{2(1-\cos\theta)},$$

and also

$$S \equiv \frac{(n+1)\sin(n-1)\theta + \sin\theta - n\sin n\theta}{2(1-\cos\theta)}.$$

It is obvious that in general if

$$f(x) = a_0 + a_1 x + a_2 x^2 + \ldots,$$

and z have the same meaning as above,

$$f(xz) + f(xz^{-1}) = 2(a_0 + a_1 x\cos\theta + \ldots + a_n x^n \cos n\theta + \ldots),$$

$$f(xz) - f(xz^{-1}) = 2i(a_1 x\sin\theta + \ldots + a_n x^n \sin n\theta + \ldots).$$

Some doubt may often arise as to the limiting values of the angle θ beyond which results found by this method may not be true, but this can always be cleared up by the use of the powers of x as coefficients as in the forms just given. Thus, to take a very well-known case, to sum the infinite series $\sin\theta - \frac{1}{2}\sin 2\theta + \frac{1}{3}\sin 3\theta \ldots$. Take

$$C \equiv x\cos\theta - \frac{x^2}{2}\cos 2\theta + \ldots, \text{ and } S \equiv x\sin\theta - \frac{x^2}{2}\sin 2\theta + \ldots$$

and we have

$$\begin{aligned} C + iS &\equiv xz - \tfrac{1}{2}x^2z^2 + \tfrac{1}{3}x^3z^3 - \ldots \\ &\equiv \log(1+xz) \\ &\equiv \log\rho(\cos\phi + i\sin\phi) \equiv \log\rho + i\phi, \end{aligned}$$

where $\rho = \sqrt{1 + 2x\cos\theta + x^2}$, and $\tan\phi = \dfrac{x\sin\theta}{1 + x\cos\theta}$.

Thus $S = \tan^{-1}\left(\dfrac{x\sin\theta}{1+x\cos\theta}\right)$, meaning by this the angle between $-\dfrac{\pi}{2}$ and $\dfrac{\pi}{2}$ whose tangent is $\dfrac{x\sin\theta}{1+x\cos\theta}$, which is free from ambiguity, (since the series manifestly vanishes when $x = 0$), and when $x = 1$, the result will be $\dfrac{\theta}{2}$ if $\dfrac{\theta}{2}$ lie between those limits, or θ between $-\pi$ and π. So also the corresponding series in cosines

$$\cos\theta - \tfrac{1}{2}\cos 2\theta + \tfrac{1}{3}\cos 3\theta - \ldots = \log\sqrt{2(1+\cos\theta)} = \tfrac{1}{2}\log\left(4\cos^2\frac{\theta}{2}\right),$$

which is sometimes written $\log\left(2\cos\dfrac{\theta}{2}\right)$ without the proper limitation that $\cos\dfrac{\theta}{2}$ must be positive. The series will be convergent only if x^2 be less than 1; and this will be generally the case.

Many series also may be summed by the same method as was explained under the corresponding head in Algebra: that is by obtaining the rth term (u_r) of the proposed series in the form $U_{r+1} - U_r$. Thus, to sum the series

$$\text{cosec}\, x + \text{cosec}\, 2x + \ldots + \text{cosec}\, 2^{n-1}x,$$

we have

$$\text{cosec}\, 2^r x \equiv \frac{\sin 2^{r-1}x}{\sin 2^{r-1}x \sin 2^r x} \equiv \frac{\sin (2^r - 2^{r-1})\, x}{\sin 2^{r-1}x \sin 2^r x} \equiv \cot 2^{r-1}x - \cot 2^r x,$$

so that $U_{r+1} = -\cot 2^{r-1}x$, and $S_n = U_{n+1} - U_1 = \cot \frac{x}{2} - \cot 2^{n-1}x$.

Such being the method, it is clear that giving the answer would, in these cases, amount to giving the whole solution.]

630. Sum the following infinite series, and the corresponding series in sines,

(1) $\cos\theta + \frac{1}{2}\cos 2\theta + \frac{1}{3}\cos 3\theta + \ldots,$

(2) $\cos\theta - \frac{\cos 3\theta}{\lfloor 3} + \frac{\cos 5\theta}{\lfloor 5} - \ldots,$

(3) $\cos\theta - \frac{1}{3}\cos 3\theta + \frac{1}{5}\cos 5\theta - \ldots,$

(4) $1 - n\cos\theta + \frac{n(n+1)}{2}\cos 2\theta - \frac{n(n+1)(n+2)}{\lfloor 3}\cos 3\theta + \ldots,$

(5) $1 - \cos\theta + \frac{\cos 2\theta}{\lfloor 2} - \frac{\cos 3\theta}{\lfloor 3} + \ldots,$

(6) $\cos\theta + \frac{1}{2}\frac{\cos 3\theta}{3} + \frac{1.3}{2.4}\frac{\cos 5\theta}{5} + \ldots,$

(7) $\cos\theta - \frac{1}{2}\frac{\cos 3\theta}{3} + \frac{1.3}{2.4}\frac{\cos 5\theta}{5} - \ldots,$

(8) $\cos\theta + \frac{2}{3}\frac{\cos 2\theta}{4} + \frac{2.4}{3.5}\frac{\cos 3\theta}{6} + \ldots,$

(9) $x\cos\theta + \frac{1}{2}x^2\cos 2\theta + \frac{1}{3}x^3\cos 3\theta + \ldots,$ when $x = \cos\theta,$

(10) $x\cos\theta + \frac{1}{2}x^3\frac{\cos 3\theta}{3} + \frac{1.3}{2.4}x^5\frac{\cos 5\theta}{5} + \ldots,$ when $x = \cos 2\theta.$

[(1) $C = -\frac{1}{2}\log\left(4\sin^2\frac{\theta}{2}\right)$, $S = \tan^{-1}\left(\frac{\sin\theta}{1-\cos\theta}\right) = \frac{\pi-\theta}{2}$ if θ lie between 0 and π,

(2) $C = \frac{1}{2}\sin(\cos\theta)\{\epsilon^{\sin\theta} + \epsilon^{-\sin\theta}\}$, $S = \frac{1}{2}\cos(\cos\theta)\{\epsilon^{\sin\theta} - \epsilon^{-\sin\theta}\}$,

(3) $C = \pm\frac{\pi}{4}$ being of the sign of $\cos\theta$, $S = \frac{1}{4}\log\left(\frac{1+\sin\theta}{1-\sin\theta}\right)$,

(4) $C = \left(2 \cos \frac{\theta}{2}\right)^{-n} \cos n\frac{\theta}{2}$, $S = \left(2 \cos \frac{\theta}{2}\right)^{-n} \sin n\frac{\theta}{2}$, θ being between $-\pi$ and π,

(5) $C = \epsilon^{-\cos\theta} \cos(\sin\theta)$, $S = -\epsilon^{-\cos\theta} \sin(\sin\theta)$,

(6) $\cos^2 C = \sin\theta$, and $S = \log\left(\cos\frac{\theta}{2} + \sin\frac{\theta}{2} + \sqrt{\sin\theta}\right)$, if $\sin\theta$ be positive,

(7) $C = \log\left(\sqrt{\cos\theta} + \sqrt{2}\cos\frac{\theta}{2}\right)$, and $\cos^2 S = \cos\theta$, if $\cos\theta$ be positive,

(8) $C = \rho^2 \cos 2\phi$, $S = \rho^2 \sin 2\phi$, where $\rho\cos\phi = \cos^{-1}\left(\sqrt{\sin\frac{\theta}{2}}\right)$, $\rho\sin\phi = \log\left(\cos\frac{\theta}{4} + \sin\frac{\theta}{4} + \sqrt{\sin\frac{\theta}{2}}\right)$, θ being between 0 and 2π,

(9) $C = \frac{1}{2}\log(\operatorname{cosec}^2\theta)$, $S = \tan^{-1}(\cot\theta)$,

(10) $\cos^2 \dfrac{C}{\sqrt{\cos 2\theta}} = \sin\theta(\sin\theta + \cos\theta)$, if $\cos 2\theta$ be positive.]

631. Sum the series

$$\sin^3\theta + \frac{\sin^3 3\theta}{3} + \ldots + \frac{\sin^3 3^{n-1}\theta}{3^{n-1}},$$

$$\cos^3\theta - \frac{\cos^3 3\theta}{3} + \ldots + (-1)^{n-1}\frac{\cos^3 3^{n-1}\theta}{3^{n-1}},$$

$$\frac{\cos\theta}{\sin 3\theta} + \frac{\cos 3\theta}{\sin 3^2\theta} + \ldots + \frac{\cos 3^{n-1}\theta}{\sin 3^n\theta},$$

$$\frac{\sin 2\theta}{1 + 2\cos 2\theta} + \frac{3\sin 6\theta}{1 + 2\cos 6\theta} + \ldots + \frac{3^{n-1}\sin 2(3^{n-1}\theta)}{1 + 2\cos 2(3^{n-1}\theta)},$$

$$\frac{2\cos\theta - \cos 3\theta}{\sin 3\theta} + \ldots + \frac{2^n\cos 3^{n-1}\theta - 2^{n-1}\cos 3^n\theta}{\sin 3^n\theta},$$

$$\frac{1 + 2\cos 2\theta}{\sin 4\theta} + \ldots + \frac{1 + 2\cos 2^{2n-1}\theta}{\sin 2^{2n}\theta},$$

$$\frac{1 - 2\cos 2\theta}{\sin 2\theta} + \ldots + 3^{n-1}\frac{1 - 2\cos 2^n\theta}{\sin 2^n\theta},$$

$$\frac{5\sin 3\theta - 3\sin 5\theta}{\cos 3\theta - \cos 5\theta} + \ldots + 4^{n-1}\frac{5\sin 3(4^{n-1}\theta) - 3\sin 5(4^{n-1}\theta)}{\cos 3(4^{n-1}\theta) - \cos 5(4^{n-1}\theta)},$$

$$\frac{1 + 4\sin\theta\sin 3\theta}{\sin 4\theta} + \ldots + 3^{n-1}\frac{1 + 4\sin 4^{n-1}\theta\sin 3(4^{n-1}\theta)}{\sin 4^n\theta},$$

$$\frac{3\sin\theta - \sin 3\theta}{\cos 3\theta} + \ldots + \frac{3\sin 3^{n-1}\theta - \sin 3^n\theta}{3^{n-1}\cos 3^n\theta},$$

632. **Prove that**

$$\sec\theta + \sec\left(\frac{2\pi}{m}+\theta\right) + \sec\left(\frac{4\pi}{m}+\theta\right) + \ldots + \sec\left\{2(m-1)\frac{\pi}{m}+\theta\right\}$$

is equal to 0, or to $(-1)^{\frac{m-1}{2}} m \sec m\theta$, according as m is even or odd: also that

$$\sec^2\theta + \sec^2\left(\frac{2\pi}{m}+\theta\right) + \ldots + \sec^2\left\{2(m-1)\frac{\pi}{m}+\theta\right\}$$

is equal to $\dfrac{m^2}{1-(-1)^{\frac{m}{2}}\cos m\theta}$, or to $m^2\sec^2 m\theta$, according as m is even or odd.

[The equation which expresses $\cos m\theta$ in terms of $\cos\theta$ will be satisfied by $\cos\theta$, $\cos\left(\frac{2\pi}{m}+\theta\right)$, ..., $\cos\left\{(m-1)\frac{2\pi}{m}+\theta\right\}$; the results of the question follow on finding the sum of the reciprocals of the roots, and the sum of their squares.]

633. Prove that

$$\cos\theta + \cos(\alpha+\theta) + \cos(2\alpha+\theta) + \ldots + \cos(\overline{m-1}\,\alpha+\theta) = 0,$$

if $m\alpha = 4\pi$, m being any positive integer except 2; that

$$\sec^2\frac{\pi}{n} + \sec^2\frac{2\pi}{n} + \ldots + \sec^2\left(\frac{n}{2}-1\right)\frac{\pi}{n} = \frac{1}{6}(n^2-4),$$

if n be any even positive integer except 2; and that

$$\sec^2\frac{\pi}{n} + \sec^2\frac{2\pi}{n} + \sec^2\frac{3\pi}{n} + \ldots + \sec^2\frac{n-1}{2}\frac{\pi}{n} = \frac{n^2-1}{2},$$

if n be any odd positive integer except 1.

634. Prove that, if n be a whole number > 4,

$$\sin^2\frac{\pi}{n}\cos^{n-4}\frac{\pi}{n} - \sin^2\frac{2\pi}{n}\cos^{n-4}\frac{2\pi}{n} + \sin^2\frac{3\pi}{n}\cos^{n-4}\frac{3\pi}{n}\ldots = 0,$$

the number of terms being $\frac{n-1}{2}$, or $\frac{n}{2}-1$, as n is odd or even.

[The roots of the equation $(1+x)^n = (1-x)^n$ are the values of $i\tan\frac{r\pi}{n}$, where r may have all integral values from 0 to $n-1$, omitting $\frac{n}{2}$ if n be even. Hence $\dfrac{nx}{(1+x)^n-(1-x)^n} = \Sigma\dfrac{A_r}{x^2+\tan^2\frac{r\pi}{n}}$, where r has all integral values from 1 to $\frac{n-1}{2}$ or $\frac{n}{2}-1$ and A_r is $(-1)^{r-1}\sin^2\frac{r\pi}{n}\cos^{n-4}\frac{r\pi}{n}$.]

635. Prove that

$$\frac{n \sin n\phi}{\cos n\phi - \cos n\theta} = \frac{\sin\phi}{\cos\phi - \cos\theta} + \frac{\sin\phi}{\cos\phi - \cos(\alpha+\theta)}$$

$$+ \frac{\sin\phi}{\cos\phi - \cos(2\alpha+\theta)} + \ldots + \frac{\sin\phi}{\cos\phi - \cos(\overline{n-1}\,\alpha+\theta)};$$

and

$$\frac{n \sin n\theta}{\cos n\phi - \cos n\theta} = \frac{\sin\theta}{\cos\phi - \cos\theta} + \frac{\sin(\alpha+\theta)}{\cos\phi - \cos(\alpha+\theta)}$$

$$+ \frac{\sin(2\alpha+\theta)}{\cos\phi - \cos(2\alpha+\theta)} + \ldots + \frac{\sin(\overline{n-1}\,\alpha+\theta)}{\cos\phi - \cos(\overline{n-1}\,\alpha+\theta)};$$

where n is a positive integer, and $n\alpha = 2\pi$.

CONIC SECTIONS, GEOMETRICAL.

I. *Parabola.*

[The focus and vertex are denoted always by S and A respectively.]

636. Two parabolas having the same focus intersect: prove that the angles between their tangents at the two points of intersection are either equal or supplementary.

637. A chord PQ of a parabola is a normal at P and subtends a right angle at the focus: prove that SQ is twice SP, and that PQ subtends a right angle at one end of the latus rectum.

638. A chord PQ of a parabola is a normal at P and subtends a right angle at the vertex: prove that SQ is three times SP.

639. Two circles each touch a parabola and touch each other at the focus of the parabola: prove that the angle between the focal distances of the points of contact is 120°.

640. Two parabolas have a common focus and axes at right angles, a circle is drawn touching both and passing through the focus: prove that the points of contact are ends of a diameter, or subtend an angle of 30° at the focus.

641. Two parabolas have a common focus, a circle is described touching both and passing through the focus: prove that the angle between the focal distances of the points of contact will be one third of the angle between the axes, or one third of the defect from four right angles of this angle.

642. Two parabolas A, B have a common focus and axes at right angles: prove that any two tangents drawn to A at right angles to each other will be equally inclined to the tangents drawn to B from the same point.

643. In a parabola AQ is drawn through the vertex A at right angles to a chord AP to meet the diameter through P in Q: prove that Q lies on a fixed straight line.

644. Through any point P of a parabola is drawn a straight line QPQ' perpendicular to the axis, and terminated by the tangents at the end of the latus rectum: prove that the distance of P from the latus rectum is a mean proportional between QP, PQ'.

645. A circle touches a parabola at a point whose distance from the focus exceeds the latus rectum, and passes through the focus: prove that it will cut the parabola in two points, and that the common chord will cut the axis of the parabola in a fixed point at a distance from the focus equal to the latus rectum.

646. A parabola is described touching a given circle, and having its focus at a given point on the circle: prove that if the distance of the point of contact from the focus be less than the radius of the circle, the circle and parabola will have two other common tangents whose common point will lie on a fixed straight line which bisects the radius drawn from the focus.

647. With a given point as focus is described a parabola touching a given circle: prove that the point of intersection of the two other common tangents lies on a fixed circle, such that the polar of the given point with respect to it passes through the centre of the given circle.

[If the given point lie on the given circle, the locus degenerates into the straight line bisecting at right angles the radius through the given point.]

648. On the tangents drawn from a point O are taken two points P, Q such that SP, SO, SQ are all equal: prove that PQ is perpendicular to the axis and its distance from O is twice its distance from A.

649. Two equal parabolas have a common focus S and axes opposite, and SPQ is any straight line meeting them in P, Q; with centres P, Q are drawn circles touching the respective tangents at the vertices: prove that these circles will have internal contact, and that the rectangle under their radii will be fixed.

650. On a focal chord PQ as diameter is described a circle which meets the parabola again in P', Q': prove that the circle $P'SQ'$ will touch the parabola.

651. A circle touches a parabola in P, passes through S and meets the parabola again in Q, Q'; a focal chord is drawn parallel to the tangent at P: prove that the circle on this chord as diameter will pass through Q, Q', and that the focal chord and QQ' will intersect on the directrix.

652. Two parabolas whose foci are S, S' have three common tangents, and the circle circumscribing the triangle formed by these tangents is drawn: prove that SS' will subtend at any point on this circle an angle equal to that between the axes of the parabolas.

653. From any point on the tangent at any point of a parabola perpendiculars are let fall on the focal distance and on the axis: prove that the sum, or the difference, of the focal distances of the feet of these perpendiculars is equal to half the latus rectum.

654. The normal at a point P is produced to O so that PO is bisected by the axis: prove that any chord through O subtends a right angle at P; and that the circle on PO as diameter will have double contact with the parabola.

655. From a fixed point O is let fall OQ perpendicular on the diameter through a point P of a parabola: prove that the perpendicular from Q on the tangent at P will pass through a fixed point, which remains the same for all equal parabolas on a common axis.

656. A circle is drawn through two fixed points R, S, and meets a fixed straight line through R again in P: prove that the tangent at P will touch a fixed parabola whose focus is S.

657. Two fixed straight lines intersect in O: prove that any circle through O and through another fixed point S meets the two fixed lines again in points such that the chord joining them touches a fixed parabola whose focus is S.

658. The perpendicular AZ on the tangent at P meets the parabola again in Q: prove that the rectangle ZA, AQ is equal to the square on the semi latus rectum and that PQ passes through the centre of curvature at A.

659. Two parabolas have a common focus and axes at right angles: prove that the directrix of either passes through the point of contact of their common tangent with the other.

660. Through any point P on a parabola is drawn PK at right angles to AP to meet the axis in K: prove that AK is equal to the focal chord parallel to AP. Explain the result when P coincides with A.

661. A circle on a double ordinate to the axis PP' meets the parabola again in Q, Q': prove that the latus rectum of the parabola which touches PQ, PQ', $P'Q$, $P'Q'$ is double that of the former, and its focus is the centre of the circle.

662. Three points A, B, C are taken on a parabola, and tangents drawn at them forming a triangle $A'B'C'$; a, b, c are the centres of the circles BCA', CAB', ABC': prove that the circle through a, b, c will pass through the focus.

663. Two points are taken on a parabola, such that the sum of the parts of the normals intercepted between the points and the axis is equal to the part of the axis intercepted between the normals: prove that the difference of the normals is equal to the latus rectum.

664. The perpendicular SY being drawn to any tangent, a straight line is drawn through Y parallel to the axis to meet in Q the straight line through S parallel to the tangent: prove that the locus of Q is a parabola.

665. If X be the foot of the directrix, SY perpendicular from the focus on a chord PP', and a circle with centre S and radius equal to XY meet the chord in QQ': prove that PP', QQ' subtend equal angles at S.

666. A given straight line meets one of a series of coaxial circles in A, B: prove that the parabola which touches the given straight line, the tangents to the circle at A, B, and the common radical axis will have another fixed tangent.

[If K be a point circle of the system, L the intersection of the given straight line with the radical axis and KO drawn at right angles to KL to meet the radical axis in O, the fixed tangent is the straight line through O perpendicular to the given straight line.]

667. Two tangents TP, TQ are drawn to a parabola, OP, OQ are tangents to the circle TPQ: prove that TO will pass through the focus.

668. A triangle ABC is inscribed in a circle, AA' is a diameter, a parabola is described touching the sides of the triangle with its directrix passing through A' and S is its focus: prove that the tangents to the circle at B, C will intersect on SA'.

669. The normals at two points P, Q meet the axis in p, q and the chord PQ meets it in O: prove that straight lines drawn through O, p, q at right angles respectively to the three lines will meet in a point.

670. Normals at P, P' meet the axis in G, G', and straight lines at right angles to the normals from G, G' meet in Q: prove that

$$PG : GQ = P'G' : G'Q.$$

671. The tangent to a parabola at P meets the tangent at Q in T and meets SQ in R; also the tangent at Q meets the directrix in K: prove that PT, TR subtend equal or supplementary angles at K.

672. Two equal parabolas have a common focus and axes inclined at an angle of 120°: prove that a tangent to either curve at a common point will meet the other in a point of contact of a common tangent.

673. The chord PR is normal at P, O is the centre of curvature at P and U the pole of PR: prove that OU will be perpendicular to SP.

674. From a fixed point O is drawn a straight line OP to any point P on a fixed straight line: prove that the straight lines drawn through P equally inclined to PO and to the fixed straight line touch a fixed parabola.

675. A parabola whose focus lies on a fixed circle and whose directrix is given, always touches two fixed parabolas whose common focus is the given centre, and whose directrices are each at a distance from the given directrix equal to the given radius; and the tangents at the points of contact are at right angles.

676. The centre of curvature at P is O, PO meets the axis in G and OL is drawn perpendicular to the axis to meet the diameter through P: prove that LG is parallel to the tangent at P.

677. The straight lines Aa, Bb, Cc are drawn perpendicular to the sides BC, CA, AB of a triangle ABC: prove that two parabolas can be drawn touching the sides of the triangles ABC, abc respectively, such that the tangent at the vertex of the former is the axis of the latter.

678. A right-angled triangle is described self-conjugate to a given parabola and with its hypotenuse in a given direction: prove that its vertex lies on a fixed straight line parallel to the axis of the parabola and its sides touch a fixed parabola.

679. Two equal parabolas have their axes in the same straight line and their vertices at a distance equal to the latus rectum; a chord of the outer touches the inner and on it as diameter is described a circle: prove that this will touch the outer parabola.

680. Tangents are drawn from a fixed point O to a series of confocal parabolas: prove that the corresponding normals envelope a fixed parabola whose directrix passes through O and is parallel to the axis of the system, and whose focus S' is such that OS' is bisected by S.

681. A point O on the directrix is joined to the focus S and SO bisected in F; with focus F is described another parabola whose axis is the tangent at the vertex of the former and from O two tangents are drawn to the latter parabola: prove that the chord of contact and the corresponding normals all touch the given parabola.

682. Prove the following construction for inscribing in a parabola a triangle with its sides in given directions:—Draw tangents in the given directions touching at A, B, C, and chords AA', BB', CC' parallel to BC, CA, AB; $A'B'C'$ will be the required triangle.

[The construction is not limited to the parabola, and a similar construction may be made for an inscribed polygon.]

683. Two fixed tangents are drawn to a parabola: prove that the centre of the nine points' circle of the triangle formed by these and any other tangent is a straight line.

684. At one extremity of a given finite straight line is drawn any circle touching the line, and from the other extremity is drawn a tangent to this circle: prove that the point of intersection of this tangent with the tangent parallel to the given line lies on a fixed parabola, and those with the tangents perpendicular to the given line on two fixed hyperbolas.

685. Two parabolas have a common focus and from any point on their common tangent are drawn other tangents to the two: prove that the distances of these from the focus are in a constant ratio.

686. Two tangents are drawn to a parabola equally inclined to a given straight line: prove that their point of intersection lies on a fixed straight line passing through the focus.

687. Two parabolas have a common focus S, parallel tangents drawn to them at P, Q meet their common tangent in P', Q': prove that the angles PSQ, $P'SQ'$ are each equal to the angle between the axes.

688. Two parabolas have parallel axes and two parallel tangents are drawn to them: prove that the straight line joining the points of contact passes through a fixed point.

[A general property of similar and similarly situate figures.]

689. On a tangent are taken two points equidistant from the focus: prove that the other tangents drawn from these points will intersect on the axis.

690. A circle is described on the latus rectum as diameter and a straight line through the focus meets the two curves in P, Q: prove that the tangents at P, Q will intersect either on the latus rectum or on a straight line parallel to the latus rectum and at a distance from it equal to the latus rectum.

691. A chord is drawn in a given direction and on it as diameter a circle is described: prove that the distance between the middle points of this chord and of the other common chord of the circle and parabola is of constant length.

692. On any chord as diameter is described a circle cutting the parabola again in two points: prove that the part of the axis of the parabola intercepted between the two common chords is equal to the latus rectum.

693. Two equal parabolas are placed with their axes in the same straight line and their vertices at a distance equal to the latus rectum; a tangent drawn to one meets the other in two points: prove that the circle of which this chord is a diameter touches the parabola of which this is a chord.

694. A parabola is described having its focus on the arc, its axis parallel to the axis, and touching the directrix, of a given parabola: prove that the two curves will touch each other.

695. Circles are described having for diameters a series of parallel chords of a parabola: prove that they will all touch another parabola related to the given one in the manner described in the last question.

696. A circle is described having double contact with a parabola and a chord QQ' of the parabola touches the circle in P: prove that QP, $Q'P$ are respectively equal to the distances of Q, Q' from the common chord.

697. The locus of the centre of the circle circumscribing the triangle formed by two fixed tangents to a parabola and any other tangent is a straight line.

698. The locus of the focus of a parabola touching two fixed straight lines one of them at a given point is a circle.

699. Two equal parabolas A, B have a common vertex and axes opposite: prove that the locus of the poles with respect to A of tangents to B is A.

700. Three common tangents PP', QQ', RR' are drawn to two parabolas and PQ, $P'Q'$ intersect in L: prove that LR, LR' are parallel to the axes. Also prove that if PP' bisect QQ' it will also bisect RR', and PP' will be divided harmonically by QQ', RR'.

701. Two equal parabolas have a common focus and axes opposite; two circles are described touching each other, each with its centre on one parabola and touching the tangent at the vertex of that parabola: prove that the rectangle under their radii is constant whether the contact be internal or external, but in the former case is four times as great as in the latter.

702. Two equal parabolas have their axes parallel and opposite, and one passes through the centre of curvature at the vertex of the other: prove that this relation is reciprocal and that the parabolas cut at right angles.

703. From the ends of a chord PP' are let fall perpendiculars PM, $P'M'$ on the tangent at the vertex: prove that the circle on PP' as diameter and the circle of curvature at the vertex have PP' for radical axis.

[The analytical proof of this is instantaneous.]

704. A parabola touches the sides of a triangle ABC in A', B', C', $B'C'$ meets BC in P, another parabola is drawn touching the sides and P is its point of contact with BC: prove that its axis is parallel to $B'C'$.

705. The directrix and one point being given, prove that the parabola will touch a fixed parabola to which the given straight line is tangent at the vertex.

706. The locus of the focus of a parabola which touches a given parabola and has a given directrix parallel to that of the given parabola is a circle.

707. A triangle is self-conjugate to a parabola, prove that the straight lines joining the mid points of its sides touch the parabola; and that the straight line joining any angular point of the triangle to the point of contact of the corresponding tangent will be parallel to the axis.

708. Four tangents are drawn to a parabola: prove that the three circles whose diameters are the diagonals of the quadrilateral will have the directrix as common radical axis.

709. A circle is drawn meeting a parabola in four points and tangents drawn to the parabola at these points: prove that the axis of the parabola will bisect the diagonals of the quadrilateral so formed.

710. The tangents at P, Q meet in T, and O is the centre of the circle TPQ: prove that OT subtends a right angle at S and that the circle OPQ passes through S.

711. Three parallels are drawn through A, B, C to meet the opposite sides of the triangle ABC in A', B', C': prove that a parabola can be drawn through $A'B'C'$ and the middle points of the sides, and that its axis will be in the same direction as the three parallels.

712. A chord LL' of a circle is bisected in O, and H is its pole; two parabolas are described with their focus at O, their directrices passing through H, and one of their common points on the circle: prove that the angle between their axes is equal to LHL'.

II. *Central Conics.*

[In these questions, unless other meanings are expressly assigned, S, S' are the foci of a central conic, C the centre, AA', BB' the major and minor axes, T, t and G, g the points where the tangent and normal at a point P meet the axes, and CD the semi-diameter conjugate to CP.]

713. If SY, SZ be perpendiculars on two tangents the straight line drawn through the intersection of the tangents perpendicular to YZ will pass through S'.

714. If SY, SZ be drawn perpendicular respectively to the tangent and normal at any point, YZ will pass through the centre.

715. A common tangent is drawn to a conic and to the circle whose diameter is the latus rectum: prove that the latus rectum bisects the angle between the focal distances of the points of contact.

716. If a triangle ABC circumscribe a conic the sum of the angles subtended by BC at the foci will exceed the angle A by two right angles.

717. Two conics U, V have a common focus S, the tangents to U at two common points meet in P and to V in Q: prove that PQ passes through S (the common points being rightly selected when there are four).

718. Perpendiculars SY, $S'Y'$ are drawn on any tangent and YP, $Y'P'$ are the other tangents from Y, Y': prove that SP, $S'P'$ will intersect on the conic.

719. A circle touches the conic at P and passes through S, PQK drawn perpendicular to the directrix meets the circle in Q: prove that QSK is a right angle.

720. On a tangent are taken two points O, O' such that $SO = SO'$ = major axis: prove that the radius of the circle $OS'O'$ is equal to the major axis.

721. Tangents OP, OQ, $O'P'$, $O'Q'$ are drawn to a certain circle: prove that the foci of the conic which touches the sides of the two triangles OPQ, $O'P'Q'$ lie on the circle.

722. In an ellipse in which $BB' = SS'$ a diameter PP' is taken and circles drawn touching the ellipse in P, P' and passing through S: their second common point will lie on the latus rectum.

723. Prove that when $SG = PG$, SP is equal to the latus rectum; and if PK drawn always at right angles to SP meet the axis major in K, SK has then its least possible length.

724. A common radius CPQ is drawn to the two auxiliary circles of an ellipse and tangents to the circles at P, Q meet the corresponding axes in U, T: prove that TU will touch the ellipse.

725. A circle has double contact with a conic: prove that the tangent from any point of the conic to the circle bears a constant ratio to the distance from the chord of contact.

726. The foot of the perpendicular from the focus on the tangent at the extremity of the farther latus rectum lies on the minor axis.

727. The common tangents to an ellipse and to a circle through the foci will touch the circle in points lying on the tangents at the ends of the minor axis; and the common tangents to an ellipse and to a circle with its centre on the major axis and dividing SS' harmonically will touch the circle in points on the tangent at one end of the major axis.

[These two cases are undistinguishable analytically.]

728. The tangent at P and normal at Q meet on the minor axis: prove that the tangent at Q and normal at P will also meet on the minor axis and PQ will always touch a confocal hyperbola.

[The data will not be possible for the ellipse unless $SS' > BB'$.]

729. Prove that at the point P where $SP = PG$,

$$SP : HP = BC^2 : AC^2.$$

730. A tangent meets the auxiliary circle in two points through which are drawn chords of the circle parallel to the minor axis: prove that the straight line drawn from the foot of the ordinate parallel to the tangent will divide either chord into segments which are as the focal distances of the points of contact.

731. Two diagonals of a quadrilateral intersect at right angles: prove that a conic can be inscribed with a focus at the intersection of the diagonals.

732. Given a focus S and two tangents, the locus of the second focus is the straight line through the intersection of the tangents perpendicular to the line joining the feet of the perpendiculars from S on the tangents.

733. Given one focus, a tangent, and a straight line on which the centre lies, prove that the conic has a second fixed tangent.

734. From the foci S, S' are drawn perpendiculars SPY, $S'P'Y'$ on any tangent to the auxiliary circle meeting the conic in P, P': prove that the rectangle SY, $S'P'$ = the rectangle $S'Y'$, $SP = BC^2$.

735. The length of the focal perpendicular on any tangent to the auxiliary circle is equal to the focal distance of the corresponding point

736. Through C is drawn a straight line parallel to either focal distance of P, and CD is the radius parallel to the tangent at P: prove that the distance of D from the former straight line is equal to BC.

737. Prove that, if an ellipse be inscribed in a given rectangle, the points of contact will be the angular points of a parallelogram of constant perimeter: and investigate the corresponding theorem when the conic is an hyperbola.

738. A straight line is drawn touching the minor auxiliary circle meeting the ellipse in P and the director circle in Q, Q': prove that QP, PQ' are equal to the focal distances of P.

739. Given a focus and two tangents of a conic, prove that the envelope of the minor axis is a parabola with its focus at the given focus: also a common tangent to this parabola and any one of the conics subtends a right angle at the given focus.

740. A perpendicular from the centre on the tangent meets the focal distances of the point of contact in two points: prove that these points are at a constant distance SC from the feet of the focal perpendiculars on the tangent.

741. The tangent at a point P meets the major axis in T; prove that

$$SP : ST :: AN : AT.$$

742. The circle passing through the feet of the perpendiculars from the foci on the tangent and through the foot of the ordinate will pass through the centre; and the angle subtended at either end of the major axis by the distance between the feet of the perpendiculars will be equal or supplementary to the angle which either focal distance makes with the corresponding perpendicular.

743. Given a focus and the length and direction of the major axis, prove that a conic will touch two fixed parabolas whose common focus is the given focus and semi latus rectum along the given line and of the given length.

744. A conic is described having one focus at the focus of a given parabola and its major axis coincident in direction with and equal to half of the latus rectum of the parabola: prove that this conic will touch the parabola.

745. A conic touches two adjacent sides of a given parallelogram and its foci lie on the two other sides one on each: prove that each directrix touches a fixed parabola. If $ABCD$ be the parallelogram, S, S' the foci on BC, CD respectively, and on AS, AS' be taken $AL = AB$, $AL' = AD$, the excentricity of the conic will be the ratio $LL' : SS'$.

746. Three points A, B, C are taken on a conic such that CA, CB are equally inclined to the tangent at C: prove that the normal at C will pass through the pole of AB.

747. Given one focus, a tangent, and the length of the major axis, prove that the locus of the second focus is a circle: and determine the portions of the locus which correspond to an ellipse, and those which correspond to an hyperbola in which the given focus belongs to the branch which touches the given straight line.

748. Given a focus, the excentricity, and a tangent: prove that the directrix will touch a fixed conic having the same focus and excentricity, and the minor axis of this envelope will lie along the given tangent.

749. A conic described with its foci at the centres of two given intersecting circles and touching a tangent drawn to either circle at a common point will touch the other tangents at the common points; its auxiliary circle will pass through the common points, and any tangent to the conic will be harmonically divided by the circles.

750. Through any point O are drawn two tangents to a conic, and on them are taken two points P, Q so that O, P, Q are equidistant from S: prove that $S'O$ is perpendicular to PQ, and, if $S'O$, PQ meet in R, that twice the rectangle $S'O$, $S'R$, together with the square on SO, is equal to the square on SS', sign being attended to.

751. In any conic if PO be taken along the normal at P equal to the harmonic mean between PG, Pg, O will be the point such that any chord through it subtends a right angle at O; and if from P perpendiculars be let fall on any two conjugate diameters the straight line joining the feet of these perpendiculars will bisect PO.

752. The triangle ABC is isosceles, A being the vertex, and conics are drawn touching the sides AB, AC and the perpendiculars from B, C on the opposite sides: prove that the foci of these conics lie on either a fixed circle or a fixed straight line, and trace the motion of the foci as the centre moves along the straight line which is its locus.

753. Two diameters PP', QQ' of a conic are drawn, and PR, PR' let fall perpendicular on $P'Q$, $P'Q'$; prove that the chord intercepted by the conic on RR' subtends a right angle at P.

754. A triangle ABC circumscribes a conic, and Sa, Sb, Sc are drawn perpendiculars on the sides; prove that

$$\triangle Sbc : \triangle S'BC = \triangle Sca : \triangle S'CA = \triangle Sab : \triangle S'AB$$
$$= \triangle abc : \triangle ABC.$$

755. From a point on an ellipse perpendiculars are let fall on the axes and *produced* to meet the corresponding auxiliary circles: prove that the straight line joining the two points of intersection passes through the centre.

756. Two conics have common foci S, S' and any straight line being taken another straight line is drawn joining the poles of the former with respect to the two conics: prove that the conic whose focus is S and which touches both these straight lines and the minor axis will be a parabola and that its directrix will pass through S'.

757. An ellipse and hyperbola are confocal, a straight line is drawn parallel to one of their common diameters and its poles with respect to the two conics joined to the centre: prove that the joining lines are at right angles, and that the polars of any point on a common diameter are also at right angles.

758. Tangents (or normals) are drawn in a given direction to a series of confocal conics: prove that the points of contact lie on a rectangular hyperbola having an asymptote in the given direction and passing through the foci.

759. An hyperbola and an ellipse are confocal, and from any point T on an asymptote are drawn TO, TP, TQ touching the hyperbola in O and the ellipse in P, Q respectively, prove that OT is a mean proportional between OP and OQ.

760. The tangents drawn to a series of confocal conics at the points where they meet a fixed straight line through S all touch a fixed parabola whose focus is S' and directrix is the given straight line, and which touches the minor axis.

761. From a point O on any ellipse are drawn tangents OP, OQ to any confocal: prove that the chord of curvature at O in direction of either tangent is double the harmonic mean between OP, OQ; tangents drawn outside the ellipse being considered negative.

762. An ellipse is described touching two given confocal conics and having the same centre: prove that the tangents at the points of contact will form a rectangle. For real contact one of the given conics must be an ellipse.

763. An ellipse is described having double contact with each of two confocals: prove that the sum of the squares on its axis is constant, and that the locus of its foci is the lemniscate in which $SP \cdot HP$ = difference of the squares on the given semi-major axes.

764. From a point O on a given ellipse are drawn two tangents OP, OQ to a given confocal ellipse and a diameter parallel to the tangent at O meets OP, OQ in the points P', Q': prove that the harmonic mean between OS, OS' bears to the harmonic mean between OP, OQ the constant ratio $OS + OS' : OP' + OQ'$.

765. On any tangent to a conic are taken two points equidistant from one focus and subtending a right angle at the other: prove that their distance from the former focus is constant.

766. The perpendicular CY on a tangent meets an ellipse in P, and Q is another point on the ellipse such that $CQ = CY$: prove that the perpendicular from C on the tangent at Q is equal to CP.

767. A tangent to a conic at P meets the minor axis in T, and TQ is drawn perpendicular to SP: prove that SQ is of constant length; and, PM being drawn perpendicular to the minor axis, that QM will pass through a fixed point.

768. Three tangents to a conic are such that their points of intersection are at equal distances from a focus: prove that each distance is equal to the major axis; and that the second focus is the centre of perpendiculars of the triangle formed by the tangents.

769. A conic is inscribed in a circle and is concentric with the nine points' circle of the triangle: prove that it will have double contact with the nine points' circle.

770. An ellipse is inscribed in an acute-angled triangle ABC with its foci S, S' at the centre of the circumscribed circle and the centre of perpendiculars respectively; SA, SB, SC meet the ellipse again in a, b, c: prove that the tangents at a, b, c are parallel to the sides of ABC, and form a triangle in which S', S are the centre of the circumscribed circle and the centre of perpendiculars respectively.

771. With the centre of perpendiculars of a triangle as centre are described two ellipses, one inscribed in the triangle the other circumscribing it: prove that these ellipses are similar and their major axes at right angles; and that the diameters of the inscribed conic parallel to the sides are as the cosines of the angles.

772. With a focus of a given conic as focus and any tangent as directrix is described a conic similar to the given conic: prove that it will touch the minor axis. If with the same focus and directrix a parabola be described it will intercept on the minor axis a segment subtending a constant angle at the focus.

[In general if the described conic have a given excentricity e' not less than that of the given conic it will intercept on the minor axis a segment subtending at the given focus a constant angle $= 2\cos^{-1}\left(\frac{e}{e'}\right)$.]

773. With a focus S of a given conic as focus and any tangent as directrix is described a conic touching a fixed straight line perpendicular to the major axis, another fixed straight line is drawn parallel and conjugate to the former: prove that the segment of this latter straight line intercepted by the variable conic subtends at S the same angle as the segment intercepted by the given conic.

774. With the vertex of a given conic as focus and any tangent as directrix is described a conic passing through one of the foci of the given conic: prove that the major axis is equal to the distance of either focus of the given conic from its directrix.

775. An ellipse is inscribed to a given triangle with its centre at the circumscribed circle of the triangle: prove that both auxiliary circles of the ellipse touch the nine points' circle of the triangle; and that the three perpendiculars of the triangle are normals to the ellipse.

776. A conic touches the sides and passes through the centre of the circumscribed circle of a triangle: prove that the director circle of the ellipse will touch the circumscribed circle of the triangle.

777. A diameter PE' being fixed, QVQ' is any chord parallel to it bisected in V, and PV intersects CQ or CQ' in R: prove that the locus of R is a parabola.

778. A chord is drawn parallel to the major axis and circles drawn through S to touch the conic at the ends of the chord: prove that the second common point of the circles is the intersection of the chord with the focal radius to its pole; and that the locus of this point is a parabola with its vertex at S.

779. With any point on a given circle as focus and a given diameter as directrix is described a conic similar to a given conic: prove that it will touch two fixed similar conics to which the given diameter is latus rectum, its points of contact lying on the radius through the focus.

780. Given the side BC of a triangle ABC and that

$$\cos A = m \cos B \cos C,$$

prove that the locus of A is an ellipse of which BC is the minor axis when m is positive, an ellipse of which BC is major axis when m is negative but $1+m$ positive, and an hyperbola of which BC is transverse axis when $1+m$ is negative.

781. Given a focus S and two tangents to a conic, prove that the envelope of the minor axis is a parabola of which S is focus.

782. A circle is drawn touching the latus rectum of a given ellipse in S the focus on the side towards the centre and also touches the tangents at the ends of the latus rectum: prove that the two other common tangents will touch the ellipse in points lying on a tangent to the circle.

783. From the foci S, S' are let fall perpendiculars SY, $S'Y'$ on any tangent to an ellipse: prove that the perimeter of the quadrilateral $SYY'S'$ will be the greatest possible when YY' subtends a right angle at the centre.

[This is only possible when SS' is greater than BB'; when SS' is equal to or less than BB' the perimeter is greatest when the point of contact is the end of the minor axis.]

784. A conic is described having one side of a triangle for directrix, the opposite vertex for centre and the centre of perpendiculars for focus: prove that the sides of the triangle which meet in the centre are conjugate.

785. The angle which a diameter of an ellipse subtends at an extremity of the minor axis is supplementary to that which its conjugate subtends at the ends of the major axis.

786. Two pairs of conjugate diameters of an ellipse are PP', DD'; pp', dd' respectively; prove that Pp, Pp' are respectively parallel to $D'd$, $D'd'$.

787. Tangents TP, TQ are drawn to a conic and chords Qq, Pp parallel to TP, TQ respectively: prove that pq is parallel to PQ. Also prove that the diameters parallel to the tangents form a harmonic pencil with CT and the diameter conjugate to CT.

788. A chord QQ' of an ellipse is parallel to one of the equal conjugate diameters and QN, $Q'N'$ are perpendiculars on an axis: prove that the triangles QCN, $Q'CN'$ are equal and that the normals at Q, Q' intersect on the diameter which is perpendicular to the other equal conjugate diameter.

789. Any ordinate NP of an ellipse is produced to meet the auxiliary circle in Q and normals to the ellipse and circle at P, Q meet in R; RK, RL are drawn perpendicular to the axes: prove that K, P, L lie on one straight line and that KP, PL are equal respectively to the semi-axes. (The point Q may be either point in which NP meets the auxiliary circle.)

790. On the normal to an ellipse at P are taken two points Q, Q' such that $QP = PQ' = CD$: prove that the cosine of the angle QCQ' is $\frac{CP^2 - CD^2}{AC^2 - BC^2}$; and if from Q or Q' be drawn a straight line normal to the ellipse at R, the parts of this straight line intercepted between R and the axes will be equal respectively to BC and AC.

791. Through any point Q of one of the auxiliary circles is drawn QPP' perpendicular to the axis of contact meeting the ellipse in P, P': prove that the normals to the ellipse at P, P' intercept on the normal to the circle at Q a length equal to the diameter of the other auxiliary circle.

792. Tangents to an ellipse at P, D ends of conjugate diameters meet in O, any other tangent meets these in P', D' respectively: prove that the rectangle under OP', OD' is double that under PP', DD'.

[The ratio of the two rectangles is constant for any two fixed points P, D, having a value depending on the area cut off by the segment PD.]

793. A chord PQ is drawn through one focus, L is its pole and O the centre of the circle LPQ: prove that the circle OPQ will pass through the second focus.

794. Through two fixed points A, B of a conic are drawn chords AP, BQ parallel to each other: prove that PQ always touches a concentric similar and similarly placed conic.

795. A parallelogram $ABCD$ circumscribes a given conic and a tangent meets AB, AD in P, Q, and CB, CD in P', Q': prove that the rectangles BP, DQ, and BP', DQ' are equal and constant.

796. An equilateral triangle PQR is inscribed in an auxiliary circle of an ellipse and P', Q', R' are the corresponding points on the ellipse: prove that the circles of curvature at P', Q', R' meet in one point lying on the ellipse and on the circle $P'Q'R'$

797. A conic, centre O, is inscribed in a triangle ABC and through B, C are drawn straight lines parallel to the diameter conjugate to OA: prove that these straight lines will be conjugates.

798. A chord EF of a given circle is divided in a given ratio in S; construct a conic of which E is one point, S a focus, and the given circle the circle of curvature at E.

799. A point P is taken on an ellipse equidistant from the minor axis and a directrix; prove that the circle of curvature at P will pass through a focus.

800. An ellipse is drawn concentric with a given ellipse, similar to it, and touching it at a point P; prove that the areas of the two are as $CP^2 : SP . S'P$; and their curvatures at P in the duplicate ratio of $SP . S'P : CP^2$.

801. Any chord PQ of an ellipse meets the circle of curvature at P in Q': prove that PQ' has to PQ the duplicate ratio of the diameters of the ellipse which are respectively parallel to the tangent at P and to the chord PQ.

802. Two circles are described with S, S' as centres and intersecting in P, P'; prove that with any point on the conic, whose foci are S, S' and which passes through P, as centre, can be described a circle touching both the former, and that all these tangent circles cut at right angles a fixed circle touching the conic in P, P'.

803. Given a focus, a point, and the length of the major axis; prove that the envelope of either directrix is a conic having its focus at the common point and excentricity equal to the ratio of the focal distances of the common point.

804. Given a point and the directrices; prove that the locus of each focus is a circle, and the envelope of the conic is a conic having the given point for focus and the distances between the directrices for major axis.

805. A circle is described having internal contact with each of two given circles one of which lies within the other, and the centre P of the moving circle describes an ellipse of which AA' is the major axis; through A is drawn a diameter of the moving circle; prove that the ends of this diameter will lie on an ellipse similar to the locus of P, and having a focus at A and centre at A'.

806. A conic has one focus in common with a given conic, touches the given conic and passes through its second focus: prove that the major axis is constant.

807. Two similar conics U, V are placed with their major axes in the same straight line, and the focus of U is the centre of V: prove that the focal distance of the point of contact with U of a common tangent is equal to the semi-major axis of V.

808. In a conic one focus, the excentricity, and the direction of the major axis are given, and tangents are drawn to it at points where it meets a given circle having its centre at the given focus: prove that these tangents all touch a fixed conic having the given excentricity and whose auxiliary circle is the given circle.

809. The tangent to a conic at P meets the axes in T, t and the central radius at right angles to CP in Q: prove that the ratio of QT to Qt is constant.

810. Through a given point O on a given conic are drawn chords OP, OQ equally inclined to a given direction: prove that PQ passes through a fixed point.

811. A chord PQ is normal at P to a given conic and a diameter LL' is drawn bisecting PQ; prove that PQ makes equal angles with LP, $L'P$ and that $LP \pm L'P$ is constant.

812. A conic is described through the foci of a given conic and touching it at the ends of a diameter: prove that the rectangle under the distances of a focus of this conic from the foci of the given conic is equal to the square on the semiminor axis of the given conic; and that the diameter of this conic which is conjugate to the major axis of the given conic is equal to the minor axis of that conic.

813. A conic is inscribed in a triangle ABC and has its focus at O; the angles BOC, COA, AOB are denoted by A', B', C'; prove that

$$\frac{OA \sin A}{\sin(A'-A)} = \frac{OB \sin B}{\sin(B'-B)} = \frac{OC \sin C}{\sin(C'-C)} = \text{major axis.}$$

Under what convention is this true if O be a point without the triangle?

814. Two conics are described having a common minor axis and such that the outer touches the directrices of the inner; MPP' is a common ordinate; prove that MP' is equal to the normal at P.

815. Two tangents OA, OB are drawn to a conic and a straight line meets the tangents in Q, Q', the chord AB in R, and the conic in P, P'; prove that

$$QP \,.\, PQ' : RP^2 = QP' \,.\, P'Q' : RP'^2,$$

and that for a given direction of the straight line each of these ratios is constant.

816. With B the extremity of the minor axis of an ellipse as centre is described a circle whose diameter is equal to the major axis, and the tangents at the end of the major axis meet the other common tangents to the ellipse and circle in P, P', Q, Q': prove that B, P, P', Q, Q' lie on a circle whose diameter is equal to the radius of curvature of the ellipse at B.

817. A chord of an ellipse subtends at S an angle equal to the angle between the equal conjugate diameters: prove that the foot of the perpendicular from C on this chord lies on a fixed circle whose diameter is equal to the radius of the director circle.

818. A parabola is drawn with its focus at S a focus of a given conic and touches the conic: prove that its directrix will touch a fixed circle whose centre is S', and that the tangent at the vertex of the parabola touches the auxiliary circle.

819. A parabola is drawn through the foci of a given ellipse with its own focus P on the ellipse; prove that the parts of the axis of the parabola intercepted between P and the axes of the ellipse are of constant length, and if through the points where the axis of the parabola meets the axes of the ellipse straight lines be drawn at right angles to the axes of the ellipse their point of intersection will lie upon the normal to the ellipse at P.

820. A given finite straight line is one of the equal conjugate diameters of an ellipse; prove that the locus of the foci is a lemniscate of Bernoulli.

821. A parallelogram is inscribed in a conic and from any point on the conic are drawn two straight lines each parallel to two sides: prove that the rectangles under the segments of these lines cut off by the sides of the parallelogram are in a constant ratio.

822. Two central conics in the same plane have two conjugate diameters of the one parallel respectively to two conjugate diameters of the other; and in general no more.

823. In two similar and similarly placed ellipses are drawn two parallel chords PP', QQ'; PQ, $P'Q'$ meet the two conics in R, S, R', S' respectively: prove that RR', SS' are parallels: also that QQ', RR' and PP', SS' intersect in points lying on a fixed straight line.

824. A circle described on the intercept of the tangent at P made by the tangents at A, A' meets the conic again in Q; prove that the ordinate of Q is to the ordinate of P as the minor axis is to the sum of the minor axis and the diameter conjugate to P. (As $BC : BC + CD$.)

825. A point P is taken on a conic and O is the centre of the circle SPS', PO is divided in O' so that $PO' : PO = BC^2 : AC^2$: prove that the circle with O' as centre and $O'P$ as radius will touch the major axis at the foot of the normal at P.

826. With a fixed point P on a given conic as focus is described a parabola touching a pair of conjugate diameters; prove that this parabola will have a fixed tangent parallel to the tangent at P and that this tangent divides CP in the ratio $CP^2 : CD^2$.

827. Through a point O are drawn two straight lines conjugates with respect to a given conic; any tangent meets them in P, Q: prove that the other tangents drawn from P, Q intersect on the polar of O.

828. A parabola is described having S for its focus and touching the minor axis; prove that a common tangent will subtend a right angle at S and that its point of contact with either conic lies on the directrix of the other.

829. Prove that the two points common to the director circles of all conics inscribed in a given quadrilateral may be constructed as follows: take aa', bb', cc' the three diagonals of the quadrilateral forming a triangle ABC and let O be the centre of the circle ABC, then if P, Q be the required points, O, P, Q lie in one straight line perpendicular to the bisector of the diagonals, PQ is bisected by this bisector and the rectangle OP, OQ is equal to the square of the radius of the circle ABC.

830. At each point P of an ellipse is drawn QPQ' parallel to the major axis so that $QP = PQ' = SP$: prove that Q, Q' will trace out ellipses whose centres are A, A' and whose areas are together double the area of the given ellipse. If QPQ' be drawn parallel to the minor axis instead of the major, the loci are ellipses whose major axes are at right angles to each other and they touch each other in S and touch the tangents at A, A'.

831. A given ellipse has its minor axis increased and major axis diminished in the ratio $\sqrt{1-e} : 1$, its centre then displaced along the minor axis through a length equal to a and the ellipse then turned about its centre through half a right angle: prove that the whole effect is equivalent to a simple shear parallel to the minor axis by which the major axis is transferred into the position of a tangent at one end of the latus rectum.

832. A point P is taken on an hyperbola such that $CP = CS$: prove that the circle PTG will touch CP at P, and, if Q, Q' be two other points such that the ordinate of P is a mean proportional between those of Q, Q', that the tangents at Q, Q' will intersect on the circle whose radius is CS.

833. An hyperbola is described through the focus of a parabola with its own foci on the parabola; prove that one of its asymptotes is parallel to the axis of the parabola.

834. A parabola passes through two given points and its axis is in a given direction: prove that its focus lies on a fixed hyperbola.

835. Two tangents of an hyperbola U are asymptotes of another V; prove that if V touch one of the asymptotes of U it will touch both.

836. In an hyperbola whose excentricity is 2, the circle on a focal chord as diameter passes through the farther vertex. Any chord of a single branch subtends at the focus S interior to that branch an angle double that which it subtends at the farther vertex A'. If RSR' be a chord, SPp, SQq chords inclined at $60°$ to the former, the circles qPR', pQR will intersect in S and A', and if the former intersect the circle on the latus rectum in U, V, the angle $A'SU$ is three times $A'SP$, and UV is a diameter of the last-mentioned circle.

837. The straight line joining two points which are conjugates with respect to a conic is bisected by the conic: prove that the line is parallel to an asymptote.

838. A conic is drawn through two given points with asymptotes in given directions: prove that the locus of its foci is an hyperbola.

839. A straight line is drawn equidistant from focus and directrix of an hyperbola, and through any point of it is drawn a straight line at right angles to the focal distance of the point: prove that the intercept made by the conic will subtend at the focus an angle equal to the angle between the asymptotes.

840. Two hyperbolas U, V are similar and have a common focus, and the directrix of V is an asymptote of U; prove that the conjugate axis of U is an asymptote of V.

841. In an hyperbola LL' is the intercept of a tangent by the asymptotes: prove that

$$SL \cdot S'L = CL \cdot LL', \text{ and } SL' \cdot S'L' = CL' \cdot LL'.$$

842. To an hyperbola the concentric circle through the foci is drawn: prove that tangents drawn from any point on this circle to the hyperbola divide harmonically the diameter of the circle which lies on the conjugate axis; and if OP, OP' the tangents meet the conjugate axis in U, U' and PM, $P'M'$ be perpendiculars on the conjugate axis, UM', $U'M$ will be divided in a constant ratio by C.

843. A circle is drawn touching both branches, prove that it intercepts on either asymptote a length equal to the major axis; the tangents to it where it meets the asymptotes pass through one or other of the foci, and those meeting in a focus are inclined at a constant angle equal to that between the asymptotes; and the straight lines joining the points where it meets the asymptotes (not being parallel to the transverse axis) will touch two fixed parabolas whose foci are the foci of the hyperbola.

III. *Rectangular Hyperbola.*

[In the questions under this head, R. H. is an abbreviation for rectangular hyperbola.]

844. Four points A, B, C, D are taken on a R. H. such that BC is perpendicular to AD: prove that CA is perpendicular to BD and AB to CD.

845. The angle between two diameters of a R. H. is equal to the angle between the conjugate diameters.

846. A point P on a R. H. is taken, and PK, PK' drawn at right angles to PA, PA' to meet the transverse axis; prove that $PK = PA'$, and $PK' = PA$, and that the normal at P bisects KK'.

847. The foci of an ellipse are ends of a diameter of a R. H.; prove that the tangent and normal to the ellipse at any one of the common points are parallel to the asymptotes of the hyperbola: and that tangents drawn from any point of the hyperbola to the ellipse are parallel to a pair of conjugate diameters of the hyperbola.

848. Prove that any chord of a R. H. subtends at the ends of a diameter angles either equal or supplementary: equal if the ends of the chord be on the same branch and on the same side of the diameter, or on opposite branches and on opposite sides; otherwise supplementary.

849. A circle and R. H. intersect in four points, two of which are ends of a diameter of the hyperbola: prove that the other two will be the ends of a diameter of the circle. Also, if AB be the diameter of the hyperbola, P any point on the circle, and PA, QB meet the hyperbola again in Q, R; prove that BQ, AR will intersect on the circle.

850. With parallel chords of a R. H. as diameters are described circles; prove that they have a common radical axis.

851. The ends of the equal conjugate diameters of a series of confocal ellipses lie on the confocal R. H.

852. The ends of a diameter of a R. H. are given; prove that the locus of its foci is a lemniscate of Bernoulli, of which the given points are foci.

853. From any point P of a R. H. perpendiculars are let fall on a pair of conjugate diameters of the hyperbola: prove that the straight line joining the feet of these perpendiculars is parallel to the normal at P.

854. The tangent at a point P of a R. H. meets a diameter QCQ' in T: prove that CQ, TQ' subtend equal angles at P.

855. Through two fixed points are drawn in a given direction two equal and parallel straight lines, and on them as diameters circles are described: prove that the locus of their common points is a R. H. Also if segments similar to a given segment be described on the two lines on opposite sides, the locus of their common points is a R. H.

856. If PP', QQ' be diameters, the angles subtended by PQ', $P'Q$ at any point of the R. H. will be equal or supplementary; and similarly for PQ, $P'Q'$.

857. Two double ordinates QQ', RR' are drawn to a diameter PP' of a R. H. on opposite branches: prove that a common tangent to the circles of which QQ', RR' are diameters will subtend a right angle at P and P'.

858. Prove that a circle drawn to touch a chord of a R. H. at one end and to pass through the centre will pass through the pole of the chord.

859. Two R. H. are such that the asymptotes of one are the axes of the other: prove that they cut each other at right angles, and that any common tangent subtends a right angle at the centre.

860. Two points are taken on a R. H. and on its conjugate such that the tangents are at right angles to each other: prove that the straight line joining them subtends a right angle at the centre.

861. Tangents to a R. H. at P, Q meet in T and intersect CQ, CP respectively in P', Q': prove that a circle can be described about $CP'TQ'$.

862. A fixed diameter PP' being taken, and Q being any other point on the curve: prove that the angles QPP', $QP'P$ differ by a constant quantity.

863. On opposite sides of any chord of a R. H. are described equal segments of circles: prove that the four points in which the completed circles again meet the hyperbola are the angular points of a parallelogram.

864. A circle and rectangular hyperbola meet in four points: prove that the diameter of the hyperbola which is perpendicular to a chord joining two of the points will bisect the chord joining the other two.

865. A point moves so that the straight lines joining it to two fixed points are equally inclined to a given direction: prove that its locus is a R. H. of which the two fixed points are ends of a diameter.

866. Circles are drawn through two given points and diameters drawn in a given direction: prove that the locus of the extremities of these diameters is a R. H. whose asymptotes make equal angles with the line of centres of the circles and with the given direction.

867. Prove that the angles which two tangents to a R. H. subtend at the centre are equal to the angles which they make with their chord of contact.

868. A parallelogram has its angular points on a R. H. and from any point on the hyperbola are drawn two straight lines parallel to the sides: prove that the four points in which these straight lines meet the sides of the parallelogram lie on a circle.

869. Two circles touch the same branch of a R. H. and touch each other in the centre: prove that the chord of the hyperbola joining the points of contact subtends an angle of 60° at the centre.

870. Two unequal parabolas have a common focus and axes opposite; a R. H. is described with its centre at the common focus touching both: prove that the chord of the hyperbola joining the points of contact subtends an angle of 60° at the centre.

871. Through any point on a R. H. are drawn two chords at right angles to each other: prove that the circle through the point and the middle points of the chords will pass through the centre.

872. A chord of a R. H. subtends a right angle at a focus: prove that the foot of the perpendicular on it from the focus lies on a fixed straight line.

873. A circle meets a R. H. in four points O, P, Q, R and OO', PP', QQ', RR' are diameters of the hyperbola: prove that O' is the centre of perpendiculars of the triangle PQR, and similarly for the others.

874. Two equal circles touch a R. H. in O and meet it again in P, Q, P', Q' respectively, O, P, Q being on one branch: prove that PP', QQ' are diameters of the hyperbola, PQ', $P'Q$ parallel to the

normal at O, and that the straight lines joining Q', P' to the centre of the circle OPQ will cut off one-third from OP, OQ respectively.

[Q', P' will be the centres of perpendiculars of the vanishing triangles O, O, P; O, O, Q respectively.]

875. The length of a chord of a R. H. which is normal at one extremity is equal to the corresponding diameter of curvature.

[Take on the hyperbola three contiguous points ultimately coincident and consider the centre of the circumscribed circle, centroid, and centre of perpendiculars of the infinitesimal triangle.]

876. A diameter PP' being taken, a circle is drawn through P' touching the hyperbola in P; prove that this circle is equal to the circle of curvature at P, and that if PI be the diameter of curvature at P, PR the common chord of the hyperbola and circle of curvature, RI will be equal and parallel to PP'.

877. A triangle is inscribed in a circle, and two parabolas drawn touching the sides with their foci at ends of a diameter of the circle: prove that their axes are asymptotes of a rectangular hyperbola passing through the centres of the four circles which touch the sides.

878. Three tangents are drawn to a R. H. such that the centre of the circle circumscribing the triangle lies on the hyperbola: prove that the centre of the hyperbola will lie on the circle; and that at any common point tangents drawn to the two curves pass through the points of contact of a common tangent.

879. A circle meets a R. H. in points P, P', Q, Q' and P, P' are ends of a diameter of the hyperbola: prove that the tangents to the hyperbola at P, P' and to the circle at Q, Q' are parallel, and the tangents to the circle at P, P' and to the hyperbola at Q, Q' all meet in one point.

880. The tangent at a point of a R. H. and the diameter perpendicular to this tangent being drawn; prove that the segments of any other tangent from its point of contact to these two straight lines subtend supplementary angles at the point of contact of the fixed tangent.

881. The normal at a point P meets the curve again in Q; RR' is a chord parallel to this normal: prove that the points of intersection of QR, PR' and of QR', PR lie on the diameter at right angles to CP.

882. A triangle ABC is inscribed in a R. H. and its sides meet one asymptote in a, b, c and the other in a', b', c' respectively: through a, b, c are drawn straight lines at right angles to the corresponding sides of the triangle: prove that these meet in a point O, and, O' being similarly found from a', b', c', that OO' is a diameter of the circle ABC.

CONIC SECTIONS, ANALYTICAL.

CARTESIAN CO-ORDINATES.

I. *Straight Line, Linear Transformation, Circle.*

[In any question relating to the intersections of a curve and two straight lines, it is generally convenient to use one equation representing both straight lines. Thus, to prove the theorem: "Any chord of a given conic subtending a right angle at a given point of the conic passes through a fixed point in the normal at the given point;" we may take the equation of the conic referred to the tangent and normal at the given point

$$ax^2 + 2hxy + by^2 = 2x;$$

the equation of any pair of straight lines through this point at right angles to each other is

$$x^2 + 2\lambda xy - y^2 = 0;$$

and at the points of intersection

$$(a + b)\, x^2 + 2\,(h + \lambda b)\, xy = 2x;$$

or, at the points other than the origin,

$$(a + b)\, x + 2\,(h + \lambda b)\, y = 2,$$

which is therefore the equation of a chord subtending a right angle at the origin. This passes through the point $y = 0$, $(a + b)\, x = 2$; a fixed point on the normal.

If two points be given as the intersections of a given straight line and a given conic the equation of the straight lines joining these points to the origin may be formed immediately, since it must be a homogeneous equation of the second degree in x, y. Thus the straight lines joining the origin to the points determined by the equations

$$ax^2 + 2hxy + by^2 = 2x,$$

$$px + qy = 1,$$

are represented by the equation

$$ax^2 + 2hxy + by^2 = 2x\,(px + qy).$$

and will be at right angles if $a+b=2p$, or if the straight line $px+qy=1$ pass through the point $\left(\frac{2}{a+b}, 0\right)$, a somewhat different mode of proving the theorem already dealt with. In general the equation of the straight lines joining the origin to the two points determined by the equations

$$ax^2+by^2+c+2fy+2gx+2hxy=0,$$

$$px+qy+r=0,$$

is $$ax^2+2hxy+by^2-\frac{2}{r}(px+qy)(gx+fy)+\frac{c}{r^2}(px+qy)^2=0.$$

The results of linear transformation may generally be obtained from the consideration that, if the origin be unaltered, the expression

$$x^2+2xy\cos\omega+y^2$$

must be transformed into

$$X^2+2XY\cos\Omega+Y^2,$$

if (x, y), (X, Y) represent the same point and ω, Ω be the angles between the co-ordinate axes in the two systems respectively. Thus if

$$u\equiv ax^2+by^2+c+2fy+2gx+2hxy$$

be transformed into

$$U\equiv AX^2+BY^2+c+2FY+2GX+2HXY,$$

then $\lambda(x^2+y^2+2xy\cos\omega)+u$ must be transformed into

$$\lambda(X^2+Y^2+2XY\cos\Omega)+U,$$

and if λ have such a value that the former be the product of two linear factors, so also must the latter; hence the two quadratic equations in λ

$$c(\lambda+a)(\lambda+b)+2fg(\lambda\cos\omega+h)$$
$$=(\lambda+a)f^2+(\lambda+b)g^2+c(\lambda\cos\omega+h)^2,$$

and $$c(\lambda+A)(\lambda+B)+2FG(\lambda\cos\Omega+H)$$
$$=(\lambda+A)F^2+(\lambda+B)G^2+c(\lambda\cos\Omega+H)^2$$

must coincide; and thus the invariants may be deduced. Also, by the same transformation,

$$\lambda(x^2+y^2+2xy\cos\omega)+ax^2+by^2+2hxy$$

must be transformed into

$$\lambda(X^2+Y^2+2XY\cos\Omega)+AX^2+BY^2+2HXY,$$

and if λ have such a value that the former is a square, so must the latter; hence the equations

$$(\lambda+a)(\lambda+b)=(\lambda\cos\omega+h)^2,$$
$$(\lambda+A)(\lambda+B)=(\lambda\cos\Omega+H)^2,$$

must coincide, whence

$$\frac{a+b-2h\cos\omega}{\sin^2\omega} = \frac{A+B-2H\cos\Omega}{\sin^2\Omega},$$

$$\frac{ab-h^2}{\sin^2\omega} = \frac{AB-H^2}{\sin^2\Omega}.$$

One special form of the equation of a circle is often useful: it is

$$(x-x_1)(x-x_2)+(y-y_1)(y-y_2)=0,$$

where (x_1, y_1), (x_2, y_2) are the ends of a diameter, and the axes rectangular. The corresponding equation when the axes are inclined at an angle ω is obtained by adding the terms

$$\{(x-x_1)(y-y_2)+(x-x_2)(y-y_1)\}\cos\omega;$$

each equation being found at once from the property that the angle in a semicircle is a right angle. In questions relating to two circles, it is generally best to take their equations as

$$x^2+y^2-2ax+k=0,$$

$$x^2+y^2-2bx+k=0,$$

the axis of x being the radical axis, and k negative when the circles intersect in real points.]

883. The equation of the straight lines which pass through the origin and make an angle a with the straight line $x+y=0$ is

$$x^2+2xy\sec 2a+y^2=0.$$

884. The equation $bx^2-2hxy+ay^2=0$ represents two straight lines at right angles respectively to the two whose equation is

$$ax^2+2hxy+by^2=0.$$

If the axes of co-ordinates be inclined at an angle ω, the equation will be

$$(a+b-2h\cos\omega)(x^2+y^2+2xy\cos\omega)=(ax^2+2hxy+by^2)\sin^2\omega.$$

885. The two straight lines

$$x^2(\tan^2\theta+\cos^2\theta)-2xy\tan\theta+y^2\sin^2\theta=0$$

make with the axis of x angles a, β such that $\tan a\sim\tan\beta=2$.

886. The two straight lines

$$(x^2+y^2)(\cos^2\theta\sin^2 a+\sin^2\theta)=(x\tan a-y\sin\theta)^2$$

include an angle a.

887. The two straight lines

$$x^2\sin^2 a\cos^2\theta+4xy\sin a\sin\theta+y^2\{4\cos a-(1+\cos a)^2\cos^2\theta\}=0$$

include an angle a.

888. Form the equation of the straight lines joining the origin to the points given by the equations

$$(x-h)^2+(y-k)^2=c^2,\ kx+hy=2hk,$$

and prove that they will be at right angles if $h^2+k^2=c^2$. Interpret geometrically.

889. The straight lines joining the points given by the equations

$$ax^2+by^2+c+2fy+2gx+2hxy=0,\ px+qy=1,$$

to the origin will be at right angles if

$$a+b+2(fq+gp)+c(p^2+q^2)=0;$$

and the locus of the foot of the perpendicular from the origin on the line $px+qy=1$ is $(a+b)(x^2+y^2)+2fy+2gx+c=0$: also the same is the locus of the foot of the perpendicular from the point

$$\left(-\frac{2g}{a+b},\ -\frac{2f}{a+b}\right).$$

890. The locus of the equation

$$y=2+\frac{x^2-1}{2+}\frac{x^2-1}{2+}\ \ldots\ldots \text{to } \infty$$

is the parts of two straight lines at right angles to each other which include one quadrant.

[The equation gives $y=1+x$ when x is positive and $y=1-x$ when x is negative.]

891. The formulæ for effecting a transformation of co-ordinates, not necessarily rectangular, are

$$x=pX+qY+r,\ y=p'X+q'Y+r';$$

prove that $$(pq-p'q')(pq'-p'q)=qq'-pp'.$$

892. The expression

$$ax^2+by^2+c+2fy+2gx+2hxy$$

is transformed into

$$Ax^2+By^2+c+2Fy+2Gx+2Hxy,$$

the origin being unchanged: prove that

$$\frac{f^2+g^2-2fg\cos\omega}{\sin^2\omega}=\frac{F^2+G^2-2FG\cos\Omega}{\sin^2\Omega},$$

and

$$\frac{2fgh-af^2-bg^2}{\sin^2\omega}=\frac{2FGH-AF^2-BG^2}{\sin^2\Omega};$$

ω, Ω being the angles between the co-ordinate axes in the two cases.

893. Prove that, ABC being a given acute-angled triangle and P any point in its plane, the three circular loci

$$PB^2 + PC^2 = n \,.\, PA^2\,;\; PC^2 + PA^2 = n \,.\, PB^2\,;\; PA^2 + PB^2 = n \,.\, PC^2,$$

have their radical centre at the centre of the circle ABC, each locus cuts the circle ABC at right angles, and the centre of any locus lies on the straight line joining an angular point to the middle point of the opposite side.

894. A certain point has the same polar with respect to each of two circles; prove that a common tangent subtends a right angle at this point.

895. A chord through A meets the tangent at B, the other end of a diameter of a given circle, in P and from any point in the chord produced are drawn two tangents to the circle: prove that the straight lines joining A to the points of contact will meet the tangent at B in points equidistant from P.

896. The radii of two circles are a, b and the distance between their centres $\sqrt{2(a^2+b^2)}$; prove that a common tangent subtends a right angle at the point which bisects the distance between their centres: and that if through the point which divides the distance between the centres in the ratio $a^2 : b^2$ be drawn two straight lines at right angles to each other equally inclined to the line of centres these straight lines will pass through the points of contact of the common tangents.

897. From a point O on a fixed straight line are drawn two tangents to a given circle meeting in P, Q the tangent at A which is parallel to the tangent at either point where the fixed straight line meets the circle: prove that $AP + AQ$ is constant.

898. Three circles U, V, W have a common radical axis and from any point on U two straight lines are drawn to touch V, W respectively: prove that the squares on these tangents will be in the ratio of the distances of the centre of U from those of V, W.

899. Tangents drawn from a point P to a given circle meet the tangent at a given point A in Q, Q'; prove that if the distance of P from the fixed tangent be given, the rectangle QA, AQ' will be constant.

900. Given two circles, a tangent to one at P meets the polar of P with respect to the other in P'; prove that the circle on PP' as diameter will pass through two fixed points which will be imaginary or real as the given circles intersect in real or imaginary points.

901. One circle lies entirely within another, a tangent to the inner meets the outer in P, P' and the radical axis in Q: prove that, if S be the internal point-circle of the system, the ratio $\sin \frac{PSP'}{2} : \cos \frac{SQP}{2}$ is constant.

902. On two circles are taken two points such that the tangents drawn each from one point to the other circle are equal: prove that the points are equidistant from the radical axis.

903. The equation of a circle in which (x_1, y_1), (x_2, y_2) are ends of the chord of a segment containing an angle θ is

$$(x-x_1)(x-x_2)+(y-y_1)(y-y_2)\pm\cot\theta\{(x-x_1)(y-y_2) - (x-x_2)(y-y_1)\}=0.$$

904. On the sides AB, AC of a triangle ABC are described two segments of circles each containing an angle $\theta\left(>\frac{\pi}{2}\right)$ and on the side BC a segment containing an angle $\frac{3\pi}{2}-\theta$: prove that the centre of the last circle lies on the radical axis of the other two; each segment being towards the same parts as the opposite angle.

905. There are two systems of circles such that any circle of one system cuts any circle of the other system at right angles; prove that the circles of either system have a common radical axis which is the line of centres of the other system.

906. On a fixed chord AB of a given circle is taken a point O such that, P being any point on the circle, $OA \,.\, OB = \pm PA \,.\, PB$: prove that the straight line which bisects PO at right angles will pass through one end of the diameter conjugate to AB; and, if Q be the other point in which the straight line meets the circle, that $QO^2 = QA \,.\, QB$.

907. A circle U lies altogether within another circle V; prove that the ratio of the segments intercepted by U, V on any straight line cannot be greater than

$$\sqrt{a^2-(b-c)^2}-\sqrt{a^2-(b+c)^2} : \sqrt{(a+c)^2-b^2}-\sqrt{(a-c)^2-b^2},$$

where a, b are the radii and c the distance between the centres.

908. An equilateral triangle is drawn with its sides passing through three given points A, B, C: prove that the locus of its centre is a circle having its centre at the centroid of ABC, and that the centres of two equilateral triangles whose sides are at right angles will be at the ends of a diameter of the locus.

[The radius of the locus is the difference of the axes of the minimum ellipse about ABC, the altitude of the maximum equilateral triangle is equal to three-fourths the sum of the axes of the minimum ellipse, and is also equal to the minimum sum of the distances of any point from A, B, C.]

909. Prove that the equation

$$\{x\cos(\alpha+\beta)+y\sin(\alpha+\beta)-a\cos(\alpha-\beta)\}\{x\cos(\gamma+\delta)+y\sin(\gamma+\delta)-a\cos(\gamma-\delta)\}$$
$$=\{x\cos(\alpha+\gamma)+y\sin(\alpha+\gamma)-a\cos(\alpha-\gamma)\}\{x\cos(\beta+\delta)+y\sin(\beta+\delta)-a\cos(\beta-\delta)\}$$

is equivalent to the equation $x^2 + y^2 = a^2$: and state the property of the circle expressed by the equation in this form.

910. Four fixed tangents to a circle form a quadrilateral whose diagonals are aa', bb', cc', and perpendiculars p, p'; q, q'; r, r' are let fall from these points on any other tangent: prove that

$$pp' \cos\frac{\beta-\gamma}{2}\cos\frac{\alpha-\delta}{2} = qq'\cos\frac{\gamma-\alpha}{2}\cos\frac{\beta-\delta}{2} = rr'\cos\frac{\alpha-\beta}{2}\cos\frac{\gamma-\delta}{2}$$

$$= 4a^2 \sin\frac{\alpha-\theta}{2}\sin\frac{\beta-\theta}{2}\sin\frac{\gamma-\theta}{2}\sin\frac{\delta-\theta}{2};$$

the co-ordinates of the points of contact being $(a\cos\alpha, a\sin\alpha)$, and the like in β, γ, δ, θ.

911. The radii of two circles are R, ρ, the distance between their centres is $\sqrt{R^2 + 2\rho^2}$ and $\rho < 2R$: prove that an infinite number of triangles can be inscribed in the first which are self-conjugate with respect to the second; and that an infinite number can be circumscribed to the second which are self-conjugate to the first.

[In general, if δ denote the distance between the centres, and the polar of a point A on the first circle with respect to the second meet the first in B, C, the chords AB, AC will touch the conic

$$y^2(2R^2 + 2\rho^2 - \delta^2) + (R^2 + \rho^2 - \delta^2)(2x^2 - 2\delta x + \delta^2 - R^2 - \rho^2) = 0,$$

and BC will touch the conic

$$\{(x-\delta)^2 + y^2\} R^2 = (\delta x + \rho^2 - \delta^2)^2;$$

and these two will coincide if $\delta^2 = R^2 + 2\rho^2$.]

912. A triangle is inscribed in the circle $x^2 + y^2 = R^2$, and two of its sides touch the circle $(x-\delta)^2 + y^2 = \rho^2$: prove that the third side will touch the circle

$$\left\{x - \frac{4R^2r^2\delta}{(R^2-\delta^2)^2}\right\}^2 + y^2 = R^2\left\{\frac{2r^2(R^2+\delta^2)}{(R^2-\delta^2)^2} - 1\right\}^2,$$

which coincides with the second circle if $\delta^2 = R^2 \pm 2Rr$. Also prove that the three circles have always a common radical axis.

913. Two given polygons of n sides are similar and similarly situated: prove that in general only two polygons can be drawn of the same number of sides circumscribing one of the two given polygons and inscribed in the other; but that if the ratio of homologous sides in the two be $\cos^2\frac{r\pi}{2n} : \cos^2\frac{r\pi}{2n} - \sin^2\frac{\pi}{n}$, where r is any whole number less than $\frac{n}{2}$, there will be an infinite number.

II. *Parabola referred to its axis.*

[The equation of the parabola being taken $y^2 = 4ax$, the co-ordinates of any point on it may be represented by $\left(\frac{a}{m^2}, \frac{2a}{m}\right)$, and with this notation the equation of the tangent is $y = mx + \frac{a}{m}$; of the normal $my + x = 2a + \frac{a}{m^2}$; and of the chord through two points (m_1, m_2), $2m_1m_2x - y(m_1 + m_2) + 2a = 0$. The equation of the polar of a point (XY) is $yY = 2a(x + X)$, and that of the two tangents drawn from (X, Y) is $(Y^2 - 4aX)(y^2 - 4ax) = \{yY - 2a(x + X)\}^2$.

As an example, we may take the following, "To find the locus of the point of intersection of normals to a parabola at right angles to each other."

If (X, Y) be a point on the locus, the points on the parabola to which normals can be drawn from (X, Y) are given by the equation

$$m^3Y + m^2(X - 2a) - a = 0;$$

so that, if m_1, m_2, m_3 be the three roots of the equation

$$m_1 + m_2 + m_3 = \frac{2a - X}{Y}, \quad m_2m_3 + m_3m_1 + m_1m_2 = 0, \quad m_1m_2m_3 = \frac{a}{Y};$$

and since two normals meet at right angles in (X, Y) the product of two of the roots is -1; let then $m_2m_3 = -1$. Then

$$m_1 = -\frac{a}{Y}, \quad m_2 + m_3 = \frac{3a - X}{Y} = -\frac{Y}{a},$$

or the locus is the parabola $y^2 = a(x - 3a)$.

Again, "The sides of a triangle touch a parabola and two of its angular points lie on another parabola with its axis in the same direction, to find the locus of the third angular point."

Let the equations of the parabolas be $y^2 = 4ax$, $(y - k)^2 = 4b(x - h)$, and let the three tangents to the former be at the points m_1, m_2, m_3. The point of intersection of (1), (2) is $\frac{a}{m_1m_2}$, $a\left(\frac{1}{m_1} + \frac{1}{m_2}\right)$, and this will lie on the second parabola if

$$\left\{a\left(\frac{1}{m_1} + \frac{1}{m_2}\right) - k\right\}^2 = 4b\left(\frac{a}{m_1m_2} - h\right),$$

and similarly for m_1, m_3. Hence m_2, m_3 are the roots of the quadratic in Z,

$$\left\{a\left(\frac{1}{m_1} + \frac{1}{Z}\right) - k\right\}^2 = 4b\left(\frac{a}{m_1Z} - h\right),$$

hence, if (X, Y) be the point of intersection of the tangents at m_2, m_3,

$$Y = a\left(\frac{1}{m_2} + \frac{1}{m_3}\right) = 2\left(k - \frac{a}{m_1}\right) + \frac{4b}{m_1},$$

$$X = \frac{a}{m_2 m_3} = \frac{\left(k - \frac{a}{m_1}\right)^2 + 4bh}{a};$$

so that $$(aY - 4bk)^2 = 4(2b - a)^2 (aX - 4bh),$$

or the third point lies on another parabola with its axis in the same direction as the two given parabolas, and which coincides with the second if $a = 4b$.]

914. Two parabolas have a common vertex A and a common axis, an ordinate NPQ meets them, the tangent at P meets the outer parabola in R, R' and AR, AR' meet the ordinate in L, M; prove that NP, NQ are respectively harmonic and geometric means between NL, NM.

915. A triangle is inscribed in a parabola and a similar and similarly placed triangle circumscribes it: prove that the sides of the latter triangle are respectively four times the corresponding sides of the latter.

916. Two tangents p, q being drawn to a given parabola U, through their point of intersection are drawn the two parabolas confocal with U, and A', A'' are their vertices: prove that

$$\left(\frac{p-q}{p+q}\right)^2 = \frac{SA' \,.\, A''A}{SA'' \,.\, A'A} = [SA'A''A];$$

A, A' being taken on opposite sides of S.

917. An equilateral triangle is inscribed in a parabola: prove that the ordinates y_1, y_2, y_3 of the angular points satisfy the equations

$$3(y_2 + y_3)(y_3 + y_1)(y_1 + y_2) + 32a^2(y_1 + y_2 + y_3) = 0,$$

$$(y_1 + y_2 + y_3)^2 + y_2y_3 + y_3y_1 + y_1y_2 + 48a^2 = 0;$$

and that its centre lies on the parabola $9y^2 = 4a(x - 8a)$.

918. An equilateral triangle circumscribes a parabola: prove that the ordinates y_1, y_2, y_3 of its angular points satisfy the equations

$$(y_1 + y_2 + y_3)^2 = 4(y_2y_3 + y_3y_1 + y_1y_2 + 3a^2),$$

$$4a^2(y_1 + y_2 + y_3) + 3(y_2 + y_3 - y_1)(y_3 + y_1 - y_2)(y_1 + y_2 - y_3) = 0.$$

[The simplest way of expressing the conditions for an equilateral triangle is to equate the co-ordinates of the centroid and of the centre of perpendiculars.]

919. The pole O is taken of a chord PQ of a parabola: prove that the perpendiculars from O, P, Q on any tangent to the parabola are in geometric progression.

920. Four fixed tangents are drawn to a parabola, and from the angular points taken in order of a quadrangle formed by them are let fall perpendiculars p_1, p_2, p_3, p_4 on any other tangent: prove that

$$p_1p_3 = p_2p_4.$$

921. The perpendiculars from the angular points of a triangle ABC, whose sides touch a parabola, on the directrix are p, q, r, and on any other tangent are x, y, z: prove that

$$\frac{p \tan A}{x(y-z)} = \frac{q \tan B}{y(z-x)} = \frac{r \tan C}{z(x-y)}.$$

[Of course the algebraical sign must be regarded.]

922. The distance of the middle point of any one of the three diagonals of a quadrilateral from the axis of the inscribed parabola is one-fourth of the sum of the distances of the four points of contact from the axis.

923. Through the point where the tangent to a given parabola at P meets the axis is drawn a straight line meeting the parabola in Q, Q' which divides the ordinate at P in a given ratio: prove that PQ, PQ' will both touch a fixed parabola having the same vertex and axis as the given one.

[If the ratio of the part cut off to the whole ordinate be $k : 1$, the ratio of the latus rectum of the envelope to that of the given parabola will be $2k : 1+k$.]

924. Two equal parabolas have axes in one straight line, and from any point on the outer tangents are drawn to the inner: prove that they will intercept a constant length on any fixed tangent to the inner equal to half the chord of the outer intercepted on the fixed tangent.

925. A tangent is drawn to the circle of curvature at the vertex and the ordinates of the points where it meets the parabola are y_1, y_2: prove that

$$\frac{1}{y_1} \sim \frac{1}{y_2} = \frac{1}{a}.$$

926. On the diameter through a point O of a parabola are taken points P, P' so that the rectangle OP, OP' is constant: prove that the four points of intersection of the tangents drawn from P, P' lie on two fixed straight lines parallel to the tangent at O and equidistant from it.

927. The points P, P' are taken on the diameter through a fixed point O of a parabola so that the mid-point of PP' is fixed: prove that the tangents drawn from P, P' to the parabola will intersect on another parabola of half the linear dimensions.

[In general if tangents to the parabola $y^2 = 4ax$ divide a given segment LL' on the axis of x harmonically, their point of intersection lies on the conic

$$(x-c)^2 + \frac{cy^2}{a} = m^2,$$

where $OL + OL' = 2c$ and $LL' = 2m$.]

928. A chord of a parabola passes through a point on the axis (outside the parabola) at a distance from the vertex equal to half the latus rectum: prove that the normals at its extremities intersect on the parabola.

929. The sum of the angles which three normals drawn from one point make with the axis exceeds the angle which the focal distance of the point makes with the axis by a multiple of π.

930. Normals are drawn at the extremities of any chord passing through a fixed point on the axis of a parabola: prove that their point of intersection lies on a fixed parabola.

[More generally, if a chord pass through (X, Y), the locus of the point of intersection of the normals at its ends is the parabola

$$2\{2ay + Y(x - X - 2a)\}^2 + (Y^2 - 4aX)\{Yy + 2X(x - X - 2a)\} = 0.]$$

931. Two normals to a parabola meet at right angles, and from the foot of the perpendicular let fall from their point of intersection on the axis is measured towards the vertex a distance equal to one-fourth of the latus rectum: prove that the straight line joining the end of this distance with the point of intersection of the normals is also a normal.

932. Two equal parabolas have their axes coincident but their vertices separated by a distance equal to the latus rectum; through the centres of curvature at the vertices are drawn chords PQ, $P'Q'$ equally inclined in opposite senses to the axis, P, P' being on the same side of the axis: prove that (1) PQ', $P'Q$ are normals to the outer parabola; (2) their common point R' lies on the inner; (3) the normals at P', Q', R' meet in a point which lies on a third equal parabola.

933. From a point O are drawn three normals OP, OQ, OR and two tangents OL, OM to a parabola: prove that the latus rectum

$$= 4\frac{OP \,.\, OQ \,.\, OR}{OL \,.\, OM}.$$

934. The normals to the parabola $y^2 = 4ax$ at points P, Q, R meet in the point (X, Y): prove that the co-ordinates of the centre of perpendiculars of the triangle PQR are $X - 6a$, $-\frac{1}{2}Y$.

935. Three tangents are drawn to a parabola so that the sum of the angles which they make with the axis is π: prove that the circle round the triangle formed by the tangents touches the axis (in the focus of course).

936. The locus of a point from which two normals can be drawn making complementary angles with the axis is the parabola

$$y^2 = a(x - a).$$

937. Two (equal) parabolas have the same latus rectum and from any point of either two tangents are drawn to the other: prove that the centres of two of the four circles which touch the sides of the triangle formed by the tangents and their chord of contact lie on the parabola to which the tangents are drawn. Also, if two points be taken conjugate to each other with respect to one of the parabolas and from them tangents drawn to the other at points L, M; N, O, respectively, the rectangle under the perpendiculars from any point of the second parabola on the chords LN, MO will be equal to that under the perpendiculars from the same point on MN, LO.

938. Prove that the two parabolas $y^2 = ax$, $y^2 = 4a(x + a)$ are so related that if a normal to the latter meet the former in P, Q and A be the vertex of the former, either AP or AQ is perpendicular to the normal.

939. The normals at three points of the parabola $y^2 = 4ax$ meet in the point (X, Y): prove that the equation of the circle through the three points is

$$2(x^2 + y^2) - 2x(X + 2a) - yY = 0;$$

and that of the circle round the triangle formed by the three tangents is

$$(x - a)(x - 2a + X) + y(y + Y) = 0.$$

[Hence if O be the point from which the normals are drawn and OO' be bisected by S, SO' is a diameter of the circle round the triangle formed by the tangents.]

940. In the two parabolas $y^2 = 2c(x \pm c)$ a tangent drawn to one meets the other in two points and on the chord intercepted as diameter is described a circle: prove that this circle will touch the second parabola.

941. On a focal chord as diameter is described a circle cutting the parabola again in P, Q: prove that the circle PSQ will touch the parabola.

942. On a chord of a given parabola as diameter a circle is described and the other common chord of the circle and parabola is conjugate to the former with respect to the parabola: prove that each chord touches a fixed parabola.

943. Two tangents OL, OM to a parabola meet the tangent at the vertex in P, Q: prove that

$$\pm PQ = OL \cos QPL = OM \cos PQM.$$

944. Two parabolas have a common focus and direction of axis, a chord QVQ' of the outer is bisected by the inner in V, VP parallel to the axis meets the outer in P: prove that QV is a mean proportional between the tangents drawn from P to the inner.

945. Prove that the parabolas

$$y^2 = 4ax, \quad y^2 + 4cy + 4ax = 8a^2$$

cut each other at right angles in two points and that each passes through the centre of curvature at the vertex of the other. If the origin be taken at the mid-point of their common chord their equations will be

$$y^2 - c^2 - 4a^2 = \pm(2cy + 4ax).$$

[The general orthogonal trajectory of the system of parabolas

$$y^2 + 2\lambda y + 4ax = 8a^2$$

for different values of λ is $\quad y^2 - 4ax = Ce^{\frac{x}{2a}}$.]

946. On a focal chord PSQ of a parabola are taken points p, q on opposite sides of S so that $qS \cdot Sp = QS \cdot SP$, and another parabola is drawn with parallel axis and passing through q, p: prove that the common chord of the two parabolas will pass through S.

947. A chord PQ of a parabola meets the axis in T, U is the mid-point and O the pole of the chord, a normal to PQ through U meets the axis in G and OK is perpendicular from O on the directrix: prove that SO is parallel to TK and SK to GU.

948. Through each point of the straight line $x = my + h$ is drawn a chord of the parabola $y^2 = 4ax$, which is bisected in the point: prove that this chord touches the parabola

$$(y - 2am)^2 = 8a(x - h).$$

949. Prove that the triangle formed by three normals to a parabola is to the triangle formed by the three corresponding tangents in the ratio

$$(t_1 + t_2 + t_3)^2 : 1,$$

where t_1, t_2, t_3 are the tangents of the angles which the normals make with the axis.

950. Three tangents to the parabola $y^2 = 4a(x + a)$ make angles α, β, γ with the axis: prove that the co-ordinates of the centre of the circle circumscribing the triangle formed by them are

$$+\tfrac{1}{2}a\frac{\sin(\alpha+\beta+\gamma)}{\sin\alpha\sin\beta\sin\gamma}, \quad -\tfrac{1}{2}a\frac{\cos(\alpha+\beta+\gamma)}{\sin\alpha\sin\beta\sin\gamma}.$$

951. Three confocal parabolas have their axes in A. P., a normal is drawn to the outer and a tangent perpendicular to this normal to the inner: prove that the chord which the middle parabola intercepts on this tangent is bisected in the point where it meets the normal.

952. Two normals OP, OQ are drawn to a parabola, and α, β are the angles which the tangents at P, Q make with the axis: prove that

$$\frac{OP}{\sin\alpha + \sin\beta\cos(\alpha-\beta)} = \frac{OQ}{\sin\beta + \sin\alpha\cos(\beta-\alpha)} = \frac{a}{\sin^2\alpha\sin^2\beta}.$$

953. From any point on the outer of two equal parabolas with a common axis tangents are drawn to the inner: prove that the part of the axis intercepted bears to the ordinate of the point from which the tangents are drawn a constant ratio equal to that which the chord intercepted on the tangent at the vertex of the inner parabola bears to the semilatus rectum.

954. Prove that the common tangent to the two parabolas

$$x^2\cos^2\alpha = 4a(x\cos\alpha + y\sin\alpha),$$
$$y^2\sin^2\alpha = -4a(x\cos\alpha + y\sin\alpha),$$

subtends a right angle at the origin.

955. Two parabolas have a common focus S and axes in the same straight line, and from a point P on the outer are drawn two tangents PQ, PQ' to the inner: prove that the ratio

$$\cos\tfrac{1}{2}QPQ' : \cos\tfrac{1}{2}ASP$$

is constant, A being the vertex of either parabola.

956. A parabola circumscribes a triangle ABC and its axis makes with CB an angle θ (measured from CB towards CA): prove that its latus rectum is

$$2R \sin\theta \sin(C-\theta)\sin(B+\theta);$$

and that for an inscribed parabola the latus rectum is four times as large.

957. A triangle ABC is inscribed in a given parabola and the focus is the centre of perpendiculars of the triangle: prove that

$$(1-\cos A)(1-\cos B)(1-\cos C) = 2\cos A\cos B\cos C;$$

and that each side of the triangle touches a fixed circle which passes through the focus and whose diameter is equal to the latus rectum.

958. A parabola is drawn touching the sides AB, AC of a triangle ABC at B, C and passing through the centre of perpendiculars: prove that the centre of perpendiculars is the vertex of the parabola and that the centre of curvature at the vertex is a point on BC.

959. The latus rectum of a parabola which touches the sides of a triangle ABC and whose focus is S is equal to $SA \,.\, SB \,.\, SC \div R^2$.

960. A chord LL' of a given circle has its mid-point at O and its pole at P; a parabola is drawn with its focus at O and its directrix passing through P: prove that the tangent to this parabola at any point where it meets the circle passes through either L or L'.

961. A triangle, self-conjugate to a given parabola, has one angular point O given: prove that the circle circumscribing the triangle passes through another fixed point Q such that OQ is parallel to the axis and bisected by the directrix.

962. A triangle is inscribed in a parabola, its sides are at distances x, y, z from the focus and subtend at the focus angles θ, ϕ, ψ (always measured in the same sense so that the sum is 2π): prove that

$$\frac{\sin\theta}{x^2} + \frac{\sin\phi}{y^2} + \frac{\sin\psi}{z^2} = \frac{\sin\theta + \sin\phi + \sin\psi + 2\tan\frac{\theta}{2}\tan\frac{\phi}{2}\tan\frac{\psi}{2}}{l^2};$$

where $2l$ is the latus rectum.

963. Two points L, L' are taken on the directrix of a parabola conjugate to each other with respect to the parabola: prove that any other conic through LSL' having its focus on LL' will have for the corresponding directrix a tangent to the parabola.

964. An ellipse of given excentricity $\frac{2\lambda}{1+\lambda^2}$ is described passing through the focus of a given parabola $y^2 = 4ax$ and with its own foci on

the parabola: prove that its major axis touches one of the parabolas, confocal with the given parabola,

$$y^2 = 4a(1-\lambda^2)(x-a\lambda^2),$$

and that its minor axis is normal to one of the two

$$y^2 = 4a(1+\lambda^2)(x+a\lambda^2).$$

965. An ellipse is described with its focus at the vertex of a given parabola; its minor axis and the distance between its foci are each double of the latus rectum of the parabola: prove that the pole with respect to the ellipse of that ordinate of the parabola with which the minor axis in one position coincides always lies on the parabola and also on an equal parabola whose axis coincides with that of the ellipse.

966. A parabola touches the sides of a triangle ABC in the points A', B', C' and O is the point of concourse of AA', BB', CC': prove that, under a certain convention as to sign,

$$OA \operatorname{cosec} BOC + OB \operatorname{cosec} COA + OC \operatorname{cosec} AOB = 0:$$

also, if P be a point such that PA' bisects the angle BPC and PB', PC' respectively bisect the external angles between PC, PA, and PA, PB,

$$PA = PB + PC.$$

967. A triangle circumscribes the circle $x^2 + y^2 = a^2$, and two angular points lie on the circle $(x-2a)^2 + y^2 = 2a^2$: prove that the third angular point lies on the parabola $y^2 = a(x - \frac{3}{4}a)$. Prove also that the three curves have two real and two impossible common tangents.

968. Two parabolas have a common focus, axes inclined at an angle α, and are such that triangles can be inscribed in one whose sides touch the other: prove that $l_2 = 2l_1(1+\cos\alpha)$, l_1, l_2 being their latera recta.

969. A circle is described with its centre at a point P of a parabola and its radius equal to twice the normal at P: prove that triangles can be inscribed in the parabola whose sides touch the circle.

970. Two parabolas A, B have their axes parallel and the latus rectum of A is four times that of B: prove that triangles can be inscribed in B whose sides touch A. If the axes be in the same straight line the normals to B at the angular points of such a triangle will all meet in one point, as will the normals to A at the points of contact, and the loci of these points of concourse are straight lines perpendicular to the axis.

[Taking the equations of the parabolas to be

$$y^2 = 16ax, \quad y^2 = 4a(x+h),$$

the straight lines will be $x = 2a$, $x = 8a + h$.]

971. The circle of curvature of a parabola at P meets the parabola again in Q and QL, QM are drawn tangents to the circle and parabola at Q, each terminated by the other curve: prove that when LM subtends a right angle at P, PL is parallel to the axis, and that this is the case when the focal distance of P is one-third of the latus rectum.

972. If the tangent at P make an angle θ with the axis, the tangent to the circle at Q will make an angle $\pi - 3\theta$ with the axis; also the angle between the tangent at P and the other common tangent to the parabola and circle will be $2\tan^{-1}(\frac{1}{2}\tan\theta)$, and if ϕ be the angle which this common tangent makes with the axis

$$\tan\frac{\phi}{2}\tan\frac{3\theta}{2} = 1.$$

973. From a point O on the normal at P are drawn two tangents to a parabola making angles α, β with OP: prove that the radius of curvature at P is $2OP\tan\alpha\tan\beta$.

974. The normal at a point of a parabola makes an angle θ with the axis: prove that the length of the chord intercepted on the normal bears to the latus rectum the ratio $1 : \sin\theta\cos^2\theta$, and the length of the common chord of the parabola and the circle of curvature at the point bears to the latus rectum the ratio $2\sin\theta : \cos^2\theta$.

975. At a point P of a parabola is drawn a circle equal to the circle of curvature and touching the parabola externally; the other common tangents to this circle and the parabola intersect in Q: prove that, if QK be let fall perpendicular on the directrix,

$$4\frac{SQ - QK}{SQ + QK} = \frac{AS}{AS + SP}.$$

III. *Ellipse referred to its axes.*

[The equation of the ellipse in the following questions is always supposed to be $\frac{x^2}{a^2} + \frac{y^2}{b^2} = 1$, and the axes to be rectangular, unless otherwise stated. The point whose excentric angle is θ is called the point θ. The excentricity is denoted by e. The tangent and normal at the point θ are respectively

$$\frac{x}{a}\cos\theta + \frac{y}{b}\sin\theta = 1, \quad \frac{ax}{\cos\theta} - \frac{by}{\sin\theta} = a^2 - b^2;$$

the chord through the two points α, β is

$$\frac{x}{a}\cos\frac{\alpha+\beta}{2} + \frac{y}{b}\sin\frac{\alpha+\beta}{2} = \cos\frac{\alpha-\beta}{2};$$

and the intersection of tangents at α, β, (the pole of this chord)

$$\frac{a\cos\frac{\alpha+\beta}{2}}{\cos\frac{\alpha-\beta}{2}}, \quad \frac{b\sin\frac{\alpha+\beta}{2}}{\cos\frac{\alpha-\beta}{2}}.$$

The polar of a point (X, Y) is $\frac{xX}{a^2}+\frac{yY}{b^2}=1$; and the equation of the two tangents from (X, Y) is

$$\left(\frac{x^2}{a^2}+\frac{y^2}{b^2}-1\right)\left(\frac{X^2}{a^2}+\frac{Y^2}{b^2}-1\right)=\left\{\frac{xX}{a^2}+\frac{yY}{b^2}-1\right\}^2.$$

It follows from the equation of the tangent that if the equation of any straight line be $lx+my=1$, and l, m satisfy the equation $a^2l^2+b^2m^2=1$, the straight line touches the ellipse $\frac{x^2}{a^2}+\frac{y^2}{b^2}=1$, a result often useful.

The equation of the tangent in the form

$$x\cos\theta+y\sin\theta=\sqrt{a^2\cos^2\theta+b^2\sin^2\theta}$$

may be occasionally employed with advantage.

The points α, β will be extremities of conjugate diameters if $\alpha\sim\beta=\frac{\pi}{2}$. Any two points are called conjugate if either lies on the polar of the other, and any two straight lines if either passes through the pole of the other.

If (X, Y) be the pole of the chord through (α, β) it will be found that

$$\frac{\sin\alpha\sin\beta}{1-\frac{X^2}{a^2}}=\frac{\cos\alpha\cos\beta}{1-\frac{Y^2}{b^2}}=\frac{\sin\alpha+\sin\beta}{\frac{2Y}{b}}=\frac{\cos\alpha+\cos\beta}{\frac{2X}{a}}=\frac{1}{\frac{X^2}{a^2}+\frac{Y^2}{b^2}},$$

which enable us to find the locus of (X, Y) when α, β are connected by some fixed equation. Thus, "If a triangle be circumscribed about an ellipse $\frac{x^2}{a^2}+\frac{y^2}{b^2}=1$ and two angular points lie on the ellipse $\frac{x^2}{a'^2}+\frac{y^2}{b'^2}=1$, to find the locus of the third angular point."

If α, β, γ be the three points of contact and (α, β), (α, γ) be the pairs of points whose tangents intersect on the second ellipse, we have

$$\frac{a^2}{a'^2}\cos^2\frac{\alpha+\beta}{2}+\frac{b^2}{b'^2}\sin^2\frac{\alpha+\beta}{2}=\cos^2\frac{\alpha-\beta}{2},$$

and the like equation with γ in place of β. Hence β, γ are the two roots of the equation

$$A\cos\alpha\cos\theta+B\sin\alpha\sin\theta=C;$$

where $\quad A\equiv-\frac{a^2}{a'^2}+\frac{b^2}{b'^2}+1,\ B\equiv\frac{a^2}{a'^2}-\frac{b^2}{b'^2}+1,\ C\equiv\frac{a^2}{a'^2}+\frac{b^2}{b'^2}-1;$

and we have therefore

$$\frac{\cos\frac{\beta+\gamma}{2}}{A\cos\alpha}=\frac{\sin\frac{\beta+\gamma}{2}}{B\sin\alpha}=\frac{\cos\frac{\beta-\gamma}{2}}{C},$$

so that the co-ordinates of the third angular point are

$$\frac{Aa}{C}\cos\alpha,\ \frac{Bb}{C}\sin\alpha,$$

so that its locus is the ellipse

$$\frac{x^2}{A^2a^2}+\frac{y^2}{B^2b^2}=\frac{1}{C^2}.$$

This locus will be found to coincide with the second ellipse if $\frac{a}{a'}\pm\frac{b}{b'}\pm 1=0$, and if we so choose the signs of a', b' that the relation is $\frac{a}{a'}+\frac{b}{b'}=1$, $A\equiv 2\frac{b}{b'}$, $B=2\frac{a}{a'}$, $C=-2\frac{ab}{a'b'}$, so that the co-ordinates of the third point are $-a'\cos\alpha$, $-b'\sin\alpha$, or its excentric angle is $\pi+\alpha$, and similarly the excentric angles of the other points are $\pi+\beta$, $\pi+\gamma$. Hence the ellipses

$$\frac{x^2}{a^2}+\frac{y^2}{b^2}=1,\quad \frac{x^2}{a'^2}+\frac{y^2}{b'^2}=1,$$

will be such that an infinite number of triangles can be inscribed in the second whose sides touch the first, if with any signs to a', b' the relation $\frac{a}{a'}+\frac{b}{b'}=1$ is satisfied and the excentric angle of any corner of such a triangle exceeds that of the corresponding point of contact by π.

If this condition be not satisfied the two given ellipses and the locus will be found to have four common tangents real or impossible.

Again, for the reciprocal problem, "If a triangle be inscribed in the ellipse $\frac{x^2}{a'^2}+\frac{y^2}{b'^2}=1$ and two of its sides touch the ellipse $\frac{x^2}{a^2}+\frac{y^2}{b^2}=1$, to find the envelope of the third side."

Taking α, β, γ for the angular points and (α,β), (α,γ) for the sides which touch the second ellipse, we have

$$\frac{a^2}{a'^2}\cos^2\frac{\alpha+\beta}{2}+\frac{b^2}{b'^2}\sin^2\frac{\alpha+\beta}{2}=\cos^2\frac{\alpha-\beta}{2},$$

and a like equation with γ in place of β. Hence, as before,

$$\frac{C^2}{A^2}\frac{\cos^2\frac{\beta+\gamma}{2}}{\cos^2\frac{\beta-\gamma}{2}}+\frac{C^2}{B^2}\frac{\sin^2\frac{\beta+\gamma}{2}}{\cos^2\frac{\beta-\gamma}{2}}=1,$$

which, since the third side is

$$\frac{x}{a'}\frac{\cos\frac{\beta+\gamma}{2}}{\cos\frac{\beta-\gamma}{2}}+\frac{y}{b'}\frac{\sin\frac{\beta+\gamma}{2}}{\cos\frac{\beta-\gamma}{2}}=1,$$

proves that the envelope is the ellipse

$$A^2\frac{x^2}{a'^2}+B^2\frac{y^2}{b'^2}=C^2,$$

which coincides with the second ellipse if $\frac{a}{a'} \pm \frac{b}{b'} \pm 1 = 0$, or if with any signs of a', b', $\frac{a}{a'} + \frac{b}{b'} = 1$. The excentric angles of the points of contact will be $\alpha - \pi$, $\beta - \pi$, $\gamma - \pi$ (or $\alpha + \pi$, $\beta + \pi$, $\gamma + \pi$, which are practically the same). If this condition be not satisfied the three conics intersect in the same four points real or impossible.

The relations between the excentric angles corresponding to normals drawn from (X, Y) may be found from the equation

$$\frac{aX}{\cos\theta} - \frac{bY}{\sin\theta} = a^2 - b^2,$$

a biquadratic whose roots give the excentric angles of the points to which normals can be drawn from (X, Y). If $\tan\frac{\theta}{2} \equiv Z$, this equation becomes

$$Z^4 bY + 2Z^3(aX + a^2 - b^2) + 2Z(aX - a^2 + b^2) - bY = 0.$$

This equation having four roots, there must be two relations independent of X, Y between the roots, as is also obvious geometrically. These relations are manifest on inspection of the equation; they are

$$Z_1Z_2Z_3Z_4 = -1,\ Z_2Z_3 + \ldots = 0;$$

and the relation between Z_1, Z_2, Z_3 is therefore

$$Z_2Z_3 + Z_3Z_1 + Z_1Z_2 = \frac{1}{Z_2Z_3} + \frac{1}{Z_3Z_1} + \frac{1}{Z_1Z_2},$$

which is equivalent to $\sin(\beta + \gamma) + \sin(\gamma + \alpha) + \sin(\alpha + \beta) = 0$, if α, β, γ be the corresponding values of θ.

Since $1 - (Z_2Z_3 + \ldots) + Z_1Z_2Z_3Z_4 = 0$, it follows that

$$\tan\frac{\alpha + \beta + \gamma + \delta}{2} = \infty,$$

or $\alpha + \beta + \gamma + \delta$ is an odd multiple of π.

The following is another method of investigating the same question. If the normal at (x, y) to the ellipse pass through (X, Y),

$$a^2xY - b^2yX = (a^2 - b^2)xy. \quad \text{(A)}$$

Now if $\frac{lx}{a} + \frac{my}{b} = 1$, and $\frac{l'x}{a} + \frac{m'y}{b} = 1$, be the equations of two lines joining the four points to which normals can be drawn from (X, Y), the equation

$$\frac{x^2}{a^2} + \frac{y^2}{b^2} - 1 + \lambda\left(\frac{lx}{a} + \frac{my}{b} - 1\right)\left(\frac{l'x}{a} + \frac{m'y}{b} - 1\right) = 0$$

can be made to coincide with (A). The identification of the two gives

$$\lambda = 1, \quad ll' + 1 = 0, \quad mm' + 1 = 0,$$

whence it follows that normals at the points where the two straight lines

$$\frac{lx}{a} + \frac{my}{b} = 1, \quad \frac{x}{al} + \frac{y}{bm} = -1$$

meet the ellipse all meet in a point. The point is given by

$$\frac{ax}{l(1 - m^2)} = \frac{-by}{m(1 - l^2)} = \frac{a^2 - b^2}{l^2 + m^2}.$$

If α, β be the two points on the former, and γ one of the points on the latter,

$$l = \frac{\cos\frac{\alpha+\beta}{2}}{\cos\frac{\alpha-\beta}{2}}, \quad m = \frac{\sin\frac{\alpha+\beta}{2}}{\cos\frac{\alpha-\beta}{2}}, \quad \frac{\cos\gamma}{l} + \frac{\sin\gamma}{m} + 1 = 0,$$

whence

$$\sin\frac{\alpha+\beta}{2}\cos\frac{\alpha+\beta}{2} + \left(\cos\gamma\sin\frac{\alpha+\beta}{2} + \sin\gamma\cos\frac{\alpha+\beta}{2}\right)\cos\frac{\alpha-\beta}{2} = 0,$$

or $$\sin(\beta+\gamma) + \sin(\gamma+\alpha) + \sin(\alpha+\beta) = 0.$$

The equation formed from this by replacing γ by δ must also hold, whence

$$\frac{\cos\frac{\gamma+\delta}{2}}{\sin\alpha+\sin\beta} = \frac{\sin\frac{\gamma+\delta}{2}}{\cos\alpha+\cos\beta} = \frac{\cos\frac{\gamma-\delta}{2}}{\sin(\alpha+\beta)},$$

and $\tan\frac{\gamma+\delta}{2} = \cot\frac{\alpha+\beta}{2}$, or $\alpha+\beta+\gamma+\delta$ is an odd multiple of $\frac{\pi}{2}$.]

976. A chord AP is drawn from the vertex of an ellipse of excentricity e, along PA is taken a length PR equal to $PA \div e^2$, and RQ is drawn at right angles to the chord to meet the straight line through P parallel to the axis: the locus of Q is a straight line perpendicular to the axis. Similarly if BP be a chord through a vertex on the minor axis and along BP be taken a length BR equal to $BP \div e^2$, and RQ be drawn at right angles to BR to meet the straight line through P parallel to the minor axis, the locus of Q is a straight line parallel to the major axis.

[The equations of the loci, with the centre as origin, are

$$\frac{x}{a} = \pm\frac{a^2+b^2}{a^2-b^2}, \quad \frac{y}{b} = \pm\frac{a^2+b^2}{a^2-b^2}.]$$

977. Tangents drawn from a point P to a given ellipse meet a given tangent whose point of contact is O in Q, Q': prove that if the distance of P from the given tangent be constant, the rectangle OQ, OQ' will be constant. Also if the length QQ' be given the locus of Q will be a conic having contact of the third order with the given ellipse at the other end of the diameter through O; and the conic will be an ellipse, parabola, or hyperbola according as the given length QQ' is less than, equal to, or greater than the diameter parallel to the given tangent.

978. Two ellipses have the same major axis and an ordinate NPQ is drawn, the tangent at P meets the other ellipse in points the lines joining which to either extremity of the major axis meet the ordinate in L, M: prove that NP is a harmonic and NQ a geometric mean between NL, NM.

979. The equation giving t the length of the tangent from (X, Y) to the ellipse $\frac{x^2}{a^2} + \frac{y^2}{b^2} = 1$ is

$$\frac{X}{a^2}\left(\frac{1}{b^2} - \frac{U}{t^2}\right)^{\frac{1}{2}} + \frac{Y}{b^2}\left(\frac{U}{t^2} - \frac{1}{a^2}\right)^{\frac{1}{2}} + \frac{U}{t}\left(\frac{1}{b^2} - \frac{1}{a^2}\right)^{\frac{1}{2}} = 0,$$

where $U \equiv \frac{X^2}{a^2} + \frac{Y^2}{b^2} - 1$.

980. The major and minor axes of an ellipse being AA', BB', another similar ellipse is described with BB' for its major axis, P is any point on the former ellipse and L the centre of perpendiculars of the triangle PBB': prove that L will lie on the second ellipse and that the normals at L, P will intersect on another ellipse whose minor axis is $4b$, and major axis $2\frac{a^2 + b^2}{a}$.

981. A given ellipse subtends a right angle at O, and OO' is drawn perpendicular to and bisected by the polar of O: prove that OO' is divided by the axes in a constant ratio, CO' is a constant length, the middle point of OO' is the point of contact of the polar of O with its envelope, and the rectangle under the perpendiculars from O, C on the polar of O is constant.

982. The rectangle under the perpendiculars let fall on a straight line, from its pole with respect to a given ellipse and from the centre of the ellipse, is constant $(=\lambda)$: prove that the straight line touches the confocal $\frac{x^2}{a^2+\lambda} + \frac{y^2}{b^2+\lambda} = 1$.

983. The rectangle under the perpendiculars drawn to the normal at a point P from the centre and from the pole of the normal is equal to the rectangle under the focal distances of P.

984. The sum or the difference of the rectangles under the perpendiculars upon any straight line (1) from its pole with respect to a given ellipse and from the centre, (2) from the foci of the given ellipse, is constant $(=b^2)$; the sum when the straight line intersects the ellipse in real points, otherwise the difference; or with proper regard to sign in both cases, the rectangle (2) always exceeds the rectangle (1) by b^2.

985. Through a point are drawn two straight lines at right angles to each other and conjugate with respect to a given ellipse: prove that the arithmetical difference between the rectangles under the perpendiculars on these lines each from the centre and from its own pole is equal to the sum of the rectangles under the focal perpendiculars, and to the rectangle under the focal distances of the point.

986. On the focal distances of any point of an ellipse as diameters are described two circles: prove that the excentric angle of the point is equal to the angle which a common tangent to the circles makes with the minor axis.

987. The ordinate NP at a point P of an ellipse is produced to Q so that $NQ : NP :: CA : CN$, and from Q two tangents are drawn to the ellipse: prove that they intercept on the minor axis produced a length equal to the minor axis.

988. A circle of radius r is described with its centre on the minor axis of a given ellipse at a distance er from the centre: prove that the tangent to this circle at a point where it meets the ellipse will touch the minor auxiliary circle.

989. A point P on the auxiliary circle is joined to the ends of the major axis and the joining lines meet the ellipse again in Q, Q': prove that the equation of QQ' is

$$(a^2 + b^2)\, y \sin\theta + 2b^2 x \cos\theta = 2ab^2,$$

where θ is the angle ACP, and if the ordinate to P meet QQ' in R, R is the point of contact of QQ' with its envelope.

990. From a point P of an ellipse two tangents are drawn to the circle on the minor axis: prove that these tangents will meet the diameter at right angles to CP in points lying on two fixed straight lines parallel to the major axis.

991. Two tangents are drawn to an ellipse from a point P: prove that the angle between them is

$$\cos^{-1}\left(\frac{CP^2 - AC^2 - BC^2}{SP \,.\, S'P}\right).$$

992. If p, q be the lengths of two tangents at right angles to each other

$$\frac{4\,(a^2 + b^2)^3}{p^2 + q^2} = \left\{a^2 + b^2 + a^2b^2\left(\frac{1}{p^2} + \frac{1}{q^2}\right)\right\}^2.$$

993. If p, q be the lengths of two tangents and $2ma$, $2mb$ the axes of the concentric similar and similarly situated ellipse drawn through their point of intersection

$$\frac{p^2 + q^2}{a^2 + b^2} = 4\left(\frac{m^2 - 1}{m^2}\right)^2.$$

994. The lengths of two tangents drawn to an ellipse from a point on one of the equal conjugate diameters are p, q: prove that

$$(a^2 + b^2)\,(p^2 - q^2)^2\,(p^2 + q^2 + a^2 + b^2)^2 = 4\,(p^2 + q^2)^3\,(a^2 - b^2)^2.$$

995. If p, q be the lengths of two tangents drawn from a point on the hyperbola $\frac{x^2}{a} - \frac{y^2}{b} = a - b$ to the ellipse $\frac{x^2}{a^2} + \frac{y^2}{b^2} = 1$, and r the central distance of the point, then will

$$pq = r^2 - a^2 + ab - b^2,\quad p - q = 2\,(a - b)\,\frac{pq}{ab + pq},$$

and

$$(p + q)^2 = 4pq\,\frac{(pq + a^2)\,(pq + b^2)}{(pq + ab)^2}.$$

996. If two tangents be drawn from any point of the hyperbola $\frac{x^2}{a} - \frac{y^2}{b} = a - b$ to the ellipse $\frac{x^2}{a^2} + \frac{y^2}{b^2} = 1$, the difference of their lengths will be $2(a-b)\left(1 - \frac{ab}{r^2 - (a-b)^2}\right)$, where r is the central distance of the point: and if a parallelogram be inscribed in the hyperbola whose sides touch the ellipse and r_1, r_2 be the central distances of two adjacent angular points, then will

$$(r_1^2 - a^2 + ab - b^2)(r_2^2 - a^2 + ab - b^2) = a^2b^2;$$

the lengths of the sides of the parallelogram will be

$$\sqrt{r_1^2 + r_2^2 - (a-b)^2} \pm (a-b),$$

and the point of contact on any side will divide that side in the ratio

$$r_1^2 + ab - a^2 - b^2 : ab.$$

997. A circle is described on a chord of the ellipse lying on the straight line $p\frac{x}{a} + q\frac{y}{b} = 1$ as diameter: prove that the equation of the straight line joining the other two common points of the ellipse and circle is

$$p\frac{x}{a} - q\frac{y}{b} = \frac{a^2 + b^2}{a^2 - b^2}.$$

998. In an ellipse whose axes are in the ratio $\sqrt{2} + 1 : 1$, a circle whose diameter joins the ends of two conjugate diameters of the ellipse will touch the ellipse.

999. Normals to an ellipse at P, Q meet in O and CO, PQ are equally inclined to the axes: prove that the part of PQ intercepted between the axes is of constant length and that the other normals drawn from O will be at right angles to each other.

1000. If O be the point in the normal at P such that chords drawn through O subtend a right angle at P, and O' be the corresponding point for another point P', OO', PP' will be equally inclined to the axes and their lengths in a constant ratio.

1001. A circle is described having for diameter the part of the normal at P intercepted between the axes, and from any point on the tangent at P two tangents are drawn to this circle: prove that the chord of the ellipse which passes through the points of contact subtends a right angle at P.

1002. The normals at three points of an ellipse whose excentric angles are α, β, γ will meet in a point, if

$$\sin(\beta + \gamma) + \sin(\gamma + \alpha) + \sin(\alpha + \beta) = 0,$$

which is equivalent to

$$\tan\frac{\beta+\gamma}{2}\cot\alpha = \tan\frac{\gamma+\alpha}{2}\cot\beta = \tan\frac{\alpha+\beta}{2}\cot\gamma.$$

1003. If four normals to an ellipse meet in a point the sum of the corresponding excentric angles will be an odd multiple of π. Also two tangents drawn to the ellipse parallel to two chords through the four points will intersect on one of the equal conjugate diameters.

1004. The normals to the ellipse at the points where it is met by the straight lines

$$\frac{px}{a}+\frac{qy}{b}=1,\quad \frac{x}{ap}+\frac{y}{bq}=-1,$$

will all intersect in one point,

$$\left(\frac{ax}{p(1-q^2)}=\frac{by}{-q(1-p^2)}=\frac{a^2-b^2}{p^2+q^2}\right).$$

1005. From a point P of an ellipse PM, PN are let fall perpendicular upon the axes and MN produced meets the ellipse in Q, q: prove that the normals at Q, q intersect in the centre of curvature at p, Pp being a diameter.

1006. From a point O are drawn normals OP, OQ, OR, OS, and p, q, r, s are taken such that their co-ordinates are equal to the intercepts on the axes made by the tangents at P, Q, R, S: prove that p, q, r, s lie in one straight line. Also, if through the centre C be drawn straight lines at right angles to CP, CQ, CR, CS to meet the corresponding tangents, the four points so determined will lie in one straight line.

[If X, Y be the co-ordinates of O, the two straight lines will be

$$xX-yY=a^2-b^2,\quad a^2Xx+b^2Yy+a^2b^2=0.]$$

1007. The normals to an ellipse at P, Q, R, S meet in a point and the circles QRS, RSP, SPQ, PQR meet the ellipse again in the points P', Q', R', S' respectively: prove that the normals at P', Q', R', S' meet in a point.

1008. Normals are drawn at the extremities of a chord parallel to the tangent at the point α: prove that the locus of their intersection is the curve

$$2(ax\sin\alpha+by\cos\alpha)(ax\cos\alpha+by\sin\alpha)=(a^2-b^2)^2\sin 2\alpha\cos^2 2\alpha.$$

1009. Normals are drawn at the extremities of a chord drawn through a fixed point on the major axis: prove that the locus of their intersection is an ellipse whose axes are

$$\frac{a^2-b^2}{a}\left(c+\frac{1}{c}\right),\quad \frac{a^2-b^2}{b}\left(c-\frac{1}{c}\right);$$

the distance of the given point from the centre being ca.

1010. The normal at a point P of an ellipse meets the curve in Q and any other chord PP' is drawn; QP' and the straight line through P at right angles to PP' meet in R: prove that the locus of R is the straight line

$$\frac{x}{a}\cos\phi - \frac{y}{b}\sin\phi = \frac{a^2+b^2}{a^2-b^2},$$

where ϕ is the excentric angle of P. The part of any tangent intercepted between this straight line and the tangent at P is divided by the point of contact into two parts which subtend equal or supplementary angles at P.

1011. A chord PQ is normal at P, PP' is a chord perpendicular to the axis, the tangent at P' meets the axes in T, T', the rectangle $TCT'R$ is completed and CR meets PQ in U: prove that

$$CR \,.\, CU = a^2 - b^2.$$

1012. Along the normal at P is measured PO inwards equal to CD, and the other normals OL, OM, ON are drawn: prove that the parts of LP, MP, NP intercepted between the axes are equal to $a+b$; the tangents at L, M, N form a triangle whose circumscribed circle is fixed; and if r_1, r_2, r_3 be the lengths LP, MP, NP,

$$r_1 + r_2 + r_3 = 2\,(a-b),$$

$$r_2 r_3 + r_3 r_1 + r_1 r_2 = PO^2 - 4ab,$$

$$r_1 r_2 r_3\,(a-b) = 2ab\,(ab - PO^2);$$

any of the three r_1, r_2, r_3 being reckoned negative when drawn from a point whose distance from the major axis is greater than $\sqrt{\dfrac{b^3}{a+b}}$. Corresponding results may be found when PO is measured outwards, but in that case two of the normals will always be impossible unless $a > 2b$.

1013. The chord PQ is normal at P, and O is the pole of PQ: prove that

$$PQ = \frac{2a^2b^2}{p\,(a^2+b^2-p^2)},\quad PO = \frac{a^2b^2}{p\sqrt{(a^2-p^2)(p^2-b^2)}},$$

where p is the perpendicular from the centre on the tangent at P.

1014. Perpendiculars p_1, p_2 are let fall from the ends of a given chord on any tangent, and a perpendicular p_3 from the pole of the chord: prove that

$$p_1 p_2 = p_3^2 \cos^2\frac{\alpha-\beta}{2},$$

where α, β are the excentric angles of the given points.

1015. Two circles have each double contact with an ellipse and touch each other: prove that

$$4b^2 = \frac{(r_1 \pm r_2)^2}{e^2} + \frac{(r_1 \mp r_2)^2}{1-e^2},$$

r_1, r_2 being the radii; also the point of contact of the two circles is equidistant from the chords of contact with the ellipse.

[Only the upper sign applies when the circles are real; the corresponding equation for the hyperbola is formed by putting $-b^2$ for b^2, as usual, when the circles touch only one branch, but for circles touching both branches the equation is

$$\frac{(r_1 + r_2)^2}{e^2} - (r_1 - r_2)^2 = \frac{4a^2}{e^2-1}.]$$

1016. Two ellipses have common foci S, S', and from a point P on the outer are drawn two tangents PQ, PQ' to the inner: prove that $\cos\frac{QPQ'}{2} : \cos\frac{SPS'}{2}$ is a constant ratio.

1017. The sides of a parallelogram circumscribing an ellipse are parallel to conjugate diameters: prove that the rectangle under the perpendiculars let fall from two opposite angles on any tangent is equal to the rectangle under those from the other two angles.

1018. The diagonals of a quadrilateral circumscribing an ellipse are aa', bb', cc', and from b, b', c, c' are let fall perpendiculars p_1, p_2, p_3, p_4 on any tangent to the ellipse: prove that the ratio $p_1p_2 : p_3p_4$ is constant and equal to λ, where

$$\frac{\lambda+1}{\lambda-1} = \frac{\frac{x_1x_2}{a^2} + \frac{y_1y_2}{b^2} + 1}{\left(\frac{x_1^2}{a^2} + \frac{y_1^2}{b^2} - 1\right)^{\frac{1}{2}} \left(\frac{x_2^2}{a^2} + \frac{y_2^2}{b^2} - 1\right)^{\frac{1}{2}}},$$

and (x_1, y_1), (x_2, y_2) are the points a, a'. If the points of contact of the tangents from b be L, L', from b' be M, M', from c be L, M', and from c' be L', M, the value of λ is equal to the ratio of $[LL'M'M]$ at any point of the ellipse to its value at the centre.

1019. Prove that the equation

$$\left\{\frac{x}{a}\cos(\alpha-\beta) + \frac{y}{b}\sin(\alpha-\beta) - 1\right\}\left\{\frac{x}{a}\cos(\alpha+\beta) + \frac{y}{b}\sin(\alpha+\beta) - 1\right\}$$
$$= \left\{\frac{x}{a}\cos\alpha + \frac{y}{b}\sin\alpha - \cos\beta\right\}^2$$

is true at any point of the ellipse $\frac{x^2}{a^2} + \frac{y^2}{b^2} = 1$; and hence that the locus of a point from which if two tangents be drawn to the ellipse the centre of the circle inscribed in the triangle formed by the two tangents and the chord of contact shall lie on the ellipse is the confocal

$$\frac{x^2}{a^2} - \frac{y^2}{b^2} = \frac{a^2 - b^2}{a^2 + b^2}.$$

1020. Two tangents are drawn to an ellipse from a point (X, Y): prove that the rectangle under the perpendiculars from any point of the ellipse on the tangents bears to the square on the perpendicular from the same point on the chord of contact the ratio $1 : \lambda$; where

$$(\lambda^2 - 1)\left(\frac{X^2}{a^4} + \frac{Y^2}{b^4}\right)^2 = \left(\frac{1}{b^4} - \frac{1}{a^4}\right)\left(\frac{X^2}{a^2} + \frac{Y^2}{b^2} - 1\right)\left(\frac{X^2}{a^2} - \frac{Y^2}{b^2} - \frac{a^2 - b^2}{a^2 + b^2}\right).$$

1021. Four points A, B, C, D are taken on an ellipse, and perpendiculars p_1, p_2, p_3, p_4 let fall from any point of the ellipse upon the chords AB, CD; AC, BD respectively: express the constant ratio $p_1p_2 : p_3p_4$ in terms of the co-ordinates (X_1, Y_1), (X_2, Y_2) of the poles of BC, AD, and prove that the value of the ratio will be unity if

$$\frac{X_1X_2}{a^4} - \frac{Y_1Y_2}{b^4} = \frac{a^2 - b^2}{a^4 + b^4}.$$

1022. A tangent is drawn to an ellipse and with the point of contact as centre is described another ellipse similar and similarly situated but of three times the area: prove that if from any point of this latter ellipse two other tangents be drawn to the former, the triangle formed by the three tangents will be double of the triangle formed by joining their points of contact.

1023. Two tangents TP, TQ meet any other tangent in P', Q': prove that

$$PP' \,.\, QQ' = TP' \,.\, TQ' \cos^2 \frac{\alpha - \beta}{2};$$

where α, β are the excentric angles of P, Q.

1024. Two sides of a triangle are given in position and the third in magnitude: prove that the locus of the centre of the nine points' circle of the triangle is an ellipse; which reduces to a limited straight line if the acute angle between the given directions be 60°. If c be the given length and 2α the given angle, the axes of the ellipse will be equal to

$$\frac{c \sin 3\alpha}{4 \sin^2 \alpha \cos \alpha}, \quad \frac{c \cos 3\alpha}{4 \sin \alpha \cos^2 \alpha}.$$

1025. The tangent at a point P meets the equal conjugate diameters in Q, Q': prove that tangents from Q, Q' will be parallel to the straight line joining the feet of the perpendiculars from P on the axes.

1026. The excentric angles of the corners of an inscribed triangle are α, β, γ: prove that the co-ordinates of the centre of perpendiculars are

$$\frac{a^2+b^2}{2a}(\cos\alpha+\cos\beta+\cos\gamma)-\frac{a^2-b^2}{2a}\cos(\alpha+\beta+\gamma),$$

$$\frac{a^2+b^2}{2b}(\sin\alpha+\sin\beta+\sin\gamma)-\frac{a^2-b^2}{2b}\sin(\alpha+\beta+\gamma);$$

and those of the centre of the circumscribed circle are

$$\frac{a^2-b^2}{4a}\{\cos\alpha+\cos\beta+\cos\gamma+\cos(\alpha+\beta+\gamma)\}$$

$$-\frac{a^2-b^2}{4b}\{\sin\alpha+\sin\beta+\sin\gamma-\sin(\alpha+\beta+\gamma)\}.$$

The loci of these points when the triangle is of maximum area are respectively

$$4(a^2x^2+b^2y^2)=(a^2-b^2)^2,$$

$$16(a^2x^2+b^2y^2)=(a^2-b^2)^2.$$

1027. The centre of perpendiculars of the triangle formed by tangents at the points α, β, γ is the point given by the equations

$$4ax\cos\frac{\beta-\gamma}{2}\cos\frac{\gamma-\alpha}{2}\cos\frac{\alpha-\beta}{2}$$

$$=a^2\{\cos\alpha+\cos\beta+\cos\gamma-\cos(\alpha+\beta+\gamma)\}+2(a^2+b^2)\cos\alpha\cos\beta\cos\gamma,$$

$$4by\cos\frac{\beta-\gamma}{2}\cos\frac{\gamma-\alpha}{2}\cos\frac{\alpha-\beta}{2}$$

$$=b^2\{\sin\alpha+\sin\beta+\sin\gamma+\sin(\alpha+\beta+\gamma)\}+2(a^2+b^2)\sin\alpha\sin\beta\sin\gamma.$$

1028. Two points H, H' are conjugate with respect to an ellipse, P is any point on the ellipse, and PH, PH' meet the ellipse again in Q, Q': prove that QQ' passes through the pole of HH'.

1029. The lines

$$l_1\frac{x}{a}+m_1\frac{y}{b}=1,\quad l_2\frac{x}{a}+m_2\frac{y}{b}=1,\quad l_3\frac{x}{a}+m_3\frac{y}{b}=1$$

form a triangle self-conjugate to the ellipse: prove that

$$l_2l_3+m_2m_3=l_3l_1+m_3m_1=l_1l_2+m_1m_2=1,$$

and that the co-ordinates of the centre of perpendiculars of the triangle are

$$\frac{a^2-b^2}{a}l_1l_2l_3,\quad \frac{b^2-a^2}{b}m_1m_2m_3.$$

1030. A triangle is self-conjugate to a given ellipse and one corner of the triangle O is fixed: prove that the circle circumscribing the triangle passes through another fixed point O', that C, O, O' are in one straight line, and that $CO\,.\,CO'=a^2+b^2$.

1031. In the ellipses

$$\frac{x^2}{a^2}+\frac{y^2}{b^2}=1,\quad \frac{x^2}{a}+\frac{y^2}{b}=a+b,$$

a tangent to the former meets the latter in P, Q: prove that the tangents at P, Q are at right angles to each other.

1032. Two tangents OP, OQ are drawn at the points α, β: prove that the co-ordinates of the centre of the circle circumscribing the triangle OPQ are

$$\frac{\cos\frac{\alpha+\beta}{2}}{\cos\frac{\alpha-\beta}{2}}\,\frac{a^2+(a^2-b^2)\cos\alpha\cos\beta}{2a},$$

$$\frac{\sin\frac{\alpha+\beta}{2}}{\cos\frac{\alpha-\beta}{2}}\,\frac{b^2+(b^2-a^2)\sin\alpha\sin\beta}{2b}.$$

If this point lie on the axis of x, the locus of O is a circle (or the axis of x).

1033. Two points P, Q are taken on an ellipse such that the perpendiculars from Q, P on the tangents at P, Q intersect on the ellipse: prove that the locus of the pole of PQ is the ellipse

$$a^2x^2+b^2y^2=(a^2+b^2)^2,$$

and that if R be another point similarly related to P, the same relation will hold between Q, R; the centre of perpendiculars of the triangle formed by the tangents at P, Q, R will be the centre of the ellipse, and the centre of perpendiculars of the triangle P, Q, R lies on the ellipse

$$a^2x^2+b^2y^2=(a^2-b^2)^2.$$

1034. Three points (x_1, y_1), (x_2, y_2), (x_3, y_3) on an ellipse are such that $x_1+x_2+x_3=0$, $y_1+y_2+y_3=0$: prove that the circles of curvature at these points will pass through a point on the ellipse whose co-ordinates are

$$\frac{4x_1x_2x_3}{a^2},\quad \frac{4y_1y_2y_3}{b^2}.$$

1035. At a point P of an ellipse is drawn a circle touching the ellipse and of radius equal to n times the radius of curvature, and the two other common tangents to the circle and ellipse intersect in (X, Y) and include an angle ϕ: prove that

$$\frac{X^2}{a^2-\lambda^2}+\frac{Y^2}{b^2-\lambda^2}=1,$$

and $$4na^2b^2\tan^2\frac{\phi}{2}=\frac{(n\lambda^2-ab)^2(n\lambda^2+ab)^2}{(a^2-n\lambda^2)(n\lambda^2-b^2)},$$

n being reckoned negative when the circle has external contact and λ being the semidiameter parallel to the tangent at P.

1036. A triangle of minimum area circumscribes an ellipse, O is its centre of perpendiculars and OM, ON perpendiculars on the axes: prove that MN is a normal to the ellipse at the point of concourse of the three circles of curvature drawn at the points of contact of the sides.

1037. The tangent at the point whose excentric angle is ϕ touches the circle of curvature at the point whose excentric angle is θ: prove that

$$\frac{\sin\frac{\phi+\theta}{2}}{\sin\frac{\phi-\theta}{2}} = \frac{1-e^2\cos^4\theta}{2\cos\theta\,(1-e^2\cos^2\theta)}.$$

If P be the point θ, and T the pole of the normal at P, PT will be the least possible when the point ϕ lies on the normal at P.

1038. The hyperbola which osculates a given ellipse at a point and has its asymptotes parallel to the equal conjugate diameters meets the ellipse again in the same point as the common circle of curvature; and if P be the point of osculation and O the centre of the hyperbola, PO is the tangent at O to the locus of O and is normal to the ellipse

$$a^2x^2 + b^2y^2 = \frac{4a^6b^6}{(a^2-b^2)^2}.$$

1039. A rectangular hyperbola osculates a given ellipse at a point P and meets the ellipse again in the same point as the common circle of curvature: prove that, if O be its centre, PO will be the tangent at O to the locus of O and will be normal to the ellipse

$$\frac{x^2}{a^2} + \frac{y^2}{b^2} = \left(\frac{a^2+b^2}{a^2-b^2}\right)^2.$$

1040. An hyperbola is described with two conjugate diameters of a given ellipse for asymptotes: prove that, if the curves intersect, the tangent to the ellipse at any common point is parallel to the tangent to the hyperbola at an adjacent common point, and the parallelogram formed by the tangents to the hyperbola will be to that formed by the tangents to the ellipse as $m^2\sin^2\theta : 1$, the equation of the hyperbola being

$$\frac{x^2}{a^2} + \frac{2xy}{ab}\cot\theta - \frac{y^2}{b^2} = m.$$

If the common points be impossible the points of contact of the common tangents will lie on two diameters, and the parallelograms formed by joining the points of contact will be for the ellipse and hyperbola respectively in the ratio $m^2 : \sin^4\theta$.

1041. A triangle circumscribes the ellipse and its centroid lies in the axis of x at a distance c from the centre: prove that its angular points will lie on the conic

$$\frac{(x-3c)^2}{a^2} + \frac{y^2(a^2-9c^2)}{a^2b^2} = 4.$$

1042. A triangle is inscribed in the ellipse and its centroid lies in the axis of x at a distance c from the centre: prove that its sides will touch the conic

$$\frac{4a^2y^2}{b^2(a^2-9c^2)}+\frac{(2x-3c)^2}{a^2}=1.$$

[In this and the preceding question the axes need not be rectangular.]

1043. A triangle is inscribed in the ellipse and the centre of perpendiculars of the triangle is one of the foci: prove that the sides of the triangle will touch one of the circles

$$\left(x\pm\frac{a^2\sqrt{a^2-b^2}}{a^2+b^2}\right)^2+y^2=\frac{a^2b^4}{(a^2+b^2)^2}.$$

1044. A triangle circumscribes the circle $x^2+y^2=a^2$ and two of its angular points lie on the circle $(x-c)^2+y^2=b^2$: prove that the locus of the third angular point is a conic touching the common tangents of the two circles; that this conic becomes a parabola if $(c\pm a)^2=b^2-a^2$; and that the chords intercepted on any tangent to this conic by the two circles are in the constant ratio

$$2a^2:\sqrt{(2ab+b^2-c^2)(2ab-b^2+c^2)}.$$

1045. A triangle circumscribes an ellipse and two of its angular points lie on a confocal ellipse: prove that the third angular point lies on another confocal and that the perimeter of the triangle is constant.

1046. Two conjugate radii CP, CD being taken, PO is measured along the normal at P equal to k times CD: prove that the locus of O is the ellipse

$$\frac{x^2}{(a-kb)^2}+\frac{y^2}{(b-ka)^2}=1;$$

and this ellipse touches the evolute of the ellipse in four points which are real only when k lies between $\frac{b}{a}$ and $\frac{a}{b}$: k being negative when PO is measured outwards.

1047. The ellipses

$$\frac{x^2}{a^2}+\frac{y^2}{b^2}=1,\quad a^2x+b^2y^2=\frac{a^4b^4}{(a^2+b^2)^2},$$

are so related that (1) an infinite number of triangles can be inscribed in the former whose sides touch the latter; (2) the central distance of any angular point of such a triangle will be perpendicular to the opposite side; (3) the normals to the first ellipse at the angles of any such triangle, and to the second at the points of contact, will severally meet in a point.

1048. The ellipses

$$\frac{x^2}{a^6} + \frac{y^2}{b^6} = \frac{1}{(a^2 - b^2)^2}, \qquad \frac{x^2}{a^2} + \frac{y^2}{b^2} = 1, \qquad (a^2 > 2b^2)$$

are such that the normals to the latter at the corners of any inscribed triangle whose sides touch the former meet on the latter.

1049. The semi-axes of an ellipse U are CA, CB; LCL' is the major axis and C the focus of another ellipse V, $LC = BC$, $CL' = CA$: prove that the auxiliary circle of V touches both the auxiliary circles of U; one of the common tangents, PP', of U and V is such that P lies on the auxiliary circle of V; and PL, PL' are parallel to CA, CB; CP', CL are equally inclined to CA, CB; if the auxiliary circle of V meet U also in Q, R, S, the triangle QRS has the centre of its inscribed circle at C, and the straight lines bisecting its external angles touch V and form a triangle whose nine points' circle is the auxiliary circle of V, and whose circumscribed circle has its centre at the second focus of V; also if the three other common tangents to U, V form a triangle $Q'R'S'$, the centre of its circumscribed circle is C and its nine points' circle is the auxiliary circle of U; the sum of the excentric angles of Q, R, S is equal to that of the points of contact of the triangle $Q'R'S'$, and if this sum be δ the excentric angles of Q, R, S are the roots of the equation $\tan\frac{\delta - \theta}{2} = \frac{b}{a}\tan\theta$, and those of the points of contact of $Q'R'S'$ are the roots of the equation $\tan\frac{\delta - \theta}{2} = \frac{a}{b}\tan\theta$; the three perpendiculars of the triangle $Q'R'S'$ are normals to U and meet in O the second focus of V, OP is normal at P and a circle goes through P and the other three points of contact. The straight lines through Q, R, S at right angles to CQ, CR, CS will touch V in points q, r, s such that Cq, Cr, Cs make with CA angles respectively equal to the excentric angles of Q, R, S. The normals at Q, R, S meet in a point O' from which, if the fourth normal $O'p$ be drawn, Pp is a diameter of U; and the normals at the points of contact of $Q'R'S'$ meet in a point o on the same normal $O'p$ such that

$$op : O'p = ab : a^2 - ab + b^2.$$

1050. A triangle LMN is inscribed in the ellipse $\frac{x^2}{a^2} + \frac{y^2}{b^2} = 1$ so that the normals at L, M, N meet in a point O, and from O the fourth normal OP is drawn: prove the following theorems.

(1) OP will bear to the semi-diameter conjugate to CP the ratio $k : 1$ where k is given by either of the equations

$$X = (kb - a)\cos(\alpha + \beta + \gamma), \quad Y = (b - ka)\sin(\alpha + \beta + \gamma),$$

where X, Y are the co-ordinates of O and α, β, γ the excentric angles of L, M, N.

(2) The sides of the triangle LMN will touch the ellipse

$$\frac{x^2}{a'^2} + \frac{y^2}{b'^2} = 1$$

in points whose excentric angles are $\pi+\alpha$, $\pi+\beta$, $\pi+\gamma$, if

$$\frac{a'}{a^2(a-kb)}=\frac{b'}{b^2(ka-b)}=\frac{1}{a^2-b^2}.$$

(3) The tangents at L, M, N will form a triangle whose corners lie on the ellipse $\frac{x^2}{A^2}+\frac{y^2}{B^2}=1$, at points whose excentric angles are $\pi+\alpha$, $\pi+\beta$, $\pi+\gamma$; where $Aa'=a^2$, $Bb'=b^2$.

(4) An infinite number of such triangles LMN can be inscribed in the ellipse $\frac{x^2}{a^2}+\frac{y^2}{b^2}=1$ and circumscribed to the ellipse $\frac{x^2}{a'^2}+\frac{y^2}{b'^2}=1$, the excentric angles α, β, γ satisfying the two independent equations

$$\cos\alpha+\cos\beta+\cos\gamma=\left(\frac{b'}{b}-\frac{a'}{a}\right)\cos(\alpha+\beta+\gamma),$$

$$\sin\alpha+\sin\beta+\sin\gamma=\left(\frac{b'}{b}-\frac{a'}{a}\right)\sin(\alpha+\beta+\gamma),$$

and the relation between the axes being $\frac{a'}{a}+\frac{b'}{b}=1$. The ratio $k:1$ remains the same for all such triangles, and if L', M', N' be the points of contact of the sides, the ratio of the areas of the triangles $L'M'N'$, LMN is always the same, being $a'b':ab$, the ratio of the areas of the corresponding conics.

(5) Four points related to each triangle LMN: (α) the centroid, (β) the centre of perpendiculars, (γ) the centre of the circumscribed circle, (δ) the point of concourse of the normals, lie each on a fixed ellipse co-axial with the original, and the excentric angle is always the excess of the sum of the excentric angles of L, M, N above π, while the several semiaxes are

$$(\alpha)\ \left(\frac{a'}{a}-\frac{b'}{b}\right)\frac{a}{3},\ \left(\frac{a'}{a}-\frac{b'}{b}\right)\frac{b}{3};\quad (\beta)\ \frac{aa'-bb'}{a},\ \frac{aa'-bb'}{b};$$

$$(\gamma)\ \frac{a^2-b^2}{2a}\frac{b'}{b},\ \frac{a^2-b^2}{2b}\frac{a'}{a};\qquad (\delta)\ \frac{a'}{a^2}(a^2-b^2),\ \frac{b'}{b^2}(a^2-b^2).$$

[The results here given include all cases of triangles inscribed in the ellipse $\frac{x^2}{a^2}+\frac{y^2}{b^2}=1$ with sides touching a *co-axial* ellipse.]

1051. Triangles are circumscribed to an ellipse such that the normal at each point of contact passes through the opposite angular point: prove that the angular points lie on the ellipse

$$\frac{a^2x^2}{(\lambda-a^2)^2}+\frac{b^2y^2}{(\lambda-b^2)^2}=1,$$

λ being the greater root of the equation

$$\frac{a^2}{\lambda - a^2} + \frac{b^2}{\lambda - b^2} = 1;$$

the locus of the centre of perpendiculars of the triangles is the ellipse

$$(a^2 - \lambda)^2 \frac{x^2}{a^2} + (b^2 - \lambda)^2 \frac{y^2}{b^2} = (a^2 - b^2)^2,$$

and the perimeter of the triangle formed by joining the points of contact is constant.

1052. The two similar and similarly situated conics

$$\frac{x^2}{a^2} + \frac{y^2}{b^2} = 1, \qquad \frac{(x-h)^2}{a^2} + \frac{(y-k)^2}{b^2} = m^2,$$

will be capable of having triangles circumscribing the first and inscribed in the second, if

$$m^2 \pm 2m = \frac{h^2}{a^2} + \frac{k^2}{b^2}.$$

1053. A circle has its centre in the major axis of an ellipse and triangles can be inscribed in the circle whose sides touch the ellipse: prove that the circle must touch the two circles

$$x^2 + (y \pm b)^2 = a^2.$$

1054. A triangle LMN is inscribed in a given ellipse and its sides touch a fixed concentric ellipse: prove that the excentric angles α, β, γ must satisfy two equations of the form

$$\sin(\beta + \gamma) + \sin(\gamma + \alpha) + \sin(\alpha + \beta) = m,$$
$$\cos(\beta + \gamma) + \cos(\gamma + \alpha) + \cos(\alpha + \beta) = n,$$

where m, n are constant; and that the equation of the ellipse touching the sides is

$$\frac{x^2}{a^2}(m^2 + \overline{n+1}^2) + \frac{y^2}{b^2}(m^2 + \overline{n-1}^2) + 4m\frac{xy}{ab} = \left(\frac{m^2 + n^2 - 1}{2}\right)^2.$$

Also prove that the area of the triangle LMN bears to the area of the triangle formed by joining the points of contact a constant ratio equal to that of the area of the ellipses. If (x_0, y_0) be the centroid, (x_1, y_1) the centre of perpendiculars, and (x_2, y_2) the centre of the circumscribed circle, and $\alpha + \beta + \gamma = \theta$,

$$\frac{3x_0}{a} = m\sin\theta + n\cos\theta, \quad \frac{3y_0}{b} = n\sin\theta - m\cos\theta;$$

$$2ax_1 = m(a^2 + b^2)\sin\theta + \{n(a^2 + b^2) - a^2 + b^2\}\cos\theta,$$

$$2by_1 = \{n(a^2 + b^2) - a^2 + b^2\}\sin\theta - m(a^2 + b^2)\cos\theta;$$

$$4ax_2 = (a^2 - b^2)\{m\sin\theta + (n+1)\cos\theta\},$$

$$4by_2 = (a^2 - b^2)\{m\cos\theta - (n-1)\sin\theta\};$$

from which the loci can easily be found.

1055. Triangles are inscribed in a given ellipse such that their sides touch a fixed concentric ellipse of given area $\frac{\pi ab}{4(k^2 \sim 1)}$: prove that this ellipse will have double contact with each of the ellipses

$$\frac{x^2}{a^2}+\frac{y^2}{b^2}=\left(\frac{k \pm 1}{2}\right)^2.$$

1056. A triangle LMN is circumscribed about a given ellipse of focus S such that the angles SMN, SNL, SLM are all equal $(=\theta)$: prove that $\sin\theta=\frac{b}{2a}$, and that L, M, N lie on one of two fixed circles whose common radius is $\frac{2a^2}{b}$: also $\tan\theta=\frac{\sin L \sin M \sin N}{1+\cos L \cos M \cos N}$, and the point of contact of MN lies on the straight line joining L to the point of intersection of the tangents at M, N to the circle LMN.

1057. A triangle is formed by tangents to the ellipse at points whose excentric angles α, β, γ satisfy the poristic system

$$\cos(\beta+\gamma) \pm (\sin\beta+\sin\gamma)+n=0, \text{ \&c.}:$$

prove that the locus of the angular points is

$$\frac{x^2}{a^2}(n+1)+\frac{y^2}{b^2}(n-1) \pm \frac{2y}{b}=0;$$

the envelope of the sides of the triangle formed by joining the points of contact is the parabola

$$\frac{y^2}{b^2}=(n-1)\left(n+1 \pm \frac{2x}{a}\right);$$

and the centroid of this latter triangle, its centre of perpendiculars, and centre of circumscribed circle lie on three fixed straight lines parallel to the axis of x.

1058. A triangle is formed by tangents to the ellipse at points whose excentric angles α, β, γ satisfy the poristic system

$$\cos\beta\cos\gamma+m(\sin\beta+\sin\gamma)+m^2=0, \text{ \&c.}:$$

prove that the locus of the angular points is

$$m^2\left(\frac{x^2}{a^2}+\frac{y^2}{b^2}\right)-\frac{y^2}{b^2}+2m\frac{y}{b}+1=0;$$

the envelope of the sides of the triangle formed by joining the points of contact is the hyperbola $\frac{x^2}{m^2a^2}-\left(\frac{y}{b}+m\right)^2+1=0$; the locus of the centroid is the ellipse $\frac{9x^2}{a^2}+\left(\frac{3y}{b}+m\right)^2=1$; that of the centre of perpendiculars the ellipse $a^2x^2+(by+m\,\overline{a^2+b^2})^2=b^4$; and that of the centre of the circumscribed circle the straight line $2by=m(a^2-b^2)$.

1059. A triangle is formed by tangents drawn to a given ellipse at points whose excentric angles satisfy the equations

$$l\cos(\alpha+\beta+\gamma)+m(\cos\overline{\beta+\gamma}+\cos\overline{\gamma+\alpha}+\cos\overline{\alpha+\beta})+n(\cos\alpha+\cos\beta+\cos\gamma)=p,$$

$$l\sin(\alpha+\beta+\gamma)+m(\sin\overline{\beta+\gamma}+\ldots+\ldots)+n(\sin\alpha+\ldots)=q:$$

prove that the angular points will lie on the conic whose equation is

$$\frac{x^2}{a^2}(\overline{l+n}^2-q^2-\overline{m-p}^2)+\frac{y^2}{b^2}(\overline{l-n}^2-q^2-\overline{m+p}^2)+4(m^2-n^2)$$

$$+4qn\frac{y}{b}+4(lm+np)\frac{x}{a}+4qm\frac{xy}{ab}=0;$$

(which, since its equation involves four independent constants, will be the general equation of the conic in which can be inscribed triangles whose sides touch the given ellipse $\frac{x^2}{a^2}+\frac{y^2}{b^2}=1$.) The loci of the centroid, &c. of the triangle whose angular points are α, β, γ can easily be formed, and it will be found that the centroid lies on an ellipse similar and similarly situated to the given ellipse; that the locus of the centre of perpendiculars is similar to the given ellipse but turned through a right angle; and that each locus reduces to a straight line when $m^2=n^2$, in which case $\alpha+\beta+\gamma$ is constant.

1060. The maximum perimeter of any triangle inscribed in a given ellipse is

$$2\sqrt{3}\,\frac{a^2+b^2+\sqrt{a^4-a^2b^2+b^4}}{\sqrt{a^2+b^2+2\sqrt{a^4-a^2b^2+b^4}}};$$

and if $2X$, $2Y$, $2Z$ be diameters parallel to its sides

$$X^2+Y^2+Z^2=a^2+b^2+\sqrt{a^4-a^2b^2+b^4}.$$

1061. A parallelogram of maximum perimeter is inscribed in a given ellipse and $2X$, $2Y$ are its diagonals: prove that

$$\frac{1}{a^2+b^2-X^2}+\frac{1}{a^2+b^2-Y^2}=\frac{1}{a^2}+\frac{1}{b^2};$$

and that the perimeter is $4\sqrt{a^2+b^2}$.

1062. A hexagon $AB'CA'BC'$ of maximum perimeter is inscribed in an ellipse: prove that its perimeter is $4\frac{a^2+ab+b^2}{a+b}$; the tangents at A, B, C and A', B', C' form triangles inscribed in the same fixed circle of radius $a+b$; also, if a triangle be inscribed in the ellipse with sides each parallel to two sides of the hexagon, the sides of this triangle will touch a fixed circle of radius $\frac{ab}{a+b}$ and its area will be half that of the hexagon. Also, if X, Y, Z be radii of the ellipse each parallel to two sides of the hexagon,

$$X^2 + Y^2 + Z^2 = a^2 + ab + b^2, \quad 2XYZ = ab\,(a + b);$$

and if X', Y', Z' be radii each parallel to the tangents at two corners of the hexagon

$$\frac{1}{X'^2} + \frac{1}{Y'^2} + \frac{1}{Z'^2} = \frac{1}{a^2} + \frac{1}{ab} + \frac{1}{b^2}, \quad \frac{2}{X'Y'Z'} = \frac{1}{ab}\left(\frac{1}{a} + \frac{1}{b}\right).$$

1063. A hexagon $AB'CA'BC'$ is inscribed in the ellipse $\frac{x^2}{a^2} + \frac{y^2}{b^2} = 1$, and its sides touch the ellipse $\frac{x^2}{a'^2} + \frac{y^2}{b'^2} = 1$; a triangle abc is inscribed in the former ellipse so that bc is parallel to $B'C$, and BC', &c.: prove that a, A will be at the ends of conjugate diameters, the area of the triangle will be half that of the hexagon, the tangents at ABC and those at $A'B'C'$ form triangles inscribed in the ellipse

$$x^2\,(b^2 - b'^2) + y^2\,(a^2 - a'^2) = a^2b^2,$$

and the sides of the triangle abc touch the ellipse $\frac{x^2}{a^2 - a'^2} + \frac{y^2}{b^2 - b'^2} = 1$.

[The relation $\frac{(a^2 - a'^2)^{\frac{1}{2}}}{a} + \frac{(b^2 - b'^2)^{\frac{1}{2}}}{b} = 1$ must be satisfied.]

1064. The tangent to a conic at P meets the directrices in K, K', and from K, K' are drawn two other tangents intersecting in Q; prove that PQ is normal at P and is bisected by the conjugate axis.

1065. Two straight lines are drawn parallel to the major axis at a distance be^{-1} from it: prove that the part of any tangent intercepted between them will be divided by the point of contact into two parts subtending equal angles at the centre.

1066. The part of any tangent intercepted between the two straight lines

$$ab\left(\frac{x}{a} + \frac{y}{b} - 1\right)^2 = (x + y - a - b)^2$$

is divided by the point of contact into two parts subtending equal angles at the point (a, b).

1067. Two tangents to the ellipse $\frac{x^2}{a^2} + \frac{y^2}{b^2} = \frac{a^2}{a^2 - b^2}$ intersect in a point T on the axis of x: prove that the part of any tangent to the ellipse $\frac{x^2}{a^2} + \frac{y^2}{b^2} = 1$ intercepted between them is divided by the point of contact into two parts subtending equal angles at the point on the axis of x which is conjugate to T with respect to the latter ellipse.

1068. The value of λ is so determined that the equation

$$\frac{x^2}{a^2} + \frac{y^2}{b^2} - 1 + \lambda\{(x - X)^2 + (y - Y)^2\} = 0$$

represents two straight lines: prove that the part of any tangent to the ellipse intercepted between these two straight lines is divided by the point of contact into two parts which subtend equal (or supplementary) angles at (X, Y). If O be the point (X, Y) the two values of $\frac{1}{\lambda}$ are $\frac{CO^2 \pm SO \,.\, S'O - CA^2 - CB^2}{2}$. Discuss the case when O coincides with S or S'.

1069. Two conjugate diameters of a given ellipse meet the fixed straight line $p\frac{x}{a} + q\frac{y}{b} = 1$ in P, P', and the straight lines drawn through P, P' respectively at right angles to these diameters intersect in Q: prove that the locus of Q is the straight line

$$apx \pm bqy = a^2 + b^2;$$

and the locus of the intersection of straight lines drawn through P, P' perpendicular respectively to CP', CP is the straight line

$$qax - pby = 0.$$

1070. A parallelogram circumscribes a given ellipse, and the ends of one of its diagonals lie on the given straight lines $p\frac{x}{a} + q\frac{y}{b} = \pm 1$: prove that the ends of the other diagonal lie on the conic

$$\frac{x^2}{a^2} + \frac{y^2}{b^2} - 1 = \left(q\frac{x}{a} - p\frac{y}{b}\right)^2.$$

1071. In the ellipses

$$\frac{x^2}{a^2} + \frac{y^2}{b^2} = \frac{a^2 - b^2}{a^2 + b^2}, \quad \frac{x^2}{a^2} + \frac{y^2}{b^2} = \frac{a^2 + b^2}{a^2 - b^2},$$

CPQ is drawn to meet the curves, and QQ' is a double ordinate of the outer: prove that PQ' is normal at Q'.

1072. From any point on the normal to a given ellipse at a fixed point $(a \cos \alpha, b \sin \alpha)$ are drawn the three other normals to the ellipse at points P, Q, R: prove that the centroid, the centre of perpendiculars, and the centre of the circumscribed circle of the triangle PQR lie respectively on the straight lines

$$ax \sin \alpha - by \cos \alpha = 0, \quad bx \sin \alpha + ay \cos \alpha = 0, \quad 2(ax \sin \alpha - by \cos \alpha) + (a^2 - b^2) \sin \alpha \cos \alpha = 0.$$

1073. A triangle is inscribed in an ellipse and its centre of perpendiculars is at the point (X, Y): prove that the locus of the poles of its sides is the conic

$$\frac{b^2}{a^2}x^2(Y^2-b^2)+\frac{a^2}{b^2}y^2(X^2-a^2)-2xyXY+(xX+yY-a^2-b^2)^2=0.$$

1074. A fixed point O is taken within a given circle, a pair of parallel tangents drawn to the circle, and AOA' is a straight line meeting the tangents at right angles. An ellipse is described with focus O and axis AA', and the other two common tangents to this ellipse and the circle meet in P: prove that P lies on a fixed straight line bisecting at right angles the distance between O and the centre of the circle.

1075. With the focus of an ellipse as centre is described a circle touching the directrix; two tangents drawn to the circle from a point P on the ellipse meet the ellipse again in Q, Q': prove that QQ' is parallel to the minor axis, and that tangents drawn from Q, Q' to the circle will intersect in a point P' on the ellipse so that PP' is also parallel to the minor axis. The tangents to the circle at the real common points pass through the further extremity of the major axis, and the points of contact with the ellipse of the (real) common tangents are at a distance from the focus equal to the latus rectum.

1076. At the ends of the equal conjugate diameters of an ellipse whose foci are given are drawn circles equal to the circle of curvature and touching the ellipse externally: prove that the common tangents to the ellipse and one of these circles intersect on the rectangular hyperbola which is confocal with the ellipse.

1077. From any point P on the ellipse $\frac{x^2}{a^2}+\frac{y^2}{b^2}=1$, tangents are drawn to the ellipse $\frac{x^2}{a^3}+\frac{y^2}{b^3}=\frac{1}{a+b}$: prove that they meet the former ellipse in points Q, Q' at the ends of a diameter, and that the tangents at Q, Q' will touch the circle which touches the ellipse externally at P and has a diameter equal to the diameter conjugate to CP.

1078. A circle is drawn through the foci of a given ellipse and common tangents drawn to the ellipse and circle: prove that one pair of straight lines through the four points of contact with the circle will envelope the hyperbola

$$\frac{x^2}{a^2-2b^2}-\frac{y^2}{b^2}=1,$$

confocal with the ellipse.

1079. From a fixed point (X, Y) are drawn tangents OP, OQ to a conic whose foci are given: prove that the locus of the centre of the circle OPQ is the straight line

$$\frac{2xX}{X^2+Y^2+c^2}+\frac{2yY}{X^2+Y^2-c^2}=1,$$

and the locus of the centre of perpendiculars is a rectangular hyperbola of which one asymptote is parallel to CO, reducing to two straight lines if O lie on the lemniscate of which the given foci are vertices.

[The equation of the rectangular hyperbola is

$$(Xx+Yy)(Xy-Yx)=c^2(Xy+Yx-2XY);$$

and if O' be its centre, $CO'.CO=CS^2$, and the angle SCO' is three times the angle SCO.]

1080. Two tangents OP, OQ being drawn to a given conic, prove that two other conics can be drawn confocal with the given conic and having for their polars of O the normals at P, Q.

1081. Two conics have common foci S, S', a point O is taken such that the rectangle under its focal distances is equal to that under the tangents to the director circles: prove that the polars of O will be normals to a third confocal conic at points lying on the polar of O with respect to that conic.

1082. A diameter PP' of a given ellipse being taken, the normal at P' intersects the ordinate at P in Q: prove that the locus of Q is the ellipse

$$\frac{x^2}{a^2}+\frac{b^2y^2}{(2a^2-b^2)^2}=1;$$

and that the tangents from Q meet the tangent at P in points on the auxiliary circle.

1083. A chord PQ of an ellipse is normal to the ellipse at P, and p, q are perpendiculars from the centre on the tangents at P, Q: prove that

$$\frac{4q^2}{p^2-q^2}=\frac{(a^2+b^2-p^2)^2}{(a^2-p^2)(p^2-b^2)}.$$

1084. The locus of the centre of an equilateral triangle inscribed in a given ellipse is the ellipse

$$\frac{x^2}{a^2}(a^2+3b^2)^2+\frac{y^2}{b^2}(3a^2+b^2)^2=(a^2-b^2)^2.$$

1085. From two points on the polar of a point O are drawn two pairs of tangents at right angles to each other to a given ellipse: prove that the four other points of intersection of these tangents lie upon the tangents at O to the confocals through O: and the tangents drawn from a pair of these points to the corresponding confocal will be parallel to each other.

[The latter proposition is more readily proved geometrically.]

1086. An ellipse is described passing through the foci of a given ellipse and having the tangents at the end of the major axis for directrices: prove that it will have double contact with the given ellipse, and that its foci will lie on two circles touching the given ellipse at the ends of its major axis and having diameters equal to half the latus rectum.

1087. The least distance between two points lying respectively on the fixed ellipses

$$\frac{x^2}{a^2}+\frac{y^2}{b^2}=1, \quad \frac{x^2}{a'^2}+\frac{y^2}{b'^2}=1,$$

is

$$\sqrt{\frac{(a'^2b^2-a^2b'^2)(a^2-a'^2-b^2+b'^2)}{(a^2-b^2)(a'^2-b'^2)}}.$$

Explain how it comes to pass that this vanishes for confocal and for similar ellipses.

1088. Prove that if the ellipse $\frac{x^2}{a^2-\lambda}+\frac{y^2}{b^2-\lambda}=1-\frac{\lambda}{\mu}$ touch a parallel to the ellipse $\frac{x^2}{a^2}+\frac{y^2}{b^2}=1$, the distance between the ellipse and its parallel will be $\sqrt{\mu}$, and the ratio of the curvatures at the point of contact will be

$$\lambda\mu(a^2-\lambda)(b^2-\lambda) : (a^2b^2\mu-\lambda^3)(\lambda-\mu).$$

IV. *Hyperbola, referred to its axes or asymptotes.*

[The equation of the hyperbola, referred to its axes, only differing from that of the ellipse by having $-b^2$ instead of b^2, many theorems which have been stated for the ellipse are obviously also true for the hyperbola. It is convenient still to use the notation of the excentric angle and denote any point on the hyperbola by $a\cos\alpha$, $bi\sin\alpha$, and all the corresponding equations, but the excentric angle is imaginary. A point on the hyperbola may be denoted by $a\sec\alpha$, $b\tan\alpha$, but the resulting equations are not nearly so symmetrical as the corresponding equations in terms of the excentric angle are for the ellipse. The angle α so used is sometimes called the excentric angle in the case of the hyperbola. When referred to its asymptotes the equation of the hyperbola is $4xy=a^2+b^2$, but the axes are not generally rectangular, and questions involving perpendicularity should not be referred to such axes. The equation is often written $xy=c^2$: in this form the equation of the polar of (X, Y) is $xY+yX=2c^2$, and that of the two tangents from (X, Y) is $4(xy-c^2)(XY-c^2)=(xY+yX-2c^2)^2$.]

1089. Prove that the four equations

$$b(x\pm\sqrt{x^2-a^2})=a(y\pm\sqrt{y^2+b^2})$$

represent respectively the portions of an hyperbola referred to its axes which lie in the four quadrants.

1090. The equation of the chord of an hyperbola referred to its axes which is bisected in the point (X, Y) is

$$b^2X(x-X)=a^2Y(y-Y);$$

and the corresponding equation when referred to the asymptotes is

$$\frac{x}{X}+\frac{y}{Y}=2.$$

1091. The equation of the chord of the hyperbola $xy = c^2$ whose extremities are the points (x_1, y_1), (x_2, y_2) is

$$\frac{x}{x_1 + x_2} + \frac{y}{y_1 + y_2} = 1.$$

1092. The locus of points whose polars with respect to a given parabola touch the circle of curvature at the vertex is a rectangular hyperbola.

1093. The normals to any hyperbola $xy = c^2$ at any point where it is met by the ellipse $x^2 + y^2 = c^2 (1 + \sec \omega)$, ω being the angle between the asymptotes, are parallel to one of the asymptotes.

1094. A circle described on a chord AB of an ellipse as diameter meets the ellipse again in C, D, and AB, CD are conjugates with respect to the ellipse: prove that AB touches the hyperbola $\frac{x^2}{a^2} - \frac{y^2}{b^2} = \frac{a^2 - b^2}{a^2 + b^2}$, and CD the hyperbola $\frac{x^2}{a^2} - \frac{y^2}{b^2} = \frac{a^2 + b^2}{a^2 - b^2}$.

1095. A double ordinate PP' is drawn to the ellipse $\frac{x^2}{a^2} + \frac{y^2}{b^2} = 1$, and the tangent at P meets the hyperbola $\frac{x^2}{a^2} - \frac{y^2}{b^2} = 1$ in Q, Q': prove that $P'Q$, $P'Q'$ are tangents to the hyperbola; and, if R, R' be the points in which these lines again meet the ellipse, that RR' divides PP' in the ratio 1 : 2.

1096. Two circles are drawn, one having double contact with a single branch of a given hyperbola and the other having single contact with each branch, and their chords of contact with the hyperbola meet on an asymptote: prove that the pole of either asymptote with respect to one circle is the pole of the other asymptote with respect to the other circle, and that its locus is a rectangular hyperbola passing through the foci of the given hyperbola and having one asymptote in common with it.

1097. A circle is drawn with its centre on the transverse axis to touch the asymptotes of an hyperbola: prove that the tangents drawn to it at the points where it meets the hyperbola will also touch the auxiliary circle of the hyperbola. If the circle have its centre on the conjugate axis and common tangents be drawn to it and the hyperbola, the locus of their points of contact with the circle is the curve

$$(b^2x^2 - a^2y^2)(x^2 + y^2 + b^2)^2 = 4(a^2 + b^2)x^2(b^2x^2 - a^2y^2 - a^2b^2).$$

1098. The tangent to an hyperbola at P meets the asymptotes in L, L': prove that the circle LCL' passes through the points where the normal meets the axes, that the points where the tangent meets the axes are conjugate with respect to the circle, and the pole of LL' is the point through which pass all chords of the hyperbola subtending a right angle at P.

1099. Two hyperbolas have the same asymptotes, and NPQ is drawn parallel to one asymptote meeting the other in N and the curves in P, Q; a tangent at Q meets the outer hyperbola in two points and the straight lines joining these to the centre meet the ordinate NQ in L, M: prove that NQ is a geometric mean between NL, NM and that NP is a harmonic mean between NQ and the harmonic mean between NL, NM.

1100. The axes of an ellipse are the asymptotes of an hyperbola which does not meet the ellipse in real points: prove that the difference of the excentric angles of the points of contact of tangents to the ellipse drawn from a point on the hyperbola will have the least possible value when the point is on one of the equal conjugate diameters; also the locus of the points of contact with the hyperbola of the common tangents is the curve

$$\frac{a^2}{x^2} + \frac{b^2}{y^2} = 4.$$

1101. The locus of the equation

$$y = x + \frac{c^2}{x +} \frac{c^2}{x +} \frac{c^2}{x +} \dots \text{ to } \infty$$

is that part of the hyperbola $y^2 - xy = c^2$ which starting from the axis of y goes to infinity along the line $y = x$.

1102. The locus of a point from which can be drawn two straight lines at right angles to each other, each of which touches one of the rectangular hyperbolas $xy = \pm c^2$, is also the locus of the feet of the perpendiculars let fall from the origin on tangents to the hyperbolas

$$x^2 - y^2 = \pm 4c^2.$$

1103. An ellipse is described confocal with a given hyperbola, and the asymptotes of the hyperbola are the equal conjugate diameters of the ellipse: prove that, if from any point of the ellipse tangents be drawn to the hyperbola, the centres of two of the circles which touch these tangents and the chord of contact will lie on the hyperbola.

1104. The centre of perpendiculars of the triangle whose angular points are $\left(cm_1, \dfrac{c}{m_1}\right)$, $\left(cm_2, \dfrac{c}{m_2}\right)$, $\left(cm_3, \dfrac{c}{m_3}\right)$ is the point $\left(c\mu, \dfrac{c}{\mu}\right)$, where $\mu m_1 m_2 m_3 = -1$.

1105. Denoting by the point m the point whose co-ordinates are cm, $\dfrac{c}{m}$, prove that if a circle meet the rectangular hyperbola $xy = c^2$ in the four points m_1, m_2, m_3, m_4,

$$m_1 m_2 m_3 m_4 = 1;$$

and if, of four points m_1, m_2, m_3, m_4, any one is the centre of perpendiculars of the triangle formed by joining the other three

$$m_1 m_2 m_3 m_4 = -1.$$

1106. The rectangular hyperbola $x^2 - y^2 = a^2$ is cut orthogonally by all the ellipses represented by the equations

$$x^2 + 3y^2 + \lambda y - 3a^2 = 0,\quad 3x^2 + y^2 + \lambda x + 3a^2 = 0;$$

and the rectangular hyperbola $xy = a^2$ by the ellipses

$$x^2 - xy + y^2 + \lambda(x + y) - 3a^2 = 0,\quad x^2 + xy + y^2 + \lambda(x - y) + 3a^2 = 0.$$

1107. Normals are drawn to a rectangular hyperbola at the ends of a chord whose direction is given: the locus of their intersection is another rectangular hyperbola, whose asymptotes make with the asymptotes of the given hyperbola angles equal and opposite to those made by the given direction.

1108. The normal to a rectangular hyperbola at P meets the curve again in Q, and θ, ϕ are the angles which the central radii to P, Q make with either asymptote: prove that

$$\tan^3\theta \tan\phi = 1,\quad \tan PCQ + 2\tan CPQ = 0,$$

and that the least value of the angle CQP is $\sin^{-1}\frac{1}{3}$. Also if the diameter PP' be drawn, QP will subtend a right angle at P'.

1109. In a rectangular hyperbola the rectangle under the distances of any point of the curve from two fixed tangents is to the square on the distance from their chord of contact as $\cos\phi : 1$, where ϕ is the angle between the tangents.

1110. A circle is described on a chord of an ellipse as diameter which is parallel to the straight line $\frac{x}{a}\cos\alpha + \frac{y}{b}\sin\alpha = 0$: prove that the locus of the pole with respect to the circle of the straight line joining the two other common points is the hyperbola

$$\frac{a^2x^2}{\cos^2\alpha} - \frac{b^2y^2}{\sin^2\alpha} = a^4 - b^4.$$

[If the diameter be the straight line $\frac{x}{a}\cos\alpha + \frac{y}{b}\sin\alpha = \cos\beta$, the pole of the other common chord is the point (X, Y), where

$$\frac{2aX}{\cos\alpha} = \frac{a^2 - b^2}{\cos\beta} + (a^2 + b^2)\cos\beta,$$

$$\frac{2bY}{\sin\alpha} = \frac{b^2 - a^2}{\cos\beta} + (a^2 + b^2)\cos\beta.]$$

1111. An ellipse and an hyperbola are so related that the asymptotes of the hyperbola are conjugate diameters of the ellipse: prove that by a proper choice of axes their equations may be expressed in the forms

$$\frac{x^2}{a^2} + \frac{y^2}{b^2} = 1,\quad \frac{x^2}{a^2} - \frac{y^2}{b^2} = m.$$

1112. An hyperbola is described with a pair of conjugate diameters of a given ellipse as asymptotes: prove that the angle at which the curves cut each other at any common point is

$$\tan^{-1}\left(\frac{2a^2b^2\sin(\theta-\theta')}{a^4\sin\theta\sin\theta'+b^4\cos\theta\cos\theta'}\right),$$

where θ, θ' are the angles which the common diameters make with the major axis of the ellipse. The equation of the hyperbola will be $\frac{x^2}{a^2}+2h\frac{xy}{ab}-\frac{y^2}{b^2}=m$, the axes of the ellipse being the co-ordinate axes, and if tangents be drawn to both curves at the common points, the parallelogram formed by the tangents to the hyperbola will bear to that formed by the tangents to the ellipse the ratio

$$m^2 : 1+h^2.$$

1113. From two points (x_1, y_1), (x_2, y_2) are drawn tangents to the rectangular hyperbola $xy=c^2$: prove that the conic passing through the two points and through the four points of contact will be a circle if

$$x_1y_2+x_2y_1=4c^2, \text{ and } x_1x_2=y_1y_2.$$

1114. A triangle circumscribes a given circle and its centre of perpendiculars is a given point: prove that its angular points lie on a fixed conic which is an ellipse, parabola, or hyperbola, as the fixed point lies within, upon, or without the given circle.

[For the co-ordinates of the centre of perpendiculars of a triangle formed by tangents to an ellipse at points whose excentric angles are α, β, γ, see Question 1027].

1115. Three tangents are drawn to the rectangular hyperbola $xy=a^2$ at the points (x_1, y_1), (x_2, y_2), (x_3, y_3) and form a triangle whose circumscribed circle passes through the centre of the hyperbola: prove that

$$\frac{x_1+x_2+x_3}{x_1x_2x_3}+\frac{y_1+y_2+y_3}{y_1y_2y_3}=0;$$

and that the co-ordinates of the centre of the circle are $\frac{x_1x_2x_3}{a^2}$, $\frac{y_1y_2y_3}{a^2}$, a point on the hyperbola.

1116. A fixed point O whose coordinates are X, Y being taken, a chord PQ of the hyperbola $xy=c^2$ is drawn so that the centroid of the triangle OPQ lies on the hyperbola: prove that PQ touches the conic

$$(xY+yX+XY-9c^2)^2=36c^2xy,$$

which is an ellipse when $XY>9c^2$, and an hyperbola when $XY<9c^2$, but degenerates when $XY=9c^2$. If OO' be bisected by the centre of the given hyperbola, the centre of the envelope divides OO' in the ratio $3:1$.

1117. A triangle is inscribed in the hyperbola $xy = c^2$ so that its centroid is the fixed point $\left(ca, \frac{c}{a}\right)$, (a point on the hyperbola): prove that its sides will touch the ellipse

$$\left(3\frac{x}{a} + 3ay - 8c\right)^2 = 4xy,$$

which touches the asymptotes and the hyperbola (at the fixed point), the curvatures of the two curves at the point of contact are as 4 : 1, and the tangents to the ellipse where it again meets the hyperbola are parallel to the asymptotes.

1118. The circle of curvature of the rectangular hyperbola at the point (a cosec θ, $a \cot \theta$) meets the curve again in the point (a cosec ϕ, $a \cot \phi$): prove that

$$\tan \frac{\phi}{2} \tan^3 \frac{\theta}{2} = 1.$$

1119. Circles of curvature are drawn to an hyperbola and its conjugate at the ends of conjugate diameters: prove that their radical axis is parallel to one of the asymptotes.

[If X, Y be the co-ordinates of one of the points, the equation of the radical axis will be

$$2\left(\frac{x}{a} \pm \frac{y}{b}\right)\left(\frac{X^3}{a^3} - \frac{Y^3}{b^3}\right) + \frac{a^2 - b^2}{a^2 + b^2} = 0,$$

the upper sign being taken when the straight line joining the two points is parallel to the asymptote $\frac{x}{a} = \frac{y}{b}$.]

1120. Triangles are inscribed to the circle $x^2 + y^2 = 2ax$ whose sides touch the rectangular hyperbola $x^2 - y^2 = a^2$: prove that the locus of the centres of perpendiculars is the circle

$$x^2 + y^2 + 4ax = 0.$$

1121. On any hyperbola P, Q, R are three contiguous points and L the centre of perpendiculars of the triangle PQR; find the limiting position of L when Q, R move up to P; and prove that its locus for different positions of P is

$$4\{S + (a^2 - b^2)U\}^3 + 27a^2b^2(a^2 - b^2)^2 S = 0,$$

where $\quad S \equiv b^2x^2 - a^2y^2 - a^2b^2$, and $U = x^2 + y^2 - a^2 + b^2$.

V. *Polar Co-ordinates.*

1122. The equation of the normal drawn to the circle $r = 2a \cos \theta$ at the point where $\theta = \alpha$ is

$$a \sin 2\alpha = r \sin (2\alpha - \theta).$$

1123. The equation of the straight line which joins the two points of the circle $r = 2a \cos\theta$ at which $\theta = \alpha$, $\theta = \beta$, is

$$2a \cos\alpha \cos\beta = r \cos(\alpha + \beta - \theta).$$

1124. A chord AP of a conic through the vertex A meets the latus rectum in Q, and a parallel chord $P'SQ'$ is drawn through a focus S: prove that the ratio $AP \,.\, AQ : Q'S \,.\, SP'$ is constant.

1125. Prove that the equations

$$c = r(e \cos\theta \pm 1)$$

represent the same conic. If (r, θ) denote a point on the curve when the upper sign is taken, $(-r, \pi + \theta)$ will denote the same point when the lower sign is taken.

1126. A chord PQ of a parabola is drawn through a fixed point on the axis, and a straight line bisecting the angle PSQ meets the directrix in O; from O perpendiculars OP', OQ' are let fall on SP, SQ: prove that SP' and SQ' will be of constant length.

1127. Two circles are described touching a parabola at the ends of a focal chord and passing through the focus: prove that they intersect at right angles and that their second point of intersection lies on a fixed circle. Also prove that the straight line joining the centres of the circles touches an ellipse whose excentricity is $\frac{1}{3}$ and which has the same focus and directrix as the parabola.

[The equation of the parabola being $2a = r(1 + \cos\theta)$, those of the circles may be taken to be

$$r \cos^3\alpha = a\cos(\theta - 3\alpha), \quad r \sin^3\alpha = a \sin(\theta - 3\alpha).]$$

1128. A conic is described having a common focus with the conic $c = r(1 + e\cos\theta)$, similar to it, and touching it where $\theta = \alpha$: prove that its latus rectum is $\dfrac{2c(1 - e^2)}{1 + 2e\cos\alpha + e^2}$, and that the angle between the axes of the two conics is $2\tan^{-1}\left(\dfrac{e + \cos\alpha}{\sin\alpha}\right)$. If $e > 1$, the conics will intersect again in two points lying on the straight line

$$\frac{c}{r}(1 + e\cos\alpha) + c\sin\alpha\,\{e\sin\theta + \sin(\theta - \alpha)\} = 0.$$

1129. Two chords QP, PR of a conic subtend equal angles at the focus: prove that the chord QR and the tangent at P intersect on the directrix.

1130. Two conjugate points Q, Q' are taken on a straight line through the focus S of a conic, and the straight line meets the conic in P: prove that the latus rectum is equal to

$$\frac{2SP \,.\, SQ}{SQ - SP} + \frac{2SP \,.\, SQ'}{SQ' - SP}.$$

1131. Through a point O on the axis of an ellipse at a distance $\sqrt{a^2 - \frac{b^4}{a^2}}$ from the centre is drawn a straight line YOP meeting the ellipse in P and a tangent at right angles in Y: prove that the rectangle $PO \cdot OY$ is equal to the square on the semi latus rectum.

1132. The points of an ellipse at which the circle of curvature passes through the other ends of the respective focal chords are given by the equation

$$2r^2 - r(3a + c) + 2ac = 0,$$

where $2a$ is the major axis, r the focal distance, and $2c$ the latus rectum.

1133. The two circles which are touched by any circle whose diameter is a focal chord of a given conic have the directrix for their radical axis and the focus for one of their point circles.

[The equations of the two circles are

$$r^2(1 \pm e) + cer\cos\theta = c^2,$$

that of the conic being $c = r(1 + e\cos\theta)$.]

1134. The radii of two circles are a, b and the distance between their centres is c, where $c(a + b + c) = 2(a - b)^2$; the centre of a circle which always touches them both traces out an ellipse whose vertex (the nearer to the centre of the smaller circle) is A: prove that the ends of the diameter of the moving circle drawn through A lie on a fixed ellipse with its focus at A.

1135. Prove that any chord of the conic

$$c = r(1 + e\cos\theta)$$

which is normal at a point where the conic is met by the straight lines

$$\frac{c}{r}\left(e + \frac{1}{e}\right) = \pm\sin\theta + (e^2 - 1)\cos\theta$$

will subtend a right angle at the pole.

1136. A conic with given excentricity and direction of axes is described with its focus at the centre of a given circle: prove that the tangents to this conic at the points where it meets the circle touch a fixed conic of which the given circle is auxiliary circle.

1137. Two parabolas have a common focus and axes opposite, a circle is drawn through the focus touching both parabolas: prove that

$$3r^{\frac{2}{3}} = a^{\frac{2}{3}} - a^{\frac{1}{3}}b^{\frac{1}{3}} + b^{\frac{2}{3}},$$

a, b being the latera recta and r the radius of the circle.

1138. Four tangents to the parabola $2a = r(1 + \cos\theta)$ are drawn at the points $2\theta_1$, $2\theta_2$, $2\theta_3$, $2\theta_4$: prove that the centres of the circles circumscribing the four triangles formed by them lie on the circle

$$2r\cos\theta_1\cos\theta_2\cos\theta_3\cos\theta_4 = a\cos(\theta_1 + \theta_2 + \theta_3 + \theta_4 - \theta).$$

1139. Through a fixed point is drawn any straight line, and on it are taken two points such that their distances from the fixed point are in a constant ratio and the line joining them subtends a constant angle at another fixed point: prove that their loci are circles.

1140. Two circles intersect, a straight line is drawn through one of their common points, and tangents are drawn to the circles at the points where this line again meets them: prove that the locus of the point of intersection of these tangents is the cardioid

$$cr = 2ab\{1 + \cos(\theta + \alpha - \beta)\};$$

the second common point of the circles being the pole, the common chord (c) the initial line, a, b the radii, and α, β the angles subtended by c in the segments of the two circles which lie each without the other circle.

1141. The equation of the circle which touches the conic

$$c = r(1 + e\cos\theta)$$

at the point where $\theta = \alpha$, and passes through the pole, is

$$\frac{r}{c}(1 + e\cos\alpha)^2 = \cos(\alpha - \theta) + e\cos(2\alpha - \theta);$$

and the equation of the chord joining their points of intersection is

$$\frac{c}{r}(1 + 2e\cos\alpha + e^2) = e^2\cos\theta + e^2\cos(\theta - \alpha).$$

1142. Two ellipses have a common focus S, a common excentricity e, axes in the same straight line, and the axis of the outer (U) is to that of the inner (V) as $2 - e^2 : 1 - e^2$; on a chord of U, which touches V, as diameter is described a circle meeting U again in the points P, Q: prove that the circle PSQ will touch U and that PQ will touch a fixed similar ellipse having the same focus S and its centre at the foot of the directrix of U.

1143. Two similar ellipses U, V have one focus S common, and the centre of V is at the foot of that directrix of U which is the polar of S; a tangent drawn to V at a point P meets U in two points: prove that the circle through these points and S will touch U at a point Q such that SP, HQ are equally inclined to the axis, H being the second focus of U.

1144. A conic is described having the focus of a given conic for its focus, any tangent for directrix, and touching the minor axis: prove that it will be similar to the given conic.

[Also easily proved by reciprocation.]

1145. Any point P is taken on a given conic, A is the vertex, S the nearer focus, and on AP is taken a point Q such that PQ exceeds SP by the sum of the distances of A, S from the directrix: prove that the locus of Q is a conic whose focus is A, similar to the given conic and having its centre at the farther vertex.

1146. The point S is a focus of an ellipse $c = r(1 + e\cos\theta)$, O, O' are two points on any tangent such that $SO = SO' = mc$, and SO, SO' meet the ellipse in P, Q: prove that PQ touches the conic

$$c = mr\{1 + e(1+m)\cos\theta\};$$

and the tangents at P, Q intersect on the conic

$$m^2(1-e^2)\left(\frac{c}{r} - e\cos\theta\right)^2 + 2em\cos\theta\left(\frac{c}{r} - e\cos\theta\right) = 1.$$

[The latter conic is a circle with its centre at the second focus and radius equal to the major axis, when $m(1-e^2) = 2$.]

1147. With the vertex of a given conic as focus and any tangent as directrix is described a conic passing through the nearer focus: prove that its major axis is of constant length equal to the distance between the focus and directrix of the given conic, and that the second directrix envelopes a conic similar to the given conic and having a focus in common with it.

1148. Two straight lines bisect each other at right angles: prove that the locus of the points at which they subtend equal angles is

$$\frac{r^2}{ab} = \frac{a\cos\theta - b\sin\theta}{b\cos\theta - a\sin\theta};$$

$2a$, $2b$ being the lengths of the lines, their point of intersection the pole and the initial line along the length $2a$.

1149. The focal distances of three points on a conic being r_1, r_2, r_3 and the angles between them α, β, γ, prove that the latus rectum $(2l)$ is given by the equation

$$\frac{4}{l}\sin\frac{\alpha}{2}\sin\frac{\beta}{2}\sin\frac{\gamma}{2} = \frac{1}{r_1}\sin\alpha + \frac{1}{r_2}\sin\beta + \frac{1}{r_3}\sin\gamma;$$

the angles α, β, γ being always taken so that their sum is 2π.

1150. An ellipse circumscribes a triangle ABC and the centre of perpendiculars of the triangle is a focus: prove that the latus rectum will be

$$\frac{2R\cos A\cos B\cos C}{\sin\frac{A}{2}\sin\frac{B}{2}\sin\frac{C}{2}};$$

R being the radius of the circle ABC.

1151. Two ellipses have a common focus and axes inclined at an angle α, and triangles can be inscribed in one whose sides touch the other: prove that

$$c_1^2 + 2c_1c_2 = e_1^2c_2^2 + e_2^2c_1^2 - 2e_1e_2c_1c_2\cos\alpha,$$

c_1, c_2 being the latera recta, and e_1, e_2 the excentricities. Also if θ, ϕ, ψ be the angles subtended by the sides of any such triangle at the focus

$$c_2 = 4c_1\cos\frac{\theta}{2}\cos\frac{\phi}{2}\cos\frac{\psi}{2}, \text{ or } 4c_1\cos\frac{\theta}{2}\sin\frac{\phi}{2}\sin\frac{\psi}{2}, \text{ \&c.}$$

1152. Two ellipses have a common focus and axes inclined at the angle

$$\cos^{-1}\left\{-\frac{c^2(1-e'^2)+c'^2(1-e^2)}{2cc'ee'}\right\},$$

where $2c$, $2c'$ are the latera recta and e, e' the excentricities: prove that any common tangent subtends a right angle at the focus.

VI. *General Equation of the Second Degree.*

[The general equation of a conic, in Cartesian Co-ordinates, being

$$u \equiv ax^2 + by^2 + c + 2fy + 2gx + 2hxy = 0,$$

the equations giving its centre are

$$\frac{du}{dx} = 0, \quad \frac{du}{dy} = 0.$$

The equation determining its excentricity may be found at once from the consideration that

$$\frac{a+b-2h\cos\omega}{\sin^2\omega}, \quad \frac{ab-h^2}{\sin^2\omega}$$

are unchanged by transformation of co-ordinates; and therefore that

$$\frac{(a+b-2h\cos\omega)^2}{(ab-h^2)\sin^2\omega} = \frac{(\alpha^2+\beta^2)^2}{\alpha^2\beta^2},$$

where ω is the angle between the co-ordinate axes, and 2α, 2β the axes.

The excentricity e is thus given by the equation

$$\frac{e^4}{1-e^2} + 4 = \frac{(a+b-2h\cos\omega)^2}{(ab-h^2)\sin^2\omega}.$$

The area of the conic $u=0$ is

$$\frac{\pi\Delta\sin\omega}{(ab-h^2)^{\frac{3}{2}}},$$

Δ being used to denote the discriminant

$$\begin{vmatrix} a, & h, & g \\ h, & b, & f \\ g, & f, & c \end{vmatrix},$$

or

$$abc + 2fgh - af^2 - bg^2 - ch^2.$$

The foci may be determined from the condition that the rectangle under the perpendiculars from them on any tangent is constant. Thus, taking the simple case when the origin is the centre and the axes rectangular, if the equation of the conic be

$$ax^2 + by^2 + 2hxy + c = 0,$$

and (X, Y), $(-X, -Y)$ two conjugate foci, we must have in order that the straight line $px+qy=1$ may be a tangent

$$\frac{(1-pX-qY)(1+pX+qY)}{p^2+q^2} = \text{a constant} \equiv \mu,$$

or $$p^2(\mu+X^2)+q^2(\mu+Y^2)+2pqXY-1=0 \;\ldots\ldots\ldots\ldots (A).$$

But the straight line will be a tangent if the quadratic equation

$$ax^2+by^2+c(px+qy)^2+2hxy=0,$$

found by combining the equations, have equal roots; that is, if

$$(a+cp^2)(b+cq^2)=(h+cpq)^2,$$

or $$p^2bc+q^2ac-2pqch+ab-h^2=0 \;\ldots\ldots\ldots\ldots\ldots\ldots (B).$$

Now (A), (B) expressing the same geometrical fact must be coincident, so that

$$\frac{\mu+X^2}{bc}=\frac{\mu+Y^2}{ac}=\frac{-XY}{ch}=\frac{1}{h^2-ab}.$$

The equations for X, Y are then

$$\frac{X^2-Y^2}{a-b}=\frac{XY}{h}=\frac{c}{ab-h^2}.$$

Also we can obtain for μ the equation

$$\left(\mu+\frac{bc}{ab-h^2}\right)\left(\mu+\frac{ac}{ab-h^2}\right)=\frac{c^2h^2}{(ab-h^2)^2},$$

equivalent to

$$\left(\frac{c}{\mu}+a\right)\left(\frac{c}{\mu}+b\right)=h^2,$$

whose roots are the squares of the semiaxes. To each root correspond two foci, real for one and unreal for the other.

The same method applies to all cases; and, the foci being found, the directrices are their polars.

The more useful special forms of the general equation are

(1) $ax^2+by^2+2hxy=2x$,

where a normal and tangent are co-ordinate axes;

(2) $\dfrac{x^2}{aa'}+2\lambda xy+\dfrac{y^2}{bb'}-x\left(\dfrac{1}{a}+\dfrac{1}{a'}\right)-y\left(\dfrac{1}{b}+\dfrac{1}{b'}\right)+1=0$,

which, for different values of λ, represents a series of conics all passing through four fixed points, a pair of joining lines being the co-ordinate axes;

(3) $\dfrac{x}{h}+\dfrac{y}{k}-1=2\lambda\left(\dfrac{xy}{hk}\right)^{\frac{1}{2}}$,

the equation of a conic touching the co-ordinate axes at distances h, k from the origin. It is sometimes convenient to use this as the equation of a conic touching four given straight lines, h, k, λ being then parameters connected by two equations

$$\left(\frac{1}{h}-\frac{1}{a}\right)\left(\frac{1}{k}-\frac{1}{b}\right)=\frac{\lambda^2}{hk}=\left(\frac{1}{h}-\frac{1}{a'}\right)\left(\frac{1}{k}-\frac{1}{b'}\right);$$

the equations of the two other given straight lines being

$$\frac{x}{a}+\frac{y}{b}=1,\quad \frac{x}{a'}+\frac{y}{b'}=1.$$

When (3) represents a parabola, $\lambda=1$; and the equation may be written

$$\left(\frac{x}{h}\right)^{\frac{1}{2}}+\left(\frac{y}{k}\right)^{\frac{1}{2}}=1.$$

The equation of the polar of (X, Y), when the equation is in the most general form $u=0$, is

$$P\equiv x(aX+hY+g)+y(hX+bY+f)+gX+fY+c=0,$$

and this may be adapted to any special case. The equation of the two tangents from (X, Y) is

$$(ax^2+by^2+c+2fy+2gx+2hxy)(aX^2+bY^2+c+2fY+2gX+2hXY)$$
$$=\{x(aX+hY+g)+y(hX+bY+f)+gX+fY+c\}^2,$$

or $$uU=P^2.$$

The equation of a tangent at (X, Y) to the parabola

$$\left(\frac{x}{h}\right)^{\frac{1}{2}}+\left(\frac{y}{k}\right)^{\frac{1}{2}}=1$$

is $$\frac{x}{(hX)^{\frac{1}{2}}}+\frac{y}{(kY)^{\frac{1}{2}}}=1,$$

the signs of the radicals in the equation of the tangent being determined by those of the corresponding radicals in the equation of the curve at the point (X, Y). Of course the equation of the polar cannot be expressed in any such form.

The condition that the straight line $px+qy+r=0$ may touch the conic $u=0$ is

$$Ap^2+Bq^2+Cr^2+2Fqr+2Grp+2Hpq=0;$$

where $A=bc-f^2$, $F=gh-af$, &c. The systems (a, b, c, f, g, h), (A, B, C, F, G, H) are of course reciprocal.]

1153. Trace the conic, any point of which is (1) $a\sin(\theta-\alpha)$, $b\sin(\theta+\alpha)$; (2) $\frac{a}{2}\left(mz+\frac{1}{mz}\right)$, $\frac{b}{2}\left(\frac{m}{z}+\frac{z}{m}\right)$, where a, b, m, and α are given.

1154. Prove that all conics represented by the equation

$$x^2(a^2+b^2-2ab\cos\theta)+y^2(a^2+b^2+2ab\cos\theta)-4xyab\sin\theta=(a^2-b^2)^2,$$

whatever the value of θ, are equal and similar, the lengths of the axes being $2(a\pm b)$.

1155. On two parallel fixed straight lines are taken points A, P; B, Q respectively, A, B being fixed and P, Q variable, subject to the condition that the rectangle under AP, BQ is constant: prove that PQ touches a fixed conic which will be an ellipse or hyperbola according as P, Q are on the same or opposite sides of AB.

1156. One side AB of a rectangle $ABCD$ slides between two rectangular axes: prove that the elliptic loci of C, D have equal areas independent of the length of AB, and that the angle between their axes is $\tan^{-1}\left(\frac{AB}{2BC}\right)$.

1157. If in any position AB make an angle θ with the axis of x and α, β be the angles which the tangents at C, D to their loci make with the axes of y, x respectively,

$$\cot\alpha + \cot\theta = \cot\beta + \tan\theta = \frac{AB}{BC}.$$

1158. A straight line of given length slides between two fixed straight lines and from its extremities two straight lines are drawn in given directions: prove that the locus of their intersection is an ellipse.

1159. A circle being traced on a plane, the locus of the vertex of all cones on that base whose principal elliptic sections have an excentricity e is the surface generated by the revolution about its conjugate axis of an hyperbola of excentricity e^{-1}.

1160. Trace the conics

$$2x^2 - 2xy + 2ay - a^2 = 0,\ 2y^2 - 2xy - 2ay + a^2 = 0,\ 2xy - 2ay + a^2 = 0;$$

proving that they touch each other two and two.

1161. Trace the following conics:

(1) $5x^2 + 20y^2 + 8xy - 35x - 80y + 60 = 0$,

(2) $36x^2 - 20y^2 + 33xy - 105x + 7y - 23 = 0$,

(3) $36x^2 + 29y^2 + 24xy - 72x + 126y + 81 = 0$,

(4) $144x^2 - 144y^2 - 120xy + 120x - 24y + 1 = 0$,

(5) $369x^2 + 481y^2 - 384xy - 2628x + 3654y + 2484 = 0$,

(6) $16x^2 + 9y^2 - 24xy - 96x - 72y + 144 = 0$,

(7) $4x^2 + y^2 - 4xy - 24x + 22y + 61 = 0$,

(8) $7x^2 - 7y^2 - 48xy + 175x + 175y - 1050 = 0$,

(9) $84x^2 + 40y^2 - 116xy - 5460x + 3780y + 88641 = 0$.

[To reduce the equation $u = 0$, if $ab - h^2$ be finite first move the origin to the point whose co-ordinates satisfy the equations

$$\frac{du}{dx} = 0,\ \frac{du}{dy} = 0;$$

the equation will become

$$ax^2 + by^2 + 2hxy = \frac{\Delta}{ab - h^2},$$

and if the axes be then turned through the acute angle θ determined by the equation $(a - b) \tan 2\theta = 2h$, the equation becomes

$$a'x^2 + b'y^2 = \frac{\Delta}{ab - h^2},$$

where a', b' are the roots of the equation $(z - a)(z - b) = h^2$, and the sign of $a' - b'$ is the same as that of h. When $ab = h^2$, we may suppose u to be $(\alpha x + \beta y)^2 + 2fy + 2gx + c$, and arranging it in the form

$$(\alpha x + \beta y - k)^2 + 2(f + k\beta) y + 2(g + k\alpha) x + c - k^2,$$

if we determine k by the equation

$$\alpha(g + k\alpha) + \beta(f + k\beta) = 0,$$

the straight line $\alpha x + \beta y = k$ will be the axis and the straight line

$$2(f + k\beta) y + 2(g + k\alpha) x + c - k^2 = 0,$$

the tangent at the vertex.]

1162. Prove that in general two parabolas can be drawn through the points of intersection of the conics

$$u \equiv ax^2 + by^2 + c + 2fy + 2gx + 2hxy = 0,$$
$$u' \equiv a'x^2 + b'y^2 + c' + 2f'y + 2g'x + 2h'xy = 0;$$

and that their axes will be at right angles if

$$\frac{h}{a - b} = \frac{h'}{a' - b'}.$$

1163. The equation of the director circle of the conic $u = 0$ is

$$C(x^2 + y^2) - 2Gx - 2Hy + A + B = 0,$$

A, B, &c. being the reciprocal coefficients.

1164. The equation of the asymptotes of the conic $u = 0$ is

$$Cu = \Delta.$$

1165. The foci of the conic $u = 0$ are given by the two equations

$$Fx + Gy - H = Cxy,$$
$$2Gx - 2Fy - A + B = C(x^2 - y^2).$$

1166. The equation of the chord of the conic $u = 0$ which is bisected by the point (X, Y) is

$$(x - X)(aX + hY + g) + (y - Y)(hX + bY + f) = 0,$$

or

$$(x - X)\frac{dU}{dX} + (y - Y)\frac{dU}{dY} = 0.$$

1167. Prove that the origin of co-ordinates lies on one of the equal conjugate diameters of the conic $u=0$, if

$$2\frac{f^2+g^2}{a+b}=\frac{af^2+bg^2-2fgh}{ab-h^2}.$$

1168. The rectangle under the distances of the origin from two conjugate foci of the conic $u=0$ is

$$\frac{1}{C}\sqrt{(A-B)^2+4H^2}.$$

1169. The equation of the asymptotes of the conic $u=0$ is

$$\frac{d^2u}{dx^2}\left(\frac{du}{dy}\right)^2+\frac{d^2u}{dy^2}\left(\frac{du}{dx}\right)^2-2\frac{d^2u}{dxdy}\frac{du}{dx}\frac{du}{dy}=0,$$

that of the axes is

$$\frac{d^2u}{dxdy}\left\{\left(\frac{du}{dx}\right)^2-\left(\frac{du}{dy}\right)^2\right\}=\left(\frac{d^2u}{dx^2}-\frac{d^2u}{dy^2}\right)\frac{du}{dx}\frac{du}{dy},$$

and the foci are determined by the equations

$$\frac{\left(\frac{du}{dx}\right)^2-\left(\frac{du}{dy}\right)^2}{\frac{d^2u}{dx^2}-\frac{d^2u}{dy^2}}=\frac{\frac{du}{dx}\frac{du}{dy}}{\frac{d^2u}{dxdy}}=2u.$$

[The co-ordinates are supposed rectangular in the two latter, but not necessarily in the first.]

1170. The equation of the equal conjugate diameters of any conic $u=0$ is

$$\left(\frac{d^2u}{dx^2}+\frac{d^2u}{dy^2}\right)\left\{\frac{d^2u}{dy^2}\left(\frac{du}{dx}\right)^2+\frac{d^2u}{dx^2}\left(\frac{du}{dy}\right)^2-2\frac{d^2u}{dxdy}\frac{du}{dx}\frac{du}{dy}\right\}$$
$$=2\left\{\frac{d^2u}{dx^2}\frac{d^2u}{dy^2}-\left|\frac{d^2u}{dxdy}\right|^2\right\}\left\{\left(\frac{du}{dx}\right)^2+\left(\frac{du}{dy}\right)^2\right\}.$$

1171. The rectangle under the distances of any point (x, y) from two conjugate foci of the conic $u=0$ is

$$\sqrt{\frac{\left\{2u\left(\frac{d^2u}{dx^2}-\frac{d^2u}{dy^2}\right)-\left(\frac{du}{dx}\right)^2+\left(\frac{du}{dy}\right)^2\right\}^2+4\left\{2u\frac{d^2u}{dxdy}-\frac{du}{dx}\frac{du}{dy}\right\}^2}{\frac{d^2u}{dx^2}\frac{d^2u}{dy^2}-\left(\frac{d^2u}{dxdy}\right)^2}}.$$

1172. The equation of a conic, confocal with $u=0$, and having its asymptotes along the equal conjugate diameters, is

$$u\left\{\left(\frac{d^2u}{dx^2}-\frac{d^2u}{dy^2}\right)^2+4\left|\frac{d^2u}{dxdy}\right|^2\right\}$$
$$=\left(\frac{d^2u}{dx^2}-\frac{d^2u}{dy^2}\right)\left(\left|\frac{du}{dx}\right|^2-\left|\frac{du}{dy}\right|^2\right)+4\frac{d^2u}{dxdy}\frac{du}{dx}\frac{du}{dy}.$$

1173. Prove that the general equation of a conic confocal with $u=0$ is

$$\lambda^2\Delta+\lambda\left(\frac{d^2u}{dx^2}+\frac{d^2u}{dy^2}\right)\left(\overline{\frac{du}{dx}}^2+\overline{\frac{du}{dy}}^2\right)=2(\lambda-8)\left(\frac{d^2u}{dx^2}+\frac{d^2u}{dy^2}\right)^2u.$$

If $\lambda=16$, this is the confocal whose asymptotes lie along the equal conjugate diameters of u; and, if $\lambda=8$, it is the confocal about which can be described parallelograms inscribed in u.

1174. The equation

$$2u\left(l^2\frac{d^2u}{dx^2}+2lm\frac{d^2u}{dxdy}+m^2\frac{d^2u}{dy^2}\right)=\left(l\frac{du}{dx}+m\frac{du}{dy}\right)^2$$

represents a pair of parallel tangents to the conic $u=0$, and the equation

$$lx\frac{du}{dy}+my\frac{du}{dx}=0$$

represents a pair of conjugate diameters of the conic

$$u\equiv ax^2+by^2+2hxy=c.$$

1175. The axes of the conic $ax^2+by^2+2hxy=c$ make with the lines bisecting the angles between the co-ordinate axes angles θ; prove that

$$\tan 2\theta=\frac{(a-b)\sin\omega}{(a+b)\cos\omega-2h}.$$

1176. If $\frac{x^2}{a'^2}+\frac{y^2}{b'^2}=1$ be the equation of a conic referred to conjugate diameters, the condition that the circle $x^2+y^2+2xy\cos\omega=r^2$ may touch the conic is

$$\left(\frac{1}{r^2}-\frac{1}{a'^2}\right)\left(\frac{1}{r^2}-\frac{1}{b'^2}\right)=\frac{\cos^2\omega}{r^4}.$$

Hence determine the relations between any conjugate diameters and the axes.

1177. The locus of the foot of the perpendicular let fall from a point (X, Y) on any tangent to the ellipse $ax^2+2hxy+by^2-1\equiv v=0$,

$$b(x-X)^2-2h(x-X)(y-Y)+a(y-Y)^2$$
$$=(ab-h^2)(x^2+y^2-xX-yY)^2,$$

the axes being rectangular. Prove that this reduces to a circle and a point-circle when

$$\frac{X^2-Y^2}{a-b}=\frac{XY}{h}=\frac{1}{h^2-ab}.$$

1178. The equations determining the foci of the conic $v=0$ are

$$\frac{y(x+y\cos\omega)}{a\cos\omega-h}=\frac{x(y+x\cos\omega)}{b\cos\omega-h}=\frac{1}{ab-h^2}.$$

[The equation $\dfrac{x^2-y^2}{a-b}=\dfrac{1}{h^2-ab}$ is true whatever be the inclination of the axes.]

1179. The general equation of a conic confocal with the conic $v=0$ is

$$(x^2+y^2)(ab-h^2)+\lambda(ax^2+by^2+2hxy)=\frac{(a+\lambda)(b+\lambda)-h^2}{\lambda};$$

and the given conic is also cut orthogonally by any conic whose equation is

$$\lambda\left(\frac{x^2-y^2}{a-b}-\frac{xy}{h}\right)+ax^2+2\frac{ab}{h}xy+by^2+1=0.$$

1180. The equations of the equal conjugate diameters of the conic $v=0$ is

$$\frac{v}{ab-h^2}=\frac{2(x^2+y^2)}{a+b},$$

when the axes are rectangular; or

$$\frac{v}{ab-h^2}=\frac{2(x^2+y^2+2xy\cos\omega)}{a+b-2h\cos\omega},$$

when the axes are oblique.

1181. The tangents to the two conics

$$\frac{x^2}{a}+\frac{y^2}{b}=1,\qquad \frac{x^2}{a}+2\lambda xy-\frac{y^2}{b}=\frac{a-b}{a+b},$$

at any common point are at right angles; and if both curves be hyperbolas they will have four real common tangents.

1182. An ellipse and an hyperbola are confocal and the asymptotes of one lie along the equal conjugate diameters of the other: prove that any conic drawn through the ends of the axes of the ellipse will cut the hyperbola orthogonally.

1183. Each common tangent drawn to the two conics

$$\frac{x^2}{a^2}+\frac{y^2}{b^2}=1,\quad \frac{x^2}{a^2}+2\lambda\frac{xy}{ab}-\frac{y^2}{b^2}+(1+\lambda^2)\frac{a^2+b^2}{a^2-b^2}=0,$$

will subtend a right angle at the centre.

1184. Two common tangents to the circle $x^2+y^2=2ax$ and the conic

$$x^2+(y-\lambda x)^2+2ax=0$$

subtend each a right angle at the origin: also the tangents are parallel to each other, and the straight lines joining the origin to the points of contact with either curve are parallel to the axes of the conic. Hence prove that, if at a point on an ellipse where the rectangle under the focal distances is equal to that under the semi-axes a circle equal to the circle of curvature be drawn touching the ellipse externally, and PP', QQ' be the other common tangents, $P'Q'$, $P'Q$ will pass through the point of contact and be parallel to the axes.

1185. Two parabolas are so situated that a circle can be described through their four common points: prove that the distance of the centre of this circle from the axis of one parabola is equal to half the latus rectum of the other.

1186. An hyperbola is drawn touching the axes of an ellipse and the asymptotes of the hyperbola touch the ellipse: prove that the centre of the hyperbola lies on one of the equal conjugate diameters of the ellipse.

1187. On two fixed straight lines are taken fixed points $A, B; C, D$: prove that the parabola which touches the two fixed straight lines and the asymptotes of any conic through A, B, C, D will also touch the straight line which bisects AB and CD.

1188. With two conjugate semi-diameters CP, CD of an ellipse as asymptotes is described an hyperbola, and pd is a common chord parallel to PD and bearing to it the ratio $n : 1$: the curvatures of the two curves at any common point will be as $1 : 1 - n^2$.

1189. Five fixed points are taken, no three of which are in one straight line, and five conics are described each bisecting all the lines joining four of the points, two and two: prove that these conics will have one common point.

1190. A conic is drawn touching a given conic at P and passing through its foci S, S': prove that the pole of SS' with respect to this conic will lie on the common normal at P, and will coincide with the common centre of curvature when the conics osculate.

1191. A parabola is drawn having its axis parallel to a given straight line and having double contact with a given ellipse: prove that the locus of its focus is an hyperbola confocal with the ellipse and having one asymptote in the given direction.

[If the given direction be that of the diameter of the ellipse through the point P $(a\cos\alpha, b\sin\alpha)$ and the latus rectum of the parabola be $2k\,a^2b^2 \div CP^2$, the co-ordinates of the focus are

$$\tfrac{1}{2}a\cos\alpha\left(\frac{k(a^2-b^2)}{a^2\cos^2\alpha+b^2\sin^2\alpha}+\frac{1}{k}\right),\quad \tfrac{1}{2}b\sin\alpha\left(\frac{k(b^2-a^2)}{a^2\cos^2\alpha+b^2\sin^2\alpha}+\frac{1}{k}\right),$$

and the equation of the directrix is

$$ax\cos\alpha+by\sin\alpha=\frac{a^2\cos^2\alpha+b^2\sin^2\alpha}{2k}+\frac{k(a^2+b^2)}{2}.]$$

1191*. An hyperbola is drawn touching a given ellipse, passing through its centre, and having its asymptotes parallel to the axes: prove that the centre of curvature of the ellipse at the point of contact lies on the hyperbola, and that the chord of intersection of the two curves touches the locus of the centre of the hyperbola at a point whose distance from the centre of the hyperbola is bisected by the centre of the ellipse. At the point of contact the curvature of the hyperbola is two-thirds of that of the ellipse.

1192. Taking the equation of a conic to be $u=0$, if λ be so determined that the equation $u+\lambda(x^2+y^2+2xy\cos\omega)=0$ represent two straight lines, the part of any tangent to the conic intercepted between these straight lines will be divided by the point of contact into two parts subtending equal (or supplementary) angles at the origin. If the coordinates be rectangular $\left(\omega=\frac{\pi}{2}\right)$ and $px+qy=1$ be one of the two straight lines, then will

$$cpq+fp+gq+h=0,\quad a+2gp+cp^2=b+2fq+cq^2.$$

1193. The part of any tangent to the ellipse $a^2y^2+b^2x^2=a^2b^2$ intercepted between two fixed straight lines at right angles to each other is divided by the point of contact into two parts subtending equal angles at the point (X, Y): prove that

$$\frac{X^2}{a^2}-\frac{Y^2}{b^2}=\frac{a^2-b^2}{a^2+b^2},$$

and that the two straight lines intersect in the point (X', Y'), where

$$\frac{X'}{X}=-\frac{Y'}{Y}=\frac{a^2+b^2}{a^2-b^2}.$$

1194. The tangent at a point P to the parabola $y^2=4ax$ meets the tangent at the vertex (A) in Q and the straight line $x+4a=0$ in Q': prove that the angles QAP, $Q'AP$ are supplementary; and, generally, that the two straight lines $y=mx+\frac{a}{m}$, $y+m(x+4a)+\frac{a}{m}=0$, have a similar property with respect to the point $\left(\frac{a}{m^2}, \frac{2a}{m}\right)$.

1195. An ordinate MP is drawn to the ellipse $\frac{x^2}{a^2}+\frac{y^2}{b^2}=1$ and the tangent at P meets the axis of x in O; from O are drawn two tangents to the ellipse $\frac{x^2}{a^2}+\frac{y^2}{b^2}=\frac{1}{e^2}$; prove that the parts of any tangent to the first ellipse intercepted between these two will be divided by the point of contact into two parts subtending equal angles at M. For the two lines to be real, P must lie between the latera recta.

1196. The straight lines AA', BB', CC' are let fall from A, B, C perpendicular to the opposite sides of the triangle ABC, and conics are described touching the sides CA, AB and the perpendiculars on them: prove that the locus of the foci is

$$(x^2+y^2)\{(b^2+ac)x+b(a-c)y-(b^2+ac)(a-c)\}$$
$$=ac\{(b^2+ac)x-b(a-c)y\},$$

reducing to

$$x(x^2+y^2-a^2)=0,\ \text{when } c=a.$$

(The origin is A', $A'A$ the axis of y, and the lengths $A'B$, $A'A$, CA' are denoted by a, b, c.) In the last case trace the positions of the foci for all different positions of the centre on $A'A$.

1197. Two conics have four-point contact at O, their foci are S, H, S', H' respectively, and the circles OSH, $OS'H'$ are drawn: prove that the poles of SH, $S'H'$ with respect to the corresponding circles lie on the common chord of the two circles.

1198. Two conics osculate at a and intersect in O, the tangents at O meet the curves again in b, c, the tangents at b, c meet the tangent at a in C, B and each other in A: prove that Aa, Bb, Cc, meet in a point and that A, O, a lie on one straight line.

1199. Two conics osculate at O and intersect in P, any straight line drawn through P meets the conics again in Q, Q': prove that the tangents at Q, Q' intersect in a point whose locus is a conic touching the other two at O and also touching them again, and the curvature at O of this locus is three-fourths of the curvature of either of the former, and that the straight lines joining O with the other two points of contact form with OP and the tangent at O a harmonic pencil. If one conic be a circle and the angle POQ a right angle, OP, OQ will be parallel to the axes of the other.

[The equations of the two conics being

$$x^2 + by^2 + hxy = ax,$$
$$x^2 + by^2 + h'xy = ax,$$

that of the locus is

$$\{(h-h')\,x + by\}^2 = 4b\,(x^2 + by^2 + hxy - ax).]$$

1200. A given conic turns in one plane about (1) its centre, (2) a focus: prove that the locus of the pole of a fixed straight line with respect to the conic is (1) a circle, (2) a conic, which is a parabola when the minor axis can coincide with the fixed straight line.

[The locus in general is to be found from the equations

$$x = p\cos\theta - q\sin\theta + \frac{a^2\cos^2\theta + b^2\sin^2\theta}{h - p\cos\theta + q\sin\theta},$$

$$y = p\sin\theta + q\cos\theta + \frac{(a^2 - b^2)\sin\theta\cos\theta}{h - p\cos\theta + q\sin\theta};$$

the fixed point about which the conic turns being origin, the fixed straight line being $x = h$; a, b the semi-axes, and p, q the co-ordinates of the centre when its axis ($2b$) is parallel to the fixed straight line.]

1201. A conic has double contact with a given conic: prove that its real foci lie on a conic confocal with the given conic, and its excentricity is given by the equation

$$\frac{e^2}{2-e^2} = \left\{X^2\frac{b^2-c^2}{a^2-c^2} + Y^2\frac{a^2-c^2}{b^2-c^2}\right\} \div \left\{X^2\frac{b^2+c^2}{a^2-c^2} + Y^2\frac{a^2+c^2}{b^2-c^2}\right\},$$

where $\dfrac{x^2}{a^2} + \dfrac{y^2}{b^2} = 1$ is the given conic, (X, Y) the pole of the chord of contact, and $a^2 - c^2$, $b^2 - c^2$ the squares on the semi-axes of the confocal through the foci. The foci are given by the equations

$$\frac{xX}{a^2-c^2} + \frac{yY}{b^2-c^2} = 1, \quad \frac{x^2}{a^2-c^2} + \frac{y^2}{b^2-c^2} = 1,$$

the equation of the conic of double contact being

$$\left(\frac{x^2}{a^2}+\frac{y^2}{b^2}-1\right)\left\{\frac{X^2}{a^2(a^2-c^2)}+\frac{Y^2}{b^2(b^2-c^2)}\right\}+\frac{1}{c^2}\left(\frac{xX}{a^2}+\frac{yY}{b^2}-1\right)^2=0.$$

1202. Tangents are drawn to the conic $ax^2+by^2+2hxy=2x$ from two points on the axis of x equidistant from the origin: prove that their four points of intersection lie on the conic $by^2+hxy=x$.

1203. On any diameter of a given ellipse is taken a point such that the tangents from it intercept on the tangent at one end of the diameter a length equal to the diameter: prove that the locus of the point is the curve

$$\left(\frac{x^2}{a^2}-\frac{y^2}{b^2}\right)^2=\left(\frac{a^2+b^2}{a^2-b^2}\right)^2\left(\frac{x^2}{a^2}+\frac{y^2}{b^2}\right).$$

1204. On the diameter through any point P of a parabola is taken a point Q such that the tangents from Q intercept on the tangent at P a length equal to the focal chord parallel to the tangent at P: prove that the locus of Q is the parabola

$$3y^2+4a(x+4a)=0.$$

1205. Tangents are drawn to the conic $ax^2+by^2+2hxy=2x$ from two points on the axis of x, dividing harmonically the segment whose extremities are at distances p, q from the origin: the locus of their points of intersection will be the conic

$$\{p(ax+hy-1)-x\}\{q(ax+hy-1)-x\}$$
$$=(apq-p-q)(ax^2+by^2+2hxy-2x\}.$$

1206. From P, P' ends of a diameter of a given conic are drawn tangents to another given concentric conic: prove that their other points of intersection lie on a fixed conic touching the four common tangents of the given conics; so that if the two given conics be confocal the locus is a third confocal and the tangents form a parallelogram of constant perimeter. In this last case, if Q, Q' be the points of intersection and tangents be drawn at P, P', Q, Q', their points of intersection will lie on a fixed circle.

[The equations in the latter case are $\frac{x^2}{a^2+\lambda}+\frac{y^2}{b^2+\lambda}=1$, $\frac{x^2}{a^2}+\frac{y^2}{b^2}=1$, $\frac{x^2}{a^2+\mu}+\frac{y^2}{b^2+\mu}=1$, where $\lambda\mu=a^2b^2$, $x^2+y^2=\frac{(a^2+\lambda)(b^2+\lambda)}{\lambda}$, and the perimeter is twice the diameter of the circle. If λ be negative, one of the conics must be an hyperbola for real tangents, in which case the locus will be an hyperbola and the difference of the sides of the parallelogram will be constant.]

1207. From two points O, O' are drawn tangents to a given conic whose centre is C: prove that if the conic drawn through the four points of contact and through O, O' be a circle, CO, CO' will be equally inclined to the axes and O, O' will be conjugate with respect to the rectangular hyperbola whose vertices are the foci of the given conic.

1208. The area of the ellipse of minimum excentricity which can be drawn touching two given straight lines at distances h, k from their point of intersection is

$$\pi hk\,(h^2+k^2)\,\frac{(h^2+k^2-2hk\cos\omega)^{\frac{1}{2}}}{(h^2+k^2+2hk\cos\omega)^{\frac{3}{2}}}\sin\omega\,;$$

and, if e be the minimum excentricity,

$$\frac{e^4}{1-e^2}=\frac{(h^2-k^2)^2}{h^2k^2\sin^2\omega}.$$

[Since $\dfrac{a+b-2h\cos\omega}{\sin^2\omega}$ and $\dfrac{ab-h^2}{\sin^2\omega}$ are invariants, so is

$$\frac{(a+b-2h\cos\omega)^2}{(ab-h^2)\sin^2\omega},$$

which is thus equal to $\dfrac{(\alpha^2+\beta^2)^2}{\alpha^2\beta^2}$ if α, β be the semi-axes, or to $\dfrac{(e^2-2)^2}{1-e^2}$ if e be the excentricity. This function of e continually increases with e^2 so long as e^2 is less than 1, hence in a system of ellipses the excentricity will be a minimum when $\dfrac{e^4}{1-e^2}$, and when therefore

$$\frac{(a+b-2h\cos\omega)^2}{ab-h^2}$$

has its least value.]

1209. The conic of four-pointic contact with a given ellipse at the point $(a\cos\theta,\ b\sin\theta)$ has its minimum excentricity (e') given by the equation

$$\frac{e'^4}{1-e'^2}=\frac{e^4}{1-e^2}\sin^2 2\theta\,;$$

and the locus of its centre for different values of θ is the curve

$$(x^2+y^2)^2\left(\frac{x^2}{a^2}+\frac{y^2}{b^2}\right)=\left(\frac{x^2}{a^2}-\frac{y^2}{b^2}\right)^2(a^2-b^2)^2.$$

1210. The axes of any conic through four given points make with the bisectors of the angle θ between the axes of the two parabolas through the four points an angle ϕ: prove that the excentricity is given by the equation

$$\frac{e^4}{1-e^2}=\frac{4\cos^2\theta}{\sin^2\theta-\sin^2 2\phi};$$

and that the minimum excentricity is $\sqrt{\dfrac{2}{1+\sec\theta}}$, when $\sin 2\phi=0$, so that either axis of the ellipse of least excentricity is equally inclined to the axes of the two parabolas.

1211. Three points A, B, C are taken on an ellipse, the circle about ABC meets the ellipse again in P, and PP' is a diameter: prove that of all ellipses through A, B, C, P' the given ellipse is that of least excentricity.

1212. Of all ellipses circumscribing a parallelogram, the one of least excentricity has its equal conjugate diameters parallel to the sides.

1213. The ellipse of least excentricity which can be inscribed in a given parallelogram is such that any point of contact divides a side into segments which are as the squares on the respective adjacent diagonals.

1214. Four points are such that ellipses can be drawn through them, and e is the least excentricity of any such ellipse, e' the excentricity of the hyperbola on which the centres of the ellipses lie: prove that

$$\frac{e'^4}{e'^2-1}+\frac{e^4}{e^2-1}=4;$$

also that the equal conjugate diameters of the ellipse are parallel to the asymptotes of the hyperbola.

1215. The equation of the conic of least excentricity through the four points $(a, 0)$, $(a', 0)$, $(0, b)$, $(0, b')$ is

$$\frac{x^2}{aa'}+\frac{4xy\cos\omega}{aa'+bb'}+\frac{y^2}{bb'}-x\left(\frac{1}{a}+\frac{1}{a'}\right)-y\left(\frac{1}{b}+\frac{1}{b'}\right)+1=0,$$

and its axes are parallel to the asymptotes of the rectangular hyperbola through the four points.

1216. The axes of the conic which is the locus of the centres of all conics through four given points are parallel to the asymptotes of the rectangular hyperbola through the four points.

1217. The equation of the director circle of the conic

$$\frac{x}{h}+\frac{y}{k}-1=2\lambda\left(\frac{xy}{hk}\right)^{\frac{1}{2}}$$

is $\quad(\lambda^2-1)(x^2+y^2+2xy\cos\omega)+hx+ky+(kx+hy-hk)\cos\omega=0.$

1218. The equation of a conic, having the centre of the ellipse $a^2y^2+b^2x^2=a^2b^2$ for focus and osculating the ellipse at the point θ, is

$$(x^2+y^2)(a^2\cos^2\theta+b^2\sin^2\theta)^2=\{(a^2-b^2)(ax\cos^3\theta-by\sin^3\theta)+a^2b^2\}^2.$$

1219. A rectangular hyperbola has double contact with a parabola: prove that the centre of the hyperbola and the pole of the chord of contact will be equidistant from the directrix of the parabola.

1220. A conic is drawn to touch four given straight lines, two of which are parallel: prove that its asymptotes will touch a fixed hyperbola and that this hyperbola touches the diagonals of the quadrilateral, formed by the given lines, at their middle points.

1221. A parabola has four-pointic contact with a conic: prove that the axis of the parabola is parallel to the diameter of the conic through the point of contact, and that the latus rectum of the parabola is $\dfrac{2a^2b^2}{r^3}$, where a, b are the semi-axes of the conic and r the central distance of the point of contact. If the conic be a rectangular hyperbola, the envelope of the directrix of the parabola is

$$a^{\frac{2}{3}}=(2r)^{\frac{2}{3}}\cos\frac{2\theta}{3}.$$

1222. The locus of the centre of a rectangular hyperbola having four-pointic contact with the ellipse $a^2y^2 + b^2x^2 = a^2b^2$ is the curve

$$\left(\frac{x^2+y^2}{a^2+b^2}\right)^3 = \frac{x^2}{a^2} + \frac{y^2}{b^2}.$$

1223. The locus of the foci of all conics which have four-pointic contact with a given curve at a given point is a curve whose equation, referred to the normal and tangent at the given point, is of the form

$$(mx + y)(x^2 + y^2) = axy.$$

1224. The excentricity e of the conic whose equation, referred to axes inclined at an angle ω, is $u = 0$, satisfies the equation

$$\frac{e^4}{1-e^2} = \frac{(a-b)^2 \sin^2\omega + (\overline{a+b}\cos\omega - 2h)^2}{(ab-h^2)\sin^2\omega}.$$

1225. The co-ordinates of the focus of the parabola

$$\left(\frac{x}{h}\right)^{\frac{1}{2}} + \left(\frac{y}{k}\right)^{\frac{1}{2}} = 1$$

are given by the equations

$$hx = ky = \frac{h^2k^2}{h^2 + k^2 + 2hk\cos\omega},$$

and the equation of its directrix is

$$x(h + k\cos\omega) + y(k + h\cos\omega) = hk\cos\omega.$$

1226. In the parabola $\left(\frac{x}{h}\right)^{\frac{1}{2}} + \left(\frac{y}{k}\right)^{\frac{1}{2}} = 1$, a tangent meets the axes of co-ordinates in P, Q and perpendiculars are drawn from P, Q to the opposite axes respectively: prove that the locus of the point of intersection is

$$\frac{x + y\cos\omega}{k} + \frac{y + x\cos\omega}{h} = \cos\omega.$$

1227. The asymptotes of the conic

$$\frac{x}{h} + \frac{y}{k} - 1 = 2\lambda\left(\frac{xy}{hk}\right)^{\frac{1}{2}}$$

always touch the parabola

$$\left(\frac{x}{2h}\right)^{\frac{1}{2}} + \left(\frac{y}{2k}\right)^{\frac{1}{2}} = 1.$$

1228. One angular point of a triangle self-conjugate to a given conic is given: prove that the circles on the opposite sides as diameters will have a common radical axis which is normal at the point of contact to the similar concentric and similarly situate conic touching the polar of the given point.

1229. Two circles of radii a, b $(a > b)$ touch each other, and a conic is described having real double contact with both: prove that, when the points of contact are not on different branches of an hyperbola, the excentricity $< \frac{1}{2}\left(1+\frac{a}{b}\right)$, and the latus rectum is $>$, $=$, or $< a-b$, according as the conic is an ellipse, parabola, or hyperbola. If the contacts are on different branches, the excentricity $< \frac{a+b}{a-b}$, and the asymptotes always touch a fixed parabola.

1230. A triangle ABC circumscribes an ellipse whose foci are S, S' and $SA = SB = SC$: prove that

$$\frac{S'A \,.\, S'B \,.\, S'C}{SA \,.\, SB \,.\, SC} = 1 - e^2;$$

and that each angle of the triangle ABC lies between the acute angles $\cos^{-1}\frac{1 \pm e}{2}$.

[When the conic is an hyperbola, $e^2 - 1$ replaces $1 - e^2$, and one angle of the triangle will be obtuse and $> \pi - \cos^{-1}\frac{e-1}{2}$.]

1231. On every straight line can be found two real points conjugate to each other with respect to a given conic and the distance between which subtends a right angle at a given point not on the straight line.

1232. Prove that the axis of a parabola, which passes through the feet of the four normals drawn to a given ellipse from a given point, will be parallel to one of the equal conjugate diameters of the ellipse.

[If (X, Y) be the given point and $a^2y^2 + b^2x^2 = a^2b^2$ the given ellipse, the equation of the axis of the parabola will be

$$\frac{1}{a}\left(x + \frac{a^4X}{a^4-b^4}\right) \pm \frac{1}{b}\left(y + \frac{b^4Y}{a^4-b^4}\right) = 0.]$$

1233. A conic is drawn through four given points lying on two parallel straight lines: prove that the asymptotes touch the parabola which touches the other four joining straight lines.

[The equation of the conic being taken to be

$$\left(\frac{x}{a}+\frac{y}{b}-1\right)\left(\frac{x}{a}+\frac{y}{b}-m\right) + (\lambda - 2)\frac{xy}{ab} = 0,$$

that of the asymptotes will be found by adding $\frac{(1+m)^2}{\lambda} - m$ to the sinister so as to give for the asymptotes the equation

$$\left(\frac{x}{a}-\frac{y}{b}\right)^2 - (1+m)\left(\frac{x}{a}+\frac{y}{b}\right) + \lambda\frac{xy}{ab} + \frac{(1+m)^2}{\lambda} = 0;$$

and for the envelope the equation

$$\left\{\left(\frac{x}{a}-\frac{y}{b}\right)^2 - (1+m)\left(\frac{x}{a}+\frac{y}{b}\right)\right\}^2 = 4\frac{xy}{ab}(1+m)^2.$$

The student should observe, and account for, the factor $\left(\frac{x}{a}-\frac{y}{b}\right)^2$.]

1234. An ellipse of constant area πc^2 is described having four-pointic contact with a given parabola whose latus rectum is $2m$: prove that the locus of the centre of the ellipse is an equal parabola whose vertex is at a distance $\left(\frac{c^4}{m}\right)^{\frac{1}{3}}$ from the vertex of the given parabola; also that when $c = m$ the axes of the ellipse make with the axis of the parabola angles

$$\tfrac{1}{2}\tan^{-1}(2\tan\phi),$$

where ϕ is the angle which the tangent at the point of contact makes with the axis.

1235. An ellipse of constant area πc^2 is described having four-pointic contact with a given ellipse whose axes are $2a$, $2b$: prove that the locus of its centre is an ellipse, concentric similar and similarly situate with the given ellipse, the linear ratio of the two being $1-\left(\frac{c^2}{ab}\right)^{\frac{2}{3}} : 1$. Also the described ellipse will be similar to the given ellipse when the point of contact P is such that

$$CP^2 : CD^2 = c^{\frac{4}{3}} : (ab)^{\frac{2}{3}}.$$

VII. *Envelopes (of the second class).*

[The equation of the tangent to a parabola, in the form

$$y = mx + \frac{a}{m},$$

gives as the condition of equal roots in m, $y^2 = 4ax$; and the equation of the tangent to an ellipse

$$\frac{x}{a}\cos\alpha + \frac{y}{b}\sin\alpha = 1,$$

written in the form

$$z^2\left(1+\frac{x}{a}\right) - 2z\frac{y}{b} + 1 - \frac{x}{a} = 0, \quad \left(z \equiv \tan\frac{\alpha}{2}\right),$$

gives as the condition of equal roots in z

$$1 - \frac{x^2}{a^2} = \frac{y^2}{b^2}.$$

So in general if the equation of a line in one plane, straight or curved, involve a parameter in the second degree, it follows that through any proposed point can be drawn two lines of the series represented by the equation. These two lines will be the tangents (rectilinear or curvilinear) from the proposed point to the curve which is the envelope of the system. If the proposed point lie on this envelope the two tangents will coincide, hence the equation of the envelope may be found as the condition of equal roots.

Thus, "To find the envelope of a system of circles each having for its diameter a focal chord of a given conic."

If LSL' be a focal chord, $ASL = \alpha$, O the mid point of LL', and P any point on the circle, we shall have

$$SL = \frac{c}{1 + e\cos\alpha},\quad SL' = \frac{c}{1 - e\cos\alpha},\quad OL = \frac{c}{1 - e^2\cos^2\alpha},\quad SO = \frac{ce\cos\alpha}{1 - e^2\cos^2\alpha},$$

and $OP^2 = SO^2 + SP^2 - 2SO\,.\,SP\cos PSO$, whence the equation of the circle

$$c^2 = r^2(1 - e^2\cos^2\alpha) - 2cer\cos\alpha\cos(\theta - \alpha),$$

or, if $\tan\alpha = \lambda$,

$$(r^2 - c^2)(1 + \lambda^2) - e^2r^2 - 2cer(\cos\theta + \lambda\sin\theta) = 0,$$

whence, as the condition for equal roots in λ,

$$\{r^2(1 - e^2) - c^2 - 2cer\cos\theta\}(r^2 - c^2) = c^2e^2r^2\sin^2\theta,$$

equivalent to

$$(r^2 - c^2 - cer\cos\theta)^2 = e^2r^4,$$

or

$$r^2(1 \pm e) - cer\cos\theta = c^2,$$

so that the envelope is two circles, one of which degenerates into a straight line when $e = 1$, being the directrix of the parabola; and one degenerates into a point when $e = 2$, being the farther vertex of the hyperbola. In general, if A, A' be the nearer and farther vertices, the two circles will have for diameters the segments ASM, $A'SM'$, where $M'S = SM = c$ the semi latus rectum. In this case every one of the system of curves has real contact in two points with the envelope, but it frequently happens that the contact becomes impossible for a part of the series.

A method which is often the best is exemplified in the following: "To find the envelope of a chord of a conic which subtends a right angle at a given point."

Move the origin to the given point, and let the equation of the conic be

$$u \equiv ax^2 + by^2 + c + 2fy + 2gx + 2hxy = 0,$$

and let $px + qy = 1$ be the equation of a chord. The equation of the straight lines joining to the origin the ends of this chord will then be

$$ax^2 + by^2 + 2hxy + 2(gx + fy)(px + qy) + c(px + qy)^2 = 0,$$

which will be at right angles if

$$a + b + 2(pg + qf) + c(p^2 + q^2) = 0.$$

Hence the equation

$$(a + b)(px + qy)^2 + 2(pg + qf)(px + qy) + c(p^2 + q^2) = 0$$

represents two parallel chords of the series and involves the parameter $p : q$ in the second degree; whence, for the envelope,

$$(\overline{a+b}\,x^2 + 2gx + c)(\overline{a+b}\,y^2 + 2fy + c) = (\overline{a+b}\,xy + fx + gy)^2,$$

or

$$(f^2 - bc + g^2 - ca)(x^2 + y^2) = (gx + fy + c)^2,$$

or the envelope is a conic having a focus at the given point, directrix the polar of the given point with respect to the given conic, and excentricity

$$\sqrt{\frac{f^2 + g^2}{f^2 + g^2 - c(a + b)}}.$$

The envelope of any series of lines is to be found from the condition that the equation shall give two equal values of the parameter, but in all the following examples it will be found that the equation of the line can be written in the form

$$U + 2\lambda V + \lambda^2 W = 0,$$

where λ is the parameter, so that the envelope is $UV = W^2$. A common form is $W = U\cos\theta + V\sin\theta$ when the envelope is, as already seen,

$$W^2 = U^2 + V^2.]$$

1236. A conic has a given focus and given length and direction of major axis: the envelope is two parabolas whose common focus is the given focus and whose common latus rectum is in the given direction and of twice the given length.

[This is obvious geometrically.]

1237. The envelope of the circles

$$x^2 + (y - p)(y - q) = 0,$$

where p, q are connected by the equation $pq \pm a(p - q) = 0$, is the two circles

$$x^2 + y^2 \pm 2ax = 0.$$

[It will be found best to make the ratio $p : q$ the parameter, so that the equation will be

$$pq(x^2 + y^2) \pm ay(p^2 - q^2) + a^2(p - q)^2 = 0.]$$

1238. The envelope of the circles

$$x^2 + (y - p)(y - q) = 0,$$

where p, q are connected by the equation

$$(p + a)(q - a) + b^2 = 0,$$

is the two circles

$$(x \pm b)^2 + y^2 = a^2.$$

1239. The ellipse $q^2x^2 + p^2y^2 = p^2q^2$ has its axes connected by the equation $a^2p^2 = q^2(p^2 - q^2)$: prove that the envelope is the two circles

$$x^2 + y^2 \pm 2ax = 0;$$

and, if the relation between the axes be $a^2p^2 = q^2(m^2p^2 - n^2q^2)$, the envelope will be

$$m^2x^2 + n^2y^2 \pm 2nax = 0.$$

1240. The envelope of the ellipse

$$x^2 + y^2 - 2(ax\cos\alpha + by\sin\alpha)\left(\frac{x}{a}\cos\alpha + \frac{y}{b}\sin\alpha\right)$$

$$+ (a^2 + b^2 - c^2)\left(\frac{x}{a}\cos\alpha + \frac{y}{b}\sin\alpha\right)^2 = a^2\sin^2\alpha + b^2\cos^2\alpha - c^2$$

is the two confocal ellipses

$$\frac{x^2}{a^2} + \frac{y^2}{b^2} = 1, \quad \frac{x^2}{a^2 - c^2} + \frac{y^2}{b^2 - c^2} = 1.$$

[The parameter $\tan \alpha$ is involved in the second degree.]

1241. The director circle of a conic and one point of the conic are given: prove that the envelope is a conic whose major axis is a diameter of the given circle.

[The equation may be taken to be $\frac{x^2}{a^2} + 2\lambda\frac{xy}{ab} + \frac{y^2}{b^2} = 1$, b, λ being parameters connected by the equation

$$a^2 + b^2 = c^2(1 - \lambda^2),$$

and a, c, given. The envelope is $\frac{x^2}{c^2} + \frac{y^2}{c^2 - a^2} = 1$.]

1242. Through a fixed point O is drawn a straight line meeting two fixed straight lines parallel to each other in L, M: the envelope of the circle whose diameter is LM is a conic whose focus is O and whose transverse axis has its ends on the two fixed straight lines.

1243. A variable tangent to a given parabola meets two fixed tangents, and another parabola is drawn touching the fixed tangents in these points: prove that the directrix of this last envelopes a third parabola touching straight lines, drawn at right angles to the two fixed tangents through their common point, in the points where they are met by the directrix of the given parabola.

1244. A variable tangent to a given parabola meets two fixed tangents, and on the intercepted segment as diameter a circle is described: the envelope is a conic touching the two fixed tangents in the points where they are met by the directrix of the given parabola.

[The given parabola being $\left(\frac{x}{a}\right)^{\frac{1}{2}} + \left(\frac{y}{b}\right)^{\frac{1}{2}} = 1$, the envelope is

$$\{x(a + b\cos\omega) + y(b + a\cos\omega) - ab\cos\omega\}^2 = 4abxy\sin^2\omega.]$$

1245. Through a point P are drawn two circles each touching two fixed equal circles which touch each other at A: prove that the angle at which the two circles intersect at P is $2\sec^{-1}(n)$, where n is the ratio of the radius of the circle drawn through P to touch the two at A to the radius of either. If P lie on BB', a common tangent to the two fixed circles, the circle through P touching the two will make with BB' an angle $4PAB$.

1246. Through each point of the straight line $\frac{lx}{a} + \frac{my}{b} = 1$ is drawn a chord of the ellipse $a^2y^2 + b^2x^2 = a^2b^2$ bisected in the point: prove that the envelope is the parabola whose focus is the point

$$\frac{x}{al} = \frac{-y}{bm} = \frac{a^2 - b^2}{l^2a^2 + m^2b^2},$$

and directrix the straight line

$$lax + mby = a^2 + b^2.$$

1247. An hyperbola has a focus at the centre of a given circle and its asymptotes in given directions: prove that tangents drawn to it at the points where it meets the given circle envelope an hyperbola to which the given circle is the auxiliary circle.

[The equation of a tangent will be $\frac{c}{r} = e \cos\theta + \cos(\theta - \alpha)$ where

$$\frac{c}{a} = 1 + e\cos\alpha,$$

or we may write the equation

$$\frac{a}{r}(1 + e\cos\alpha) = e\cos\theta + \cos(\theta - \alpha),$$

involving only the parameter α, and giving the envelope

$$\left(\frac{a}{r} - e\cos\theta\right)^2 = \left(\cos\theta - \frac{ae}{r}\right)^2 + \sin^2\theta,$$

or

$$\frac{x^2}{a^2} - \frac{y^2}{a^2(e^2 - 1)} = 1.]$$

1248. A circle subtends the same given angle at each of two given points: prove that its envelope is an hyperbola whose foci are the two given points and whose asymptotes include an angle supplementary to the given one.

1249. A circle has its centre on a fixed straight line and intercepts on another straight line a segment of constant length: prove that its envelope is an hyperbola of which the first straight line is the conjugate axis and the second straight line is an asymptote.

1250. A conic is drawn having its focus at A, the vertex of a given conic, passing through a focus S of the given conic, and having for directrix a tangent to the given conic: prove that its envelope is a conic having its focus at the given vertex and excentricity $\frac{e}{2 \mp e}$. Also the envelopes of the minor axis, the second latus rectum, and the second directrix are each conics similar to the given conic, and bearing to it the linear ratios $1 \mp e : 2e$, $1 \mp e : e$; and $1 + e \mp e : e$, the upper or lower signs being taken according as S is the nearer or farther focus.

1251. A triangle is inscribed in the hyperbola $xy = c^2$ whose centroid is the fixed point (cm, cm^{-1}): prove that its sides envelope the ellipse

$$\left(3my + \frac{3x}{m} - 8c\right)^2 = 4xy,$$

which touches the asymptotes of the hyperbola; and also touches the hyperbola at the point (cm, cm^{-1}), the curvatures at the point being as $4 : 1$. Where the ellipse again meets the hyperbola, its tangents are parallel to the asymptotes.

1252. The centre and directrix of an ellipse are given: the envelope is two parabolas having their common focus at the given centre.

[The ellipse may be taken $\frac{x^2}{p}+\frac{y^2}{q}=1$, where $p^2=c^2(p-q)$; or $(qx^2+py^2)\,p=c^2q\,(p-q)$, involving only the parameter $p:q$.]

1253. One extremity of the minor axis, and the directrix, of a conic are given: the envelope is a circle with centre at the given point and touching the given line.

[The equation is $\frac{x^2}{a^2}+\frac{y^2}{b^2}=\frac{2y}{b}$, where $\frac{a^4}{a^2-b^2}=c^2$, or may be written $(b^2x^2+a^2y^2)^2=4c^2y^2\,(a^2b^2-b^4)$, involving only the parameter $a^2:b^2$. This theorem is easily proved geometrically.]

1254. Find the envelope of the circle $(x-h)^2+y^2=r^2$, when h, r are connected by the equation $h^2=2\,(r^2+a^2)$, and a is given.

1255. A circle rolls with internal contact upon a circle of half the radius: prove that the envelope of any chord of the rolling circle is a circle which reduces to a point when the chord is a diameter.

1256. A parabola rolls on an equal parabola, similar points being always in contact: prove that the envelope of any straight line perpendicular to the axis of the moving parabola is a circle.

[Also obvious geometrically.]

1257. A parabola has a given focus and intercepts on a given straight line a segment subtending a constant angle (2α) at the focus: prove that the envelope of its directrix is an ellipse having the given point for focus, the given straight line for minor axis, and excentricity $\cos\alpha$.

1258. The envelope of the straight line

$$\frac{x}{\cos\theta}-\frac{y\sin\phi}{\sin\theta}=a\cos\phi,$$

where θ, ϕ are parameters connected by the equation

$$h\cos\theta-\frac{k\sin\theta}{\sin\phi}=\frac{a}{\cos\phi},$$

is the parabola

$$(hx+ky-a^2)^2=4hkxy.$$

[By combining the two equations we may obtain the equation in the form

$$hx+ky-a^2-\lambda kx-\frac{1}{\lambda}hy=0.]$$

1259. The envelope of the conic

$$\frac{x^2}{a\,(a-kb)}+\frac{y^2}{b\,(ka-b)}=1$$

is the four straight lines $(x\pm y)^2=a^2-b^2$.

1260. A parabola is drawn touching a given straight line at a given point O, also the point on the normal at O, chords through which subtend a right angle at O, is given: the envelope is a circle in which the two given points are ends of a diameter, and each parabola touches the envelope at the point opposite to O in the parabola.

[The equation of the parabola may be taken $\sqrt{1-\lambda}\,x+\sqrt{1+\lambda}\,y=\sqrt{ax}$.]

1261. Through each point P of a given circle is drawn a straight line PQ of given length and direction (a given vector), and a circle is described on PQ as diameter: prove that the envelope of the common chord of the two circles is a parabola. The envelope of the circle is obviously two circles.

[If $x^2+y^2=a^2$ be the given circle, $2h$ the given length, axis of y the given direction, the envelope is $x^2+(y-h)^2=\left(a-\frac{h}{a}y\right)^2$.]

1262. Two points are taken on a given ellipse such that the normals intersect in a point lying on a fixed normal: prove that the envelope of the chord joining the two points is a parabola whose directrix passes through the centre and whose focus is the foot of the perpendicular from the centre on the tangent which is perpendicular to the given normal.

[If α be the excentric angle of the foot of the given normal, the equation of one of the chords will be

$$p\frac{x}{a}\cos\alpha+q\frac{y}{b}\sin\alpha=1,$$

where p, q are connected by the equation $pq+p+q=0$.]

1263. A chord PQ is drawn through a fixed point (X, Y) to the ellipse $\frac{x^2}{a^2}+\frac{y^2}{b^2}=1$, the normals at P, Q meet in O and from O are drawn OP', OQ' also normals: prove that the envelope of $P'Q'$ is the parabola

$$\left(\frac{xX}{a^2}+\frac{yY}{b^2}+1\right)^2=\frac{4xyXY}{a^2b^2},$$

whose focus is the foot of the perpendicular from the origin on the line $\frac{xX}{a^2}+\frac{yY}{b^2}+1=0$, and directrix the straight line $a^2xY+b^2yX=0$.

1264. Three points are taken on a given ellipse so that their centroid is a fixed point, the straight lines joining them two and two will touch a fixed conic.

[Refer the ellipse to conjugate diameters so that the fixed point is $\left(\frac{m}{3}a, 0\right)$, then taking the three points $(a\cos\alpha, b\sin\alpha)$, &c., we may take the chord joining two to be $p\frac{x}{a}+q\frac{y}{b}=\frac{1}{2}$, where

$$(m^2-1)(p^2+q^2)-2mp+1=0.$$

The envelope is

$$\left(\frac{2x}{a}-m\right)^2+\frac{4y^2}{b^2(1-m^2)}=1.]$$

1265. A triangle inscribed in the ellipse $a^2y^2+b^2x^2=a^2b^2$ has its centroid at the point $\left(\frac{X}{3}, \frac{Y}{3}\right)$: prove that its sides touch the conic

$$\left\{\frac{4x(x-X)}{a^2}+\frac{X^2}{a^2}+\frac{Y^2}{b^2}-1\right\}\left\{\frac{4y(y-Y)}{b^2}+\frac{X^2}{a^2}+\frac{Y^2}{b^2}-1\right\}$$
$$=\frac{4}{a^2b^2}(2xy-xY-yX)^2.$$

[The poles of the sides lie on the conic

$$\left(\frac{X^2}{a^2}+\frac{Y^2}{b^2}-1\right)\left(\frac{x^2}{a^2}+\frac{y^2}{b^2}\right)=4\left(\frac{xX}{a^2}+\frac{yY}{b^2}-1\right).]$$

1266. Three points are taken on a given ellipse such that the centre of perpendiculars of the triangle is a fixed point the envelope of the chords will be a fixed conic whose asymptotes are perpendicular to the tangents from the fixed point to the given conic.

[If (X, Y) be the given point, α, β, γ excentric angles of the corners of one of the triangles, then (1026)

$$(a^2+b^2)\cos\alpha-(a^2-b^2)\cos(\alpha+\beta+\gamma)=2aX-(a^2+b^2)(\cos\beta+\cos\gamma),$$
$$(a^2+b^2)\sin\alpha-(a^2-b^2)\sin(\alpha+\beta+\gamma)=2bY-(a^2+b^2)(\sin\beta+\sin\gamma);$$

square and add, and we obtain the relation

$$(a^2X^2+b^2Y^2)(p^2+q^2)-a^4q^2-b^4p^2-2(a^2+b^2)(paX+qbY)+(a^2+b^2)^2=0,$$

connecting the parameters in the equation $p\frac{x}{a}+q\frac{y}{b}=1$ of one of the sides. The equation of the envelope is

$$\left(2xX-\frac{a^2X^2+b^2Y^2-b^4}{a^2+b^2}\right)\left(2yY-\frac{a^2X^2+b^2Y^2-a^4}{a^2+b^2}\right)$$
$$+x^2(X^2-a^2)+y^2(Y^2-b^2)=2xyXY.]$$

1267. The envelope of a chord of the conic $\frac{x^2}{a^2}+\frac{y^2}{b^2}=1$ which subtends a right angle at the point (X, Y) is

$$\{(x-X)^2+(y-Y)^2\}(a^2+b^2-X^2-Y^2)=a^2b^2\left(\frac{xX}{a^2}+\frac{yY}{b^2}-1\right)^2;$$

and thence that if F be the point (X, Y), F' the second focus of this envelope, FF' is divided by either axis of the given ellipse into segments in the ratio $a^2+b^2 : a^2-b^2$, and that the major axis of the envelope bears to the minimum chord of the director circle through F the ratio $2ab : a^2+b^2$.

1268. A parabola has its focus at the focus of a given conic and touches the conic: prove that its directrix and the tangent at its vertex both envelope circles, the former one of radius equal to the major axis and with its centre at the second focus of the given conic, the latter the auxiliary circle of the given conic.

1269. Each diameter of a given parabola meets a fixed straight line, and from their common point is drawn a straight line making a given angle with the tangent corresponding to the diameter: the envelope is a parabola, degenerating when the given angle is equal to the angle which the fixed straight line makes with the axis.

1270. From the point where a diameter of a given conic meets a fixed straight line is drawn a straight line inclined at a given angle to the conjugate diameter: the envelope is a parabola. If the constant angle be made with the first diameter, the envelope is another parabola.

[The equations of the conic and straight line being

$$\frac{x^2}{a^2}+\frac{y^2}{b^2}=1,\ p\frac{x}{a}+q\frac{y}{b}=1,$$

and t the tangent of the given angle, the envelopes are respectively

$$\{pa(y-tx)+qb(x+ty)+t(a^2-b^2)\}^2=4ab\{p(x+ty)-a\}\{q(y-tx)-b\},$$

and $\{bp(x+ty)-aq(y-tx)\}^2+4ab\{q(x+ty)-bt\}\{p(y-tx)+at\}=0.$]

1271. A fixed point A is taken on a given circle, and a chord of the circle PQ is such that $PQ=e(AP\pm AQ)$: prove that the envelope of PQ is a circle of radius $a(1-e^2)$ touching the given circle at A, a negative value of the radius meaning external contact.

1272. A fixed point A is taken within a given circle, and a chord of the circle PQ is such that $PQ=e(AP+AQ)$: prove that the envelope of PQ is a circle coaxial with the given circle and the point A and whose radius is $\sqrt{(a^2-e^2c^2)(1-e^2)}$, where a is the radius of the given circle and c the distance of A from its centre.

1273. One given circle U lies within another V, and PQ is a chord of V touching U, S the interior point circle coaxial with U and V, PSP' a chord of V: prove that the envelope of QP' is a third coaxial circle such that the tangents drawn from any point of it to U, V are in the ratio $a^2-c^2 : a^2$, where a is the radius of U and c the distance between their centres.

1274. A circle passes through two fixed points A, B, and a tangent is drawn to it at the second point where it meets a fixed straight line through A: the envelope of this tangent is a parabola, whose focus is B, whose directrix passes through A, and whose axis makes with BA an angle double that which the fixed straight line makes with BA.

[If the two points be $(\pm a, 0)$, and $y\cos\alpha=(x-a)\sin\alpha$ the given straight line, the envelope is

$$(x+a)^2+y^2=\{(x-a)\cos 2\alpha+y\sin 2\alpha\}^2.]$$

1275. Through any point O on a fixed tangent to a given parabola is drawn a straight line $OPTP'$ meeting the parabola in P, P' and a given straight line in T, and OT is a mean proportional between OP, OP': prove that PP' either passes through a fixed point or envelopes a parabola.

1276. The envelope of the polar of the origin with respect to any circle circumscribing a maximum triangle inscribed in the ellipse $a^2y^2 + b^2x^2 = a^2b^2$ is

$$\frac{x^2}{a^2} + \frac{y^2}{b^2} = 4\left(\frac{a^2+b^2}{a^2-b^2}\right)^2.$$

[If θ be the fourth point in which the circumscribing circle meets the ellipse, the equation of the polar will be

$$-\frac{x}{a}\cos\theta + \frac{y}{b}\sin\theta = 2\frac{a^2+b^2}{a^2-b^2}.]$$

1277. A chord of a given conic is drawn through a given point, another chord is drawn conjugate to the former and equally inclined to a given direction: prove that the envelope of this latter chord is a parabola.

1278. A triangle is self-conjugate to the circle $(x-c)^2 + y^2 = b^2$, and two of its sides touch the circle $x^2 + y^2 = a^2$: prove that if the equation of the third side be $p(x-c) + qy = 1$, p, q will be connected by the equation

$$p^2(a^2+b^2) + q^2(a^2+b^2-c^2) + 2pc + c^2 - 2a^2 = 0;$$

and find the Cartesian equation of the envelope.

1279. A triangle ABC is inscribed in the circle $x^2 + y^2 = a^2$, and A is the pole of BC with respect to the circle $(x-c)^2 + y^2 = b^2$: prove that AB, AC envelope the conic

$$\frac{(2x-c)^2}{2a^2+2b^2-c^2} + \frac{2y^2}{a^2+b^2-c^2} = 1.$$

1280. From a fixed point O are drawn tangents OP, OP' to one of a series of conics whose foci are given points S, S': prove that (1) the envelope of the normals at P, P' is the same as the envelope of PP', (2) the circle OPP' will pass through another fixed point, (3) the conic $OPP'SS'$ will pass through another fixed point.

1281. A chord of a parabola is drawn through a fixed point and on it as diameter a circle is described: prove that the envelope of the polar of the vertex with respect to this circle is a conic which degenerates when the fixed point is on the tangent at the vertex.

[This conic will be a circle for the point $\left(-\frac{a}{2}, 0\right)$, and a rectangular hyperbola when the point lies on the parabola $y^2 = 4a(2x-a)$.]

1282. The centre of a given circle is C and a diameter is AB, chords AP, PB are drawn and perpendiculars let fall on these chords from a fixed point O: prove that the envelope of the straight line joining the feet of these perpendiculars is a conic whose directrices are AB and a parallel through O, whose excentricity is $CP : CO$, and whose focus corresponding to the directrix through O lies on CO.

1283. From the centres A, B of two given circles are drawn radii AP, BQ whose directions include a constant angle 2α: prove that the envelope of PQ is a conic whose excentricity is

$$\frac{c}{\sqrt{a^2+b^2-2ab\cos 2\alpha}},$$

where a, b are the radii and c the distance AB.

[The conic is always an ellipse when one circle lies within the other, and always an hyperbola when each lies entirely without the other; when the circles intersect, the conic is an ellipse if 2α be greater than the angle subtended by AB at a common point, reduces to the two common points when 2α has that critical value, and is an hyperbola for any smaller numerical value. When $2\alpha = 0$ or π, the envelope degenerates to a point. For different values of α, the foci of the envelope lie on a fixed circle of radius $\frac{abc}{a^2-b^2}$, and whose centre divides AB externally in the ratio $a^2 : b^2$.]

1284. Two conjugate chords AB, CD of a conic are taken, P is any point on the conic, PA, PB meet CD in a, b, O is another fixed point, and Oa, Ob meet PB, PA in Q, R: prove that QR envelopes a conic which degenerates if O lie on AB or on the conic.

1285. The point circles coaxial with two given equal circles are S, S', a straight line parallel to SS' meets the circles in H, H' so that $SH = S'H'$, and with foci H, H' is described a conic passing through S, S': prove that its directrices are fixed and that its envelope is a conic having S, S' for foci.

1286. Two conics U, V osculate in O and PP' is the remaining common tangent, PQ, $P'Q'$ are drawn tangents to V, U respectively: prove that PP', QQ' and the tangent at O meet in a point, and that, if from any point on PP' be drawn other tangents to U, V, the straight line joining their points of contact envelopes a conic touching both curves at O and touching U, V again in Q', Q respectively.

1287. Normals PQ, PQ' are drawn to the parabola $y^2 = 4ax$ from a point P on the curve: prove that the envelope of the circle PQQ' is the curve

$$y^2\left(x+\frac{a}{4}\right) = x^2(2a-x),$$

which is the pedal of the parabola $y^2 = a(x-2a)$ with respect to the origin A.

[The chord QQ' always passes through a fixed point $C(-2a, 0)$, and if S be the focus of the given parabola, C, S are single foci and A a double focus of the envelope: for a point P on the loop, $CP = AP + 2SP$, and for a point on the sinuous branch, $CP = 2SP - AP$, so that AP may be regarded as changing sign in vanishing.]

VIII. *Areal Co-ordinates.*

[In this system the position of a point P with respect to three fixed points A, B, C not in one straight line is determined by the values of the ratios of the three triangles PBC, PCA, PAB to the triangle ABC, any one of them PBC being esteemed positive or negative according as P and A are on the same or on opposite sides of BC. These ratios being denoted by x, y, z will always satisfy the equation $x+y+z=1$. A point is completely determined by the ratios of its areal co-ordinates $(X:Y:Z)$ or by two equations, as $lx=my=nz$. It is sometimes convenient to use trilinear co-ordinates, x, y, z being then the distances of the point from the sides of the triangle of reference ABC and connected by the equation $ax+by+cz=2K$, where a, b, c are the sides and K the area of the triangle ABC. A point would obviously be equally well determined by x, y, z being any fixed multiples of its areal or trilinear co-ordinates, a relation of the form $Ax+By+Cz=1$ always existing. In the questions under this head areal co-ordinates will generally be taken for granted.

The general equation of a straight line is $px+qy+rz=0$, and p, q, r are proportional to the perpendiculars from A, B, C on the straight line, sign of course being always regarded. When p, q, r are the actual perpendiculars, $px+qy+rz$ is the perpendicular distance from the line of the point whose areal co-ordinates are x, y, z.

The condition that the straight lines $p_1x+q_1y+r_1z=0$, $p_2x+q_2y+r_2z=0$ shall be parallel is

$$\begin{vmatrix} p_1, & q_1, & r_1 \\ p_2, & q_2, & r_2 \\ 1, & 1, & 1 \end{vmatrix} = 0,$$

and that they may be at right angles is

$$p_1p_2\sin^2 A + \ldots + \ldots = (q_1r_2+q_2r_1)\sin B\sin C\cos A + \ldots + \ldots$$

If $p_1=q_1=r_1$, or if $p_2=q_2=r_2$, both these equations are true. The straight line $x+y+z=0$, *the line at infinity*, may then be regarded as both parallel and perpendicular to every finite straight line, the fact being that the direction of the line at infinity is really indeterminate.

In questions relating to four points it is convenient to take the points to be $(X, \pm Y, \pm Z)$, or given by the equations $lx=\pm my=\pm nz$; and similarly to take the equations of four given straight lines to be $px\pm qy\pm rz=0$.

The general equation of a conic is

$$u\equiv ax^2+by^2+cz^2+2fyz+2gzx+2hxy=0;$$

and the polar of any point $(X : Y : Z)$ is

$$X\frac{du}{dx} + Y\frac{du}{dy} + Z\frac{du}{dz} = 0,$$

or (the same thing)

$$x\frac{dU}{dX} + y\frac{dU}{dY} + z\frac{dU}{dZ} = 0.$$

The special forms of this equation most useful are

(1) circumscribing the triangle of reference $(a, b, c = 0)$

$$fyz + gzx + hxy = 0;$$

(2) inscribed in the triangle of reference

$$(lx)^{\frac{1}{2}} + (my)^{\frac{1}{2}} + (nz)^{\frac{1}{2}} = 0;$$

(3) touching the sides AB, AC at the points B, C

$$kx^2 = yz;$$

but when this form is used it is often better to take such a multiple of the ratio $\Delta PBC : \Delta ABC$ for x as to reduce the equation to the form

$$x^2 = yz;$$

(4) to which the triangle is self-conjugate $(f, g, h = 0)$

$$lx^2 + my^2 + nz^2 = 0;$$

here again it is often convenient to use such multiples of the triangle ratios as to give us the equation

$$x^2 + y^2 + z^2 = 0.$$

As a general rule, when metrical results are wanted, it will be found simpler to keep to the true areal or trilinear co-ordinates.

The form (4) is probably the most generally useful. We may denote any point on such a conic by a single variable, as with the excentric angle in the case of a conic referred to its axes, which is indeed a particular case of this form. Thus any point on the conic $x^2 + y^2 + z^2 = 0$ may be represented by the equations

$$\frac{x}{\cos\theta} = \frac{y}{\sin\theta} = iz;$$

and we may call this the point θ.

The equation of the tangent at θ is

$$x\cos\theta + y\sin\theta = iz,$$

that of the chord through θ, ϕ is

$$x\cos\tfrac{1}{2}(\theta + \phi) + y\sin\tfrac{1}{2}(\theta + \phi) = iz\cos\tfrac{1}{2}(\theta - \phi);$$

and the intersection of the tangents at θ, ϕ is

$$\frac{x}{\cos\frac{1}{2}(\theta+\phi)} = \frac{y}{\sin\frac{1}{2}(\theta+\phi)} = \frac{iz}{\cos\frac{1}{2}(\theta-\phi)}.$$

The equations of any two conics may be taken to be

$$x^2+y^2+z^2=0,\ ax^2+by^2+cz^2=0,$$

their common points being $\frac{x^2}{b-c}=\frac{y^2}{c-a}=\frac{z^2}{a-b}$. The multipliers of the areal co-ordinates will not all be real when the triangle of reference is real, and when the conics have two real and two impossible common points the triangle will be imaginary.

Any point on the conic $x^2=yz$ may be denoted by the co-ordinates $(\lambda : 1 : \lambda^2)$ and called the point λ. The tangent at this point is

$$\lambda^2 y - 2\lambda x + z = 0,$$

and the chord through λ, μ is

$$\lambda\mu y - (\lambda+\mu)x + z = 0.$$

Any point on the conic $fyz+gzx+hxy=0$ may be taken to be

$$\frac{x\cos^2\theta}{f} = \frac{y\sin^2\theta}{g} = \frac{z}{-h},$$

the tangent at the point being

$$\frac{x}{f}\cos^4\theta + \frac{y}{g}\sin^4\theta + \frac{z}{h} = 0;$$

and any point on the conic $(lx)^{\frac{1}{2}}+(my)^{\frac{1}{2}}+(nz)^{\frac{1}{2}}=0$ to be

$$\frac{lx}{\cos^4\theta} = \frac{my}{\sin^4\theta} = nz,$$

with the corresponding tangent $\frac{lx}{\cos^2\theta}+\frac{my}{\sin^2\theta}-nz=0$; but these equations are not often required.]

1288. The equations of the straight lines each bisecting two of the sides of the triangle formed by joining the feet of the perpendiculars of the triangle ABC are

$$x = y\cot^2 B + z\cot^2 C, \text{ \&c.},$$

and the perpendiculars from (x, y, z) on them are

$$2R\sin^2 B\sin^2 C\,(y\cot^2 B + z\cot^2 C - x), \text{ \&c.}$$

1289. The sides of the triangle of reference are bisected in the points A_1, B_1, C_1; the triangle $A_1B_1C_1$ is treated in the same way and so on n times: prove that the equation of B_nC_n is

$$y+z = \frac{2^{n+1}+(-1)^n}{2^n+(-1)^{n+1}}\,x.$$

1290. The equation of the straight line passing through the centres of the inscribed and circumscribed circles is

$$\frac{x}{\sin A}(\cos B - \cos C) + \frac{y}{\sin B}(\cos C - \cos A) + \frac{z}{\sin C}(\cos A - \cos B) = 0;$$

and the point $\sin A(m + n\cos A) : \sin B(m + n\cos B) : \sin C(m + n\cos C)$ lies on this straight line for all values of $m : n$.

1291. If x, y, z be perpendiculars from any point on three straight lines which meet in a point and make with each other angles A, B, C, the equation $lx^2 + my^2 + nz^2 = 0$ will represent two straight lines which will be real, coincident, or imaginary, according as

$$mn\sin^2 A + nl\sin^2 B + lm\sin^2 C$$

is negative, zero, or positive.

1292. The perpendiculars from A, B, C on a straight line are p, q, r, and the areal co-ordinates of any point on the line are x, y, z: prove that $px + qy + rz = 0$, and the perpendicular distance of any point (x, y, z) from the line is $px + qy + rz$.

1293. The perpendiculars from A, B, C on the straight line joining the centres of the inscribed and circumscribed circles are p, q, r: prove that

$$p^2 = \frac{2R^2(1 - \cos\overline{B - C})(1 - \cos B)(1 - \cos C)}{3 - 2\cos A - 2\cos B - 2\cos C},$$

and two similar equations for q, r.

1294. The straight lines bisecting the external angles at A, B, C meet the opposite sides in A', B', C', and p, q, r are the perpendiculars from A, B, C upon the straight line $A'B'C'$: prove that

$$p\sin A = q\sin B = r\sin C = \frac{2R\sin A\sin B\sin C}{\sqrt{3 - 2\cos A - 2\cos B - 2\cos C}}.$$

1295. Within a triangle ABC are taken two points O, O'; AO, BO, CO meet the opposite sides in A', B', C', and the points of intersection of $O'A, B'C'$; $O'B, C'A'$; $O'C, A'B'$ are respectively D, E, F: prove that $A'D, B'E, C'F$ will meet in a point which remains the same if O, O' be interchanged in the construction.

[If $(x_1 : y_1 : z_1)$ and $(x_2 : y_2 : z_2)$ be the points O, O' the point is determined by the equations

$$\frac{x}{x_1x_2(y_1z_2 + y_2z_1)} = \frac{y}{y_1y_2(z_1x_2 + z_2x_1)} = \frac{z}{z_1z_2(x_1y_2 + x_2y_1)}.]$$

1296. The perpendiculars p, q, r are let fall from A, B, C on any tangent (1) to the inscribed circle, (2) to the circumscribed circle, (3) to the nine-points' circle, and (4) to the polar circle: prove that

(1) $p\sin A + q\sin B + r\sin C = 2R\sin A\sin B\sin C$;

(2) $p\sin 2A + q\sin 2B + r\sin 2C = 4R\sin A\sin B\sin C$;

(3) $p\sin A\cos(B-C) + \ldots + \ldots = 2R\sin A\sin B\sin C$;

(4) $p\tan A + q\tan B + r\tan C = \dfrac{2R\sin A\sin B\sin C}{\sqrt{-\cos A\cos B\cos C}}$;

and also that in (4) $p^2\tan A + q^2\tan B + r^2\tan C = 0$.

[The general relation for the tangent to any circle, whose centre is $(x_1 : y_1 : z_1)$ and radius ρ, is

$$px_1 + qy_1 + rz_1 = (x_1 + y_1 + z_1)\,\rho.]$$

1297. **The feet of the perpendiculars let fall from $(x_1 : y_1 : z_1)$ on the sides of the triangle of reference are A', B', C': prove that straight lines drawn through A, B, C perpendicular to $B'C'$, $C'A'$, $A'B'$ respectively meet in the point**

$$\frac{xx_1}{a^2} = \frac{yy_1}{b^2} = \frac{zz_1}{c^2}.$$

1298. A triangle LMN has its angular points on the sides of the triangle ABC, and AL, BM, CN meet in a point $(x_1 : y_1 : z_1)$; a straight line $px + qy + rz = 0$ is drawn meeting the sides of LMN in three points which are joined to the corresponding angular points of ABC: prove that the joining lines meet the sides of ABC in points lying on the straight line

$$\frac{x}{(q+r)\,x_1} + \frac{y}{(r+p)\,y_1} + \frac{z}{(p+q)\,z_1} = 0.$$

1299. The two points at which the escribed circles of the triangle of reference subtend equal angles lie on the straight line

$$(b-c)\,x\cot A + (c-a)\,y\cot B + (a-b)\,z\cot C = 0.$$

1300. Four straight lines form a quadrilateral, and from the middle points of the sides of the triangle formed by three of them perpendiculars are let fall on the straight line which bisects the diagonals: prove that these perpendiculars are inversely proportional to the perpendiculars from the angular points of the triangle on the fourth straight line.

[The three being taken to form the triangle of reference, and the fourth being $px + qy + rz = 0$, the equation of the bisector of the diagonals will be found to be

$$2\,(qrx + rpy + pqz) = (qr + rp + pq)\,(x + y + z).]$$

1301. A straight line meets the sides of the triangle ABC in A', B', C', the straight line joining A to the point (BB', CC') meets BC in a, and b, c are similarly determined: prove that if any point O be taken the straight lines joining a, b, c to the intersections of OA, OB, OC with $A'B'C'$ will pass through a point O'; and that OO' will pass through a point whose position is independent of O.

[If $px+qy+rz=0$ be the line $A'B'C'$, and $(x_1:y_1:z_1)$ the point O, the straight line OO' is

$$px(qy_1-rz_1)+qy(rz_1-px_1)+rz(px_1-qy_1)=0,$$

passing through the point $px=qy=rz$.]

1302. The two points whose distances from A, B, C are as BC, CA, AB respectively both lie on the straight line joining the centroid G and the centre of perpendiculars L of the triangle.

[The two points are given by the equations

$$\frac{S-b^2z-c^2y}{a^2}=\frac{S-c^2x-a^2z}{b^2}=\frac{S-a^2y-b^2x}{c^2},$$

where $$S=a^2yz+b^2zx+c^2xy.]$$

1303. The distances of L from A, B, C are as $\cos A:\cos B:\cos C$: prove that the other point P whose distances from A, B, C are as $\cos A:\cos B:\cos C$ also lies on the straight line GL and is reciprocal to L with respect to the circumscribed circle. Also

$$\frac{AP^2}{AL^2}=\frac{BP^2}{BL^2}=\frac{CP^2}{CL^2}=\frac{OP}{OL}=\frac{1}{1-8\cos A\cos B\cos C},$$

where O is the centre of the circle ABC.

1304. Each of the straight lines

$$x\sin^2 A+y\sin^2 B+z\sin^2 C=0,$$
$$x\cos^2 A+y\cos^2 B+z\cos^2 C=0,$$

is perpendicular to the straight line joining the centroid and the centre of perpendiculars.

1305. The equation of the straight line bisecting the diagonals of the quadrilateral, whose four sides are $px\pm qy\pm rz=0$, is

$$p^2x+q^2y+r^2z=0;$$

and that of the radical axis of the three circles whose diameters are the diagonals is

$$(q^2-r^2)(b^2z+c^2y)+\ldots+\ldots=0.$$

1306. One of the sides of a quadrilateral passes through the centre of one of the four circles which touch the diagonals: prove that each of the other three passes through one of the other three centres, and that the circles whose diameters are the diagonals touch each other in a point lying on the circle circumscribing the triangle formed by the diagonals, the common tangent being normal to the circumscribed circle.

[Generally the three circles intersect in real points if each of the four sides has two of the four centres on each side of it. If the four sides be $px\pm qy\pm rz=0$, the two common points are such that

$$\frac{AP}{p}=\frac{BP}{q}=\frac{CP}{r}.]$$

1307. The equation of a circle passing through B and C and whose segment on BC (on the same side as A) contains an angle θ, is

$$a^2yz + b^2zx + c^2xy = x(x+y+z)\frac{bc\sin(\theta - A)}{\sin\theta}.$$

1308. The locus of the radical centre of three circular arcs on BC, CA, AB, respectively, containing angles $A+\theta$, $B+\theta$, $C+\theta$, for different values of θ is the straight line

$$\frac{x}{a}\sin(B-C) + \frac{y}{b}\sin(C-A) + \frac{z}{c}\sin(A-B) = 0;$$

when $\theta = \frac{\pi}{2}$, the radical centre is the centre of the circumscribed circle, when $\theta = 0$, the radical centre is the point of concourse of the three straight lines joining A, B, C respectively to the points of intersection of the tangents to the circumscribed circle; and generally the radical centre divides the distance between these two points in the ratio

$$\cos\theta\,(1 + \cos A\cos B\cos C) : \sin\theta\sin A\sin B\sin C.$$

1309. A straight line drawn through the centre of the inscribed circle meets the sides of the triangle ABC in a, b, c, and these points are joined to the centres of the corresponding escribed circles: prove that the joining lines meet two and two on the sides of the triangle; and, if a', b', c' be their points of intersection, the circles on aa', bb', cc' as diameters will touch each other in one point lying on the circumscribed circle, their common tangent being normal to the circumscribed circle.

1310. The equation of a circle which passes through the centres of the escribed circles of the triangle of reference is

$$bcx^2 + cay^2 + abz^2 + (a+b+c)(ayz + bzx + cxy) = 0;$$

and, if we change the sign of one of the three a, b, c throughout, we get the equation of the circle through the centres of the inscribed circle and two of the escribed circles.

1311. A circle meets the sides of the triangle ABC in P, P'; Q, Q'; R, R' respectively, and AP, BQ, CR meet in the point $X : Y : Z$: prove that AP', BQ', CR' meet in the point $X' : Y' : Z'$, where

$$\frac{XX'(Y+Z)(Y'+Z')}{a^2} = \frac{YY'(Z+X)(Z'+X')}{b^2} = \frac{ZZ'(X+Y)(X'+Y')}{c^2}.$$

1312. The lines joining the feet of the perpendiculars of the triangle ABC meet the corresponding sides in A', B', C': prove that the circles whose diameters are AA', BB', CC' will touch each other if

$$\sec^2 A + \sec^2 B + \sec^2 C - 2\sec A\sec B\sec C + 7 = 0.$$

1313. Two circles cut each other at right angles and from three points A, B, C on one are drawn tangents whose lengths are p, q, r to the other: prove that

$$p^2 \sin 2A + q^2 \sin 2B + r^2 \sin 2C = 8\Delta ABC.$$

1314. The two point-circles coaxial with the circumscribed circle and the nine-points' circle of the triangle of reference are

$$mbc \sin A\,(x \cot A + y \cot B + z \cot C)(x + y + z) = a^2yz + b^2zx + c^2xy,$$

m being a root of the equation

$$m^2(1 - 8 \cos A \cos B \cos C) - 2m(1 - 2 \cos A \cos B \cos C) + 1 = 0.$$

1315. The equation of a circle in which the centroid and the centre of perpendiculars of the triangle of reference are ends of a diameter is

$$2bc \sin A\,(x \cot A + \cot B + z \cot C)(x + y + z) = 3\,(a^2yz + b^2zx + c^2xy);$$

and the tangent to it from an angular point bears to the tangent from the same point to the nine-points' circle the ratio $2 : \sqrt{3}$.

1316. The conic $lx^2 + my^2 + nz^2 = 0$ will represent the polar circle of the triangle of reference if

$$l \tan A = m \tan B = n \tan C.$$

1317. The necessary and sufficient conditions that the equation

$$lx^2 + my^2 + nz^2 + 2pyz + 2qzx + 2rxy = 0$$

may represent a circle are

$$\frac{m + n - p}{a^2} = \frac{n + l - q}{b^2} = \frac{l + m - r}{c^2}.$$

1318. The lengths of the tangents from A, B, C to a certain circle are p, q, r: prove that the equation of the circle is

$$(p^2x + q^2y + r^2z)(x + y + z) - a^2yz - b^2zx - c^2xy = 0,$$

and that the square of the tangent from (x, y, z) is the left-hand member.

1319. The nine-points' circle of the triangle of reference touches the inscribed and escribed circles in the points P, P_1, P_2, P_3: prove that (1) the equations of the tangents at these points are

$$\frac{x}{b - c} + \frac{y}{c - a} + \frac{z}{a - b} = 0,$$

and the three equations formed from this by changing a into $-a$, b into $-b$, or c into $-c$; (2) PP_1, P_2P_3 meet BC in the same points as the straight lines bisecting the internal and external angles at A; (3) PP_1, P_2P_3 intersect in the point

$$\frac{-x}{b^2 - c^2} = \frac{y}{c^2 - a^2} = \frac{z}{a^2 - b^2};$$

and (4) the tangents at P, P_1, P_2, P_3 all touch the maximum ellipse inscribed in the triangle.

1320. The straight line $lx + my + nz = 0$ meets the sides of the triangle ABC in A', B', C': prove that the circles on AA', BB', CC' have the common radical axis

$$l(m-n)\,x\cot A + m(n-l)\,y\cot B + n(l-m)\,z\cot C = 0;$$

and the circles will touch each other if

$$(mn + nl + lm)^2 \sin A \sin B \sin C = 2lmn\,(l\sin 2A + m\sin 2B + n\sin 2C).$$

1321. A circle having its centre at the point $(X : Y : Z)$ cuts at right angles a given circle $u = 0$: prove that its equation is

$$(X + Y + Z)\,u = (x + y + z)\left(X\frac{du}{dx} + Y\frac{du}{dy} + Z\frac{du}{dz}\right).$$

1322. A conic touches the sides of the triangle ABC in the points a, b, c and Aa meets the conic again in A': prove that the equation of the tangent at A' is

$$\frac{x}{2X} = \frac{y}{Y} + \frac{z}{Z},$$

where $(X : Y : Z)$ is the point of concourse of Aa, Bb, Cc.

1323. The two conics circumscribing the triangle of reference, passing through the point $(X : Y : Z)$, and touching the straight line $px + qy + rz = 0$ will be real if

$$\frac{pqr\,(pX + qY + rZ)}{XYZ} < 0.$$

Interpret this result geometrically.

1324. Find the two points in which the straight line $y = kz$ meets the conic $(lx)^{\frac{1}{2}} + (my)^{\frac{1}{2}} + (nz)^{\frac{1}{2}} = 0$, and from the condition that one of the points may be at infinity determine the direction of the asymptotes. Prove that the conic will be a rectangular hyperbola if

$$l^2a^2 + m^2b^2 + n^2c^2 + 2mnbc\cos A + 2nlca\cos B + 2lmab\cos C = 0.$$

1325. A conic touches the sides of the triangle ABC, any point is taken on the straight line which passes through the intersections of the chords of contact with the corresponding sides, and the straight lines joining this point to A, B, C respectively meet BC, CA, AB in a, b, c: prove that corresponding sides of the triangles ABC, abc intersect in points lying on one tangent to the conic.

1326. A conic touches the sides of the triangle ABC, and the straight lines joining A, B, C to the points of contact meet in O; through O is drawn a straight line meeting BC, CA, AB respectively in a, b, c, and from a, b, c are drawn respectively three other tangents: prove that the intersections of these tangents two and two lie upon a fixed conic circumscribing the triangle.

1327. If e be the excentricity of a conic inscribed in the triangle ABC, and $x : y : z$ the *trilinear* co-ordinates of a focus,

$$1 - e^2 = \frac{xyz\,(ax + by + cz)\,(ayz + bzx + cxy)}{R^2\,(y^2 + z^2 + 2yz\cos A)\,(z^2 + x^2 + 2zx\cos B)\,(x^2 + y^2 + 2xy\cos C)}.$$

1328. Two parabolas are inscribed in the triangle of reference such that triangles can be inscribed in one whose sides touch the other; and $(x : y : z)$, $(x' : y' : z')$ are their foci: prove that

$$\left(\frac{x}{x'}\right)^{\frac{1}{2}} + \left(\frac{y}{y'}\right)^{\frac{1}{2}} + \left(\frac{z}{z'}\right)^{\frac{1}{2}} = 0.$$

1329. The equation of the axis of a parabola inscribed in the triangle ABC is $px + qy + rz = 0$: prove that

$$\frac{a^2p}{q-r} + \frac{b^2q}{r-p} + \frac{c^2r}{p-q} = 0.$$

1330. Prove that the equation of a conic, inscribed in the triangle ABC and having $px + qy + rz = 0$ for an asymptote, is

$$p\sqrt{(q-r)\,x} \pm q\sqrt{(r-p)\,x} \pm r\sqrt{(p-q)\,x} = 0.$$

1331. The condition that the straight line $px + qy + rz = 0$ may be an asymptote of a rectangular hyperbola circumscribing the triangle ABC is

$$p\,(q-r)^2 \cot A + q\,(r-p)^2 \cot B + r\,(p-q)^2 \cot C = 0.$$

1332. Prove that, at any point P on the minimum ellipse circumscribing the triangle ABC,

$$\frac{AP}{\sin BPC} + \frac{BP}{\sin CPA} + \frac{CP}{\sin APB} = 0,$$

and $$\cot BPC + \cot CPA + \cot APB = \cot A + \cot B + \cot C,$$

the angles BPC, CPA, APB being so measured that their sum is 360°.

1333. A conic $lyz + mzx + nxy = 0$ is such that the normals to it at the points A, B, C meet in a point, prove that

$$bcl\,(m^2 - n^2) + cam\,(n^2 - l^2) + abn\,(l^2 - m^2) = 0;$$

the point of concourse of the normals must lie on the curve

$$x\,(y^2 - z^2)\,(\cos A - \cos B \cos C) + \ldots + \ldots = 0;$$

and the centre of the conic on the curve

$$bcx\,(y^2 - z^2) + cay\,(z^2 - x^2) + abz\,(x^2 - y^2) = 0.$$

[*Trilinear* co-ordinates are here employed.]

1334. The conic $(lx)^{\frac{1}{2}} + (my)^{\frac{1}{2}} + (nz)^{\frac{1}{2}} = 0$ is such that the normals at its points of contact with the sides meet in a point: prove that

$$\frac{l^2}{a^2}\,(m-n) + \frac{m^2}{b^2}\,(n-l) + \frac{n^2}{c^2}\,(l-m) = 0,$$

that the centre lies on the same curve as in the last question; and that when with a point on this curve as centre are described two conics, one touching the sides and the other passing through the angular points, the directions of their axes will be the same.

1335. An ellipse inscribed in a given triangle passes through the centre of the circumscribed circle: prove that the locus of its centre is the conic whose foci are the centres of the circumscribed and nine-points circles and whose major axis is $\frac{1}{2}R$. Also if an inscribed ellipse pass through the centre of perpendiculars, the locus of its centre is a conic whose centre is the centre of the nine-points circle, to which the perpendiculars of the triangle are normals, the sum of whose axes is R and the difference is the distance between the centre of the circumscribed circle and the centre of perpendiculars.

1336. Find all the common points of the two conics

$$(lx)^{\frac{1}{2}} + (my)^{\frac{1}{2}} + (nz)^{\frac{1}{2}} = 0,$$

$$(lx + my + nz)(x + y + z) = 2(m + n)yz + 2(n + l)zx + 2(l + m)xy;$$

and prove that their areas are as

$$\left(\frac{mn + nl + lm}{l + m + n}\right)^{\frac{3}{2}} : \frac{(m + n)(n + l)(l + m)}{8(lmn)^{\frac{1}{2}}}.$$

1337. A conic passes through the corners of a triangle and through its centroid, prove that the pole of the mid point of any side with respect to this conic is the straight line bisecting the other two sides.

[For, when $l + m + n = 0$, the equations $lyz + mzx + nxy = 0$,

$$l(y + z - x)^2 + m(z + x - y)^2 + n(x + y - z)^2 = 0,$$

coincide.]

1338. The minimum excentricity of any conic through the four points $\frac{x^2}{X^2} = \frac{y^2}{Y^2} = \frac{z^2}{Z^2}$, is given by the equation

$$\frac{e^4}{e^2 - 1} = 4\frac{(X^2 \cot A + Y^2 \cot B + Z^2 \cot C)^2}{(X + Y + Z)(-X + Y + Z)(X - Y + Z)(X + Y - Z)}.$$

1339. Two conics have a common director circle, one being inscribed in the triangle ABC and the triangle self-conjugate to the other: prove that the common centre must lie on the line

$$x \cot A + y \cot B + z \cot C = 0.$$

1340. Two concentric conics are drawn one circumscribing the triangle of reference and to the other the triangle bisecting the sides is self-conjugate: prove that the two conics are similar and similarly situated, and their areas in the ratio

$$-8xyz : (-x + y + z)(x - y + z)(x + y - z),$$

where $(x : y : z)$ is their common centre.

[Also very easily proved by orthogonal projection.]

1341. A conic is inscribed in the triangle of reference such that the triangle formed by joining the feet of the perpendiculars let fall from its centre on the sides is an arithmetic mean between the corresponding triangles for the foci: prove that the locus of its centre is the curve (a bicircular quartic)

$$(x^2 \cot A + y^2 \cot B + z^2 \cot C)^2$$
$$= (x + y + z)(-x + y + z)(x - y + z)(x + y - z).$$

1342. Prove that, if α, β be the semi-axes of the conic

$$(lx)^{\frac{1}{2}} + (my)^{\frac{1}{2}} + (nz)^{\frac{1}{2}} = 0,$$

$$\alpha^2 + \beta^2 = \frac{l^2a^2 + m^2b^2 + n^2c^2 + 2mnbc \cos A + \ldots + \ldots}{4(l + m + n)^2},$$

$$\alpha\beta = \tfrac{1}{2} bc \sin A \sqrt{\frac{lmn}{(l + m + n)^3}}.$$

1343. Of all ellipses inscribed in a given circle, that has the least director circle whose centre is the centre of perpendiculars, the radius of its director circle being $2R\sqrt{\cos A \cos B \cos C}$ and the area of the ellipse

$$2\pi R^2 \sqrt{(\cos A - \cos B \cos C)(\cos B - \cos C \cos A)(\cos C - \cos A \cos B)};$$

provided that of the segments into which each perpendicular is divided by the centre of perpendiculars the segment next the base is less than the one to the vertex. (R denotes the radius of the circle ABC.)

1344. A conic is inscribed in the triangle ABC with its centre at the centre of perpendiculars, and θ is the angle which its axis makes with the side BC: prove that

$$\frac{\tan 2\theta}{\tan B - \tan C} = \frac{\cos A - \cos B \cos C}{\cos A - 2 \cos B \cos C};$$

also that, if A', B', C' be its points of contact with the sides, the centre of perpendiculars of the triangle $A'B'C'$ will lie on the conic and the tangent there will be parallel to BC.

1345. If $(x : y : z)$ be the centre of an inscribed conic, the sum of the squares on its semi-axes is

$$4R^2 \cos A \cos B \cos C + p^2;$$

where p is the distance of its centre from the centre of perpendiculars.

1346. The centre of a conic is the point $(x : y : z)$, its excentricity is e, the radius of its director circle ρ, and K denotes twice the area of the triangle of reference ABC: prove that (1) for a conic inscribed in the triangle ABC

$$\frac{(e^2 - 2)^2}{1 - e^2} = \frac{(x^2 \cot A + y^2 \cot B + z^2 \cot C)^2}{(x + y + z)(-x + y + z)(x - y + z)(x + y - z)},$$

$$\rho^2 = K \frac{x^2 \cot A + y^2 \cot B + z^2 \cot C}{(x + y + z)^2};$$

(2) for a circumscribed conic

$$\frac{(e^2-2)^2}{1-e^2} = \frac{(a^2yz + \dots + \dots - x^2bc\cos A - \dots - \dots)^2}{K^2(x+y+z)(-x+y+z)(x-y+z)(x+y-z)},$$

$$\rho^2 = \frac{4xyz\,(a^2yz + \dots - x^2bc\cos A - \dots - \dots)}{(x+y+z)^2(-x+y+z)(x-y+z)(x+y-z)};$$

(3) for a conic to which the triangle is self-conjugate,

$$\frac{(e^2-2)^2}{1-e^2} = \frac{(a^2yz + b^2zx + c^2xy)^2}{K^2xyz\,(x+y+z)},$$

$$\rho^2 = -\frac{a^2yz + b^2zx + c^2xy}{(x+y+z)^2}.$$

1347. Two similar conics have a common centre, one is inscribed and the other circumscribed to the triangle of reference: prove that their common centre lies either on the circumscribed circle or on the circle of which the centroid and centre of perpendiculars are ends of a diameter.

1348. The equation of an asymptote of the conic $yz = kx^2$ is

$$2\mu kx - ky - \mu^2 z = 0,$$

where μ is given by the equation $\mu^2 + \mu + k = 0$; and the asymptotes, for different values of k, envelope the parabola

$$(y-z)^2 + 4x\,(x+y+z) \equiv (y+z+2x)^2 - 4yz = 0.$$

1349. A conic is inscribed in the triangle ABC and its centre lies on a fixed straight line parallel to BC $(y+z=kx)$: prove that its asymptotes envelope the conic

$$(k-1)\,(x+y+z)^2 = 16yz.$$

1350. The radius of curvature of the conic $x^2 = kyz$ at the point B is $\dfrac{kR\sin^2 C}{\sin A \sin B}$.

1351. Prove that the equation $x^2 = 4yz$ represents a parabola; and that the tangential equation of the same parabola is $qr = p^2$.

1352. The tangents to a given conic at two fixed points A, B meet in C, and the tangent at any point P meets CA, CB in B', A' respectively: find the locus of the point of intersection of AA', BB'; and, if AP, BP meet CB, CA in a, b respectively, find the envelope of ab.

[Taking the original conic to be $z^2 = kxy$, the locus and envelope are respectively $4z^2 = kxy$, $z^2 = 4kxy$.]

1353. The tangents to a given conic at A, B meet in C and a, b are two other fixed points on the conic; a tangent to the conic meets CA, CB in B', A': prove that the locus of the intersection of aB', bA' is a conic passing through a, b and the intersections of Ca, Bb, and of Cb, Aa.

[Taking the given conic to be $z^2 = xy$, and $(x_1 : y_1 : z_1)$, $(x_2 : y_2 : z_2)$ the points a, b, the locus is

$$\left(\frac{x}{x_1} - \frac{y}{y_1}\right)\left(\frac{x}{x_2} - \frac{y}{y_2}\right) = 4\left(\frac{z}{x_1} - \frac{y}{z_1}\right)\left(\frac{z}{y_2} - \frac{x}{z_2}\right).]$$

1354. The tangents to a given conic at A, B meet in C; P is any other point on the conic and AP, BP meet CB, CA in a, b: prove that the triangle abP is self-conjugate to another fixed conic touching the former at A, B.

[Taking the given conic to be $z^2 = xy$ and $(x_1 : y_1 : z_1)$ the point P, the fixed conic is $z^2 = 2xy$, which is equivalent to

$$\left(\frac{x}{x_1} - \frac{z}{z_1}\right)^2 + \left(\frac{y}{y_1} - \frac{z}{z_1}\right)^2 = \left(\frac{x}{x_1} + \frac{y}{y_1} - \frac{z}{z_1}\right)^2.]$$

1355. Prove that the locus of the foci of the conic $x^2 = kyz$, for different values of k, is the circular cubic

$$x(y^2 - z^2) + 2yz(y \cos B - z \cos C) = 0,$$

trilinear co-ordinates being used.

1356. Three tangents touch a conic in A, B, C and form a triangle abc; BC, CA, AB meet a fourth tangent in α, β, γ, and $A\alpha$, $B\beta$, $C\gamma$ meet the conic again in A', B', C': prove that $B'C'$, bc; $C'A'$, ca, $A'B'$, ab intersect in points on the fourth tangent, and aA', bB', cC' meet in a point. If a family of conics be inscribed in the quadrilateral formed by the four tangents, the centre of homology of the two triangles abc, $A'B'C'$ lies on the curve

$$(px)^{\frac{1}{3}} + (qy)^{\frac{1}{3}} + (rz)^{\frac{1}{3}} = 0,$$

x, y, z and $px + qy + rz$ being the four tangents.

1357. Three points A, B, C are taken on a conic and the tangents form a triangle abc, a fourth point O $(X : Y : Z)$ is taken on the conic and Oa, Ob, Oc meet the tangents at A, B, C in α, β, γ from which points other tangents are drawn forming a triangle $A'B'C'$: prove that AA', BB', CC' will meet in O and that the axis of the two triangles ABC, $A'B'C'$ envelopes the curve

$$\left(\frac{X}{x}\right)^{\frac{1}{2}} + \left(\frac{Y}{y}\right)^{\frac{1}{2}} + \left(\frac{Z}{z}\right)^{\frac{1}{2}} = 0.$$

[If the points be fixed and the conic variable, the straight lines $B'C'$, $C'A'$, $A'B'$ each envelope a fixed tricusp, two of the cusps (for $B'C'$) being B, C and the tangents OB, OC, and the third cusp, lying on OA and having OA for tangent, being at a point O' such that, if AO meet BC in D, $\{AO'OD\} = \frac{4}{3}$.]

1358. In the last question, if ABC and the conic be fixed and the point O vary, the axis envelopes the curve $(lx)^{-\frac{1}{3}}+(my)^{-\frac{1}{3}}+(nz)^{-\frac{1}{3}}=0$, the conic being $(lx)^{-1}+(my)^{-1}+(nz)^{-1}=0$.

1359. Prove that the general equation of a conic, with respect to which the conic $lx^2+my^2+nz^2=0$ is its own polar reciprocal, is

$$(lx^2+my^2+nz^2)(p^2mn+q^2nl+r^2lm)=2lmn(px+qy+rz)^2.$$

1360. Every hyperbola is its own polar reciprocal with respect to a parabola having double contact with it at the ends of a chord which touches the conjugate hyperbola.

1361. An ellipse is its own reciprocal polar with respect to a rectangular hyperbola which has double contact with it at the ends of a chord touching the hyperbola which is confocal with the ellipse and has its asymptotes along the equal conjugate diameter of the ellipse.

1362. A parabola is its own reciprocal with respect to any rectangular hyperbola which has double contact with it at the ends of a chord touching the other parabola which has the same latus rectum.

1363. An hyperbola is its own reciprocal with respect to either circle which touches both branches of the hyperbola and intercepts on the transverse axis a length equal to the conjugate axis.

1364. Each of two conics U, V is its own reciprocal with respect to the other, prove that they must have double contact and that each is its own reciprocal with respect to any conic which has double contact with both U and V provided the contacts are different.

1365. Through the fourth common point of the two conics

$$lyz+mzx+nxy=0,\quad l'yz+m'zx+n'xy=0,$$

is drawn a straight line meeting the conics again in P, Q: prove that the locus of the intersection of the tangents at P, Q is the curve (tricusp quartic)

$$\{ll'(mn'-m'n)yz\}^{\frac{1}{2}}+\{mm'(nl'-n'l)zx\}^{\frac{1}{2}}+\{nn'(lm'-l'm)xy\}^{\frac{1}{2}}=0.$$

1366. From any point on the fourth common tangent to the two conics $x^{\frac{1}{2}}+y^{\frac{1}{2}}+z^{\frac{1}{2}}=0$, $(lx)^{\frac{1}{2}}+(my)^{\frac{1}{2}}+(nz)^{\frac{1}{2}}=0$ are drawn two other tangents to the conics: prove that the envelope of the straight line joining their points of contact is the curve

$$\{l(m-n)x\}^{\frac{1}{3}}+\{m(n-l)y\}^{\frac{1}{3}}+\{n(l-m)z\}^{\frac{1}{3}}=0.$$

1367. The sides of the triangle ABC touch a conic U; O, O_1, O_2, O_3 are the centres of the inscribed and escribed circles of ABC, a conic V is described through B, C, O, O_1, and one focus of U, and a conic W through B, C, O_2, O_3, and the same focus of U: prove that the fourth common point of V, W will be the conjugate focus of U; also that, if the conic W be fixed, the major axis of the conic U will always pass through a fixed point on the internal bisector of the angle A, and if the conic V be fixed, through a fixed point on the external bisector of the angle A.

1368. Four conics are described with respect to each of which three of the four straight lines $px \pm qy \pm rz = 0$ form a self-conjugate triangle and the fourth is the polar of a fixed point $(X : Y : Z)$; prove that all four will have two common tangents, meeting in $(X : Y : Z)$, whose equation is

$$\frac{p^2X}{yZ - zY} + \frac{q^2Y}{zX - xZ} + \frac{r^2Z}{xY - yX} = 0.$$

1369. The triangle ABC is self-conjugate to a given conic and on the tangent to the conic at any point P is taken a point Q such that the pencil $Q\{ABCP\}$ is constant: prove that the locus of Q is a quartic having nodes at A, B, C and touching the conic in four points.

[If the conic be $lx^2 + my^2 + nz^2 = 0$, and k the given anharmonic ratio, the locus of Q is $\frac{(k-1)^2}{lx^2} + \frac{1}{my^2} + \frac{k^2}{nz^2} = 0$.]

1370. Two given conics intersect in A, B, C, D and from any point O on AB are drawn tangents OP, OQ to one conic, Op, Oq to the other: prove that Pp, Qq intersect in one fixed point and Pq, Qp in another; that these points remain the same if A, B be interchanged with C, D; and that the six such points corresponding to all the common chords lie on four straight lines.

[Taking the conics to be $x^2 + y^2 + z^2 = 0$, $ax^2 + by^2 + cz^2 = 0$, the six points are $x = 0$, $b(c-a)y^2 = c(a-b)x^2$, &c.]

1371. A triangle $A'B'C'$ is drawn similar to the triangle of reference ABC and with its sides passing respectively through A, B, C; another similar triangle abc is drawn with its sides parallel to those of the former and its angular points upon the sides BC, CA, AB respectively: prove that the triangle ABC is a mean proportional between the triangles abc, $A'B'C'$; and that the straight lines $A'a$, $B'b$, $C'c$ meet in the point

$$\frac{x}{\sin 2A + \sin 2(B-\theta) + \sin 2(C+\theta)} = \frac{y}{\sin 2B + \sin 2(C-\theta) + \sin 2(A+\theta)}$$
$$= \frac{z}{\sin 2C + \sin 2(A-\theta) + \sin 2(B+\theta)};$$

where θ is the angle between the directions of BC, $B'C'$. Prove also that the locus of this point is a conic having an axis along the straight line joining the centroid and the centre of perpendiculars.

[The equation of the conic is $u^2 + v^2 = w^2$, where

$$u = x\cos 2A + y\cos 2B + z\cos 2C + 4(x+y+z)\cos A\cos B\cos C,$$

$$v = x(\sin 2B - \sin 2C) + y(\sin 2C - \sin 2A) + z(\sin 2A - \sin 2B),$$

$$w = 2(x\cos 2A + y\cos 2B + z\cos 2C) + x + y + z,$$

of which u and w are parallel to each other and perpendicular to v.]

1372. Four fixed tangents are drawn to a given conic forming a quadrilateral whose diagonals are aa', bb', cc'; three other conics are drawn osculating the given conic at the same point P and passing through a, a'; b, b'; and c, c' respectively: prove that the tangents at a, a', b, b', c, c' all meet in one point; that the locus of this point as P moves is the envelope of the straight line joining it to P and is a fourth class sextic having two cusps on each diagonal and touching the given conic at the points of contact of the four tangents.

[If the four tangents be $px \pm qy \pm rz = 0$ and the conic

$$lx^2 + my^2 + nz^2 = 0,$$

the locus is

$$(p^4mnx^2)^{\frac{1}{3}} + (q^4nly^2)^{\frac{1}{3}} + (r^4lmz^2)^{\frac{1}{3}} = 0.]$$

1373. A conic is drawn through B, C osculating in P the conic

$$(lx)^{\frac{1}{2}} + (my)^{\frac{1}{2}} + (nz)^{\frac{1}{2}} = 0:$$

prove that the locus of the pole of BC with respect to this conic is the cubic

$$(lx + 4my + 4nz)^3 = 27lx\,(my - nz)^2:$$

also if A' be the point of contact of BC and another conic be drawn also osculating the given conic in P but passing through A, A', the tangents at B, C, A, A' will meet in a point.

1374. A parabola touches the sides of the triangle ABC and the straight line $B'C'$ joining the feet of the perpendiculars from B, C on the opposite sides: prove that its focus lies on the straight line joining A to the intersection of BC, $B'C'$.

1375. A triangle is self-conjugate to a parabola: prove that the straight lines each bisecting two of the sides are tangents to the parabola, and thence that the focus lies on the nine-points' circle and the directrix passes through the centre of perpendiculars.

1376. A triangle is self-conjugate to a parabola and the focus of the parabola lies on the circle circumscribing the triangle: prove that the poles of the sides of the triangle with respect to the circle lie on the parabola.

1377. A rectangular hyperbola is inscribed in the triangle ABC: prove that the locus of the pole of the straight line which bisects the two sides AB, AC is the circle

$$x^2(a^2 + b^2 + c^2) + (y^2 + 2xy)(a^2 + b^2 - c^2) + (z^2 + 2zx)(a^2 - b^2 + c^2) = 0;$$

that this circle is equal to the polar circle of the triangle and its centre is the point of the circle ABC opposite to A.

1378. A conic is drawn touching the four straight lines

$$px \pm qy \pm rz = 0;$$

prove that its equation is $lx^2 + my^2 + nz^2 = 0$, where l, m, n are connected by the equation $p^2mn + q^2nl + r^2lm = 0$, and investigate the species of this conic with respect to the position of its centre on its rectilinear locus.

[If the middle points of the internal diagonals of the convex quadrangle be L, M, and that of the external diagonal be N, L, M, N being in order, the conic is an hyperbola when the centre lies between $-\infty$ and L, an ellipse from L to M, an hyperbola from M to N, and an ellipse from N to $+\infty$. Hence there are two true minimum excentricities.]

1379. A conic is drawn touching the four straight lines

$$px \pm qy \pm rz = 0,$$

prove that any two straight lines $p_1x + q_1y + r_1z = 0$, $p_2x + q_2y + r_2z = 0$, will be conjugate with respect to this conic if

$$\frac{p_1p_2}{p^2} = \frac{q_1q_2}{q^2} = \frac{r_1r_2}{r^2}.$$

1380. The straight lines $p_1x + q_1y + r_1z = 0$, $p_2x + q_2y + r_2z = 0$ will be conjugate with respect to all parabolas inscribed in the triangle of reference if

$$q_1r_2 + q_2r_1 = r_1p_2 + r_2p_1 = p_1q_2 + p_2q_1.$$

[A particular case of the last with different notation.]

1381. The two points $(x_1 : y_1 : z_1)$, $(x_2 : y_2 : z_2)$, will be conjugate with respect to any conic through the four points $(X : \pm Y : \pm Z)$, if

$$\frac{x_1x_2}{X^2} = \frac{y_1y_2}{Y^2} = \frac{z_1z_2}{Z^2}.$$

1382. A triangle is self-conjugate to a rectangular hyperbola: prove that the foci of any conic inscribed in the triangle will be conjugate with respect to the hyperbola.

[A particular case of the last.]

1383. The locus of the foci of all conics touching the four straight lines $px \pm qy \pm rz = 0$ is the cubic whose equation is, if

$$(l, m, n) \equiv l^2 \sin^2 A + m^2 \sin^2 B + n^2 \sin^2 C - 2mn \sin B \sin C \cos A$$
$$- 2nl \sin C \sin A \cos B - 2lm \sin A \sin B \cos C,$$

$$\frac{(l, m, n)}{lx + my + nz} = \frac{(-l, m, n)}{-lx + my + nz} + \frac{(l, -m, n)}{lx - my + nz} + \frac{(l, m, -n)}{lx + my - nz};$$

and this equation may be reduced to the form

$$(x + y + z)(l^2x^2 \cot A + m^2y^2 \cot B + n^2z^2 \cot C)\, bc \sin A$$
$$= (l^2x + m^2y + n^2z)(a^2yz + b^2zx + c^2xy).$$

1384. Of all the conics inscribed in a given quadrilateral there are only two which have an axis along the straight line which is the locus of the centres of the conics, and the two conics will be real and the axis the major axis when the centre of perpendiculars of the triangle formed by the diagonals of the quadrilateral is on the opposite sides of the locus of centres to the three corners of the triangle.

1385. The equations determining the foci of the conic

$$lx^2 + my^2 + nz^2 = 0$$

are

$$\frac{1}{a^2}\left\{\frac{(y+z)^2}{l} + \frac{x^2}{m} + \frac{x^2}{n}\right\} = \frac{1}{b^2}\left\{\frac{(z+x)^2}{m} + \frac{y^2}{n} + \frac{y^2}{l}\right\} = \frac{1}{c^2}\left\{\frac{(x+y)^2}{n} + \frac{z^2}{l} + \frac{z^2}{m}\right\}.$$

1386. One directrix of the conic $lx^2 + my^2 + nz^2 = 0$ passes through A: prove that

$$mn = l\,(m \cot^2 C + n \cot^2 B);$$

and that the conjugate focus lies on the straight line joining the feet of the perpendiculars from B, C on the opposite sides.

1387. A conic is described to which the triangle ABC is self-conjugate and its centre lies on the straight line bisecting two of the sides of the triangle formed by joining the feet of the perpendiculars of the triangle ABC, prove that one of its foci is a fixed point.

[It is at the foot of the perpendicular from A on BC.]

1388. Given a point O on a conic and a triangle ABC self-conjugate to the conic; AO, BO, CO meet the opposite sides in three points and the straight lines joining these two and two meet the corresponding sides in A', B', C': prove that the intersections of BB', CC'; CC', AA'; and AA', BB' also lie on the conic.

1389. Any tangent to a conic meets the sides of the triangle ABC which is self-conjugate to the conic in a, b, c; the straight line joining A to the intersection of Bb, Cc meets BC in A', and B', C' are similarly determined: prove that $B'C'$, $C'A'$, $A'B'$ are also tangents to the conic.

1390. Two conics U, V have double contact and from a point O on the chord of contact are drawn tangents OP, OQ; Op, Oq; another conic W is drawn through p, q touching OP, OQ: prove that the tangent to W at any point where it meets U will touch V.

1391. A conic passes through four given points: prove that the locus of tangents drawn to it from a given point is in general a cubic, which degenerates into a conic if the point be in the same straight line with two of the former and in that case the locus passes through the other two points and the tangents to it at them pass through the fifth point.

1392. A conic is inscribed in a given quadrilateral and tangents are drawn to it from a given point: prove that the locus of their points of contact is a cubic passing through the ends of the diagonals of the quadrilateral, through the given point, and through any point where the straight line joining the given point to the intersection of two diagonals meets the third: there is a node at the given point the tangents at which form a harmonic pencil with the straight lines to the ends of any diagonal.

[The node is a crunode when the given point lies within the convex quadrangle or in any of the portions of space vertically opposite any angle of the convex quadrangle.]

1393. Prove that, if $px+qy+rz=0$ be the equation of the axis of a parabola inscribed in the triangle ABC, or the asymptote of a rectangular hyperbola to which the triangle is self-conjugate,

$$\frac{a^2p}{q-r}+\frac{b^2q}{r-p}+\frac{c^2r}{p-q}=0.$$

1394. A parabola is inscribed in the triangle ABC and S is its focus (a point on the circle ABC), the axis meets the circle ABC again in O: prove that, if with centre O a rectangular hyperbola be described to which the triangle is self-conjugate, one of its asymptotes will coincide with OS.

1395. The conics passing through two given points and touching three given straight lines are either all four real or all four impossible.

[If the three given straight lines form the triangle of reference and $(x_1 : y_1 : z_1)$ $(x_2 : y_2 : z_2)$ be the two given points, the conics will be

$$(lx)^{\frac{1}{2}}+(my)^{\frac{1}{2}}+(nz)^{\frac{1}{2}}=0,$$

where

$$\frac{l}{y_1z_2+y_2z_1\pm 2\sqrt{y_1y_2z_1z_2}}=\frac{m}{z_1x_2+z_2x_1\pm 2\sqrt{z_1z_2x_1x_2}}=\frac{n}{x_1y_2+x_2y_1\pm 2\sqrt{x_1x_2y_1y_2}},$$

in which ambiguities an odd number of negative signs must be taken. If the points of contact with BC be A_1, A_2, A_3, A_4 these can always be taken so that

$$BA_1 . BA_2 : CA_1 . CA_2 = BA_3 . BA_4 : CA_3 . CA_4.]$$

1396. The locus of the foci of a rectangular hyperbola, to which the triangle ABC is self-conjugate, is the tricyclic sextic

$$\frac{a^2}{x\{U-x(x+y+z)\,bc\cos A\}}+\ldots+\ldots=0,$$

where $U\equiv a^2yz+b^2zx+c^2xy$.

1397. A triangle circumscribes the conic $x^2+y^2+z^2=0$ and two of its angular points lie on the conic $lx^2+my^2+nz^2=0$: prove that the locus of the third angular point is the conic

$$\frac{x^2}{(-l+m+n)^2}+\frac{y^2}{(l-m+n)^2}+\frac{z^2}{(l+m-n)^2}=0;$$

that this will coincide with the second if $l^{\frac{1}{2}}+m^{\frac{1}{2}}+n^{\frac{1}{2}}=0$; and that the three conics have always four common tangents.

1398. The angular points of a triangle lie on the conic

$$lx^2+my^2+nz^2=0$$

and two of its sides touch the conic $x^2+y^2+z^2=0$; prove that the envelope of the third side is the conic

$$l(-l+m+n)^2x^2+m(l-m+n)^2y^2+n(l+m-n)^2z^2=0;$$

that this will coincide with the second if $l^{\frac{1}{2}}+m^{\frac{1}{2}}+n^{\frac{1}{2}}=0$, and that the three conics have always four common points.

1399. A triangle is self-conjugate to the conic $x^2+y^2+z^2=0$ and two of its angular points lie on the conic $lx^2+my^2+nz^2=0$; prove that the locus of the third angular point is the conic

$$(m+n)x^2+(n+l)y^2+(l+m)z^2=0;$$

that this will coincide with the second if $l+m+n=0$; and that the three have always four common points. Also prove that the straight line joining the two angular points will touch the conic

$$\frac{x^2}{m+n}+\frac{y^2}{n+l}+\frac{z^2}{l+m}=0.$$

1400. A triangle is self-conjugate to the conic $lx^2+my^2+nz^2=0$ and two of its sides touch the conic $x^2+y^2+z^2=0$; prove that the envelope of the third side is the conic

$$\frac{lx^2}{m+n}+\frac{my^2}{n+l}+\frac{nz^2}{l+m}=0;$$

that this will coincide with the second if $l+m+n=0$; and that the three have always four common tangents. Also prove that the locus of the intersection of the two sides is the conic

$$l(m+n)x^2+m(n+l)y^2+n(l+m)z^2=0.$$

IX. *Anharmonic Ratio. Homographic Pencils and Ranges. Involution.*

[The anharmonic ratio of four points A, B, C, D in one straight line, denoted by $\{ABCD\}$, means the ratio $\frac{AB}{BD} : \frac{AC}{CD}$, or $\frac{AB \cdot CD}{AC \cdot BD}$; the order of the letters marking the direction of measurement of the segments and segments measured in opposite directions being affected with opposite signs. So, if A, B, C, D be any four points in a plane and P any other point in the same plane, $P\{ABCD\}$ denotes $\frac{\sin APB \sin CPD}{\sin APC \sin BPD}$, the same rules being observed as to direction of measurement and sign for the angles in this expression as for the segments in the other.

Either of these ratios is called harmonic when its value is -1; in which case AD is the harmonic mean between AB and AC, and DA is the harmonic mean between DB and DC. The anharmonic ratio of four points or four straight lines can never be equal to 1; as that value leads immediately to the result $AD \cdot BC = 0$ or $\sin APD \sin BPC = 0$ making two of the points or two of the lines coincident.

A series of points on a straight line is called a range, and a series of straight lines through a point is called a pencil, the straight line or point being the axis or vertex of the range or pencil respectively. If two ranges $abcd \ldots$, $a'b'c'd' \ldots$ be so connected that each point a of the first determines one point a' of the second and each point a' of the second determines one point a of the first, the ranges are homographic. So also two pencils, or a range and a pencil, may be homographic; and in all such cases the anharmonic ratio of any range or pencil is equal to the anharmonic ratio of the corresponding range or pencil in any homographic system.

If four fixed points A, B, C, D be taken on any conic and P be any other point on the same conic, $P\{ABCD\}$ is constant for all positions of P and is harmonic when BC, AD are conjugates with respect to the conic. Also, if the tangent at P meet the tangents at A, B, C, D in the points a, b, c, d, the range $\{abcd\}$ is constant and equal to the former pencil.

A range of points on any straight line is homographic with the pencil formed by their polars with respect to any conic.

If the equations of four straight lines can be put in the form $u = \mu_1 v$, $u = \mu_2 v$, $u = \mu_3 v$, $u = \mu_4 v$, the anharmonic ratio of the pencil formed by them or of the range in which any straight line meets them is

$$\frac{(\mu_1 - \mu_2)(\mu_3 - \mu_4)}{(\mu_1 - \mu_3)(\mu_2 - \mu_4)}.$$

A very great number of loci and envelopes can be determined immediately from the following theorems: (1) The locus of the intersection of corresponding rays of two homographic pencils is a conic passing through the vertices of the pencils (O, O') and the tangents at O, O' are the rays corresponding to $O'O$, OO' respectively: (2) The envelope of a straight line which joins corresponding points of two homographic ranges is a conic touching the axes of the two ranges in the points which correspond to the common point of the axes.

A series of pairs of points on a straight line is said to be in involution when there exist two fixed points (f, f') on the line such that, a, a' being any pair, $\{aff'a'\} = -1$. The points f, f' are called the foci or double points of the range, since when a is at f, a' will also be at f. The middle point (C) of ff' is called the centre and $Ca \,.\, Ca' = Cf^2$. The foci may be either both real or both impossible, but the centre is always real; and when two corresponding points are on the same side of the centre the foci are real. Similarly a series of pairs of straight lines, or rays, drawn from a point is in involution when there exist two fixed rays forming with any pair of corresponding rays a harmonic pencil in which the two fixed rays are conjugate. This pair of fixed straight lines is called the focal lines or double rays.

Any straight line is divided in involution by the six straight lines joining the points of a quadrangle, and any two corresponding points of the involution will lie on a conic round the quadrangle.

The pencil formed by joining any point to the six points of intersection of the sides of a quadrilateral is in involution and any pair of corresponding rays touch a conic inscribed in the quadrilateral.

The locus of the intersection of two tangents to a given conic drawn from corresponding points of an involution is the conic which passes through the double points of the involution and through the points of contact of tangents to the given conic drawn from the double points.

The envelope of a chord of a given conic whose ends lie on corresponding rays of a pencil in involution is a conic touching the double rays of the involution and also touching the tangents drawn to the given conic at the points where the double rays meet it.

These two theorems will be found to include as particular cases many well-known loci and envelopes.

It may be mentioned that a large proportion of the questions which are given under this head might equally well have appeared in the next division: Reciprocal Polars and Projection.]

1401. Two fixed straight lines meet in A; B, C, D are three fixed points on another straight line through A; any straight line through D meets the two former straight lines in b, c and Bb, Cc meet in P, Bc, Cb in Q: prove that the loci of P, Q are straight lines through A which make with the two former a pencil whose ratio is $(\{ABCD\})^2$.

1402. On a straight line are taken points O, A, B, C, A', B', C' such that

$$\{OABC\} = \{OAB'C'\} = \{OA'BC'\} = \{OA'B'C\};$$

prove that each $= \{OA'B'C'\}$, and that the ranges $\{OBCA'\}$, $\{OCAB'\}$, and $\{OABC'\}$ will each be harmonic.

1403. Two fixed straight lines intersect in a point O on the side BC of a triangle ABC; any point P being taken on AO the straight lines PB, PC meet the two fixed straight lines in B_1, B_2, C_1, C_2 respectively: prove that B_1C_2 and B_2C_1 pass each through a fixed point on BC.

1404. From a fixed point are let fall perpendiculars on conjugate rays of a pencil in involution: prove that the straight line through the feet of these perpendiculars passes through a fixed point.

1405. Two conjugate points a, a' of a range in involution being joined to a fixed point O, straight lines drawn through a, a' at right angles to aO, $a'O$ meet in a point which lies on a fixed straight line.

1406. Chords are drawn through a fixed point of a conic equally inclined to a given direction; prove that the straight line joining their extremities passes through a fixed point.

1407. Through a given point are drawn chords PP', QQ' of a given conic so as both to touch a confocal conic: prove that the points of intersection of PQ, $P'Q'$, and of PQ', $P'Q$ are fixed.

1408. A circle is described having for ends of a diameter two conjugate points of a pencil in involution: prove that this circle will be cut orthogonally by any circle through the two double points of the range.

1409. Two triangles are formed each by two tangents to a conic and their chord of contact; prove that their angular points lie on one conic.

1410. Four points A, B, C, D being taken on a conic, any straight line through D meets the conic again in D' and the sides of the triangle ABC in A', B', C': prove that the range $\{A'B'C'D'\}$ is equal to the pencil $\{ABCD\}$ at any point on the conic.

1411. The sides of a triangle ABC touch a conic in the points A', B', C' and the tangent at any point O meets the sides of the two triangles in a, b, c, a', b', c' respectively: prove that $\{Oabc\} = \{Oa'b'c'\}$.

1412. Four chords of a conic are drawn through a point, and two other conics are drawn through the point, one passing through four extremities of the four chords and the other through the other four extremities: prove that these conics will touch each other at the point of concourse.

[Also very easily proved by projection.]

1413. Through a given point O is drawn any straight line meeting a given conic in Q, Q', and a point P is taken on this line such that the range $\{OQQ'P\}$ is constant: prove that the locus of P is an arc of a conic having double contact with the given conic.

1414. Given two points A, B of a given conic; the envelope of a chord PQ such that the pencil $\{APQB\}$ at any point of the conic has a given value is a conic touching the given conic at A, B.

1415. Through a fixed point is drawn any straight line meeting two fixed straight lines in Q, R respectively; E, F are two other fixed points: prove that the locus of the point of intersection of QE, RF is a conic passing through E, F, and the common point of the two fixed straight lines.

1416. Three fixed points A, B, C being taken on a given conic, two other points P, P' are taken on the conic such that the pencils $\{PABC\}$, $\{P'ABC\}$ are equal at any point on the conic: prove that PP', CA, and the tangent at B, meet in a point.

1417. Six fixed points A, B, C, A', B', C' are taken on a given conic such that, at any point on the conic, $\{B'ABC\} = \{BA'B'C'\}$; and P, P' are two other points on the conic such that, at any point on the conic, $\{PABC\} = \{P'A'B'C'\}$: prove that PP', AA', BB', CC' all intersect in one point.

[Of course the six points subtend a pencil in involution at any point of the conic, and conjugate rays pass through P, P'.]

1418. A conic passes through two given points A, A', and touches a given conic at a given point O; prove that their other common chord will pass through a fixed point B on AA', and that if the straight line through A, A' meet the given conic in C, C' and the tangent at O in B', the points A, A'; B, B'; C, C' will be in involution.

1419. Two chords AB, CD of a conic being conjugate, the angle ACB is a right angle, and any chord DP through D meets AB in Q; prove that the angle PCQ is bisected by CA or CB.

1420. Three fixed points A, B, C being taken on a conic, and P being any other point on the conic, through P is drawn a straight line meeting the sides of the triangle ABC in points a, b, c such that $\{Pabc\}$ has a given value: prove that the straight line passes through a fixed point O on the conic such that the pencil $\{OABC\}$ at any point of the conic has the same given value.

1421. Prove that the two points in which a given straight line meets any conic through four given points are conjugate with respect to the conic which is the locus of the pole of the given straight line with respect to the system of conics.

1422. Three fixed tangents to a given conic form a triangle ABC, and on the tangent at any point P is taken a point O such that the pencil $O\{PABC\}$ has a given value: prove that the locus of O is a straight line which touches the conic.

1423. Two conics circumscribe a triangle ABC, any straight line through A meets them again in P, Q: prove that the tangents at P, Q divide BC in a constant anharmonic ratio.

1424. Conics are described touching four given straight lines, of which two meet in A, and the other two in B; on the two meeting in A are taken two fixed points C, D, and the tangents drawn from them to one of the conics meet in P: prove that the locus of P is a straight line through B which forms with BC, BD and one of the tangents through B a pencil equal to that formed by BA, BC, BD and the other tangent through B.

1425. The diagonals of a given quadrilateral are AA', BB', CC', and on them are taken points a, a'; b, b'; c, c', so that each diagonal is divided harmonically: prove that if a, b, c be collinear, so also will a', b', c', and their common point will be the point where either of them is touched by a conic inscribed in the quadrilateral.

[This is also a good example of the use of Projection.]

1426. Two fixed tangents OA, OB are drawn to a given conic and a fixed point C taken on AB; through C is drawn a straight line meeting the fixed tangents in A', B': prove that the remaining tangents from A', B' intersect in a point whose locus is a fixed straight line through O.

1427. Two fixed points A, B are taken on a given conic and a fixed straight line drawn conjugate to AB; any point P being taken on this last straight line chords APQ, BPQ' are drawn; prove that QQ' passes through a fixed point on AB.

1428. A conic is inscribed in a triangle ABC, the polar of A meets BC in a, and aP is drawn to touch the conic; prove that if from any point Q on aP another tangent be drawn, this tangent and QA will form with QB, QC a harmonic pencil.

1429. Two chords AO, BC of a conic are conjugate, any chord OP meets the sides of the triangle ABC in a, b, c: prove that the range $\{abcP\}$ is harmonic.

1430. Two fixed points A, B are taken on a given conic, P is any other point on the conic: prove that the envelope of the straight line joining the points where PA, PB meet two fixed tangents to the conic is a conic which touches at A, B the straight lines joining these points to the points of contact of the corresponding fixed tangents, and which also touches the two fixed tangents.

1431. One diagonal of a quadrilateral circumscribing a conic is AA': prove that another conic can be described touching two of the sides of the quadrilateral in A, A' and passing through the points of contact of the other two.

1432. On the normal to an ellipse at a point P are taken two points O, O' such that the rectangle $PO \,.\, PO'$ is equal to that under the focal distances of P, and from these points tangents are drawn to the ellipse: prove that their points of intersection lie on the circle whose diameter is QQ', where Q, Q' are the points in which the tangent at P meets the director circle.

1433. A range of points in involution lie on a fixed straight line and a homographic system on another fixed straight line; a, a' are conjugate points of the former and A, A' the corresponding points on the latter: prove that the locus of the intersection of aA, $a'A'$ or of aA', $a'A$ is a straight line.

1434. A pencil in involution has a point O for its vertex, and a homographic pencil is drawn from another point O', corresponding rays of the two intersect in P and the conjugate rays in P': prove that PP' passes through a fixed point.

1435. In any conic the tangent at A meets the tangents at C, B in b, c which are joined to a point O by straight lines meeting BC in b', c': prove that AC, cb' intersect on the polar of O, as also AB, bc'.

1436. The triangle ABC is self-conjugate to a given conic U, a conic V is inscribed in the triangle and its points of contact are A', B', C': prove that, if $B'C'$ touch U, so also will $C'A'$, $A'B'$, and the straight line in which lie the points $(BC, B'C')$, $(CA, C'A')$, and $(AB, A'B')$.

1437. A variable tangent to a conic meets two fixed tangents in P, Q; A, B are two fixed points: prove that the locus of the intersection of AP, BQ is a conic, passing through A, B, and the intersections of (OA, Bb) and (OB, Aa); Oa, Ob being the fixed tangents.

1438. Parallel tangents are drawn to a given conic and the point where one meets a given tangent is joined to the point where the other meets another given tangent: prove that the envelope of the joining line is a conic to which the two given tangents are asymptotes.

1439. Through a fixed point O of an hyperbola is drawn a straight line parallel to an asymptote, and on it are taken two points P, P' such that the rectangle $OP . OP'$ is constant; the locus of the intersection of tangents drawn from P, P' is two fixed straight lines passing through the common point of the tangent at O and the asymptote, and forming with them an harmonic pencil.

1440. Four fixed points A, B, C, D are taken on a given conic; through D is drawn any straight line meeting the conic again in P and the sides of the triangle ABC in A', B', C': prove that the range $\{PA'B'C'\}$ is constant.

1441. The tangent to a parabola at any point P meets two fixed tangents CA, CB in a, b, the diameters through the points of contact A, B in a', b', and the chord of contact AB in c': prove that

$$Pa . Pa' : Pb . Pb' = ac' : bc'.$$

1442. A tangent to an hyperbola at P meets the asymptotes in a, b, the tangent at a point Q in c, and the straight lines drawn through Q parallel to the asymptotes in a', b': prove that

$$Pa' : Pb' = ca : bc.$$

1443. The anharmonic ratio of the pencil subtended by the four points whose excentric angles are α_1, α_2, α_3, α_4 at any point of an ellipse is

$$\frac{\sin\frac{1}{2}(\alpha_1-\alpha_2)\sin\frac{1}{2}(\alpha_3-\alpha_4)}{\sin\frac{1}{2}(\alpha_1-\alpha_3)\sin\frac{1}{2}(\alpha_2-\alpha_4)}.$$

1444. Tangents are drawn to a conic at four points A, B, C, D, and form a quadrilateral whose diagonals are aa', bb', cc', the tangents at A, B, C forming the triangle abc, and being met by the tangent at D in a', b', c'; the middle points of the diagonals are A', B', C' and the centre of the conic is O: prove that the range $\{A'B'C'O\}$ is equal to the pencil $\{ABCD\}$ at any point of the conic.

1445. A conic is drawn through four given points A, B, C, D; BC, AD meet in A'; CA, BD in B'; AB, CD in C'; and O is the centre of the conic: prove that the pencil $\{ABCD\}$ on the conic is equal to the pencil $\{A'B'C'O\}$ on the conic which is the locus of O.

1446. The anharmonic ratio of the four common points of the two conics

$$x^2+y^2+z^2=0,\quad ax^2+by^2+cz^2=0,$$

at any point on the former is one of the three

$$\frac{a-b}{a-c},\ \frac{b-c}{b-a},\ \frac{c-a}{c-b},$$

or the reciprocal of one of them, according to the order of taking the four points; also these are the values of the range formed on any tangent to the second conic by their four common tangents.

1447. Two fixed tangents are drawn to a given conic intersecting each other in O and a fixed straight line in L, M; from any point on LM are drawn two tangents to the conic meeting the two fixed tangents in A, B; A', B', respectively: prove that a conic drawn to touch the two fixed tangents at points where they are met by LM, and touching one of the straight lines AB', $A'B$, will also touch the other.

1448. A quadrilateral circumscribes a conic and AA', BB' are two of its diagonals; any point P being taken on the conic, BP, $B'P$ and the tangent at P meet AA' in the points t', t, p respectively: prove that

$$Ap^2 : A'p^2 = At\,.\,At' : A't\,.\,A't'.$$

[Also easily proved by projecting A, A' into foci.]

1449. Four tangents TP, TQ, $T'P'$, $T'Q'$ are drawn to a parabola: prove that the conic $TPQT'P'Q'$ will be a circle if TT' be bisected by the focus.

[A parabola can be drawn with its focus at T' touching PQ and the normals at P, Q; and another with its focus at T touching $P'Q'$ and the normals at P', Q'; and the axis of the given parabola will be the tangent at the vertex of either of these.]

1450. Four tangents TP, TQ, $T'P'$, $T'Q'$ are drawn to a given ellipse: prove that the conic $TPQT'P'Q'$ will be a circle when CT, CT' being equally inclined to the major axis and T, T' on the same side of the minor axis, $CT \,.\, CT' = CS^2$, where C is the centre and S a focus of the given ellipse.

[A parabola can be drawn with T' for focus and CT for directrix touching PQ and the normals at P, Q; and another parabola with T for focus and CT' for directrix will touch $P'Q'$ and the normals at P', Q'; and these parabolas are the same for a series of conics confocal with the given ellipse.]

1451. The locus of the intersection of tangents to the ellipse $ax^2 + by^2 + 2hxy = 1$ drawn parallel to conjugate diameters of the ellipse $a'x^2 + b'y^2 + 2h'xy = 1$ is

$$(ab - h^2)(a'x^2 + b'y^2 + 2h'xy) = ab' + a'b - 2hh'.$$

1452. Through each point of the conic $ax^2 + by^2 + 2hxy = 1$ is drawn a pencil in involution whose double rays are parallel to the co-ordinate axes: prove that the chord cut off by a pair of conjugate rays passes through a fixed point whose locus is the conic

$$ax^2 + by^2 + 2hxy = \frac{ab}{h^2}.$$

1453. Two conjugate rays of a pencil in involution meet the conic

$$u \equiv ax^2 + by^2 + c + 2fy + 2gx + 2hxy = 0$$

in the points P, P'; Q, Q', the double rays of the pencil being the axes of co-ordinates: prove that the conic enveloped by PQ, PQ', $P'Q$, $P'Q'$ is

$$4(fg - ch)\,xy = (fy + gx + c)^2.$$

[If $fg = ch$, the double rays are conjugate with respect to the conic u, and the chords pass through the two fixed points where the double rays meet the polar of the vertex: if $c = 0$, the vertex is on the curve and the chord determined by conjugate rays passes through the point

$$\left(-\frac{f}{h}, -\frac{g}{h}\right).]$$

1454. Four fixed tangents to a conic form a quadrilateral of which AA', BB' are two diagonals, any other tangent meets AA' in P and the range $\{APP'A'\}$ is harmonic: prove that the locus of the intersection of BP', or of $B'P'$, with the last tangent is a conic passing through AA' and touching the given conic where BB' meets it.

[Taking the given conic to be $x^2 = yz$, and the straight line AA' to be $px + qy + rz = 0$, the locus is

$$p^2(x^2 - yz) = (qy - rz)^2,$$

degenerating to the straight line $qy + rz = px$ when $p^2 = 4qr$; that is, when AA' is a tangent to the given conic.]

1455. Three fixed tangents are drawn to a conic and their points of intersection joined to a focus; any other tangent meets these six lines in an involution such that the distance between the double points subtends a right angle at the focus. Also the locus of the double points for different positions of the last-named tangent is the curve

$$\frac{c}{r} = e\cos\theta + \cos(\alpha + \beta + \gamma - 3\theta),$$

where $c = r(1 + e\cos\theta)$ is the equation of the given conic, and α, β, γ are the values of θ at the points of contact of the fixed tangents.

X. *Reciprocal Polars and Projections.*

[If there be a system of points, and straight lines, lying in the same plane and we take the polars of the points and the poles of the straight lines with respect to any conic in that plane, we obtain a system of straight lines and points reciprocal to the former; so that to a series of points lying on any curve in the first system correspond a series of straight lines touching a certain other curve in the second system, and *vice versâ*: and, in particular, to any number of points lying on a straight line or a conic, correspond a number of straight lines passing through a point or touching a conic. Thus from any general theorem of position may be deduced a reciprocal theorem. It is in nearly all cases advisable to take a circle for the *auxiliary conic* with respect to which the system is reciprocated; the point (p) corresponding to any proposed straight line being then found by drawing through O, the centre of the circle, OP perpendicular to the proposed straight line and taking on OP a point p such that $OP \cdot Op = k^2$, k being the radius of the circle; and similarly the straight line through p at right angles to Op is the straight line corresponding to the point P.

To draw the figure reciprocal to a triangle ABC, with respect to a circle whose centre is O or more shortly *with respect to the point* O, draw Oa perpendicular to BC and on it take any point a; through a, O draw straight lines perpendicular to OC, CA, meeting in b; and through b, O draw straight lines perpendicular to OA, AB meeting in c; then the points a, b, c will be the poles of the sides of the triangle ABC and the straight lines bc, ca, ab the polars of the points A, B, C, with respect to some circle with centre O. Now suppose we want to find the point corresponding to the perpendicular from A on BC; it must lie on bc and on the straight line through O at right angles to Oa since Oa is parallel to the straight line whose reciprocal is required; it is therefore determined. Hence to the theorem that the three perpendiculars of a triangle meet in a point corresponds the following: if through any point (O) in the plane of a triangle (abc) be drawn straight lines at right angles to Oa, Ob, Oc to meet the respectively opposite sides, the three points so determined will lie on one straight line, or be *collinear*.

So from the theorem that the bisectors of the angles meet in a point we get the following: the straight lines drawn through O bisecting the external angles (or one external and two internal angles) between Ob, Oc; Oc, Oa; Oa, Ob, respectively, will meet the opposite sides in three collinear points.

If a circle with centre A and radius R be reciprocated with respect to O, the reciprocal curve is a conic whose focus is O, major axis along OA, excentricity $OA \div R$, and latus rectum $2k^2 \div R$ or $2 \div R$ if we take the radius of the auxiliary circle to be unity. The centre A is reciprocated into the directrix. Focal properties of conics are thus deduced from theorems relating to the circle. For instance, if O be a point on the circle and OP, OQ chords at right angles, PQ will pass through the centre. Reciprocating with respect to O, to the circle corresponds a parabola and to the points P, Q two tangents to the parabola at right angles to each other; perpendicular tangents to a parabola therefore intersect on the directrix.

Again, to find the condition that two conics which have one focus common should be such that triangles can be inscribed in one whose sides touch the other. Take two circles which have this property, and let R, r be their radii, δ the distance between their centres; then

$$\delta^2 = R^2 \pm 2Rr.$$

Reciprocate the system with respect to a point at distances x, y from the centres, and let a be the angle between these distances. Then a will be the angle between the axes of the two conics, and, if $2c_1$, $2c_2$ be the latera recta, e_1, e_2 the excentricities,

$$c_1 = \frac{1}{r}, \quad c_2 = \frac{1}{R}, \quad e_1 = \frac{y}{r}, \quad e_2 = \frac{x}{R}, \quad \delta^2 = x^2 + y^2 - 2xy \cos a = R^2 \pm 2Rr,$$

whence
$$\frac{1}{c_2^2} \pm \frac{2}{c_1 c_2} = \frac{e_2^2}{c_2^2} + \frac{e_1^2}{c_1^2} - 2\frac{e_1 e_2}{c_1 c_2} \cos a,$$

or
$$c_1^2 \pm 2c_1 c_2 = e_2^2 c_1^2 + e_1^2 c_2^2 - 2e_1 e_2 c_1 c_2 \cos a;$$

the required relation.

If a system of confocal conics be reciprocated with respect to one of the foci, the reciprocal system will consist of circles having a common radical axis; the radical axis being the reciprocal of the second focus, and the first focus being a point-circle of the system.

The reciprocal of a conic with respect to any point in its plane is another conic which is an ellipse, parabola, or hyperbola according as the point lies within, upon, or without the conic. To the points of contact of tangents from the point correspond the asymptotes, and to the polar of the point the centre of the reciprocal. So also to the asymptotes and centre of the original conic correspond the points of contact and polar with respect to the reciprocal.

As an example we may reciprocate the elementary property that the tangent at any point of a conic makes equal angles with the focal distances. The theorem so obtained is that if we take any point

O in the plane of the conic there exist two fixed straight lines (reciprocals of the foci) such that if a tangent to the conic at P meet them in Q, Q', OP makes equal angles with OQ, OQ'. (More correctly there are two such pairs of straight lines, one pair only being real.) If however the point O lie on the curve the original curve was a parabola; and one of the straight lines being the reciprocal of the point at infinity on the parabola will be the tangent at O. Another property of the focus, that any two straight lines through it at right angles to each other are conjugate, shews us that if on either of the two straight lines we take two points L, L' such that LOL' is a right angle, L, L' will be conjugate.

Since the anharmonic ratio of the pencil formed by any four rays is equal to that of the range formed by their poles with respect to any conic it follows that, in any reciprocation whatever, a pencil or range is replaced by a range or pencil* having the same anharmonic ratio.

The method of Projections enables us to make the proof of any general theorem of position depend upon that of a more simple particular case of that theorem. Given any figure in a plane we have five constants disposable to enable us to simplify the projected figure, three depending on the position of the vertex and two on the direction of the plane of Projection. It is clear that relations of tangency, of pole and polar, and anharmonic ratio, are the same in the original and projected figure.

As a good example of the use of this method, we will by means of it prove the theorem that if two triangles be each self-conjugate to the same conic their angular points lie on one conic.

Let the two triangles be ABC, DEF, and abc, def their projections; project the conic into a circle with its centre at d, then e, f will be at infinity, and de, df at right angles. Draw a conic through $abcde$, then since abc is self-conjugate to a circle whose centre is d, d is the centre of perpendiculars of the triangle abc, the conic is therefore a rectangular hyperbola, and e being one of its points at infinity, f must be the other. Thus $abcdef$ lie on one conic, and therefore $ABCDEF$ also lie on one conic. Again, retaining the centre at d, take any other conic instead of a circle; de, df will still be conjugate diameters, and therefore if any conic pass through a, b, c, d, its asymptotes will be parallel to a pair of conjugate diameters of the conic whose centre is d and to which abc is self-conjugate. The same must therefore be the case with respect to the four conics each having its centre at one of the four points a, b, c, d, and the other three points corners of a self-conjugate triangle. These four conics must therefore be similar and similarly situated. Moreover if we draw the two parabolas which can be drawn through a, b, c, d their axes must be parallel respectively to coincident conjugate diameters of any one of the four conics; that is to the asymptotes. But the axes of these parabolas must be parallel to the asymptotes of the conic which is the locus of the centres of all conics through a, b, c, d, since the centre is at infinity for a parabola. Hence, finally, if we have four points in a plane, the four conics each of which has one of the four points for its centre and the other three at the corners of a self-conjugate triangle are all similar and similarly situated to each other and to the conic which is the locus of centres of all conics through the four points.

(The same results might also be proved by orthogonal projection, making d the centre of perpendiculars of the triangle abc, in which case the five conics are all circles.)

Let A, B be any two fixed points on a circle, ∞, ∞' the two impossible circular points at infinity, P any other point on the circle; then $P\{A\infty\infty'B\}$ is constant. Hence PA, PB divide the segment terminated by the two circular points in a constant anharmonic ratio. Hence two straight lines including a given angle may be projected into two straight lines dividing a given segment in a constant anharmonic ratio. In particular, if APB be a right angle, AB passes through the centre of the circle (the pole of $\infty\infty'$), and the ratio becomes harmonic.

Thus, projecting properties of the director circle of a conic, we obtain the following important theorem: the locus of the intersection of tangents to a conic which divide a given segment harmonically is a conic passing through the ends of the segment and through the points of contact of tangents to the conic drawn from the ends. If the straight line on which the segment lies touch the conic, the locus degenerates to a straight line joining the points of contact of the other tangents drawn from the ends of the segment.

Reciprocating, we get the equally important theorem: if a chord of a given conic be divided harmonically by the conic and by two given straight lines its envelope will be a conic touching the two given straight lines and also the tangents to the given conic at the points where the given straight lines meet it; but when the two given straight lines intersect on the given conic the chord which is divided harmonically will pass through a fixed point, the intersection of the tangents to the given conic at the points where the given straight lines again meet it.

If tangents be drawn to any conic through ∞, ∞' their four other points of intersection are the real and impossible foci of the conic. When the conic is a parabola the line joining ∞, ∞' is a tangent, and one of the real foci is at infinity, while the two impossible foci are the circular points. Many focal properties, especially of the parabola, may thus be generalized by projection. Thus since the locus of intersection of tangents to a parabola including a constant angle is a conic having the same focus and directrix, it follows that if a conic be inscribed in a triangle ABC, and two tangents be drawn dividing BC in a constant range, the locus of their point of intersection is a conic touching the former in the points where AB, AC touch it. Here B, C are the projections of ∞, ∞', A of the focus, and the directrix is the polar of the focus.

The circular points at infinity have singular properties in relation to many other well-known curves. All epicycloids and hypocycloids pass through them, the cardioid has cusps at them, and may be projected into a three-cusped epicycloid.]

1456. Two conics have a common focus S, and two common tangents PP', QQ'; prove that the angles PSP', QSQ' are equal or supplementary.

1457. Two conics have a common focus, and (1) equal minor axes, (2) equal latera recta: prove that (1) the common tangents are parallel, (2) one of their common chords passes through the common focus.

1458. The straight line drawn through the focus S at right angles to any straight line SO will meet the polar of O on the directrix.

1459. The fixed point O is taken on a given conic also any three other points on the conic L, M, N: straight lines drawn through O at right angles to OL, OM, ON meet MN, NL, LM in three points lying on a straight line which meets the normal at O in a fixed point (chords through which subtend a right angle at O).

1460. Given a conic and a point O: prove that there are two real straight lines such that the distance between any two points on either, which are conjugate with respect to the conic, subtends a right angle at O.

1461. A fixed point O is taken on a conic, and OR is the chord normal at O, OP, OQ any other chords: prove that a certain straight line can be drawn through the pole of OR such that the tangents at P, Q intercept on it a segment which subtends at O an angle $2POQ$.

1462. On any straight line can be found two points, conjugate to a given conic, such that the segment between them subtends a right angle at a given point.

1463. A point O being taken in the plane of a triangle ABC, straight lines drawn through O at right angles to OA, OB, OC meet the respectively opposite sides in A', B', C': prove that any conic which touches the sides of the triangle and the straight line $A'B'C'$ subtends a right angle at O.

1464. An ellipse is described about an acute-angled triangle ABC, and one focus is the centre of perpendiculars of the triangle: prove that its latus rectum is

$$2R\frac{\cos A\cos B\cos C}{\sin\frac{A}{2}\sin\frac{B}{2}\sin\frac{C}{2}}.$$

1465. A parabola and hyperbola have a common focus and axis, and the parabola touches the directrix of the hyperbola; prove that any straight line through the focus is harmonically divided by a tangent to the parabola and the two parallel tangents to the hyperbola.

1466. A series of conics are described having equal latera recta, a focus of a given conic their common focus, and tangents to the conic their directrices: prove that the common tangents of any two intersect on the directrix of the given conic at a point such that the line joining it to the focus is at right angles to one of their common chords which passes through the focus.

1467. A point S is taken within a triangle ABC such that the sides subtend at S equal angles, and four conics are drawn with S as focus circumscribing the triangle: prove that one of these will touch the other three, and that the tangent to this conic at A will meet BC in a point A' such that ASA' is a right angle.

1468. Prove that, with the centre of the circumscribed circle as focus, three hyperbolas can be described circumscribing a given triangle ABC; that their excentricities are $\operatorname{cosec} B \operatorname{cosec} C$, &c.; their latera recta $2R \cot B \cot C$, &c.; their directrices the straight lines joining the middle points of the sides; and that the fourth common point of any two lies on the straight line joining one of the points A, B, C to the mid point of the opposite side.

1469. With a given point as focus four conics can be drawn circumscribing a given triangle, and the latus rectum of one of these will be equal to the sum of the other three. Also if any conic U be drawn touching the directrices of the four conics the polar of the given point with respect to it will be a tangent to the conic V which has the given point for focus, and which touches the sides of the triangle, and the conic U will subtend a right angle at the given point.

[If l, l_1, l_2, l_3 be the four latera recta, and $l_1 + l_2 + l_3 = l$, the latus rectum of V will be $\dfrac{4ll_1l_2l_3}{(l_2+l_3)(l_3+l_1)(l_1+l_2)}$.]

1470. From a point P on the circle ABC are drawn PA', PB', PC' at right angles to PA, PB, PC respectively to meet the corresponding sides of the triangle ABC: prove that the straight line $A'B'C'$ passes through the centre of the circle.

1471. With a point on the circumscribed circle of a triangle ABC as focus are described four conics circumscribing the triangle: prove that the corresponding directrices will pass each through the centre of one of the four circles touching the sides.

1472. A triangle is inscribed in an ellipse so that the centre of the inscribed circle coincides with one of the foci; prove that the radius of the inscribed circle is $\dfrac{c}{1+\sqrt{1+e^2}}$; $2c$ being the latus-rectum, and e the excentricity.

1473. A triangle is self-conjugate to an hyperbola, and one focus is equidistant from the sides of the triangle: prove that each distance is $\dfrac{c}{\sqrt{e^2-2}}$, $2c$ being the latus rectum, and e the excentricity.

1474. Two conics have a common focus, and triangles can be inscribed in one which are self-conjugate to the other; prove that

$$2c_1^2 + c_2^2 = c_1^2 c_2^2 + e_2^2 c_1^2 - 2e_1 e_2 c_1 c_2 \cos\alpha;$$

c_1, c_2 being their latera recta, e_1, e_2 their excentricities; and α the angle between their axes. Prove also that in this case triangles can be circumscribed to the second which shall be self-conjugate to the first.

1475. A conic passes through two fixed points A, A' and touches a given conic at a fixed point O: prove that their chord of intersection meets AA' in a fixed point B; and, if the given conic meet AA' in C, C' and the tangent at O meet it in B', AA', BB', CC' will be in involution.

1476. Four points being taken on a circle, four parabolas can be drawn having a common focus, and each touching the sides of the triangle formed by joining two and two three of the four points.

1477. Three tangents to an hyperbola are so drawn that the centre of perpendiculars of the triangle formed by them is at one of the foci; prove that the polar circle and the circumscribed circle of the triangle are fixed.

1478. Three tangents to a parabola form a triangle ABC, and perpendiculars x, y, z are let fall on them from the focus S; prove that

$$yz \sin BSC + zx \sin CSA + xy \sin ASB = 0,$$

the angles at S being measured so that their sum is 360°. Also prove that if $2l$ be the latus rectum,

$$\frac{\sin 2A}{x^2} + \frac{\sin 2B}{y^2} + \frac{\sin 2C}{z^2} = \frac{8 \sin A \sin B \sin C}{l^2}.$$

1479. The minor axis of an ellipse is BB', and B is the centre of curvature at B'; a point P is taken on the circle of curvature at B', and tangents drawn from P to the ellipse meet the tangent at B in Q, Q': prove that a conic drawn to touch QB', $Q'B'$ with its focus at B and directrix passing through B' will touch the circle at P.

1480. An hyperbola is drawn osculating a given parabola at P, passing through the focus, and having an asymptote parallel to the axis: prove that the tangent to it at the focus and the asymptote aforesaid intersect in the centre of curvature at P.

1481. Given a circle and a straight line not meeting it in real points, the two point-circles S, S' have with the given circle the given straight line for radical axis; two conics are drawn osculating the circle at P and having one a focus at S and the other a focus at S': prove that the corresponding directrices coincide and pass through the point of contact of the parabola which osculates the given circle at P and touches the given straight line.

1482. An ellipse is drawn osculating a given circle at P and having one focus at a point O of the circle; a parabola is also drawn osculating at P and touching the tangent at O: prove that the directrix of the ellipse is parallel to the axis of the parabola and passes through the point of contact of the parabola with the tangent at O.

1483. A point O is taken within a circle, and with O as focus is described a parabola touching the radical axis of the circle and the point-circle O; AOA' is a chord of the circle bisected in O: prove that tangents from A, A' to the parabola touch it in points lying on the circle.

1484. A chord LL' of a given circle is bisected in O and P is its pole; a parabola is drawn with its focus at O and directrix passing through L: prove that the tangents drawn to this parabola at points where it meets the circle pass through L or L'; and, if two such parabolas intersect the circle in any the same point, the angle between their axes is constant.

1485. Two fixed points are taken on a given conic and joined to any point on a given straight line: prove that the envelope of the straight line joining the points in which these joining lines again meet the conic is a conic having double contact with the given conic at the points where the given straight line meets it and also touching the straight line joining the two fixed points.

1486. Any straight line drawn through a given point meets two fixed tangents to a given conic in two points from which are drawn other tangents to the given conic: the locus of the common point of these last tangents is a conic which touches the given conic at the points of contact of tangents from the fixed point and passes through the common point of the fixed tangents.

1487. Four fixed points O, A, B, C being taken, OB, CA meet in B', OC, AB in C', and from a fixed point on OA two tangents are drawn to any conic through O, A, B, C: prove that the points of contact and the points B, C, B', C' lie on a fixed conic.

1488. With the centre of perpendiculars of a triangle as focus are described two conics, one touching the sides and the other passing through the feet of the perpendiculars; prove that these conics will touch each other and that their point of contact will lie on the conic which touches the sides of the triangle at the feet of the perpendiculars.

1489. A conic is inscribed in a triangle and one focus lies on the polar circle of the triangle: prove that the corresponding directrix passes through the centre of perpendiculars.

1490. With the centre of the circumscribed circle of a triangle as focus are described two ellipses, one touching the sides and the other passing through the middle points of the sides: prove that they will touch each other.

1491. Four fixed straight lines form a quadrilateral whose diagonals are AA', BB', CC': prove that the envelope of tangents drawn to any conic inscribed in the quadrilateral at the points where it meets a fixed straight line through A is a conic which touches BB', CC' and the two sides of the quadrilateral which do not pass through A; and if BB', CC' meet AA' in c, b and the fixed straight line through A in b', c', that bb', cc' are also tangents to this envelope.

1492. Five points are taken no three lying in one straight line, and with one of the points as focus are described four conics each touching the sides of a triangle formed by joining two and two three of the remaining four points: prove that these four conics have a common tangent.

[If A, B, C, D, E be the five points, A the one taken for focus, AP, AQ two chords at right angles of the conic $ABCDE$, then the common tangent is the locus of the intersection of the tangents at P, Q.]

1493. Through a fixed point O are drawn two straight lines meeting a given conic in P, P'; Q, Q'; and a given straight line in R, R', and RR' subtends a right angle at another fixed point: prove that PQ, PQ', $P'Q$, $P'Q'$ all touch a certain fixed conic.

1494. Given a conic and a point in its plane O: prove that there exist two real points L, such that if any straight line through L meet the polar of L in P and P' be the pole of this straight line, PP' will subtend a right angle at O.

1495. Any conic drawn through four fixed points meets two fixed straight lines drawn through one of the points again in P, Q: prove that the envelope of PQ is a conic touching the straight lines joining the other three given points.

1496. Two equal circles U, V touch at a point S, a tangent to V meets U in P, Q, and O is its pole with respect to U: prove that the directrices of two of the conics described with focus S and circumscribing the triangle OPQ will touch the circle U.

1497. A conic touches the sides of a triangle ABC in a, b, c and Aa, Bb, Cc meet in S; three conics are drawn with S for focus osculating the former at a, b, c; prove that all four conics have one common tangent which also touches the conic having one focus at S and touching the sides of the triangle ABC.

1498. Given four straight lines, prove that two conics can be constructed so that an assigned straight line of the four is directrix and the other three form a self-conjugate triangle; and that, whichever straight line be taken for directrix, the corresponding focus is one of two fixed points.

1499. A quadrilateral can be projected into a rhombus on any plane parallel to one of its diagonals, and the vertex will be any point on a certain circle in a certain parallel plane.

1500. A conic inscribed in a triangle ABC touches BC in a and Aa again meets the conic in A'; the tangent at any point P meets the tangent at A' in T: prove that the pencil $T\{ABCP\}$ is harmonic.

1501. A conic is inscribed in a triangle ABC and OP, OQ are two other tangents; another conic is drawn through $OPQBC$ and T is the pole of BC with respect to it: prove that $A\{OBCT\}$ is harmonic. Also prove that if O lie on the straight line joining A to the point of contact of BC, T will coincide with A.

1502. A conic is inscribed in a given triangle ABC and touches BC in a fixed point a; b, c are two other fixed points on BC: prove that tangents drawn from b, c to the conic intersect in a point lying on a fixed straight line through A.

1503. A triangle is self-conjugate to a rectangular hyperbola U and its sides touch a parabola V; a diameter of U is drawn through the focus of V: prove that the conjugate diameter is parallel to the axis of V.

1504. Two tangents OP, OQ are drawn to a parabola; an hyperbola drawn through O, P, Q with one asymptote parallel to the axis of the parabola meets the parabola again in R: prove that its other asymptote is parallel to the tangent at R to the parabola.

1505. Two tangents OP, OQ are drawn to an hyperbola; another hyperbola is drawn through O, P, Q with asymptotes parallel to those of the former: prove that it will pass through the centre C of the former and that CO will be a diameter.

1506. A triangle is self-conjugate to a conic U and from any other two points conjugate to U tangents are drawn to a conic V inscribed in the triangle: prove that the other four points of intersection of these tangents are two pairs of conjugate points to U.

1507. A conic drawn through four fixed points A, B, C, D meets a fixed straight line L in P, Q: prove that the conic which touches the straight lines AB, CD, L and the tangents at P and Q will have a fourth fixed tangent which with L divides AB and CD harmonically.

1508. Through two fixed points O, O' are drawn two straight lines which are conjugate to each other with respect to a given conic U: prove that the locus of their common point is a conic V passing through O, O' and the points of contact of the tangents from O, O' to the given conic. Also, if two points be taken on the polars of O, O' which are conjugates with respect to U, the envelope of the straight line joining them is a conic V' which touches the polars of OO' and the tangents from O, O' to U.

1509. From two points O, O' are drawn tangents OP, OQ; $O'P'$, $O'Q'$ to a given conic U, and a conic V is drawn through $OPQO'P'Q'$; a triangle is inscribed in V, two of whose sides touch U: prove that the third side passes through the common point of PQ, $P'Q'$. Also the tangents to U at the points where the straight line OO' meets it meet V in the points of contact of the common tangents to U, V.

[V is the locus of the intersection of tangents to U which divide OO' harmonically, and U is the envelope of straight lines divided harmonically by V and by the tangents to U at the points where OO' meets it.]

1510. From two points O, O' are drawn tangents OP, OQ; $O'P'$, $O'Q'$ to a given conic U; a conic V is drawn through $OPQO'P'Q'$, and another conic V' touches the sides of the triangles OPQ, $O'P'Q'$: prove that V, V' are polar reciprocals of each other with respect to U. Also PQ, $P'Q'$ and the tangents to V at O, O' intersect in one point.

1511. Any conic is drawn touching four fixed straight lines and from a fixed point on one of the lines a second tangent is drawn to the conic: prove that the locus of its point of contact is a conic circumscribing the triangle formed by the other three given lines.

[If the four be the sides of a triangle ABC and a straight line meeting the sides in A', B', C' and the fixed point O be on the last, the locus passes through A, B, C and through the point of concourse of Aa, Bb, Cc, where a is the point (BB', CC'); also if any other straight line through O meet the sides of the triangle ABC in A'', B'', C'', and BB'', CC'' meet in α, &c., then $A\alpha$, $B\beta$, $C\gamma$ intersect in a point on the locus.]

1512. A conic is inscribed in a given quadrilateral and from two fixed points on one of the sides are drawn other tangents to the conic: prove that the locus of their common point is a conic passing through the two given points and the points of intersection of the other three straight lines.

1513. Two common tangents to two conics meet in A, the other two in A'; from a point O on AA' tangents OP, OQ, OR, OS are drawn to the two conics, and the conic through $OPQRS$ meets AA' again in O' and the conics again in P', Q', R', S': prove that $O'P'$, $O'Q'$, $O'R'$, $O'S'$ will be the tangents to the two conics at P', Q', R', S', and that the conic $OPQRS$ will pass through the other four points of intersection of the four common tangents.

1514. A tangent OP is drawn from a given point O to a conic inscribed in a given quadrilateral of which AA', BB', CC' are diagonals, and a straight line drawn through P which with PO divides AA' harmonically: prove that the envelope of this line is also the envelope of the polar of O and is a conic which touches the three diagonals. Also, if OP, OP' be the two tangents from O the conic through $OAA'PP'$ will pass through a fourth fixed point.

1515. A conic is inscribed in a given quadrilateral and from two given points on one of the diagonals tangents are drawn: prove that their points of intersection lie on a fixed conic which passes through the ends of the other two diagonals and divides harmonically the segment terminated by the two given points; also if tangents be drawn to the former conic at points where the second conic meets it four of their points of intersection will lie on a conic which passes through the points of contact of the given quadrilateral and through the ends of the given diagonal.

1516. A conic U is inscribed in a given quadrilateral and another conic V is drawn through the ends of two of the diagonals: prove that the tangents to U at the points where it meets V pass through the points of intersection of V with the third diagonal; and the points of contact with V of the common tangents to U, V lie on the tangents to U at the points where it meets the third diagonal.

1517. A conic is drawn through four given points: prove that the envelope of the straight line joining the points where this conic again meets two fixed straight lines through one of the points is a conic which touches the two fixed straight lines and the straight lines joining two and two the other three given points.

1518. Find the locus of a point such that one double ray of the involution determined by the tangents from the point to two given conics may pass through a fixed point; and prove that the other double ray will envelope a conic, which touches the diagonals of the quadrilateral formed by the common tangents to the two given conics.

1519. Three conics U, V, W have two and two double contact, not at the same points: prove that the chords of contact of V, W with U will pass through the intersection of the common tangents to V, W and form with the common tangents an harmonic pencil.

1520. Two ellipses have the same (impossible) asymptotes: prove that any ellipse which has double contact with both will touch them so that the chords of contact will lie along conjugate diameters.

1521. A point Q is taken on the directrix of a parabola whose focus is S, a circle is described whose centre O lies on SQ produced and whose radius is a mean proportional between OQ, OS: prove that the points of contact with the circle of the common tangents lie on the tangents drawn from Q to the parabola.

1522. Two chords OP, OQ of a given conic are at right angles, another conic is described with a focus at O and PQ has the same pole with respect to the two conics: prove that tangents to the second conic at points where it meets the first pass through P or Q.

1523. Two fixed points A, B are taken on a given conic and another fixed point O in the plane: a chord PQ of the conic such that $O\{PABQ\}$ is harmonic will have for its envelope a conic touching OA, OB and the tangents to the given conic at A, B. Also if PQ meet AB in R and R' be taken in PQ so that $\{PRR'Q\}$ is harmonic, the locus of R' is a conic through A, B and having double contact with the envelope of PQ.

1524. Two points P, Q are taken on a given hyperbola and straight lines drawn from P, Q each parallel to an asymptote meet in O; a parabola is drawn touching the sides of the triangle OPQ: prove that the tangents to the parabola at points where it meets the hyperbola pass through the two points where the hyperbola is met by a straight line through O parallel to PQ.

1525. A point L is taken on the directrix of a parabola whose focus is S, and a circle is drawn such that the radical axis of the circle and S is the straight line through L at right angles to LS: prove that the points of contact with the circle of common tangents to it and the parabola lie on the tangents drawn to the parabola from L; and that the tangents to the parabola at their common points pass through the points on the circle where the straight line through S at right angles to SL meets it.

1526. Through a fixed point O are drawn two chords PP', QQ' of a given conic such that the two bisectors of the angles at O are fixed: prove that the straight lines PQ, $P'Q$, PQ', $P'Q'$ all touch a fixed conic which degenerates when the two bisectors are conjugate with respect to the given conic.

1527. The equation of the polar reciprocal of the evolute of the ellipse $a^2y^2 + b^2x^2 = a^2b^2$ with respect to the centre is

$$\frac{a^2}{x^2} + \frac{b^2}{y^2} = \frac{(a^2 - b^2)^2}{k^4}.$$

1528. Two fixed points O, O' are taken, and on the side BC of a triangle ABC is taken a point A' such that the pencil $A'\{AOO'B\}$ is harmonic; B', C' are similarly determined on the other sides: prove that AA', BB', CC' meet in a point, that the four such points corresponding to the four triangles formed by any four straight lines are collinear, and that tangents drawn from any point on this line, to the conic which touches the four straight lines and OO', will divide OO' harmonically.

1529. Two conics touch at O, and any straight line through O meets them in P, Q; prove that the tangents at P, Q intersect in a point lying on the chord of intersection of the two conics.

1530. Four tangents a, b, c, d are drawn to a conic, and the straight line joining the points of contact of b, c meets a, d in A, D; prove that a conic drawn touching a, b, c, d so that A is its point of contact with a will also have D for its point of contact with d.

1531. Two conics U, V intersect in A, B, C, D, and the pole of AB with respect to U is the pole of CD with respect to V: prove that the pole of CD with respect to U is the pole of AB with respect to V.

1532. A conic is drawn through four given points: prove that its asymptotes meet the conic which is the locus of centres of all conics through the four points in two points at the ends of a diameter.

1533. Four points and one straight line being given, four conics are described such that with respect to any one of them three of the points are corners of a self-conjugate triangle, and the fourth is the pole of the given straight line: prove that these four conics will meet the given straight line in the same two points which are points of contact of the two conics through the four points touching the line. Also prove that any conic through the four points will divide the segment between the two common points harmonically.

1534. Four straight lines and a point being given, four conics are described such that with respect to any one of them three of the straight lines form a self-conjugate triangle and the fourth is the polar of the given point: prove that these four conics will have two common tangents from the given point, and these tangents are tangents to the two conics through the given point touching the given lines. Also prove that tangents from the given point to any conic touching the given lines form a harmonic pencil with the two common tangents.

1535. Two fixed tangents CA, CB are drawn to a given conic: prove that the envelope of the straight line, joining any point in AB to the point in which its polar meets a fixed straight line, is a conic touching the sides of the triangle ABC and the fixed straight line.

1536. The sides of the triangle ABC are met by a transversal in A', B', C'; the straight line joining A to the point (BB', CC') meets BC in a, and b, c are similarly determined: four conics are drawn touching the sides of the triangle ABC, and meeting the transversal in the same two points: prove that the other common chord of any two of these conics passes through either a, b, or c; that these six common chords intersect by threes in four points; and that these four points are the poles of the transversal with respect to the four conics which touch the sides of the triangle abc and pass through the before-mentioned two points on the transversal.

1537. Two conics U, V have double contact, and from a fixed point O on their chord of contact are drawn tangents OP, OP'; OQ, OQ'; another conic W is drawn through Q, Q' to touch OP, OP': prove that the tangents to W at the points where it meets U will touch either V or another fixed conic V' which has double contact with both U and V. The pole of QQ' with respect to W lies either on the chord of contact of U and V or on a fixed straight line through O dividing PP', and QQ', harmonically to the chord of contact of U and V: if the former, the tangents to W at the points where it meets U touch V; if the latter, they touch V', which touches V at Q, Q' and U at the points where the fixed straight line before mentioned meets it.

1538. The four conics which can pass through three given points and touch two given straight lines are drawn, and their remaining pair of common tangents drawn to every two: prove that the six points of intersection will lie by threes on four straight lines, and that the diagonals of the quadrilateral formed by these four lines pass one through each of the three given points.

1539. Two conics intersect in O, A, B, C; through O is drawn a straight line to meet the curves again in two points: prove that the locus of the point of intersection of the tangents at these two points is a quartic having cusps at A, B, C, and touching both conics.

1540. Prove that the envelope of a straight line joining points of contact of parallel tangents to two given parabolas is a fourth-class cubic, and that the tangents at its three points of inflexion are the common tangents of the parabolas.

XI. *Invariant relations between Conics. Covariants.*

[If $U=0$, $U'=0$ be the equations of two conics referred to any system of line co-ordinates and the discriminant of $kU+U'$ be $k^3\Delta+k^2\Theta+k\Theta'+\Delta'$, the coefficients Δ, Θ, Θ', Δ' are the fundamental invariants of the two conics, and most relations between the two conics which are unaltered by Projection can be expressed in terms of these invariants. The locus of points from which the tangents drawn to the two conics are harmonically conjugate is a covariant conic denoted by $F=0$, and the envelope of straight lines divided harmonically by the two is another conic denoted by $F'=0$. Any other covariant can be expressed in terms of any three of these in the form $lU+mU'+nF$, the coefficients l, m, n being functions of the invariants. For instance $F'\equiv\Theta'U+\Theta U'-F$. The forms of F and F' which we use are determined by stating that their respective discriminants are

$$\Delta\Delta'(\Theta\Theta'-\Delta\Delta'),\ (\Theta\Theta'-\Delta\Delta')^2.$$

Since any anharmonic ratio assumes the form $\frac{0}{0}$ when any three of its constituents coincide, and since from a point of contact of a common tangent to U and U' three of the four tangents to the two conics coincide, we see that the harmonic locus F must pass through the eight points of contact of the four common tangents. Similarly the harmonic envelope F' must touch the eight tangents drawn to the two conics at their common points. So also $V=0$, the reciprocal polar of U with respect to U', is the locus of the poles with respect to U' of tangents to U, and must therefore pass through the points of contact with U' of the common tangents. But these are the points where U' meets F, and thus we see beforehand that V can be expressed in the form $lU'+mF$. Again V is the envelope of the polars with respect to U' of points on U, and must therefore touch the tangents drawn to U' at the common points which are the common tangents of U' and F'. Hence if we denoted the tangential equations of U', F', and V by u', f', v, we must have $v\equiv lu'+mf'$. So of course V' the reciprocal polar of U' with respect to U must pass through the points of contact of the common tangents with U, and must touch the tangents to U at the common points. The locus of the points from which tangents drawn to U and U' form a pencil of given anharmonic ratio must be of the form

$$\lambda UU'=\left(\frac{k-1}{k+1}\right)^2F^2,$$

where k, or k^{-1}, is the given ratio; for we can see that it must touch U and U' where their common tangents touch them, that is at points on F; and when $k=1$ the tangents drawn to one of the conics must coincide, or the locus reduces to the conics themselves; and when $k=-1$ the locus is F. When $k=0$ one tangent drawn to one conic must coincide with one tangent drawn to the other, and the point must be on one of the common tangents. Hence the equation of the four common tangents is $\lambda UU' = F^2$. The value of λ is readily found, by taking

$$U \equiv x^2+y^2+z^2, \quad U' = lx^2+my^2+nz^2$$

to be $4\Delta\Delta'$. A similar equation gives the four common tangents and the tangential equation of the envelope of a straight line divided by the two conics in a given ratio.

The equation between Δ, Θ, Θ', Δ' which expresses some geometrical relation between two conics (which relation must be such as is unaltered by projection) can be generally formed by taking some particular case of that relation; as is exemplified by taking $U = b^2x^2 + a^2y^2 - a^2b^2$, and $U' = x^2+y^2-a^2-b^2$. The discriminant of $kU + U'$ is

$$(kb^2+1)(ka^2+1)(ka^2b^2+a^2+b^2),$$

so that

$$\Delta = a^4b^4, \quad \Theta = 2a^2b^2(a^2+b^2), \quad \Theta' = (a^2+b^2)^2+a^2b^2, \quad \Delta' = a^2+b^2,$$

$$\Theta^2 - 4\Theta'\Delta = -4a^6b^6, \quad \Theta(\Theta^2-4\Theta'\Delta) = -8a^8b^8(a^2+b^2) = -8\Delta^2\Delta',$$

so that the relation is

$$\Theta^3 - 4\Theta\Theta'\Delta + 8\Delta^2\Delta' = 0,$$

which is homogeneous, when we reckon Δ, Θ, Θ', Δ' to be (1) all of the same dimensions, (2) of dimensions 3, 2, 1, 0, (3) of dimensions 0, 1, 2, 3. Any invariant relation expressing a projective relation should be truly homogeneous in this way. Now U' is the director circle of U, that is the locus of tangents to U dividing a certain segment $(\infty\infty')$ harmonically, and we see that the locus, being a circle, passes through $\infty\infty'$; and it is easily proved that it passes through the points of contact of the tangents drawn to U from ∞, ∞' (the points where U meets its directrices). Hence, generally, if two conics U, U' satisfy the equation

$$\Theta^3 - 4\Theta\Theta'\Delta + 8\Delta^2\Delta' = 0,$$

U' is the locus of the intersection of tangents to U which divide a certain fixed segment harmonically, and quadrilaterals can be circumscribed to U such that the ends of two diagonals lie on U', the third diagonal being fixed (containing the fixed segment), and therefore the intersection of the other two diagonals being fixed also, the polar of the third diagonal with respect to each conic. The relation between U and U' may also be expressed as follows, U is the envelope of straight lines divided harmonically by U' and the two straight lines $\frac{x^2}{a^2}+\frac{y^2}{b^2}=0$, and it is obvious that the tangents to U' at the points where those straight lines meet it also touch U. Thus we see that if U, U' be two conics such that quadrilaterals can be circumscribed to U, of which the ends of two diagonals lie on U', U' is the locus of the intersection of tangents to U which divide harmonically a certain chord of U' lying on the third diagonal of any of the quadrilaterals, and U is the envelope of a chord of U' which is divided harmonically by two tangents to U drawn from a point which is the pole of the third diagonal with respect to either conic.

Another method of investigating such invariant relations is as follows: let U, U' be two conics such that triangles can be circumscribed to U whose angular points lie on U', then generally if any tangent to U meet U' in P, Q, the second tangents drawn from P, Q will intersect in a point R on U'. Hence, if we take P at one of the common points of U, U'; Q, R must coincide; or, if P be a common point of U and U', and PQ the tangent to U at P be a chord of U', the tangent to U' at Q will also touch U. We may therefore, by properly choosing a triangle of reference, write

$$U = x^2 + 2yz, \quad U' = y^2 + 2fyz + 2gzx,$$

and thus $\Delta = 1$, $\Theta = 2f$, $\Theta' = f^2$, $\Delta' = g^2$, and the obvious invariant relation is $\Theta^2 = 4\Theta'\Delta$. Of course a common tangent might be used instead of a common point but would give us exactly the same result. So, when U, U' are such that triangles can be circumscribed to U which are self-conjugate to U', if any tangent be drawn to U, and we take its pole with respect to U', the two tangents drawn to U from this pole will be conjugate with respect to U'. By considering this tangent to U to be a common tangent we may see that by a proper choice of the triangle of reference

$$U = y^2 + 2yz + z^2 + 2gzx, \quad U' = x^2 + 2fyz,$$

giving $\Delta = g^2$, $\Theta = 0$, $\Theta' = 2f$, $\Delta' = f^2$, or the invariant relation is $\Theta = 0$. When triangles can be inscribed in U' which are self-conjugate to U, we get in the same way, by considering a common point,

$$U = x^2 + 2yz, \quad U' = y^2 + 2gzx + 2hxy,$$

so that $\Delta = 1$, $\Theta = 0$, $\Theta' = 2gh$, $\Delta' = g^2$, or the relation is again $\Theta = 0$.]

1541. Denoting two conics by U, U', the locus of points from which tangents drawn to the two form a harmonic pencil by F, the envelope of straight lines divided harmonically by the two by F', the polar reciprocal of U with respect to U' by V, and that of U' with respect to U by V', the discriminant of F being $\Delta\Delta'(\Theta\Theta' - \Delta\Delta')$, and $(\Theta\Theta' - \Delta\Delta')^2$ that of F': prove that

$$V = \Theta U' - F = F' - \Theta' U, \quad V' = \Theta' U - F = \Theta U' - F'.$$

1542. Prove that, when U and U' are circles, their centres are the foci of F', and the excentricity of F' is

$$\sqrt{\frac{a^2 + b^2 + 2ab\cos\alpha}{a^2 + b^2 - 2ab\cos\alpha}},$$

where a, b are the radii and α the angle at which U and U' intersect.

1543. A triangle ABC is inscribed in U so that A is the pole of BC with respect to U': prove that BC envelopes V, and AB, AC envelope F'. Also if a triangle ABC circumscribe U, and A be the pole of BC with respect to U', the locus of A is V', and that of B and C is F.

1544. A triangle ABC is self-conjugate to U, and B, C lie on U'; prove that the locus of A is $\Theta' U - \Delta U' = 0$, the envelope of BC is F'' and that of AB and AC is V'.

1545. A triangle ABC is self-conjugate to U, and its sides AB, AC touch U': prove that the envelope of BC is

$$\Theta'^2 U - \Theta' F'' + \Delta\Delta' U' = 0\,;$$

the locus of A is F, and that of B, C is V'.

1546. A triangle is inscribed in U, and two of its sides touch U': prove that the envelope of the third side is

$$(\Theta'^2 - 4\Theta\Delta')\, U + 4\Delta\Delta' U' = 0.$$

1547. A triangle is circumscribed to U, and two of its angular points lie on U': prove that the locus of the third is

$$(\Theta^2 - 4\Theta'\Delta)^2\, U + 4\Delta\, (\Theta^2 - 4\Theta'\Delta)\, F + 16\Delta^2\Delta' U' = 0.$$

1548. A conic osculates U at P and U' at P', and the tangents at P, P' meet in Q: prove that the locus of Q is

$$\Delta' U^3 - \Delta U'^3 = 0,$$

and the envelope of PP' is

$$\Delta\Delta'^2 U^3 + \Delta^2\Delta' U'^3 + 3\Delta\Delta' UU'F = F^3.$$

1549. Two conics for which F and F' degenerate into two straight lines and two points respectively ($\Theta\Theta' = \Delta\Delta'$) will have either four real common points and no real common tangents, two real common points and two real common tangents, or no real common points and four real common tangents. When they have four real common points, these will be the points of contact of tangents from either of the points into which F' degenerates, to the two conics; and when they have four real common tangents these are the tangents drawn to the two conics at the points, in which either of the straight lines into which F degenerates, meets them.

1550. The condition $\Theta\Theta' = \Delta\Delta'$ is satisfied by a circle and rectangular hyperbola when one of their common chords is a diameter of the circle, and the other (therefore) a diameter of the hyperbola.

[When the second common chord is real, there are four real common points and no real common tangents, and F is two impossible straight lines, having one real common point where the two common chords meet. When the second common chord meets the conics in unreal points, there are two real common points and two real common tangents,

F' is two real straight lines forming a harmonic pencil with the two common chords, and the two points into which F'' degenerates are the poles of the two common chords.]

1551. An hyperbola is described whose asymptotes are conjugate diameters of a given ellipse: prove that the relation $\Theta\Theta' = \Delta\Delta'$ is satisfied for the two conics: that when there are four real common points the two points F'' are two real points at ∞, the poles of the common diameters; and the two straight lines F' are two impossible diameters: when there are four real common tangents, the points of contact lie on two diameters (the straight lines F') and the points F'' are impossible.

1552. The general equation of a conic, for which the relation $\Theta\Theta' = \Delta\Delta'$ is satisfied with the given conic $lx^2 + my^2 + nz^2 = 0$, is

$$(lx^2 + my^2 + nz^2)\left(\frac{pp'}{l} + \frac{qq'}{m} + \frac{rr'}{n}\right) = 2(px + qy + rz)(p'x + q'y + r'z).$$

1553. Prove that the equations of two conics, satisfying the relation $\Theta\Theta' = \Delta\Delta'$, may be always reduced to the forms

$$x^2 + y^2 - z^2 = 0, \quad x^2 - y^2 + mz^2 = 0:$$

and reduce in this manner the two pairs of circles

(1) $x^2 + y^2 = 49, \quad x^2 + y^2 - 20x + 99 = 0;$

(2) $x^2 + y^2 = 16, \quad x^2 + y^2 - 10x + 16 = 0.$

[(1) $(5x - 49)^2 - 24y^2 + 49(x - 5)^2 = 0, \quad (5x - 49)^2 + 24y^2 = (x - 5)^2;$

(2) $\{(2 + i)x - 4(1 + 2i)\}^2 + \{(2 - i)x - 4(1 - 2i)\}^2 + 6y^2 = 0,$

$\{(2 + i)x - 4(1 + 2i)\}^2 - \{(2 - i)x - 4(1 - 2i)\}^2 + 8iy^2 = 0.$]

1554. The equations of two conics, for which the relation $\Theta\Theta' = \Delta\Delta'$ is satisfied, can always be put in the forms (areal co-ordinates)

$$U \equiv lx^2 + my^2 + nz^2 = 0, \quad U' \equiv x^2 + 2pyz = 0,$$

and, if the two straight lines F' meet the two curves in P, Q, P', Q'; p, q, p', q'; the ranges $\{PP'Q'Q\}$, $\{pp'q'q\}$ will be equal; and similarly the tangents drawn to the two curves from the two points F'' will form pencils of equal ratios.

1555. If ABC be the triangle of reference in the last question, the quadrangle formed by the points of contact of the common tangents with either curve will have the same vertices as the quadrangle formed by the points in which AB, AC meet U.

1556. The tangent and normal to a rectangular hyperbola at P meet the transverse axis in T, G, and a circle is drawn with centre G and radius GP: prove that straight lines drawn through T parallel to the asymptotes will pass through the points of contact of common tangents drawn to this circle and to the auxiliary circle, and the tangents drawn to the two circles from any point on either of these straight lines will form a harmonic pencil.

1557. The harmonic locus and envelope of the conics

$$x^2 = 2pyz, \quad x^2 = 2qyz$$

are respectively $$x^2 = 2ryz, \quad x^2 = 2r'yz,$$

where r, r' are the arithmetic and geometric means between p and q.

1558. The harmonic locus and envelope of the conics

$$2lx^2 + 2\lambda xy + y^2 = 2ax, \quad 2mx^2 + 2\mu xy + y^2 = 2ax,$$

are respectively

$$\{l + m - \tfrac{1}{2}(\lambda - \mu)^2\} x^2 + (\lambda + \mu) xy + y^2 - 2ax = 0,$$
$$\{l + m + \tfrac{1}{4}(\lambda - \mu)^2\} x^2 + (\lambda + \mu) xy + y^2 - 2ax = 0.$$

1559. Prove that when two conics have contact of the third order the harmonic locus and envelope coincide.

$$[2mx^2 + y^2 = 2ax, \quad 2nx^2 + y^2 = 2ax, \quad (m + n) x^2 + y^2 = 2ax.]$$

1560. Four tangents are drawn to a circle U forming a quadrilateral such that the extremities of two of its diagonals lie on another circle U': prove that if a, a' be the radii, and b the distance between the centres,

$$b = a' \text{ or } (b^2 - a'^2)^2 = 2a^2(b^2 + a'^2).$$

[In the former case U' is the locus of the points from which tangents drawn to U divide harmonically the diameter of U' drawn from the centre of U, and U is the envelope of chords of U' divided harmonically by the radical axis and by the diameter of U' which is at right angles to the line of centres.]

1561. Two conics will be such that quadrilaterals can be circumscribed to either with the ends of two diagonals on the other, if

$$\Theta^2 = 2\Theta'\Delta, \text{ and } \Theta'^2 = 2\Theta\Delta':$$

and the curves $x^2 - y^2 = a^2$, $x^2 + y^2 \pm 2\sqrt{3}ax + 2a^2 = 0$ are so related.

1562. Prove that if a circle and rectangular hyperbola be described so that each passes through the centre of the other, and a parabola be described with its focus at the centre of the hyperbola and directrix touching the hyperbola at the centre of the circle, the three form a harmonic system, such that, if any two be taken as U and U', the covariants F, F', V, V' all coincide with the third, and thus that an infinite number of triangles can be inscribed in the first whose sides touch the second and which are self-conjugate to the third, whatever be the order in which the three are taken.

1563. A straight line is divided harmonically by two conics and its pole with respect to either lies on the other: prove that the same property is true for every other straight line divided harmonically; that, for the two conics, $\Theta = 0$, $\Theta' = 0$; and that the harmonic locus and envelope coincide and form with the two a harmonic system.

1564. The conics U_1, U_2, U_3 form a harmonic system and any triangle ABC is inscribed in U_1 whose sides touch U_2 in a, b, c; then abc will be a triangle whose sides touch U_3 (in A', B', C') and $A'B'C'$ will be a triangle whose sides touch U_1 in A, B, C.

1565. Prove that any two conics of a harmonic system have two real common points and two real common tangents; and, if A, A' be the common points, the common tangents BC, $B'C'$ can be so taken that AB, AC, $A'B'$, $A'C'$ are the tangents to the two at A, A', and the third conic of the system will touch AB, AC at B, C, and $A'B'$, $A'C'$ at B', C'.

1566. Prove that the equations of three conics forming a harmonic system can be obtained in areal co-ordinates in the forms

$$x^2 = 2pyz, \quad y^2 = 2qzx, \quad z^2 = 2rxy,$$

where $pqr + 1 = 0$, the triangle of reference being either of two triangles.

[By using multiples of areal co-ordinates, the equation may be written in the more symmetrical form

$$x^2 + 2yz = 0, \quad y^2 + 2zx = 0, \quad z^2 + 2xy = 0.]$$

1567. In a harmonic system

$$x^2 + 2yz = 0, \quad y^2 + 2zx = 0, \quad z^2 + 2xy = 0, \qquad (1, 2, 3)$$

a triangle ABC is taken whose sides touch (2) in the points a, b, c and angular points A, B, C lie on (3): prove that Aa, Bb, Cc intersect in a point lying on (3) such that if A be the point $(\frac{1}{2} : -\lambda^2 : \lambda)$, and B, C be similarly denoted by μ, ν, the point of concourse will be $(\frac{1}{2} : -k^2 : k)$ where

$$\frac{3}{k} + \frac{1}{\lambda} + \frac{1}{\mu} + \frac{1}{\nu} = 0, \quad \lambda + \mu + \nu = 0, \quad 4\lambda\mu\nu + 1 = 0,$$

and
$$\frac{\lambda}{4\lambda^2 + 1} = \frac{\mu}{4\mu^2 + 1} = \frac{\nu}{4\nu^2 + 1} = \frac{-k}{3} = \frac{\lambda\mu\nu}{\mu\nu + \nu\lambda + \lambda\mu}.$$

Also prove that bc, ca, ab touch (1) in points similarly denoted by λ, μ, ν.

1568. Prove that the three conics whose equations are

$$3x^2 - y^2 + 4(y + 2a)^2 \pm 6xy = 0, \quad 3x^2 - 3y^2 - 8ay - 8a^2 = 0,$$

form a harmonic system, which may be reduced to the standard form by either

$$X = y + 4a + x, \qquad Y = y + 4a - x, \qquad Z = 4(y + a),$$

or by
$$X = 3y + 4a + 3x, \quad Y = 3y + 4a - 3x, \quad Z = 4a.$$

1569. A circle and rectangular hyperbola are such that the centre of either lies on the other, and the angles at which they intersect (in real points) are θ, θ': prove that

$$(2(\cos\theta)^{\frac{2}{3}} + 1)(2(\cos\theta')^{\frac{2}{3}} + 1) = 3,$$

and the squares of their latera recta are as $(1 + 8\cos^2\theta)^{\frac{3}{2}} : 8\sin^3\theta\cos\theta$ (the same ratio as $(1 + 8\cos^2\theta')^{\frac{3}{2}} : 8\sin^3\theta'\cos\theta'$).

1570. A circle and parabola are such that the focus of the parabola lies on the circle and the directrix of the parabola passes through the centre of the circle, and the two intersect in two real points at angles θ, θ': prove that

$$(2(\cos\theta)^{\frac{2}{3}} + 1)(2(\cos\theta')^{\frac{2}{3}} + 1) = 3,$$

and that the latus rectum of the parabola is to the diameter of the circle as

$$8\sin^3\theta\cos\theta : (1 + 8\cos^2\theta)^{\frac{3}{2}}.$$

1571. A parabola and rectangular hyperbola are such that the focus of the parabola is the centre of the hyperbola and the directrix of the parabola touches the hyperbola, and they intersect in two real points at angles 2θ, $2\theta'$: prove that

$$(2(\sin\theta)^{\frac{2}{3}} + 1)(2(\sin\theta')^{\frac{2}{3}} + 1) = 3,$$

and that the squares of their latera recta are as

$$8\cos^3\theta\sin\theta : (1 + 8\sin^2\theta)^{\frac{3}{2}}.$$

THEORY OF EQUATIONS.

1572. The product of two unequal roots of the equation

$$ax^3 + bx^2 + cx + d = 0$$

is 1: prove that the third root is $\dfrac{a-c}{b-d}$.

1573. The roots of the equation $x^2 - px + q = 0$, when real, are the limits of the infinite continued fractions

$$\frac{q}{p-}\,\frac{q}{p-}\,\frac{q}{p-}\ldots, \quad p - \frac{q}{p-}\,\frac{q}{p-}\ldots\ldots.$$

Explain these results when $p^2 < 4q$.

1574. Prove that, when the equation $x^3 - px^2 + qx - r = 0$ has two equal roots, the third root must satisfy either of the equations

$$x(x-p)^2 = 4r, \quad (x-p)(3x+p) + 4q = 0.$$

1575. Find the relation between p, q, r in order that the roots of the equation $x^3 - px^2 + qx - r = 0$ may be (1) the tangents, (2) the cosines, (3) the sines, of the angles of a triangle.

[The results are

(1) $p = r$, (2) $p^2 - 2q + 2r = 1$, (3) $p^4 - 4p^2q + 8pr + 4r^2 = 0$.]

1576. Prove that the roots of the equations

(1) $x^3 - 5x^2 + 6x - 1 = 0$; (2) $x^3 - 6x^2 + 10x - 4 = 0$;

(3) $x^4 - 7x^3 + 15x^2 - 10x + 1 = 0$;

(4) $x^6 - 11x^5 + 45x^4 - 84x^3 + 70x^2 - 42x + 11 = 0$;

are (1) $4\cos^2\dfrac{\pi}{7}$, $4\cos^2\dfrac{2\pi}{7}$, $4\cos^2\dfrac{3\pi}{7}$;

(2) $4\cos^2\dfrac{\pi}{8}$, $4\cos^2\dfrac{2\pi}{8}$, $4\cos^2\dfrac{3\pi}{8}$;

(3) $4\cos^2\dfrac{\pi}{9}$, $4\cos^2\dfrac{2\pi}{9}$, $4\cos^2\dfrac{3\pi}{9}$, $4\cos^2\dfrac{4\pi}{9}$;

(4) $4\cos^2\dfrac{\pi}{13}$, $4\cos^2\dfrac{2\pi}{13}$, …… $4\cos^2\dfrac{6\pi}{13}$.

1577. Determine the relation between q and r necessary in order that the equation $x^3 - qx + r = 0$ may be put into the form

$$(x^2 + mx + n)^2 = x^4;$$

and solve in this manner the equation

$$8x^3 - 36x + 27 = 0.$$

1578. Find the condition necessary in order that the equation

$$ax^3 + bx^2 + cx + d = 0$$

may be put under the form $(x^2 + px + q)^2 = x^4$; and solve in this manner the equation $x^3 + 3x^2 + 4x + 4 = 0$.

[The condition is $c^3 - 4bcd + 8ad^2 = 0$; and the proposed equation may be written $(x^2 + 2x + 4)^2 = x^4$.]

1579. Prove that, if the roots of the equation $x^3 - px^2 + qx - r = 0$ are in H.P., those of the equation

$$\{p^3(1-n) + n^2(pq - nr)\}x^3 - (p^2 - 2npq + 3n^2r)x^2 + (pq - 3nr)x - r = 0$$

are also in H.P.

1580. Reduce the equation $x^3 - px^2 + qx - r = 0$ to the form $y^3 \pm 3y + m = 0$ by assuming $x \equiv ay + b$; and solve this equation by assuming $y \equiv z \mp \frac{1}{z}$. Hence prove the condition for equal roots to be

$$4(p^2 - 3q)^3 = (2p^3 - 9pq + 27r)^2.$$

1581. Prove that the roots of the auxiliary quadratic, used in solving a cubic equation by Cardan's (or Tartaglia's) rule, are

$$\frac{(2\alpha - \beta - \gamma)(2\beta - \gamma - \alpha)(2\gamma - \alpha - \beta) \pm 3\sqrt{-3}(\beta - \gamma)(\gamma - \alpha)(\alpha - \beta)}{54};$$

where α, β, γ are the roots of the cubic.

1582. Prove that any cubic equation in x can be reduced to the form $(ay + b)^3 = cy^3$ by putting $x \equiv y + z$, and the roots of the quadratic for z will be

$$\frac{\alpha(\beta - \gamma)^2 + \beta(\gamma - \alpha)^2 + \gamma(\alpha - \beta)^2 \pm \sqrt{-3}(\beta - \gamma)(\gamma - \alpha)(\alpha - \beta)}{(\beta - \gamma)^2 + (\gamma - \alpha)^2 + (\alpha - \beta)^2}.$$

1583. Prove that, if the cubic $(p_0, p_1, p_2, p_3 \between x, 1)^3 = 0$ be put in the form $A(x + \alpha)^3 + B(x + \beta)^3 = 0$, α, β will be the roots of the quadratic

$$(p_1^2 - p_0p_2)z^2 - (p_1p_2 - p_0p_3)z + (p_2^2 - p_1p_3) = 0;$$

and thence deduce the condition for equal roots.

[The true condition for two equal roots is given by making this quadratic have equal roots; yet, if $\alpha = \beta$, the equation reduces to $(A + B)(x + \alpha)^3 = 0$. The student should explain this result.]

1584. A cubic equation is solved by putting it in the form $(x+p)^3 = z(x+q)^3$: prove that the roots of the quadratic for z are $\left(\frac{\alpha+\beta\omega^2+\gamma\omega}{\alpha+\beta\omega+\gamma\omega^2}\right)^3$, $\left(\frac{\alpha+\beta\omega+\gamma\omega^2}{\alpha+\beta\omega^2+\gamma\omega}\right)^3$, where α, β, γ are the roots of the cubic, and ω an impossible cube root of 1.

Solve the equation $x^3+9x^2-33x+27=0$ in this manner.

[$5(x-1)^3 = 4(x-2)^3$.]

1585. Prove that the equation

$$(x-a)(x-b)(x-c) - f^2x^2(x-a) - g^2x^2(x-b) - h^2x^2(x-c) + 2fghx^3 = 0,$$

when a, b, c are all of the same sign, will have two equal roots only when

$$\frac{af}{f-gh} = \frac{bg}{g-hf} = \frac{ch}{h-fg}.$$

[The equation may be reduced to the form

$$\frac{\frac{gh}{fa}}{\frac{1}{x}-\frac{1}{\alpha}} + \frac{\frac{hf}{gb}}{\frac{1}{x}-\frac{1}{\beta}} + \frac{\frac{fg}{hc}}{\frac{1}{x}-\frac{1}{\gamma}} = 1,$$

where α, β, γ are the three $\frac{af}{f-gh}$, &c.]

1586. The equation $x^4-4x^3+5x^2-3=0$ can be solved as follows: $(x^4+5x^2-3)^2 = 16x^6$, therefore $(x^4-3x^2+5)^2=16$, or $x^4-3x^2+5\pm4=0$: prove that the equation $x^4-2ax^3+(a^2+1)x^2=a^2-1$ can be solved in the same way; solve it, and select the roots which belong to the original equation.

1587. Prove that the equation

$$x^3+(a+b+c)x^2+2(bc+ca+ab-a^2-b^2-c^2)x-4abc=0$$

has all its roots real for all real values of a, b, c, and that the roots are separated by the three

$$a-b-c-\frac{bc}{a},\quad b-c-a-\frac{ca}{b},\quad c-a-b-\frac{ab}{c}.$$

[If these three expressions be denoted by α, β, γ, the equation may be written

$$abc = \frac{b^2c^2}{x-\alpha} + \frac{c^2a^2}{x-\beta} + \frac{a^2b^2}{x-\gamma}.]$$

1588. Investigate whether the general cubic equation can be reduced by assuming it to coincide with either of the forms

(1) $(2x^2+(a+a')x+b+b')^2 = (2x^2+(a-a')x+(b-b')^2$;

(2) $(x^2+ax+b+c)^2 = (x^2-ax+b-c)^2$.

1589. Prove that, if $\alpha, \beta, \gamma, \delta$ be the roots of the equation

$$x^4 + qx^2 + rx + s = 0,$$

the roots of the equation

$$s^3x^4 + qs(1-s)^2x^2 + r(1-s)^3x + (1-s)^4 = 0$$

will be $\beta + \gamma + \delta + \frac{1}{\beta\gamma\delta}$, &c.

1590. Prove that the equation $x^4 - 2x^3 + m(2x-1) = 0$ has two real and two impossible roots, for all real finite values of m, except when $m = 1$.

[The equation may be written

$$(x^2 - x + z)^2 = (2z+1)x^2 - 2(m+z)x + m + z^2,$$

and the dexter is a square when $2z^3 = m(m-1)$.]

1591. Prove that the equation $x^4 + 2px^3 + 2rx + rp = 0$ has in general two real and two impossible roots: the only exception being when three roots are equal.

1592. Prove that the roots of the equation

$$x^3 - 6x^2 + 9x - 4\sin^2\alpha = 0$$

are all real and positive, and that the difference between the greatest and least lies between 3 and $2\sqrt{3}$.

[$f(0)$ is negative, $f(1)$ positive, $f(3)$ negative, and $f(4)$ positive. The actual roots are readily found, by putting $x = 4\sin^2\theta$, to be $4\sin^2\frac{\pi \pm \alpha}{3}$ and $4\sin^2\frac{\alpha}{3}$, and α may be supposed to lie between 0 and $\frac{\pi}{2}$.]

1593. In the equation $x^4 - p_1x^3 + p_2x^2 - p_3x + p_4 = 0$, prove that the sum of two of the roots will be equal to the sum of the other two, if $8p_3 - 4p_1p_2 + p_1^3 = 0$; and the product of two equal to the product of the other two, if $p_1^2p_4 = p_3^2$.

1594. The roots of a biquadratic are $\alpha, \beta, \gamma, \delta$, and it is solved by putting it in the form

$$(x^2 + ax + b)^2 = (cx - d)^2;$$

prove that the values of $2b$ are

$$\beta\gamma + \alpha\delta, \quad \gamma\alpha + \beta\delta, \quad \alpha\beta + \gamma\delta;$$

those of $\pm 2c$ are

$$\beta + \gamma - \alpha - \delta, \quad \gamma + \alpha - \beta - \delta, \quad \alpha + \beta - \gamma - \delta;$$

and those of $\pm 2d$ are

$$\beta\gamma - \alpha\delta, \quad \gamma\alpha - \beta\delta, \quad \alpha\beta - \gamma\delta.$$

1595. A biquadratic in x may be solved by putting $x \equiv my + n$ and making the equation in y reciprocal: prove that the three values of n are

$$\frac{\beta\gamma - \alpha\delta}{\beta + \gamma - \alpha - \delta}, \quad \frac{\gamma\alpha - \beta\delta}{\gamma + \alpha - \beta - \delta}, \quad \frac{\alpha\beta - \gamma\delta}{\alpha + \beta - \gamma - \delta};$$

and those of m^2 are

$$\frac{(\alpha - \gamma)(\alpha - \delta)(\beta - \gamma)(\beta - \delta)}{(\alpha + \beta - \gamma - \delta)^2}, \text{ \&c.}$$

1596. Prove that the equation $3x^4 + 8x^3 - 6x^2 - 24x + r = 0$ will have four real roots, if $r < -8 > -13$; two real roots if $r > -8 < 19$; and no real roots, if $r > 19$.

1597. Prove that, if $\frac{p_n}{q_n}$ be the nth convergent to the infinite continued fraction

$$\frac{1}{a-}\,\frac{1}{a-}\,\frac{1}{a-}\,\ldots\ldots,$$

$x^{n+1} - q_n x + p_n$ will be divisible by $x^2 - ax + 1$, and conversely.

1598. Prove that, if $\frac{p_n}{q_n}$ be the nth convergent (unreduced) to the infinite continued fraction

$$\frac{b}{a-}\,\frac{b}{a-}\,\frac{b}{a-}\,\ldots\ldots,$$

$x^{n+1} - q_n x + p_n$ will be divisible by $x^2 - ax + b$, and the quotient will be

$$x^{n-1} + q_1 x^{n-2} + q_2 x^{n-3} + \ldots + q_{n-2}\,x + q_{n-1}.$$

1599. Prove that, if $\frac{p_n}{q_n}$ be the nth convergent to the smaller root of the equation $x^2 - ax + b$, which has real roots, the convergents to the other root will be

$$\frac{a}{1}, \frac{q_2}{q_1}, \frac{q_3}{q_2}, \ldots\ldots \frac{q_n}{q_{n-1}}, \text{ or } \frac{p_2}{p_1}, \frac{p_3}{p_2}, \ldots\ldots \frac{p_{n+1}}{p_n}.$$

1600. The n roots of the equation

$$a^{n+2}\frac{(x-b)(x-c)}{(a-b)(a-c)} + b^{n+2}\frac{(x-c)(x-a)}{(b-c)(b-a)} + c^{n+2}\frac{(x-a)(x-b)}{(c-a)(c-b)} = x^{n+2},$$

different from a, b, c are given by the equation

$$x^n + H_1 x^{n-1} + H_2 x^{n-2} + \ldots + H_n = 0,$$

where H_p is the sum of the homogeneous products of powers of a, b, c of p dimensions.

1601. **The n roots of the equation**

$$\left\{\frac{a_1^{n+r}}{(x-a_1)f'(a_1)}+\frac{a_2^{n+r}}{(x-a_2)f'(a_2)}+\ldots+\frac{a_r^{n+r}}{(x-a_r)f'(a_r)}\right\}f(x)=x^{n+r},$$

different from a_1, a_2, ... a_r, the roots of $f(x)=0$, are given by the equation

$$x^n+H_1x^{n-1}+H_2x^{n-2}+\ldots+H_n=0,$$

where H_p is the sum of the homogeneous products of powers of $a_1, a_2 \ldots a_r$ of p dimensions.

1602. Prove that, if

$$(1+x+x^2+\ldots+x^{p-1})^n\equiv a_0+a_1x+a_2x^2+\ldots\ldots$$

and $S_r\equiv a_r+a_{r+p}+a_{r+2p}+\ldots$, where r may have any of the p values 0, 1, 2, ... $p-1$, then of the p quantities $S_0, S_1, S_2, \ldots S_{p-1}$, $p-1$ are equal to each other, and differ from the pth by 1.

[If $n\equiv 0 \pmod{p}$, $S_1=S_2=\ldots=S_{p-1}=S_0-(-1)^n$;

and if $n\equiv r \pmod{p}$, $S_1=S_2=\ldots=S_{p-1}=S_{p-r}-(-1)^n$.]

1603. Prove that the equation $x^n-rx^{n-p}+s=0$ will have two equal roots if

$$\left\{\frac{r}{n}(n-p)\right\}^n=\left\{\frac{s}{p}(n-p)\right\}^p.$$

1604. Prove that, if (x) have two roots equal to a, and the corresponding partial fractions in $\frac{\phi(x)}{f(x)}$ be $\frac{A}{(x-a)^2}+\frac{B}{(x-a)}$,

$$A=\frac{2\phi(a)}{f''(a)},\ B=\frac{2}{3}\,\frac{3\phi'(a)f''(a)-\phi(a)f'''(a)}{\{f''(a)\}^2}.$$

1605. The coefficients $a_0\, a_1, a_2 \ldots a_n$ can be so determined as to make the expression

$$a_0(x^{2n+2}+1)-a_1x(x^{2n}+1)+a_2x^2(x^{2n-2}+1)\ldots+(-1)^n a_nx^n(x^2+1)$$

equal to $(x-1)^{2n}\{(n+1)(x^2+1)+2nx\}$.

[The necessary value of a_{r+1} is $\frac{2(n-r)^2}{r+1}\,{}_{2n+1}C_r$.]

1606. Prove that, if $x_1, x_2, \ldots x_n$ be determined by the n simple equations

$$x_1-2^{2r}x_2+3^{2r}x_3-\ldots+(-1)^{n-1}n^{2r}x_n=(-1)^{n-1}(n+1)^{2r},$$

r having successively the values 1, 2, ... n,

$$x_n=2n+2,\quad x_{n-1}=\frac{(2n+2)(2n+1)}{\lfloor 2},\ \ldots\ x_r=\frac{\lfloor 2n+2}{\lfloor n+1-r\,\lfloor n+1+r}.$$

1607. Prove the identity (for integral values of n)

$$2^n(x-1)^n-(n-1)2^{n-2}(x-1)^{n-2}(x+1)^2+\frac{(n-2)(n-3)}{\lfloor 2}2^{n-4}(x-1)^{n-4}(x+1)^4-\dots$$

$$\equiv(n+1)\left(x^n-\frac{(2n+1)2n}{\lfloor 3}x^{n-1}+\frac{(2n+1)\,2n\,(2n-1)\,(2n-2)}{\lfloor 5}x^{n-2}-\dots\right),$$

the number of terms being n in the dexter, and $\frac{n+1}{2}$ or $\frac{n}{2}+1$ in the sinister.

1608. Prove that the expression

$$x^n-2nx^{n-1}+\frac{(2n-1)(2n-2)}{\lfloor 2}x^{n-2}-\frac{(2n-2)(2n-3)(2n-4)}{\lfloor 3}x^{n-3}+\dots$$

is unchanged, or changed in sign only, if $4-x$ be substituted for x; and deduce the identity

$$4^r-2n\frac{r}{n}4^{r-1}+\frac{(2n-1)(2n-2)}{\lfloor 2}\,\frac{r(r-1)}{n(n-1)}4^{r-2}-\dots$$

$$\equiv\frac{(2n-r+1)(2n-r)\dots(2n-2r+2)}{n(n-1)\dots(n-r+1)}.$$

[The roots of the expression are $4\cos^2\frac{a_1}{2}$, $4\cos^2\frac{a_2}{2}$,... where $a_1, a_2,\dots$ are the roots of $\frac{\sin(n+1)\theta}{\sin\theta}=0$.]

1609. Prove that the roots of the equations

$$(1)\quad x^n-(2n-1)x^{n-1}+\frac{(2n-2)(2n-3)}{\lfloor 2}x^{n-2}$$
$$-\frac{(2n-3)(2n-4)(2n-5)}{\lfloor 3}x^{n-3}+\dots=0,$$

$$(2)\quad x^{\frac{n-1}{2}}-nx^{\frac{n-3}{2}}+\frac{n(n-3)}{\lfloor 2}x^{\frac{n-5}{2}}-\frac{n(n-4)(n-5)}{\lfloor 3}x^{\frac{n-7}{2}}+\dots=0,$$

are respectively

$$(1)\quad 4\cos^2\frac{\pi}{2n+1},\ 4\cos^2\frac{2\pi}{2n+1},\dots 4\cos^2\frac{n\pi}{2n+1};$$

$$(2)\quad 4\sin^2\frac{\pi}{n},\ 4\sin^2\frac{2\pi}{n},\dots 4\sin^2\frac{(n-1)\pi}{2n}.$$

1610. Prove that, if $a_1, a_2,\dots a_n$ be the roots (all unequal) of $f(x)$, and the coefficient of x^n in $f(x)$ be 1,

$$\frac{a_1^{\,n+r-1}}{f'(a_1)}+\frac{a_2^{\,n+r-1}}{f'(a_2)}+\dots+\frac{a_n^{\,n+r-1}}{f'(a_n)}$$

will be equal to the sum of the homogeneous products of r dimensions of powers of the n quantities $a_1, a_2, \dots a_n$. Prove also that

$$\frac{f''(a_1)}{a_1 f'(a_1)} + \frac{f''(a_2)}{a_2 f'(a_2)} + \dots + \frac{f''(a_n)}{a_n f'(a_n)}$$

$$\equiv \frac{1}{a_1^2} + \frac{1}{a_2^2} + \dots + \frac{1}{a_n^2} - \left(\frac{1}{a_1} + \frac{1}{a_2} + \dots + \frac{1}{a_n}\right)^2.$$

1611. Prove that the equation in x

$$\frac{a_1}{x+a_1} + \frac{a_2}{x+a_2} + \dots + \frac{a_n}{x+b_n} = 0$$

will be an identical equation if

$$\Sigma(a) = 0, \ \Sigma(ab) = 0, \ \Sigma(ab^2) = 0, \dots \ \Sigma(ab^{n-1}) = 0;$$

but that these conditions are equivalent to $\Sigma(a) = 0$, $b_1 = b_2 = \dots = b_n$.

1612. If four quantities a, b, c, d be such that

$$bc + ad + ca + bd + ab + cd = 0,$$

and
$$\frac{1}{bc+ad} + \frac{1}{ca+bd} + \frac{1}{ab+cd} = 0;$$

while $a+b+c+d$, and $\frac{1}{a} + \frac{1}{b} + \frac{1}{c} + \frac{1}{d}$ are real and finite, two of the four will be real and two impossible.

1613. Prove that, if all the roots of the cubic $x^3 - 3px^2 + 3qx - r = 0$ be real, the difference between any two roots cannot exceed $2\sqrt{3(p^2-q)}$, and the difference between the greatest and least must exceed $3\sqrt{p^2-q}$. Also, if β be the mean root,

$$\beta^2 > \frac{3}{4}\frac{q^2-pr}{p^2-q} \text{ and } < \frac{4}{3}\frac{q^2-pr}{p^2-q}.$$

1614. Prove that the sum of the ninth powers of the roots of the equation $x^3 + 3x + 9 = 0$ is 0.

1615. The system of equations of which the type is

$$a_1^r x_1 + a_2^r x_2 + \dots + a_n^r x_n = c^r$$

is true for integral values of r from $r = 1$ to $r = n+1$: prove that they are true for all values of r.

1616. Having given the two equations

$$\cos n\alpha + p_1 \cos(n-1)\alpha + p_2 \cos(n-2)\alpha + \dots + p_n = 0,$$

$$\sin n\alpha + p_1 \sin(n-1)\alpha + p_2 \sin(n-2)\alpha + \dots p_{n-1} \sin\alpha = 0;$$

prove that

$$1 + p_1 \cos\alpha + p_2 \cos 2\alpha + \dots + p_n \cos n\alpha = 0,$$

and
$$p_1 \sin\alpha + p_2 \sin 2\alpha + \dots + p_n \sin n\alpha = 0.$$

1617. The sum of two roots of the equation

$$x^4 - 8x^3 + 21x^2 - 20x + 5 = 0$$

is equal to 4: explain why, on attempting to solve the equation from the knowledge of this fact, the method fails.

1618. The equation $x^5 - 209x + 56 = 0$ has two roots whose product is 1, determine them: also determine the roots of the equation

$$x^5 - 387x + 285 = 0$$

whose sum is 5.

1619. Prove that all the roots of the equation

$$(1-x)^m - mnx(1-x)^{m-1} + \frac{m(m-1)}{2}\frac{n(n-1)}{2}x^2(1-x)^{m-2} - \ldots = 0,$$

the number of terms being $m+1$, are all real, and that none lie beyond the limits 0, 1; m, n being whole numbers and $m > n$.

1620. Find the sum of the nth powers of the roots of the equation $x^4 - x^2 + 1 = 0$; and form the equation whose roots are the squares of the differences of the roots of the proposed equation.

[If S_r denote the sum of the rth powers, $S_{2n-1} = 0$, $S_{2n} = 4\cos\frac{n\pi}{3}$; and the required equation is

$$(x^2 + 4x + 3)(x^2 + 2x\sqrt{3} + 4)(x^2 + 7 - 4\sqrt{3}) = 0.]$$

1621. The sum of the rth powers of the roots of the equation

$$x^n + p_1x^{n-1} + p_2x^{n-2} + \ldots + p_n = 0,$$

is denoted by s_r, and $S_m = s_1 + s_2 + s_3 + \ldots + s_m$; prove that, if S_m have a finite limit when m is indefinitely increased, that limit is

$$-\frac{p_1 + 2p_2 + 3p_3 + \ldots + np_n}{1 + p_1 + p_2 + \ldots + p_n}.$$

1622. The roots of the equation

$$x^n - p_1x^{n-1} + p_2x^{n-2} - \ldots = 0$$

are $\alpha, \beta, \gamma, \delta, \ldots$: prove that

$$\Sigma(2\alpha - \beta - \gamma)(2\beta - \gamma - \alpha)(2\gamma - \alpha - \beta)$$
$$= (n-1)(n-2)p_1^3 - 3n(n-2)p_1p_2 + 3n^2p_3;$$

and determine what symmetrical functions of the differences of the roots are equal to

(1) $(n-2)(n-3)p_2^2 - 2(n-1)(n-3)p_1p_3 + 2n(n-1)p_4$.

(2) $-(n-1)(n-2)(n-3)p_1^4 + 4n(n-2)(n-3)p_1^2p_2$
$\qquad - 8n^2(n-3)p_1p_3 + 8n^3p_4$.

1623. The **roots** of the equation

$$x^5 - p_1 x^4 + p_2 x^3 - p_3 x^2 + p_4 x - p_5 = 0$$

exceed those of the equation

$$x^5 - q_1 x^4 + q_2 x^3 - q_3 x^2 + q_4 x - q_5 = 0$$

respectively, each by the same quantity : **prove that**

$$2p_1^2 - 5p_2 = 2q_1^2 - 5q_2,$$
$$4p_1^3 - 15p_1 p_2 + 25p_3 = 4q_1^3 - 15q_1 q_2 + 25q_3,$$
$$3p_2^2 - 8p_1 p_3 + 20p_4 = 3q_2^2 - 8q_1 q_3 + 20q_4,$$
$$8p_1^2 p_3 - 3p_1 p_2^2 - 50p_1 p_4 + 5p_2 p_3 + 250p_5$$
$$= 8q_1^2 q_3 - 3q_1 q_2^2 - 50q_1 q_4 + 5q_2 q_3 + 250q_5.$$

1624. **Prove that** if the n roots of an algebraical equation be $a, \beta, \gamma, \delta, \ldots,$

$$6\Sigma (a - \beta)(a - \gamma)(a - \delta) = (n - 3) \Sigma (2a - \beta - \gamma)(2\beta - \gamma - a)(2\gamma - a - \beta).$$

1625. Prove that, if H_r denote the sum of the **homogeneous** products of r dimensions of the powers of the **roots of the** equation $x^n + p_1 x^{n-1} + p_2 x^{n-2} + \ldots + p_n = 0,$

$$H_{n+r} + p_1 H_{n+r-1} + p_2 H_{n+r-2} + \ldots + p_n H_r = 0.$$

1626. Two homogeneous functions of x, y of n dimensions **are** denoted by u_n, v_n : prove that the equation found by eliminating y between the two equations $u_n = a, v_n = b$, will be a rational equation **of** the nth degree in x^n.

1627. Prove that

(1) $\begin{vmatrix} -2a, & a+b, & a+c \\ b+a, & -2b, & b+c \\ c+a, & c+b, & -2c \end{vmatrix} \equiv 4(b+c)(c+a)(a+b);$

(2) $\begin{vmatrix} (b+c)^2, & c^2, & b^2 \\ c^2, & (c+a)^2, & a^2 \\ b^2, & a^2, & (a+b)^2 \end{vmatrix} \equiv 2(bc+ca+ab)^3;$

(3) $\begin{vmatrix} a-b-c, & 2a, & 2a \\ 2b, & b-c-a, & 2b \\ 2c, & 2c, & c-a-b \end{vmatrix} \equiv (a+b+c)^3;$

(4) $\begin{vmatrix} \dfrac{-bc}{b+c}, & b, & c \\ a, & \dfrac{-ca}{c+a}, & c \\ a, & b, & \dfrac{-ab}{a+b} \end{vmatrix} \equiv \dfrac{(bc+ca+ab)^3}{(b+c)(c+a)(a+b)};$

(5) $\begin{vmatrix} (b+c)^2, & b^2, & c^2 \\ a^2, & (c+a)^2, & c^2 \\ a^2, & b^2, & (a+b)^2 \end{vmatrix} \equiv 2abc\,(a+b+c)^3;$

(6) $\begin{vmatrix} a-nb-nc, & (n+1)\,a, & (n+1)\,a \\ (n+1)\,b, & b-nc-na, & (n+1)\,b \\ (n+1)\,c, & (n+1)\,c, & c-na-nb \end{vmatrix} \equiv n^2\,(a+b+c)^3;$

(7) $\begin{vmatrix} -a\,(b^2+c^2-a^2), & 2b^3, & 2c^3 \\ 2a^3, & -b\,(c^2+a^2-b^2), & 2c^3 \\ 2a^3, & 2b^3, & -c\,(a^2+b^2-c^2) \end{vmatrix} \equiv abc\,(a^2+b^2+c^2)^3.$

1628. Prove that

(1) $$\begin{vmatrix} 1, & \cos(\beta+\gamma), & \sin^2\frac{\beta-\theta}{2}\sin^2\frac{\gamma-\theta}{2} \\ 1, & \cos(\gamma+\alpha), & \sin^2\frac{\gamma-\theta}{2}\sin^2\frac{\alpha-\theta}{2} \\ 1, & \cos(\alpha+\beta), & \sin^2\frac{\alpha-\theta}{2}\sin^2\frac{\beta-\theta}{2} \end{vmatrix}$$

$$\equiv 2\sin\frac{\beta-\gamma}{2}\sin\frac{\gamma-\alpha}{2}\sin\frac{\alpha-\beta}{2}\{\sin 2\theta+2\sin(\alpha+\beta+\gamma-\theta) - \sin(\beta+\gamma)-\sin(\gamma+\alpha)-\sin(\alpha+\beta)\};$$

(2) $$\begin{vmatrix} 1, & \cos(\alpha+\theta), & \sin^2\frac{\beta-\theta}{2}\sin^2\frac{\gamma-\theta}{2} \\ 1, & \cos(\beta+\theta), & \sin^2\frac{\gamma-\theta}{2}\sin^2\frac{\alpha-\theta}{2} \\ 1, & \cos(\gamma+\theta), & \sin^2\frac{\alpha-\theta}{2}\sin^2\frac{\beta-\theta}{2} \end{vmatrix}$$

$$\equiv 2\sin\frac{\beta-\gamma}{2}\sin\frac{\gamma-\alpha}{2}\sin\frac{\alpha-\beta}{2}\{2\sin 2\theta+\sin(\alpha+\beta+\gamma-\theta) - \sin(\alpha+\theta)-\sin(\beta+\theta)-\sin(\gamma+\theta)\}.$$

1629. Prove that the determinant

$$\begin{vmatrix} 1, & \cos\theta, & \cos 2\theta, & \ldots\ldots & \cos(n-1)\,\theta \\ \cos\theta, & \cos 2\theta, & \cos 3\theta, & \ldots\ldots & \cos n\theta \\ \cos 2\theta, & \cos 3\theta, & \cos 4\theta, & \ldots\ldots & \cos(n+1)\,\theta \\ \cdots & \cdots & \cdots & \cdots & \cdots \\ \cdots & \cdots & \cdots & \cdots & \cdots \\ \cos(n-1)\,\theta, & \cos n\theta, & \ldots\ldots & & \cos(2n-2)\,\theta \end{vmatrix}$$

and all its first 2nd, ... minors, to the $n-3^{\text{th}}$, $\equiv 0$, if n be any integer > 3.

1630. Prove that the value of the determinant

$$\begin{vmatrix} 1, & 1, & 1, & \ldots\ldots & 1 \\ 1, & 2, & 3, & \ldots\ldots & n \\ 1, & 3, & 6, & \ldots\ldots & \frac{n(n+1)}{2} \\ 1, & 4, & 10, & \ldots\ldots & \frac{n(n+1)(n+2)}{\lfloor 3} \\ \ldots & \ldots & \ldots & \ldots & \ldots \\ \ldots & \ldots & \ldots & \ldots & \ldots \\ 1, & n, & \frac{n(n+1)}{2}, & \ldots\ldots & \frac{n(n+1)\ldots(2n-1)}{\lfloor n-1} \end{vmatrix}$$

is 1, and that of its first, second, &c. principal minors are

$$n, \quad \frac{n(n+1)}{2}, \ldots\ldots$$

1631. Prove that the determinant

$$\begin{vmatrix} (s-x_1)^2, & x_2^2, & x_3^2, & x_4^2 \\ x_1^2, & (s-x_2)^2, & x_3^2, & x_4^2 \\ x_1^2, & x_2^2, & (s-x_3)^2, & x_4^2 \\ x_1^2, & x_2^2, & x_3^2, & (s-x_4)^2 \end{vmatrix} \equiv 2s^5 x_1 x_2 x_3 x_4 \left(\frac{1}{x_1} + \frac{1}{x_2} + \frac{1}{x_3} + \frac{1}{x_4} - \frac{4}{s}\right),$$

where $s \equiv x_1 + x_2 + x_3 + x_4$.

1632. The determinant of the $(n+1)^{\text{th}}$ order

$$\begin{vmatrix} a_1, & a_2, & a_3, & \ldots\ldots & a_n, & z \\ a_2, & a_3, & \ldots\ldots & a_n, & z, & a_1 \\ a_3, & a_4, & \ldots\ldots & z, & a_1, & a_2 \\ \ldots & \ldots & \ldots & \ldots & \ldots & \ldots \\ \ldots & \ldots & \ldots & \ldots & \ldots & \ldots \\ z, & a_1, & a_2, & \ldots\ldots & a_{n-1}, & a_n \end{vmatrix}$$

is equal to $(-1)^{\frac{(n-1)n}{2}} a_n^{n+1} (1-x_1^{n+1})(1-x_2^{n+1}), \ldots\ldots (1-x_n^{n+1})$; where $x_1, x_2, \ldots x_n$ are the roots of the equation

$$a_n x^n + a_{n-1} x^{n-1} + \ldots\ldots + a_1 x + z = 0.$$

1633. Prove that the determinant

$$\begin{vmatrix} 1, & \cos\alpha, & \cos(\alpha+\beta), & \cos(\alpha+\beta+\gamma) \\ \cos(\beta+\gamma+\delta), & 1, & \cos\beta, & \cos(\beta+\gamma) \\ \cos(\gamma+\delta), & \cos(\gamma+\delta+\alpha), & 1, & \cos\gamma \\ \cos\delta, & \cos(\delta+\alpha), & \cos(\delta+\alpha+\beta), & 1 \end{vmatrix}$$

and all its first minors, will vanish when $\alpha+\beta+\gamma+\delta = 2\pi$.

1634. Prove that, if u_n denote the determinant of the nth order,

$$\begin{vmatrix} a, & 1, & 0, & 0, & \dots\dots\dots & 0 \\ 1, & a, & 1, & 0, & \dots\dots\dots & 0 \\ 0, & 1, & a, & 1, & 0, \dots\dots & 0 \\ 0, & 0, & 1, & a, & 1, \dots\dots & 0 \\ \dots & \dots & \dots & \dots & \dots & \dots \\ \dots & \dots & \dots & \dots & \dots & \dots \\ 0, & 0, & 0, & \dots\dots & 1, \; a, & 1 \\ 0, & 0, & 0, & \dots\dots & 0, \; 1, & a \end{vmatrix}$$

$u_{n+1} - au_n + u_{n-1} = 0$; and thence (or otherwise) obtain its value in one of the three equivalent forms

(1) $a^n - (n-1)\,a^{n-2} + \dfrac{(n-2)(n-3)}{\underline{|2}}\,a^{n-4} - \dots\dots,$

(2) $\sin(n+1)\,\alpha \div \sin\alpha$, where $2\cos\alpha = a$,

(3) $(p^{n+1} - q^{n+1}) \div (p-q)$, where p, q are the roots of $x^2 - ax + 1 = 0$.

1635. Prove that

(1) $$\begin{vmatrix} 1-n, & 1, & 1, & \dots\dots & 1 \\ 1, & 1-n, & 1, & \dots\dots & 1 \\ 1, & 1, & 1-n, & \dots\dots & 1 \\ \dots & \dots & \dots & \dots & \dots \\ \dots & \dots & \dots & \dots & \dots \\ 1, & 1, & 1, & \dots\dots & 1-n \end{vmatrix} \equiv 0,$$ n being the order of the determinant;

(2) $$\begin{vmatrix} x, & 1, & 1, & 1, & \dots\dots & 1 \\ 1, & x, & 1, & 1, & \dots\dots & 1 \\ 1, & 1, & x, & 1, & \dots\dots & 1 \\ \dots & \dots & \dots & \dots & \dots & \dots \\ \dots & \dots & \dots & \dots & \dots & \dots \\ 1, & 1, & \dots & \dots & \dots\dots & x \end{vmatrix} \equiv (x-1)^{n-1}(x+n-1);$$

(3) $$\begin{vmatrix} x_1, & 1, & 1, & \dots\dots & 1 \\ 1, & x_2, & 1, & \dots\dots & 1 \\ 1, & 1, & x_3, & \dots\dots & 1 \\ \dots & \dots & \dots & \dots & \dots \\ \dots & \dots & \dots & \dots & \dots \\ 1, & 1, & 1, & \dots\dots & x_n \end{vmatrix} \equiv p_n - p_{n-2} + 2p_{n-3} - 3p_{n-4} + \dots + (-1)^{n-1}(n-1),$$

where $x_1, x_2, \dots x_n$ are roots of the equation

$$x^n - p_1x^{n-1} + p_2x^{n-2} - \dots + (-1)^n p_n = 0;$$

(4) $$\begin{vmatrix} x, & x^2, & x^3, & \dots\dots\dots & x^n \\ x^2, & x^3, & x^4, & \dots\dots \; x^n, & x \\ x^3, & x^4, & \dots\dots \; x^n, & x, & x^2 \\ \dots & \dots & \dots & \dots & \dots \\ x^n, & x, & x^2, & \dots\dots & x^{n-1} \end{vmatrix} \equiv (-1)^{\frac{(n-1)n}{2}}\,x^n(x^n-1)^{n-1}.$$

1636. **Prove that**

$$\begin{vmatrix} \cos\theta, & \cos 2\theta, & \cos 3\theta, & \dots\dots\dots\dots & \cos n\theta \\ \cos 2\theta, & \cos 3\theta, & \cos 4\theta, & \dots\ \cos n\theta, & \cos\theta \\ \cos 3\theta, & \cos 4\theta, & \dots\ \cos n\theta, & \cos\theta, & \cos 2\theta \\ \dots & \dots & \dots & \dots & \dots \\ \cos n\theta, & \cos\theta, & \cos 2\theta, & \dots & \cos(n-1)\theta \end{vmatrix} \equiv \frac{\{\cos\theta - \cos(n+1)\theta\}^n - (1-\cos n\theta)^n}{2(-1)^{\frac{(n-1)n}{2}}(1-\cos n\theta)}.$$

1637. Prove that

$$4\begin{vmatrix} \alpha^6, & \alpha^3, & \alpha^2, & \alpha, & 1 \\ \beta^6, & \beta^3, & \beta^2, & \beta, & 1 \\ \gamma^6, & \gamma^3, & \gamma^2, & \gamma, & 1 \\ \delta^6, & \delta^3, & \delta^2, & \delta, & 1 \\ \epsilon^6, & \epsilon^3, & \epsilon^2, & \epsilon, & 1 \end{vmatrix} - 6\begin{vmatrix} \alpha^5, & \alpha^4, & \alpha^2, & \alpha, & 1 \\ \beta^5, & \beta^4, & \beta^2, & \beta, & 1 \\ \gamma^5, & \gamma^4, & \gamma^2, & \gamma, & 1 \\ \delta^5, & \delta^4, & \delta^2, & \delta, & 1 \\ \epsilon^5, & \epsilon^4, & \epsilon^2, & \epsilon, & 1 \end{vmatrix}$$

$$\equiv (\alpha-\beta)(\alpha-\gamma)(\alpha-\delta)(\alpha-\epsilon)(\beta-\gamma)(\beta-\delta)\dots\dots\ \Sigma(\alpha-\beta)^2.$$

1638. Prove that the determinants

$$\begin{Vmatrix} 0, & 0, & 0, & a, & b, & c \\ 0, & 0, & z, & a, & b, & 0 \\ 0, & y, & 0, & a, & 0, & c \\ x, & 0, & 0, & 0, & b, & c \\ x, & y, & z, & 0, & 0, & 0 \end{Vmatrix} \equiv 0.$$

1639. **Prove that, if u_n denote the determinant of the nth order,**

$$\begin{vmatrix} a, & 1, & 0, & 0, & 0, & \dots\dots\dots\dots & & & 0 \\ a, & a, & 1, & 0, & 0, & \dots\dots\dots\dots & & & 0 \\ 1, & a, & a, & 1, & 0, & \dots\dots\dots\dots & & & 0 \\ 0, & 1, & a, & a, & 0, & \dots\dots\dots\dots & & & 0 \\ \dots & \dots & \dots & \dots & \dots & \dots & \dots & \dots & \dots \\ \dots & \dots & \dots & \dots & \dots & \dots & \dots & \dots & \dots \\ 0, & 0, & 0, & \dots\dots & 0, & 1, & a, & a, & 1 \\ 0, & 0, & 0, & \dots\dots & 0, & 0, & 1, & a, & a \end{vmatrix}$$

$u_{n+1} - au_n + au_{n-1} - u_{n-2} = 0$; and express the developed determinant in the forms

(1) $\dfrac{v_{n+1} - v_n - 1}{a-3}$, where $v_n \equiv (a-1)^n - (n-1)(a-1)^{n-2}$

$$+ \frac{(n-2)(n-3)}{\lfloor 2}(a-1)^{n-4} - \dots\dots,$$

(2) $\{p^{n+2} - p^{n+1} - p - q^{n+2} + q^{n+1} + q\} \div (p-q)(p+q-2)$,

where p, q are the roots of the equation $x^2 - (a-1)x + 1 = 0$;

(3) $$\frac{-1}{2(1+\cos\theta)}\left\{1 + (-1)^n \frac{\sin(n+1)\theta + \sin(n+2)\theta}{\sin\theta}\right\},$$

where $2\cos\theta = 1 - a$.

1640. Prove that, if u_n denote the determinant of the nth order,

$$\begin{vmatrix}
a_1, & 1, & 0, & 0, & 0, & 0, & \dots\dots & 0 \\
a_2, & a_1, & 1, & 0, & 0, & 0 & \dots\dots & 0 \\
a_3, & a_2, & a_1, & 1, & 0, & 0, & \dots\dots & 0 \\
\dots & \dots & \dots & \dots & \dots & \dots & \dots & \dots \\
\dots & \dots & \dots & \dots & \dots & \dots & \dots & \dots \\
a_r, & a_{r-1}, & a_{r-2}, & \dots & a_2, & a_1, & 1, \; 0, \; 0 \dots\dots & 0 \\
0, & a_r, & a_{r-1}, & \dots\dots & a_2, & a_1, & 1, \; 0 \dots\dots & 0 \\
\dots & \dots & \dots & \dots & \dots & \dots & \dots & \dots \\
\dots & \dots & \dots & \dots & \dots & \dots & \dots & \dots \\
0, & 0, & 0, & \dots\dots a_r, & a_{r-1}, & \dots\dots a_2, & a_1, & 1 \\
0, & 0, & 0, & \dots\dots 0, & a_r, & \dots\dots a_3, & a_2, & a_1
\end{vmatrix}$$

$$u_{n+r} - a_1 u_{n+r-1} + a_2 u_{n+r-2} - \dots\dots + (-1)^r a_r u_n = 0;$$

and that, if $x_1, x_2, x_3 \dots\dots, x_r$ be the roots of the equation

$$f(x) \equiv x^r - a_1 x^{r-1} + a_2 x^{r-2} - \dots\dots + (-1)^r a_r = 0,$$

$$u_n = \frac{x_1^{n+r-1}}{f'(x_1)} + \frac{x_2^{n+r-1}}{f'(x_2)} + \dots\dots + \frac{x_r^{n+r-1}}{f'(x_r)}.$$

Also prove that, when $a_r = a_{r-1} = \dots = a_2 = a_1 = 1$, $u_n = 0$ except when $n \equiv 0$ or 1 (mod. r), and is then equal to $(-1)^{\frac{n}{r}}$ or $(-1)^{\frac{n-1}{r}}$.

1641. Prove that

$$\begin{vmatrix}
1-x_1, & x_1(1-x_2), & x_1x_2(1-x_3), & \dots\dots & x_1x_2\dots x_{n-1}(1-x_n), & x_1x_2\dots x_n \\
-1, & 1-x_2, & x_2(1-x_3), & \dots\dots & x_2\dots x_{n-1}(1-x_n), & x_2x_3\dots x_n \\
0, & -1, & 1-x_3, & \dots\dots & x_3\dots x_{n-1}(1-x_n), & x_3\dots x_n \\
\dots & \dots & \dots & \dots & \dots & \dots \\
0, & 0, & 0, & \dots\dots & -1, & 1
\end{vmatrix}$$

is equal to 1; the second row being formed by differentiating the first with respect to x_1, the third by differentiating the second with respect to x_2, and so on.

1642. Having given

$$\sin x \sin (a+x) \sin (2a+x) \ldots \sin \{(n-1) a+x\} \equiv 2^{1-n} \sin nx,$$

where n is a whole number and $na = \pi$: prove that

(1) $\cot x + \cot (a+x) + \cot (2a+x) + \ldots + \cot \{(n-1) a + x\} \equiv n \cot nx$,

(2) $\cot^2 x + \cot^2 (a+x) + \cot^2 (2a+x) + \ldots + \cot^2 \{(n-1) a + x\} \equiv n (n-1) + n^2 \cot^2 nx$.

1643. Prove that the limit of $(\cos x)^{\cot^2 x}$, as x tends to zero, is $\epsilon^{-\frac{1}{2}}$.

1644. Prove that the equation $(1-x^2)\dfrac{dy}{dx} - xy + 1 = 0$ is satisfied either by

$$y\sqrt{1-x^2} = \cos^{-1} x, \text{ or by } y\sqrt{x^2-1} = \log (x + \sqrt{x^2-1}).$$

1645. Prove that the equation

$$(1+x^2)\frac{d^2y}{dx^2} + x\frac{dy}{dx} = \tfrac{1}{4} y$$

is satisfied by any one of the four functions

$$C(\sqrt{1+x^2} \pm 1)^{\frac{1}{2}}, \quad C'(\sqrt{1+x^2} \pm x)^{\frac{1}{2}},$$

and therefore by the sum of the four functions each with an arbitrary multiplier; and account for the apparent anomaly.

1646. Prove that, if $y = \cot^{-1} x$,

$$\frac{d^ny}{dx^n} = (-1)^n \,\underline{|n-1}\, \sin ny \sin^n y;$$

and, if $y = \tan^{-1}\left(\dfrac{x \sin a}{1 + x \cos a}\right)$,

$$\frac{d^ny}{dx^n} = (-1)^{n-1} \frac{\underline{|n-1}}{\sin^n a} \sin n (a-y) \sin^n (a-y).$$

1647. Prove that the function $4x \log x - x^2 - 2x + 3$ is positive for all values of x lying between 0 and 2.

1648. Prove that, if n be a positive integer, the expression

$$(x-n)\,\epsilon^x + \frac{x^{n-1}}{\lfloor n-1} + \frac{2x^{n-2}}{\lfloor n-2} + \frac{3x^{n-3}}{\lfloor n-3} + \dots + (n-1)\,x + n$$

will be positive for all positive values of x; and will be positive or negative for negative values of x according as n is odd or even.

1649. Having given $x^2 \dfrac{d^2y}{dx^2} + x\dfrac{dy}{dx} + y = 0$, prove that

$$x^2 \frac{d^{n+2}y}{dx^{n+2}} + (2n+1)\,x \frac{d^{n+1}y}{dx^{n+1}} + (n^2+1)\frac{d^ny}{dx^n} = 0.$$

1650. Having given $y \equiv (x + \sqrt{1+x^2})^m + (x + \sqrt{1+x^2})^{-m}$, prove that

$$(1+x^2)\frac{d^{n+2}y}{dx^{n+2}} + (2n+1)\,x\frac{d^{n+1}y}{dx^{n+1}} + (n^2 - m^2)\frac{d^ny}{dx^n} = 0.$$

1651. Assuming the expansion of $\sin(m \tan^{-1} x)$ to be

$$a_1 x + a_2 \frac{x^2}{\lfloor 2} + \dots + a_n \frac{x^n}{\lfloor n} + \dots,$$

prove that

$$a_{n+2} + (2n^2 + m^2)\,a_n + (n-1)^2\,(\overline{n-1}\,|^2 - 1)\,a_{n-2} = 0.$$

1652. Assuming the expansion of $\{\log(1+x)\}^3$ to be

$$a_3 \frac{x^3}{\lfloor 3} + a_4 \frac{x^4}{\lfloor 4} + \dots + a_n \frac{x^n}{\lfloor n} + \dots,$$

prove that

$$a_{n+2} + (n+1)\,a_{n+1} = 6\,(-1)^{n-1}\,\lfloor n \left\{\frac{1}{1} + \frac{1}{2} + \frac{1}{3} + \dots + \frac{1}{n}\right\};$$

and thence that

$$a_{n+2} = 6\,(-1)^{n+1}\,\lfloor n+1\,\{\tfrac{1}{2} + \tfrac{1}{3}(1+\tfrac{1}{2}) + \tfrac{1}{4}(1+\tfrac{1}{2}+\tfrac{1}{3}) + \tfrac{1}{5}(1+\tfrac{1}{2}+\tfrac{1}{3}+\tfrac{1}{4}) + \dots$$
$$+ \frac{1}{n+1}\left(1 + \tfrac{1}{2} + \tfrac{1}{3} + \dots + \frac{1}{n}\right)\Big\}.$$

1653. Prove that, if $y = x^n (\log x)^r$,

when $r=1$, $x\dfrac{d^{n+1}y}{dx^{n+1}} = \lfloor n$; when $r=2$, $x^2\dfrac{d^{n+2}y}{dx^{n+2}} + x\dfrac{d^{n+1}y}{dx^{n+1}} = 2\lfloor n$;

when $r=3$, $x^3\dfrac{d^{n+3}y}{dx^{n+3}} + 3x^2\dfrac{d^{n+2}y}{dx^{n+2}} + x\dfrac{d^{n+1}y}{dx^{n+1}} = \lfloor 3\,\lfloor n$;

and generally that

$$\frac{\Delta^r 0^r}{\underline{|r}} x^r \frac{d^{n+r}y}{dx^{n+r}} + \frac{\Delta^{r-1}0^r}{\underline{|r-1}} \frac{d^{n+r-1}y}{dx^{n+r-1}} + \ldots + \frac{\Delta^2 0^r}{\underline{|2}} x^2 \frac{d^{n+2}y}{dx^{n+2}} + x\frac{d^{n+1}y}{dx^{n+1}} = \underline{|r}\, \underline{|n}.$$

[It is a singular property, but easy to prove, that the sum of the coefficients of the sinister is equal to the limit of the product of the two infinite series

$$\left\{\frac{1}{\underline{|2}} - \frac{1}{\underline{|3}} + \frac{1}{\underline{|4}} - \ldots\right\}\left\{1^{r-1} + 2^{r-1} + \frac{3^{r-1}}{\underline{|2}} + \frac{4^{r-1}}{\underline{|3}} + \ldots\right\}.]$$

1654. Prove that, if $y = x^{r-1}\log(1+x)$, **where** r is **a** positive **integer,**

$$\frac{d^ry}{dx^r} = \underline{|r}\left(\frac{1}{1+x} + \frac{1}{(1+x)^2} + \ldots + \frac{1}{(1+x)^r}\right);$$

and, if $n > r$,

$$\frac{d^ny}{dx^n} = (-1)^{n-r}\frac{\underline{|r-1}\,\underline{|n-r}}{(1+x)^n}\left\{x^{r-1} + nx^{r-2} + \frac{n(n-1)}{\underline{|2}}x^{r-3} + \ldots \text{ to } r \text{ terms}\right\}$$

$$= \frac{(-1)^{n-1}\underline{|n}}{(1+x)^n}\left\{\frac{x^{r-1}}{n} - \frac{r-1}{n-1}x^{r-2}(1+x) + \frac{(r-1)(r-2)}{\underline{|2}\,(n-2)}x^{r-3}(1+x)^2 - \ldots + (-1)^r\frac{(1+x)^{r-1}}{n-r+1}\right\}.$$

1655. Prove that, if $y = (1+x)^r \log x$,

$$\frac{d^{n+r}y}{dx^{n+r}} = (-1)^{n-1}\frac{\underline{|n-1}\,\underline{|r}}{x^{n+r}}\left(x^r - nx^{r-1} + \frac{n(n+1)}{\underline{|2}}x^{r-2} - \ldots \text{ to } r+1 \text{ terms}\right);$$

and deduce the identity

$$(1+x)^r - n(1+x)^{r-1} + \frac{n(n-1)}{\underline{|2}}(1+x)^{r-2} - \ldots \text{ to } r+1 \text{ terms}$$

$$\equiv x^r - (n-r)x^{r-1} + \frac{(n-r)(n-r+1)}{\underline{|2}}x^{r-2} - \ldots \text{ to } r+1 \text{ terms}, \ (n > r).$$

1656. Prove that, in the expansion of $(1+x)^n\log(1+x)$, the coefficient of $\dfrac{x^{n+r}}{\underline{|n+r}}$ is $(-1)^{r-1}\underline{|n}\,\underline{|r-1}$; and that of $\dfrac{x^{n-r}}{\underline{|n-r}}$ is

$$\frac{\underline{|n}}{\underline{|r}}\left(\frac{1}{n} + \frac{1}{n-1} + \ldots + \frac{1}{r+1}\right),$$

n, r, **and** $n-r$ being all positive integers.

1657. Prove that the expansion of $\dfrac{\log(1+x)}{(1+x)^n}$ is

$$nx\,.\,\frac{1}{n} - n(n+1)\left(\frac{1}{n} + \frac{1}{n+1}\right)\frac{x^2}{\underline{|2}}$$

$$+ n(n+1)(n+2)\left(\frac{1}{n} + \frac{1}{n+1} + \frac{1}{n+2}\right)\frac{x^3}{\underline{|3}} - \ldots$$

and deduce the **identity**

$$\frac{1}{r} + \frac{n+1}{r-1} + \frac{(n+1)(n+2)}{\lfloor 2} \frac{1}{r-2} + \dots \text{ to } r \text{ terms}$$

$$\equiv \frac{\lfloor n+r}{\lfloor n \lfloor r} \left\{ \frac{1}{n+r} + \frac{1}{n+r-1} + \dots + \frac{1}{n+1} \right\}.$$

[When n is not a whole number, the last identity should be corrected by writing $(n+1)(n+2)\dots(n+r)$ for $\lfloor n+r \div \lfloor n$.]

1658. Prove that in the expansion of $(1+x)^{n-\frac{1}{2}} \log(1+x)$, when n is a positive integer, the coefficient of x^{2n} is 0.

1659. Prove that, in the equation $f(x+h) = f(x) + hf'(x+\theta h)$, **the** limiting value of θ, when h tends to zero, is $\frac{1}{2}$; and that, if θ **be** constant, $f(x) \equiv A + Bx + Cx^2$, where A, B, C are independent of x. Also prove that, if θ be independent of x, $f(x) \equiv A + Bx + Cm^x$, where A, B, C, m are independent of x, and find the value of θ when $f(x)$ has this form.

[The value of θ is $\frac{1}{h\log m} \log\left\{\frac{\epsilon^{h\log m} - 1}{h\log m}\right\}$.]

1660. In the equation $f(x+h) = f(x) + hf'(x+\theta h)$, prove that **the** first three terms of the expansion of θ in ascending powers of h are

$$\frac{1}{2} + \frac{h}{24}\frac{f'''(x)}{f''(x)} + \frac{h^2}{48}\frac{f''(x)f^{iv}(x) - \{f'''(x)\}^2}{\{f''(x)\}^2},$$

and calculate them when $f(x) \equiv \sin x$.

[$\frac{1}{2} + \frac{h\cot x}{24} - \frac{h^2}{48\sin^2 x}$, provided **cot x be** finite.]

1661. In the equation

$$f(a+h) = f(a) + hf'(a) + \frac{h^2}{\lfloor 2} f''(a) + \dots + \frac{h^n}{\lfloor n} f^n(a+\theta h),$$

the limiting value of θ, when h tends to zero, is $\frac{1}{n+1}$; and, if $f(x)$ **be** a rational algebraical expression of $n+1$ dimensions in x, the value of θ is always $\frac{1}{n+1}$. Also prove that, **if $f(x) = \epsilon^{mx}$**, θ is independent of x; and that **the** general form of $f(x)$ in order that θ may be independent of x, is

$$A_0 + A_1 x + A_2 x^2 + \dots + A_n x^n + B\epsilon^{mx}.$$

1662. Prove that, in the equation

$$f(x) = f(0) + xf'(0) + \frac{x^2}{\lfloor 2} f''(0) + \dots + \frac{x^n}{\lfloor n} f^n(\theta x),$$

if $f(x) \equiv (1-x)^m$ where m **is** any positive quantity, the limiting value of θ as x tends **to 1** will be $1 - \left(\frac{n}{m}\right)^{\frac{1}{m-n}}$.

1663. Prove that, in the equation

$$\frac{F(x+h)-F(x)}{f(x+h)-f(x)}=\frac{F'(x+\theta h)}{f'(x+\theta h)},$$

the limiting value of θ when h tends to zero is $\frac{1}{2}$; and that when $F(x) \equiv \sin x$ and $f(x) = \cos x$, the value of θ is always $\frac{1}{2}$.

1664. Prove that the expansion of $(\text{vers}^{-1} x)^2$ is

$$2\left(x+\frac{1}{3}\frac{x^2}{2}+\frac{1.2}{3.5}\frac{x^3}{3}+\frac{1.2.3}{3.5.7}\frac{x^4}{4}+\dots\right).$$

1665. Prove that

(1) $\frac{\pi}{4}\frac{e^\pi-1}{e^\pi+1}=\frac{1}{1+1^2}+\frac{1}{1+3^2}+\frac{1}{1+5^2}+\dots$ to ∞,

(2) $1+\frac{\pi^4}{\lfloor 4}+\frac{\pi^8}{\lfloor 8}+\dots$ to $\infty=\left(1+\frac{4}{1^2}\right)\left(1+\frac{4}{3^2}\right)\left(1+\frac{4}{5^2}\right)\dots$ to ∞.

1666. Prove that the limit of the fraction

$$\frac{2+\frac{2}{3}+\frac{2.4}{3.5}+\frac{2.4.6}{3.5.7}+\dots \text{ to } n \text{ terms}}{1+\frac{1}{2}+\frac{1.3}{2.4}+\frac{1.3.5}{2.4.6}+\dots \text{ to } n \text{ terms}},$$

when n is infinite, is $\frac{\pi}{2}$; and the limit of the ratio of the nth term of the numerator to the nth term of the denominator is π.

[This may be deduced from the equation

$$\tfrac{1}{2}(\sin^{-1}x)^2=\frac{x^2}{2}+\frac{2}{3}\frac{x^4}{4}+\frac{2.4}{3.5}\frac{x^6}{6}+\dots]$$

1667. Prove that

$$\sqrt{1}+\sqrt{2}+\sqrt{3}+\dots+\sqrt{n}>\tfrac{2}{3}n^{\frac{3}{2}}<\tfrac{2}{3}\{(n+1)^{\frac{3}{2}}-1\},$$

$$1+\tfrac{1}{2}+\tfrac{1}{3}+\dots+\frac{1}{n}>\log(1+n)<1+\log n;$$

$$1^{p-1}+2^{p-1}+3^{p-1}+\dots+n^{p-1}>\frac{1}{p}n^p<\frac{1}{p}\{(n+1)^p-1\},$$

$$\frac{1}{1^{p+1}}+\frac{1}{2^{p+1}}+\frac{1}{3^{p+1}}+\dots+\frac{1}{n^{p+1}}>\frac{1}{p}\left\{1-\frac{1}{(n+1)^p}\right\}<\frac{1}{p}\left\{p+1-\frac{1}{n^p}\right\},$$

$$\cos a+\cos 2a+\dots+\cos na>\frac{\sin(n+1)a}{a}-1<\frac{\sin na}{a};$$

p being a positive quantity, and $2na$ in the last $<\pi$.

1668. Prove that if a, h, b be eliminated by differentiation from the equation $ax^2+by^2+2hxy=1$, the resultant equation will coincide with that obtained by eliminating θ from the equations

$$\frac{d^2x}{d\theta^2}+n^2x=0,\quad \frac{d^2y}{d\theta^2}+n^2y=0.$$

[From the former equation may be deduced

$$\frac{\dfrac{d^2y}{dx^2}}{\left(x\dfrac{dy}{dx}-y\right)^3}=ab-h^2,$$

which may be interpreted to mean that the curvature varies as the cube of the perpendicular from the origin on the tangent.]

1669. Prove that the general term of the expansion of $\sin xy$ in terms of x, when x, y are connected by the equation $y=x\cos xy$, is

$$\frac{(-1)^n}{2}\frac{x^{4n+2}}{\underline{|2n+1}}\left\{(n+1)^{2n}+(2n+2)n^{2n}+\frac{(2n+2)(2n+1)}{\underline{|2}}(n-1)^{2n}+\dots\right.$$
$$\left.\text{to } n+1 \text{ terms}\right\}.$$

1670. Prove that, if $(a+x+\sqrt{a^2+2bx+x^2})^{n+1}$ be expanded in ascending powers of x, the coefficients of x^{n-1}, x^n, and x^{n+1} are respectively

$$(n+1)(a+b)(\overline{n+2}a+\overline{n-2}b)2^{n-2},\quad (n+1)(a+b)2^n,\ \text{and}\ 2^{n+1}-\left(1-\frac{b}{a}\right)^{n+1}.$$

1671. Find the limiting value, when x tends to zero, of

$$\frac{d^n}{dx^n}\left(\frac{x}{f(x)}\right)^{n+1},$$

when $f(x)$ has the values

(1) $\sin x$, (2) $\tan x$, (3) $\log(1+x)$, (4) $1+x-\sqrt{1+x^2}$, (5) ϵ^x-1,
(6) $\dfrac{\epsilon^x-\epsilon^{-x}}{2}$, (7) $\dfrac{\epsilon^x-\epsilon^{-x}}{\epsilon^x+\epsilon^{-x}}$.

n being of course a positive integer.

[When n is odd, the values are 0, 0, 1, $\frac{1}{2}\underline{|n+1}$, $\underline{|n}$, 0, 0; and when n is even, (1) $(1.3.5\dots\overline{n-1})^2$, (2) $(-1)^{\frac{n}{2}}\underline{|n}$, (3) 1, (4) $\frac{1}{2}\underline{|n+1}$, (5) $\underline{|n}$, (6) $(-1)^{\frac{n}{2}}(1.3.5\dots\overline{n-1})^2$, and (7) $\underline{|n}$.]

1672. Prove that the limiting value, when x tends to zero, of

$$\frac{d^n}{dx^n}\left\{\left(\frac{x}{\sin x}\right)^{n+1}\cos(n+1)x\right\}\ \text{is}\ 2^n\underline{|n}(-1)^{\frac{n}{2}};\ n\ \text{being even}.$$

1673. Prove that the limiting value of u^v, when x tends to a value a for which both u and v vanish, will always be 1, if the corresponding limiting value of $\frac{u}{v}$ be finite; or if the limiting value of $u\frac{dv}{dx} \div v\frac{du}{dx}$ be finite.

1674. Prove that the limiting value of u^v, when x tends to a critical value a for which $u = 1$ and $v = \infty$, is $\epsilon^{mf'(a)}$, where m is the limit of $(x - a)v$. Apply this to find the limits of $\left(\frac{\epsilon^x - \epsilon^{-x}}{2x}\right)^{\frac{1}{x^2}}$, and $\left\{\frac{\epsilon^x - \epsilon^{-x} - 2x}{\frac{x^3}{3}}\right\}^{\frac{1}{x^2}}$, when x tends to zero.

1675. Prove that the limiting values of

(1) $\frac{\sin(n+1)x + a_n \sin nx + a_{n-1} \sin(n-1)x + \ldots + a_1 \sin x}{x^{2n+1}}$,

(2) $\frac{\sin(2n+1)x + a_n \sin(2n-1)x + a_{n-1} \sin(2n-3)x + \ldots + a_1 \sin x}{x^{2n+1}}$,

when x tends to zero, and the coefficients $a_1, a_2, \ldots a_n$ have such values that both limits are finite, are $(-1)^n$, $(-4)^n$ respectively.

1676. Prove that the limiting values of

(1) $\frac{\cos(n+1)x + a_n \cos nx + a_{n-1} \cos(n-1)x + \ldots + a_1 \cos x}{x^{2n}}$,

(2) $\frac{\cos nx + a_n \cos(n-1)x + a_{n-1} \cos(n-2)x + \ldots + a_2 \cos x + a_1}{x^{2n}}$,

when x tends to zero, and the constants are so determined in each case that the limit is finite, are $(-1)^n \frac{2n+1}{n+1}$, $\frac{(-1)^n}{2}$ respectively.

1677. Having given

$$z = \epsilon^x f_1(x+y) + \epsilon^{2x} f_2(x+y) + \ldots + \epsilon^{nx} f_n(x+y);$$

prove that

$$\left(\frac{d}{dx} - \frac{d}{dy} - 1\right)\left(\frac{d}{dx} - \frac{d}{dy} - 2\right) \ldots \left(\frac{d}{dx} - \frac{d}{dy} - n\right) z = 0.$$

1678. Having given

$$z = \epsilon^x f_1\left(\frac{y}{x}\right) + \epsilon^{2x} f_2\left(\frac{y}{x}\right) + \ldots + \epsilon^{nx} f_n\left(\frac{y}{x}\right),$$

prove that

$$(p + q - 1)(p + q - 2) \ldots (p + q - n) z = 0,$$

where $p^a q^b z$ denotes $\left(\frac{y}{x}\right)^b \frac{d^{a+b}z}{dx^a dy^b}$.

1679. Having given

$$u = f(x^2 + 2yz, \quad y^2 + 2zx),$$

prove that

$$(y^2 - zx)\frac{du}{dx} + (x^2 - yz)\frac{du}{dy} + (z^2 - xy)\frac{du}{dz} = 0:$$

and having given

$$u = f\{u(x-t), \quad u(y-t), \quad u(z-t)\},$$

prove that

$$\frac{du}{dx} + \frac{du}{dy} + \frac{du}{dz} + \frac{du}{dt} = 0.$$

1680. Prove that

$$\left(\frac{d}{dx}\right)^n \left(x\frac{d}{dx} - n\right)^r y \equiv \left(x\frac{d}{dx}\right)^r \frac{d^n y}{dx^n}$$

$$\equiv x^r \frac{d^{n+r}y}{dx^{n+r}} + \frac{\Delta^{r-1}0^r}{\lfloor r-1} x^{r-1} \frac{d^{n+r-1}y}{dx^{n+r-1}} + \dots$$

$$+ \Delta^2 0^r x^2 \frac{d^{n+2}y}{dx^{n+2}} + x\frac{d^{n+1}y}{dx^{n+1}}.$$

[More generally,

$$\left(\frac{d}{dx}\right)^n \left(\overline{a+x}\frac{d}{dx} - n\right)^r y \equiv \left(\overline{a+x}\frac{d}{dx}\right)^r \frac{d^n y}{dx^n}.]$$

1681. The co-ordinates of a point referred to axes inclined at an angle ω are (x, y), and u is a function of the position of the point: prove that

$$\frac{1}{\sin^2\omega}\left(\frac{d^2u}{dx^2} + \frac{d^2u}{dy^2} - 2\cos\omega\frac{d^2u}{dx\,dy}\right), \quad \frac{1}{\sin^2\omega}\left(\frac{d^2u}{dx^2}\frac{d^2u}{dy^2} - \left|\overline{\frac{d^2u}{dx\,dy}}\right|^2\right)$$

are independent of the particular axes.

[Their values in polar co-ordinates are

$$\frac{d^2u}{dr^2} + \frac{1}{r^2}\frac{d^2u}{d\theta^2} + \frac{1}{r}\frac{du}{dr}, \quad \frac{1}{r^2}\frac{d^2u}{dr^2}\frac{d^2u}{d\theta^2} + \frac{1}{r}\frac{du}{dr}\frac{d^2u}{dr^2} - \frac{1}{r^2}\left(\frac{d^2u}{dr\,d\theta} - \frac{1}{r}\frac{du}{d\theta}\right)^2.]$$

1682. Having given $2x \equiv r(\epsilon^\theta + \epsilon^{-\theta})$, $2y \equiv r(\epsilon^\theta - \epsilon^{-\theta})$, prove that

$$\frac{d^2u}{dx^2} - \frac{d^2u}{dy^2} \equiv \frac{d^2u}{dr^2} - \frac{1}{r^2}\frac{d^2u}{d\theta^2} + \frac{1}{r}\frac{du}{dr},$$

$$\frac{d^2u}{dx^2}\frac{d^2u}{dy^2} - \left(\frac{d^2u}{dx\,dy}\right)^2 \equiv \frac{1}{r^2}\frac{d^2u}{dr^2}\frac{d^2u}{d\theta^2} - \frac{1}{r}\frac{du}{dr}\frac{d^2u}{dr^2} - \frac{1}{r^2}\left(\frac{d^2u}{dr\,d\theta} - \frac{1}{r}\frac{du}{d\theta}\right)^2.$$

1683. Having given $x + y \equiv X$, $y \equiv XY$, prove that

$$x\frac{d^2u}{dx^2} + y\frac{d^2u}{dx\,dy} - \frac{du}{dx} \equiv X\frac{d^2u}{dX^2} - Y\frac{d^2u}{dX\,dY} - \frac{du}{dX}.$$

1684. Having given $x + y \equiv \epsilon^{\theta+\phi}$, $x - y \equiv \epsilon^{\theta-\phi}$, prove that

$$\frac{d^2u}{dx^2} - \frac{d^2u}{dy^2} \equiv \epsilon^{-2\theta}\left(\frac{d^2u}{d\theta^2} - \frac{d^2u}{d\phi^2}\right).$$

1685. Having given $\epsilon^x \equiv r^{\cos\theta}$, $\epsilon^y \equiv r^{\sin\theta}$, prove that

$$x^2\frac{d^2u}{dy^2} - 2xy\frac{d^2u}{dx\,dy} + y^2\frac{d^2u}{dx^2} \equiv \frac{d^2u}{dr^2} + r\frac{du}{dr}\log r.$$

1686. Having given

$$2x^{1-n} \equiv v^{1-n} + w^{1-n},\quad 2y^{1-n} \equiv w^{1-n} + u^{1-n},\quad 2z^{1-n} \equiv u^{1-n} + v^{1-n},$$

prove that

$$u^n\frac{d\phi}{du} + v^n\frac{d\phi}{dv} + w^n\frac{d\phi}{dw} \equiv x^n\frac{d\phi}{dx} + y^n\frac{d\phi}{dy} + z^n\frac{d\phi}{dz};$$

$$u^{2n}\frac{d^2\phi}{du^2} + \dots + \dots + 2v^nw^n\frac{d^2\phi}{dv\,dw} + \dots + \dots + nu^{1-2n}\frac{d\phi}{du} + \dots + \dots$$

$$\equiv x^{2n}\frac{d^2\phi}{dx^2} + \dots + \dots + nx^{1-2n}\frac{d\phi}{dx} + \dots + \dots$$

1687. Having given $ux^2 = vy^2 = wz^2 = uvw$, prove that

$$\left(u\frac{d}{du} + v\frac{d}{dv} + w\frac{d}{dw}\right)^n\phi = \left(x\frac{d}{dx} + y\frac{d}{dy} + z\frac{d}{dz}\right)^n\phi;$$

or generally, when $x_1X_1^{n-1} = x_2X_2^{n-1} = \dots = x_nX_n^{n-1} = x_1x_2\dots x_n$,

$$\left(x_1\frac{d}{dx_1} + x_2\frac{d}{dx_2} + \dots + x_n\frac{d}{dx_n}\right)^r\phi = \left(X_1\frac{d}{dX_1} + \dots + X_n\frac{d}{dX_n}\right)^r\phi.$$

1688. Prove that, if $x_1 + x_2 + \dots + x_n = X_1 + X_2 + \dots + X_n$,

$$\frac{d\phi}{dx_1} + \frac{d\phi}{dx_2} + \dots + \frac{d\phi}{dx_n} = \frac{d\phi}{dX_1} + \frac{d\phi}{dX_2} + \dots + \frac{d\phi}{dX_n};$$

and, if

$$a_{11}x_1^2 + a_{22}x_2^2 + \dots + 2a_{rs}x_rx_s + \dots$$
$$= A_{11}X_1^2 + A_{22}X_2^2 + \dots + 2A_{rs}X_rX_s + \dots,$$

that

$$a_{11}\frac{d^2\phi}{dx_1^2} + a_{22}\frac{d^2\phi}{dx_2^2} + \dots + 2a_{rs}\frac{d^2\phi}{dx_r\,dx_s} + \dots$$
$$= A_{11}\frac{d^2\phi}{dX_1^2} + A_{22}\frac{d^2\phi}{dX_2^2} + \dots + 2A_{rs}\frac{d^2\phi}{dX_r\,dX_s} + \dots.$$

1689. Prove that, if u be a function of four independent variables x_1, x_2, x_3, x_4, and

$$x_1 = r\sin\theta\sin\phi,\quad x_2 = r\sin\theta\cos\phi,\quad x_3 = r\cos\theta\sin\psi,\quad x_4 = r\cos\theta\cos\psi,$$

$$\frac{d^2u}{dx_1^2} + \frac{d^2u}{dx_2^2} + \frac{d^2u}{dx_3^2} + \frac{d^2u}{dx_4^2} = \frac{d^2u}{dr^2} + \frac{1}{r^2}\frac{d^2u}{d\theta^2} + \frac{1}{r^2\sin^2\theta}\frac{d^2u}{d\phi^2}$$
$$+ \frac{1}{r^2\cos^2\theta}\frac{d^2u}{d\psi^2} + \frac{3}{r}\frac{du}{dr} + \frac{2}{r^2}\cot 2\theta\frac{du}{d\theta}.$$

1690. Prove that, if x, y, z be three variables connected by one equation only, and p, q, r, s, t denote $\frac{dz}{dx}$, $\frac{dz}{dy}$, $\frac{d^2z}{dx^2}$, $\frac{d^2z}{dxdy}$, $\frac{d^2z}{dy^2}$ as usual,

$$\frac{dx}{dz}=\frac{1}{p},\quad \frac{dx}{dy}=-\frac{q}{p},\quad \frac{d^2x}{dz^2}=-\frac{r}{p^3},$$

$$\frac{d^2x}{dydz}=\frac{qr-ps}{p^3},\quad \frac{d^2x}{dy^2}=-\frac{p^2t-2pqs+q^2r}{p^3}.$$

1691. The distances of any point from two fixed points are r_1, r_2, and a maximum or minimum value of $f(r_1, r_2)$ for points lying on a given curve is c: prove that the curve $f(r_1, r_2)=c$ will touch the given curve.

1692. In the straight line bisecting the angle A of a triangle ABC is taken a point P: prove that the difference of the angles APB, APC will be a maximum when AP is a mean proportional between AB, AC.

[A parabola can be drawn with its focus at A touching PB, PC at B, C.]

1693. Normals are drawn to an ellipse at the ends of two conjugate diameters: prove that a maximum distance of their common point from the centre is $(a^2+b^2)^{\frac{3}{2}} \div 3\sqrt{3}ab$, provided that $a^2 > 5b^2$; and that a minimum distance is always $(a^2-b^2)\div a\sqrt{2}$.

1694. Prove that $\phi\{f(x)\}$ is always a maximum or minimum when $f(x)$ is so; but that, if a be the maximum or minimum value of $f(x)$, $\phi(a)$ is not a maximum or minimum value of $\phi(x)$.

1695. The least area which can be included between two parabolas, whose axes are parallel and at a given distance a, and which cut each other at right angles in two points, is $a^2\frac{\sqrt{3}}{2}$.

[The included area may be proved to be $a^2 \div 3 \sin\omega\cos^2\omega$, where ω is the inclination of the common chord to the axes.]

1696. From a point O on the evolute of an ellipse are drawn the two normals OP, OQ (not touching the evolute at O): prove that if $a^2 < 2b^2$, PQ will have its minimum value when the eccentric angle of the point for which O is the centre of curvature is $\tan^{-1}\left(\frac{2a^2-b^2}{2b^2-a^2}\right)^{\frac{1}{4}}$.

1697. Prove that, if $m-1$, $n-1$, and $m-n$ be positive, the expression $(\cos x + m\sin x)^{\frac{m+1}{m-1}} \div (\cos x + n \sin x)^{\frac{n+1}{n-1}}$ will be a maximum when $x=\frac{\pi}{4}$, and a minimum when $x=\frac{\pi}{4}+\cot^{-1}(m)+\cot^{-1}(n)$. Also, if $m-n$ and $mn-1$ be positive, $(\cos x + m\sin x)^n \div (\cos x + n\sin x)^m$ will be a maximum when $x=0$, and a minimum when $x=\tan^{-1}m+\tan^{-1}n-\frac{\pi}{2}$.

1698. Prove that, if n be an odd integer or a fraction whose numerator and denominator are odd integers, the only maximum and minimum values of $\sin^n x \cos nx$ are determined by the equation $\cos(n+1)x = 0$. Also, with the same form of $n\ (>1)$, the maximum and minimum values of $\tan nx (\cot x)^n$ correspond to the values $0, \pi, 2\pi, \ldots$ of $(n-1)x$ and $\pi, 3\pi, 5\pi, \ldots$ of $2(n+1)x$, the zero value giving a maximum, and any value of x which occurs in both series being rejected.

1699. Through each point within a parabola $y^2 = 4ax$ it is obvious that at least one minimum chord can be drawn: prove that the part from which two minimum chords and one maximum can be drawn is divided from the part through which only one minimum can be drawn by the curve

$$(x-5a)^{-\frac{1}{3}} + (4x-6y+4a)^{-\frac{1}{3}} + (4x+6y+4a)^{-\frac{1}{3}} = 0;$$

and that, at any point on the parabola $y^2 = 4a(x-a)$, one minimum chord is that passing through the focus, $(a, 0)$.

[The curve has two rectilinear asymptotes $3x \pm 3\sqrt{3}y + 5a = 0$, and a parabolic asymptote $27y^2 = 32a(3x-a)$, which crosses the curve when $3x = 17a$, and is thence almost coincident with the inner branches.]

1700. The maximum value of the common chord of an ellipse and its circle of curvature at any point is

$$\frac{4}{3\sqrt{3}(a^2-b^2)}\{(a^2+b^2)(2a^2-b^2)(a^2-2b^2) + 2(a^4-a^2b^2+b^4)^{\frac{3}{2}}\}^{\frac{1}{2}}.$$

1701. A chord PQ of an ellipse is normal at P and O is its pole: prove that, when PQ is a minimum, its length will be

$$3\sqrt{3}\, a^2b^2 \div (a^2+b^2)^{\frac{3}{2}}$$

and Q will be the centre of curvature at P; and when OP is a minimum the other common tangent to the ellipse and the circle of curvature at P will pass through O.

[The minimum value of PQ here given will only exist when $a^2 > 2b^2$, the axes being both maximum values of PQ; when $a^2 < 2b^2$, one axis is the maximum and the other the minimum value of PQ, and there are no other maximum or minimum values.]

1702. In any closed oval curve, PQ, a chord which is normal at P, will have its maximum or minimum values either when Q is the centre of curvature at P, or when PQ is normal at Q as well as at P, which must always be the case for two positions at least of PQ.

1703. Prove that the expression $\dfrac{1+2x-x^2+2\sqrt{x-x^3}}{1+x^2}$ has $1 \pm \sqrt{2}$ for its maximum and minimum values corresponding to $x = -1 \pm \sqrt{2}$.

1704. Two fixed points A, B are taken on a given circle, and another given circle has its centre at B and radius greater than BA;

any point P being taken on the second circle, PA meets the first circle again in Q: prove that the maximum lengths P_1Q_1, P_2Q_2 of PQ are equally inclined to AB and each subtends a right angle at B, and the minimum lengths both lie on the straight line through A at right angles to AB: also P_1, A, B, P_2 lie on a circle which is orthogonal to the first circle.

1705. The least acute angle which the tangent at any point of an elliptic section of a cone of revolution makes with the generating line through the point is $\cos^{-1}(\cos\beta\sec\alpha)$, where 2α is the angle of the cone and β the angle which the plane of the section makes with the axis.

1706. Prove that three parabolas of maximum latus rectum can be drawn circumscribing a given triangle; and, if α, β, γ be the angles which the axis of any one of them makes with the sides, that

$$\cot\alpha + \cot\beta + \cot\gamma = 0.$$

1707. Prove that, if $x + y + z = 3c$, $f(x)f(y)f(z)$ will be a maximum or minimum when $x = y = z = c$, according as

$$f''(c) > \text{ or } < \{f'(c)\}^2 \div f'''(c).$$

1708. The minimum value of $(lx + my + nz)^2 \div (yz + zx + xy)$ is $2mn + 2nl + 2lm - l^2 - m^2 - n^2$, provided this value be positive; otherwise there is neither maximum nor minimum value.

1709. Prove that the maximum value of

$$\sin x\sqrt{a\sin^2 y + b\cos^2 y} + \cos x\sqrt{a\cos^2 y + b\sin^2 y}$$

is $\sqrt{a+b}$; and the minimum value of

$$\frac{\sqrt{a^2\sin^2 x + b^2\cos^2 x} + \sqrt{a^2\sin^2 y + b^2\cos^2 y}}{\sin(x-y)}$$

is $a + b$.

1710. Prove that, when x, y, z vary, subject to the single condition $xyz(yz + zx + xy) = x + y + z$, the minimum value of

$$\frac{(1+yz)(1+zx)(1+xy)}{(1+x^2)(1+y^2)(1+z^2)}$$

is $-\frac{1}{8}$.

1711. Find the plane sections of greatest and least area which can be drawn through a given point on a given paraboloid of revolution; proving that, if θ_1, θ_2 be the angles which the planes of maximum and minimum section make with the axis,

$$2\tan\theta_1\tan\theta_2 = 3.$$

1712. The maximum and minimum values of $f(x, y, z)$, where x, y, z are the distances of a point from three fixed points (all in one plane), are to be determined from the equations

$$\frac{1}{\sin(y, z)}\frac{df}{dx} = \frac{1}{\sin(z, x)}\frac{df}{dy} = \frac{1}{\sin(x, y)}\frac{df}{dz};$$

(y, z) denoting the angle between the distances y, z.

1713. Prove that, if A, B, C, D be corners of a tetrahedron and P a point the sum of whose distances from A, B, C, D is a minimum,

$$\frac{PA \cdot Pa}{Aa} = \frac{PB \cdot Pb}{Bb} = \frac{PC \cdot Pc}{Cc} = \frac{PD \cdot Pd}{Dd};$$

a, b, c, d being the points in which PA, PB, PC, PD respectively meet the opposite faces. Also prove that when $lPA^2 + mPB^2 + nPC^2 + rPD^2$ is a minimum,

$$\frac{\text{vol}.\,PBCD}{l} = \frac{\text{vol}.\,PCDA}{m} = \frac{\text{vol}.\,PDAB}{n} = \frac{\text{vol}.\,PABC}{r}.$$

1714. The distances of any variable point from the corners of a given tetrahedron are denoted by u, x, y, z: prove that, when $f(u, x, y, z)$ is a maximum or minimum,

$$\frac{1}{1 - a^2 - b^2 - c^2 + 2abc}\left(\frac{df}{du}\right)^2 = \frac{1}{1 - a^2 - b'^2 - c'^2 + 2ab'c'}\left(\frac{df}{dx}\right)^2$$

$$= \frac{1}{1 - a'^2 - b^2 - c'^2 + 2a'bc'}\left(\frac{df}{dy}\right)^2 = \frac{1}{1 - a'^2 - b'^2 - c^2 + 2a'b'c}\left(\frac{df}{dz}\right)^2;$$

a, b, c, a', b', c' denoting the cosines of the angles between the distances (y, z), (z, x), (x, y), (u, x), (u, y), (u, z) respectively.

1715. Prove that, if O be the point the sum of the squares of whose distances from n given straight lines, or planes, is a minimum, O will be the centre of mean position of the feet of the perpendiculars from O on the given straight lines or planes.

1716. A convex polygon of a given number of sides circumscribes a given oval, without singular points: prove that, when the perimeter of the polygon is a minimum, the point of contact of any side is the point of contact of the circle which touches that side and the two adjacent sides produced.

1717. In the curve $y^3 = 3ax^2 - x^3$, the tangent at P meets the curve again in Q: prove that

$$\tan QOx + 2\tan POx = 0,$$

O being the origin. Also prove that if the tangent at P be a normal at Q, P lies on the curve

$$4y(3a - x) = (2a - x)(16a - 5x).$$

1718. Prove that any tangent to the hypocycloid $x^{\frac{2}{3}} + y^{\frac{2}{3}} = a^{\frac{2}{3}}$, which makes an angle $\frac{1}{2}\tan^{-1}\frac{1}{2}$ with the axis of x, is also a normal to the curve.

1719. The tangent to the evolute of a parabola at a point where it meets the parabola is also a normal to the evolute.

1720. From a point on the evolute of an ellipse $a^2y^2 + b^2x^2 = a^2b^2$ the two other normals to the ellipse are drawn: prove that the straight line joining the feet of these normals will be a normal to the ellipse

$$(a^2x^2 + b^2y^2)(a^2 - b^2)^2 = a^4b^4.$$

1721. A tangent to a given ellipse at P meets the axes in two points, through which are drawn straight lines at right angles to the axes meeting in p: prove that the normal at p to the locus of p and the straight line joining the centre of the ellipse to the centre of curvature at P are equally inclined to the axes.

1722. Trace the curve $\left(\frac{x}{a}\right)^n + \left(\frac{y}{b}\right)^n = 1$ when n is an indefinitely large integer, (1) when n is even, (2) when n is odd.

[(1) the curve is undistinguishable from the sides of the rectangle formed by $x^2 = a^2$, $y^2 = b^2$; (2) when $x^2 < a^2$, $y = b$; when $y^2 < b^2$, $x = a$; and when $x^2 > a^2$ and $y^2 > b^2$, $\frac{x}{a} + \frac{y}{b} = 0$, or the curve coincides with two sides of the rectangle and with the part of one diagonal which is without the rectangle.]

1723. Trace the curve determined by the equations

$$x = a\cos\theta,\quad y = a\frac{\theta}{\sin\theta},$$

and prove that the whole curve can only be obtained by using impossible (pure imaginary) values of θ.

[The two curves (1) $x = a\cos\theta$, $y = a\dfrac{\theta}{\sin\theta}$, (2) $x = a\cos \mathrm{h}\theta$, $y = \dfrac{a\theta}{\sin \mathrm{h}\theta}$, give the same differential equation of the first order

$$(1 - x^2)\frac{dy}{dx} - xy + 1 = 0,$$

and starting from the same point (a, a) when $\theta = 0$ must coincide.]

1724. Trace the curve $4(x^2 + 2y^2 - 2ay)^2 = x^2(x^2 + 2y^2)$, proving that the area of a loop is $\frac{4\pi}{\sqrt{3}}(2 - \sqrt{3})a^2$, and that the area included between the loops is $\frac{8a^2}{3\sqrt{3}}(2\pi - 3\sqrt{3})$.

1725. A curve is given by the equations

$$x = \frac{a\cos\theta\{a^2 + (a^2 - b^2)\cos^2\theta\}}{a^2\cos^2\theta + b^2\sin^2\theta},\quad y = \frac{b\sin\theta\{b^2 + (b^2 - a^2)\sin^2\theta\}}{a^2\cos^2\theta + b^2\sin^2\theta};$$

prove that its arc is given by the equation

$$\frac{ds}{d\theta} = \frac{(a^2\sin^2\theta + b^2\cos^2\theta)^{\frac{3}{2}}}{a^2\cos^2\theta + b^2\sin^2\theta}.$$

1726. Prove that the curve whose intrinsic equation is $\frac{ds}{d\phi} = a\sec 2\phi$, if x, y, $\frac{dy}{dx}$, and ϕ vanish together, has the two rectilinear asymptotes

$$y \pm x = -\frac{a}{\sqrt{2}}\log(\sqrt{2} + 1).$$

1727. Two contiguous points P, P' on a curve being taken, PO, $P'O$ are drawn at right angles to the radius vector of each point: prove that the limiting value of PO when P' moves up to P is $\pm\frac{dr}{d\theta}$.

1728. Two fixed points S, S' being taken, a point P moves so that the rectangle SP, $S'P$ is constant: prove that straight lines drawn from S, S' at right angles respectively to SP, $S'P$ will meet the tangent at P in points equidistant from P.

1729. In a lemniscate of Bernoulli, the tangent at any point makes acute angles θ, θ' with the focal distances r, r': prove that

$$\sin\theta = \frac{r \sim 3r'}{2\sqrt{2r'}}, \quad \sin\theta' = \frac{r' \sim 3r}{2\sqrt{2r}}, \quad \sqrt{2}\sin\frac{\theta \pm \theta'}{2} = \cos\frac{\theta \mp \theta'}{2}.$$

1730. In a family of lemniscates the foci S, S' are given ($SS' = 2a$): prove that in any one in which the rectangle under the focal distances (c^2) is less than a^2, the curvature is a minimum at the points of contact of tangents drawn from the centre, and that these points all lie on a lemniscate (of Bernoulli) of which S, S' are vertices. Also, when $c^2 > a^2$, the points of inflexion lie on another such lemniscate equal to the former but with its axis at right angles to that of the former.

[In any of these curves, if p denote the perpendicular from the centre on the tangent, $2c^2pr = r^4 + c^4 - a^4$, and the radius of curvature is

$$2c^2r^3 \div (3r^4 + a^4 - c^4).$$

The points of maximum curvature are the vertices.]

1731. In the curve

$$r\left(m + n\tan\frac{\theta}{2}\right) = 1 + \tan\frac{\theta}{2},$$

the locus of the extremity of the polar subtangent is a cardioid.

1732. The tangent at any point P of a certain curve meets the tangent at a fixed point O in T, and the arc OP is always equal to $n \cdot TP$: prove that the intrinsic equation of the curve is

$$s = c\,(\sin\phi)^{\frac{1}{n-1}};$$

and that the curve is a catenary when $n = \frac{1}{2}$, the evolute of a parabola when $n = \frac{2}{3}$, a four-cusped hypocycloid when $n = \frac{3}{2}$, and a cycloid when $n = 2$.

1733. From a fixed point are let fall perpendiculars on the tangent and normal at any point of a curve, and the straight line joining the feet of the perpendiculars passes through another fixed point: prove that the curve is one of a system of confocal conics.

1734. A circle is drawn to touch a cardioid and pass through the cusp: prove that the locus of its centre is a circle. If two such circles be drawn, and through their second common point any straight line be drawn, the tangents to the circles at the points where this straight line again meets them will intersect on the cardioid.

[Of course this and many properties of the cardioid are most easily proved by inversion from the parabola.]

1735. Two circles touch the curve $r^m = a^m \cos m\theta$ in the points P, Q, and touch each other in the pole S: prove that the angle PSQ is equal to $\frac{n\pi}{1-m}$, n being a positive or negative integer.

1736. The locus of the centre of a circle touching the curve $r^m = a^m \cos m\theta$ and passing through the pole is the curve $(2r)^n = a^n \cos n\theta$, where $n(1-m) = m$.

1737. In the curve $r = a \sec^n \theta$, prove that, at a point of inflexion the radius vector makes equal angles with the prime radius and the tangent; and that the distance of the point of inflexion from the pole increases from a to $a\sqrt{\epsilon}$, as n increases from 0 to ∞. If n be negative, there is no real point of inflexion.

1738. A perpendicular, SY, is drawn from the pole S to the tangent to a curve at P: prove that, when there is a cusp at P, the circle of curvature at Y to the locus of Y will pass through S; also that, when there is a point of inflexion at Y in the locus of Y, the chord of curvature at P through S will be equal to $4SP$.

1739. The equation of the pedal of a curve is $r = f(\theta)$: prove that the equation found by eliminating a from the equations

$$r\cos a = f(\theta - a), \quad r\sin a = f'(\theta - a),$$

is that of the curve.

1740. Prove that for any cubic there exists one point such that the points of contact of tangents drawn from it to the curve lie on a circle. If the equation of the cubic be

$$ax^3 + 3gx^2y + 3fxy^2 + by^3 + Ax^2 + 2Hxy + By^2 + \ldots = 0,$$

and if $\frac{a-f}{g} = \frac{g-b}{f} = \frac{A-B}{H}$, there will be a straight line such that, if tangents be drawn to the cubic from any point of it, the points of contact will lie on a circle.

1741. The asymptotes to a cuspidal cubic are given: prove that the tangent at the cusp envelopes a curve which is the orthogonal projection of a three-cusped hypocycloid, the circle inscribed in the hypocycloid being projected into the locus of the cusp.

[The locus of the cusp is the maximum ellipse inscribed in the triangle formed by the asymptotes, and any tangent to the tricusp at a point O meets this ellipse in two points P, Q so that OQ is bisected in P; the point corresponding to O is P.]

1742. The equation of a curve of the nth order being

$$x^n\phi_1\left(\frac{y}{x}\right)+x^{n-1}\phi_2\left(\frac{y}{x}\right)+x^{n-2}\phi_3\left(\frac{y}{x}\right)+\ldots=0,$$

$\phi_1(z)$ has two roots μ and $\phi_2(\mu)=0$: prove that there will be two corresponding rectilinear asymptotes, whose equations are

$$(y-\mu x)^2\phi_1''(\mu)+2(y-\mu x)\phi_2'(\mu)+2\phi_3(\mu)=0.$$

1743. Two points P, Q describe two curves so that corresponding arcs are equal, and the radius vector of Q is always parallel to the tangent at P: show how to find P's path when Q's is given; and in especial prove that when Q describes a straight line P describes a catenary, and when Q describes a cardioid, with the cusp as pole, P describes a two-cusped epicycloid.

1744. The rectangular co-ordinates of a point on a given curve being (x, y), the radius of curvature at the point is ρ, and the angle which the tangent makes with a fixed straight line is ϕ: prove that

$$\left(\frac{d^2x}{d\phi^2}\right)^2+\left(\frac{d^2y}{d\phi^2}\right)^2=\rho^2+\left(\frac{d\rho}{d\phi}\right)^2,$$

$$\left(\frac{d^3x}{ds^3}\right)^2+\left(\frac{d^3y}{ds^3}\right)^2=\frac{1}{\rho^4}\left\{1+\left(\frac{d\rho}{ds}\right)^2\right\},$$

$$\left(\frac{d^3x}{d\phi^3}\right)^2+\left(\frac{d^3y}{d\phi^3}\right)^2=\left(\frac{d^2\rho}{d\phi^2}-\rho\right)^2+4\left(\frac{d\rho}{d\phi}\right)^2,$$

and, in general,

$$\left(\frac{d^{n+1}x}{d\phi^{n+1}}\right)^2+\left(\frac{d^{n+1}y}{d\phi^{n+1}}\right)^2=u^2+v^2,$$

where

$$2u=\left\{\left(\frac{d}{d\phi}+i\right)^n+\left(\frac{d}{d\phi}-i\right)^n\right\}\rho,\quad 2iv=\left\{\left(\frac{d}{d\phi}+i\right)^n-\left(\frac{d}{d\phi}-i\right)^n\right\}\rho.$$

1745. A curve represented by the equation $f\left(\frac{x-X}{c},\ \frac{y-Y}{c}\right)=0$ is drawn having contact of the second order with a given curve at a point P: prove that, if O be the point (X, Y), PO will be the tangent at O to the locus of O.

1746. A rectangular hyperbola whose axes are parallel to the co-ordinate axes has three-point contact with a given curve at the point (x, y): prove that the co-ordinates (X, Y) of the centre of the hyperbola are given by the equations

$$\frac{X-x}{\frac{dy}{dx}}=Y-y=\frac{\left(\frac{dy}{dx}\right)^2-1}{\frac{d^2y}{dx^2}},$$

and that the central radius to the point (x, y) is the tangent at (X, Y) to the locus of the centre. Also, when the given curve is (1) the parabola $y^2 = 4ax$, (2) the ellipse $a^2y^2 + b^2x^2 = a^2b^2$; prove that the locus of the centre of the hyperbola is

$$(1)\ 4(x+2a)^3 = 27ay^2, \quad (2)\ (ax)^{\frac{2}{3}} + (by)^{\frac{2}{3}} = (a^2+b^2)^{\frac{2}{3}}.$$

1747. An ellipse is described having four-point contact with a given ellipse at P, and with one of its equal conjugate diameters passing through P: prove that the locus of its centre is the curve

$$(x^2+y^2)^2\left(\frac{x^2}{a^2}+\frac{y^2}{b^2}\right) = (a^2-b^2)^2\left(\frac{x^2}{a^2}-\frac{y^2}{b^2}\right)^2.$$

[The curve consists of four loops, and its whole area is to that of the ellipse as

$$(a-b)^2\{(a^2+ab+b^2)^2-5a^2b^2\} : 2a^2b^2.]$$

1748. The equation of the conic of closest contact which can be described at any point of a given curve, when referred to the tangent and normal at the point as axes, is $ax^2+by^2+2hxy=2y$, where

$$a=\frac{1}{\rho}, \quad h=-\frac{1}{3\rho}\frac{d\rho}{ds}, \quad b=\frac{1}{\rho}+\frac{2}{9\rho}\left(\frac{d\rho}{ds}\right)^2-\frac{1}{3}\frac{d^2\rho}{ds^2},$$

and ρ is the radius of curvature at the point.

1749. The sum of the squares on the semi-axes of the ellipse of five pointic contact at any point of a curve is

$$9\rho^2\left\{18+2\left(\frac{d\rho}{ds}\right)^2-3\rho\frac{d^2\rho}{ds^2}\right\} \div \left\{9+\left(\frac{d\rho}{ds}\right)^2-3\rho\frac{d^2\rho}{ds^2}\right\}^2,$$

the product of the semi-axes is $27\rho^2 \div \overline{9+\left(\frac{d\rho}{ds}\right)^2-3\rho\frac{d^2\rho}{ds^2}}^{\frac{3}{2}}$, the rectangle under the focal distances is $9\rho^2 \div \left\{9+\left(\frac{d\rho}{ds}\right)^2-3\rho\frac{d^2\rho}{ds^2}\right\}$, and the excentricity e is given by the equation

$$9\frac{(e^2-2)^2}{1-e^2} = \frac{\left\{18+2\left(\frac{d\rho}{ds}\right)^2-3\rho\frac{d^2\rho}{ds^2}\right\}^2}{9+\left(\frac{d\rho}{ds}\right)^2-3\rho\frac{d^2\rho}{ds^2}}.$$

1750. A chord QQ' is drawn to a curve parallel to the tangent at P, a neighbouring point, and the straight line bisecting the external angle between PQ, PQ' meets QQ' in O: prove that the limiting value of PQ is $6\rho \div \frac{d\rho}{ds}$.

1751. In any curve a chord PQ is drawn parallel and indefinitely near to the tangent at a point O: prove that the straight line joining the middle point of the chord to O will make with the normal at O an angle whose limiting value is $\tan^{-1}\left(\frac{1}{3}\frac{d\rho}{ds}\right)$. Reconcile this result with the fact that the segments into which the chord is divided by the normal at O are ultimately in a ratio of equality. If the chord meet the normal in R, and P', Q' be respectively the mid point and the foot of the bisector of the angle QOP, the limiting value of the third proportional to RP', RQ' will be $\frac{2}{3}\rho\frac{d\rho}{ds}$.

1752. A chord PQ of a curve is drawn always parallel to the tangent at a point O: prove that the radius of curvature at O of the locus of the middle point of this chord is

$$\frac{10\rho\left(9+\overline{\frac{d\rho}{ds}}\right|^2\Big)^{\frac{3}{2}}}{9\rho^2\frac{d^3\rho}{ds^3}+\left(36+4\overline{\frac{d\rho}{ds}}\right|^2-9\rho\frac{d^2\rho}{ds^2}\Big)\frac{d\rho}{ds}}.$$

1753. A chord PQ of a curve is drawn parallel to the tangent at O and is met in R by the bisector of the angle POQ: prove that the radius of curvature at O of the locus of R is $3\rho\frac{ds}{d\rho}$.

1754. The normal chord PQ at any point P of a conic is equal to $18\rho\div\left(9+2\overline{\frac{d\rho}{ds}}\right|^2-3\rho\frac{d^2\rho}{ds^2}\Big)$; in a parabola, if I be the centre of curvature at P and O the pole of PQ, $PQ.PI=2PO^2$, and IO is perpendicular to the focal distance SP; and, in all conics, $PO=3\rho\frac{ds}{d\rho}$, and the angles IOP, IPC are equal, C being the centre.

1755. Prove that the curves $r=a\theta$, $r=\frac{3a\sin\theta}{2+\cos\theta}$ have five-point contact at the pole.

1756. The centre of curvature at a point P of a parabola is O, OQ is drawn at right angles to OP meeting the focal distance of P in Q: prove that the radius of curvature of the evolute at O is equal to $3QO$.

1757. All the curves represented by the equation

$$\frac{x^{n+1}}{a}+\frac{y^{n+1}}{b}=\left(\frac{ab}{a+b}\right)^n,$$

for different values of n, touch each other at the point $\left(x=y=\frac{ab}{a+b}\right)$, and the radius of curvature is $(a^2+b^2)^{\frac{3}{2}}\div n(a+b)^2$.

1758. At each point P of a curve is drawn the equiangular spiral of closest contact (four-point), and s, σ are corresponding arcs of the curve and of the locus of the pole S of the spiral: prove that

$$\frac{d\sigma}{ds} = \frac{\rho \frac{d^2\rho}{ds^2}}{1 + \left(\frac{d\rho}{ds}\right)^2}, \quad PS = \frac{\rho}{\sqrt{1 + \left(\frac{d\rho}{ds}\right)^2}};$$

and that the tangents at P, S to the two curves are equally inclined to SP. Prove that, when the curve is a cycloid, the locus of S is an equal cycloid, the image of the former with respect to the base; and when the curve is a catenary of equal strength, $y = a \log \sec \frac{x}{a}$, PS is constant $(= a)$, corresponding arcs are equal, and the curvatures at corresponding points are as $1 : 3$. In the curve whose intrinsic equation is

$$\frac{ds}{d\phi} = a (\cos n\phi)^{\frac{1}{n}}, \quad \frac{d\sigma}{ds} = n,$$

and the curvatures at S, P are as $2n - 1 : n$; when $2n = 1$, the locus of S is a straight line.

1759. At each point of a parabola is described the rectangular hyperbola of four-point contact: prove that the locus of its centre is an equal parabola, the image of the former with respect to the directrix.

1760. At each point of a given closed oval are drawn the parabola and rectangular hyperbola of four-point contact: prove that the arc traced out by the centre of the hyperbola exceeds twice that traced by the focus of the parabola by the arc of the oval, provided no parabola have five-point contact.

1761. At each point of a given curve is drawn the curve in which the chord of curvature through the pole bears to the radius vector the constant ratio $2 : n + 1$, having four-point contact with the given curve, and corresponding arcs of the given curve and of the locus of the pole of the osculating curve are s, σ: prove that

$$\frac{d\sigma}{ds} = n - (n - 1)\frac{(n+1)^2 \rho \frac{d^2\rho}{ds^2}}{(n-1)^2 + (n+1)^2\left(\frac{d\rho}{ds}\right)^2}.$$

Also, if S be the pole corresponding to a point P on the given curve, the tangents at P, S are equally inclined to PS,

$$PS = (n + 1)\rho \div \sqrt{1 + \left(\frac{n+1}{n-1}\frac{d\rho}{ds}\right)^2},$$

and the locus of S for different values of n is the locus of the foci of the conics which have four-point contact with the given curve at P.

1762. In the last question, prove that, when the given curve is such that corresponding arcs traced out by S and P are equal, the intrinsic equation of the given curve is either

$$\frac{ds}{d\phi} = c(\cos\phi)^{\frac{1-n}{1+n}} \text{ or } \frac{ds}{d\phi} = c\sec\frac{n-1}{n+1}\phi;$$

that in the former case PS is constant in direction and the locus of S is the image of the given curve with respect to a straight line, and in the latter that PS is constant in length, $\left(\overline{n+1}\,c \text{ or } \frac{n+1}{n}c\right)$, and the curvatures of the two curves at P, S will be as $1+n : 3-n$ or $n+1 : 3n-1$.

1763. At each point of a given curve are drawn the cardioid and lemniscate of four-point contact, and the arcs traced out by the cusp and node respectively corresponding to an arc s of the given curve are σ, σ': prove that $2\sigma + \sigma' = 3s$.

1764. At each point of the curve whose equation is

$$2\epsilon^{\frac{y}{a}} = \epsilon^{\frac{x}{a\sqrt{3}}} + \epsilon^{\frac{-x}{a\sqrt{3}}},$$

are drawn the rectangular hyperbola and parabola of four-point contact: prove that the distance from the point of osculation to the centre and focus respectively are a, $\frac{a}{2}$, and corresponding arcs of the three curves are equal.

1765. At each point of a cardioid is drawn the lemniscate of closest contact, the locus of its node will be an epicycloid, whose fixed circle is that with which the cardioid is generated as an epicycloid and whose moving circle is twice the radius.

1766. At each point of the curve $r^n = a^n \sin n\theta$ is drawn the curve similar to $r^{\frac{1}{n}} = a^{\frac{1}{n}} \sin\frac{\theta}{n}$ and having four-point contact with the former curve: prove that the locus of its pole is the curve whose intrinsic equation is

$$\frac{ds}{d\phi} = \frac{a(n+1)}{n(3n+1)}\left(\sin\frac{n\phi}{3n+1}\right)^{\frac{1-n}{n}},$$

that the radius of curvature of this curve bears to the common radius of curvature of the osculating curves at the corresponding point the ratio $(n+1)^2 : n(3n+1)$. Also the area traced out by the radius vector to the pole is

$$\frac{n+1}{8n}a^2\left(\overline{n+1}\,\theta + \sin 2\theta\right).$$

1767. At each point of an epicycloid is drawn the equiangular spiral of closest contact: prove that the locus of the pole of this spiral will be the inverse of the epicycloid with respect to its centre; and, conversely, the curves for which this property is true are those whose intrinsic equations are

$$s = a(1 - \cos m\phi),\quad 2s = a(\epsilon^{m\phi} + \epsilon^{-m\phi} - 2),$$

measuring from a cusp in each case.

1768. A curve is such that any two corresponding points of its evolute and an involute are at a constant distance: prove that the straight line joining the two points is also constant in direction.

1769. The reciprocal polar of the evolute of a parabola with respect to the focus is a cissoid, which will be equal to the pedal with respect to the vertex when the radius of the auxiliary circle is one-fourth of the latus rectum.

1770. In any epicycloid or hypocycloid the radius of curvature is proportional to the perpendicular on the tangent from the centre of the fixed circle.

1771. The co-ordinates of a point of a curve, referred to the tangent and normal at a neighbouring point as axes of co-ordinates, are

$$x = s - \frac{s^3}{6\rho^2} + \frac{s^4}{8\rho^3}\frac{d\rho}{ds} + \frac{s^5}{\underline{|5}\,\rho^4}\left(1 - 11\left|\frac{d\rho}{ds}\right|^2 + \rho\frac{d^2\rho}{ds^2}\right) + \ldots,$$

$$y = \frac{s^2}{2\rho} - \frac{s^3}{6\rho^2}\frac{d\rho}{ds} - \frac{s^4}{\underline{|4}\,\rho^3}\left(1 - 2\left|\frac{d\rho}{ds}\right|^2 + \rho\frac{d^2\rho}{ds^2}\right) + \ldots$$

where ρ, $\frac{d\rho}{ds}$, $\frac{d^2\rho}{ds^2}$, ... are the values of the radius of curvature and its differential coefficients at the origin, and s is the arc measured from the origin.

1772. Prove that, if the tangents at two points P, Q meet in O, the limiting value of $\frac{OP + OQ - \text{arc } PQ}{OP + OQ - \text{chord } PQ}$ is $\frac{2}{3}$, that of $\frac{1}{OP} - \frac{1}{OQ}$ is $\frac{2}{3\rho}\frac{d\rho}{ds}$, and that of $\frac{1}{OP^2} + \frac{1}{OQ^2} - \frac{2\cos\phi}{OP\,.\,OQ}$ is $\frac{4}{9\rho^2}\left(9 + \left|\frac{d\rho}{ds}\right|^2\right)$, ϕ being the angle between the tangents.

1773. Prove that, in the curve whose intrinsic equation is

$$\frac{ds}{d\phi} = a(1 + m\epsilon^{n\phi})^{-\frac{3}{2}},$$

the axes of the conic of closest contact at each point are inclined at constant angles to the tangent and normal. Also, at any point in the curve $\frac{ds}{d\phi} = a\sec 3\phi$, the rectangle under the focal distances in the conic of closest contact is constant.

1774. Tangents to an ellipse are drawn intercepting a given length on a fixed straight line: prove that the locus of their common point is a quartic having four-point contact with the ellipse at the points where the tangents are parallel to the fixed straight line; and trace the curve when the fixed straight line meets the ellipse, (1) in real points, (2) in impossible points; the given intercept being greater than the diameter parallel to the fixed straight line.

1775. The curvature at any point of the lemniscate of Bernoulli varies as the difference of the focal distances; and in the lemniscate in which the rectangle under the focal distances is $2a^2$, where $2a$ is the distance between the foci, the curvature varies as

$$(r_1^2+r_2^2)(r_1-r_2)^2 \div (r_1^2+r_2^2-r_1r_2)^{\frac{3}{2}}.$$

1776. Prove that the three equations

$$y=\frac{2a\cos\phi}{1-\cos\phi^2},\quad x+y\tan\phi=s,\quad y=\rho\cos\phi\,(1-\cos\phi),$$

all belong to the same curve, ρ being the radius of curvature at the point (x, y); ϕ the angle which the tangent makes with the axis of x and s the arc.

1777. The curve in which the radius of curvature at any point is n times the normal cut off by a fixed straight line (the base) is the locus of the pole of the curve $r=a\left(\cos\frac{\theta}{n-1}\right)^{n-1}$ rolling along that fixed straight line; and is also the envelope of the base of the curve, in which the radius of curvature is $n-1$ times the normal, when the curve rolls along the same straight line. The two rolling curves may be taken to have always the same point of contact P, in which case the pole of the former, Q, will always lie on the base of the latter at the point where it touches its envelope; the radius of curvature at Q of the roulette or glissette will be nQP, and the radii of curvature at P will be $(n-1)\,PG$ in the envelope curve and $\left(1-\frac{1}{n}\right)PG$ in the locus curve, PG being drawn at right angles to the fixed base to meet the moving base in G.

[All the curves involved are easily found for the values of n, -2, -1, 0, 1, 2.]

1778. The curve in which the radius of curvature is always three times the normal cut off by the base is an involute of a four-cusped hypocycloid which passes through two of the cusps: if $4a$ be the longest diameter of this curve, $2a$ will be the shortest, and the curve will lie altogether within an ellipse whose axes are $4a$, $2a$, the maximum distance cut off on any normal to the ellipse being $\frac{1}{12}a$; and the minimum normal chords in the two curves will be of lengths $1\cdot840535\,a$ and $1\cdot859032\,a$, inclined at angles $51°\,33'\,39''\cdot4$ and $61°\,52'\,28''\cdot2$ respectively to the major axis.

1779. A point P being taken on a given curve, P' is the corresponding point on an inverse to the given curve: prove that (1) a circle can be drawn touching the two curves at P, P', which will be its own inverse, (2) when the diameter of this circle is always equal to the radius of curvature at P the given curve is either an ellipse or an epicycloid, and (3) the circle of inversion is the director circle for the ellipse and the circle through the cusps for the epicycloid.

1780. The perpendicular from a fixed point on the tangent to a certain curve is $a\left(\sin\frac{\phi}{n+2}\right)^n\left(n+2\cos\frac{2\phi}{n+2}\right)$, where ϕ is the angle which the tangent makes with a fixed straight line: prove that the radius of curvature at the point of contact is $\frac{n(n-1)}{n+2}a\left(\sin\frac{\phi}{n+2}\right)^{n-2}$; and identify the curves when n is 1, 0, and -1 respectively.

[If the straight line from which ϕ is measured be the axis of x and the fixed point the origin, the curves are (1) the points $(x\pm a)^2+y^2=0$, (2) the points $x^2+(y\pm a)^2=0$, and (3) the parabola $y^2=-4a(x+3a)$.]

1781. Prove that the equation of the first negative pedal of the parabola $y^2=4a(a+x)$ is $27ay^2=(a+x)(x-8a)^2$, and that the equation of the evolute of this curve is

$$\left(\frac{y}{4a}\right)^{\frac{2}{3}}+\left(\frac{3a-2x}{3a}\right)^{\frac{2}{3}}=1.$$

[The intrinsic equation of this evolute is $s=\frac{6a}{(1+\cos\phi)^2}-\frac{3a}{2}$.]

1782. The radius of curvature ρ of a curve at a point whose areal co-ordinates are (x, y, z) is given by the equation

$$\frac{9}{\kappa^2\rho^2}+\frac{\left\{\frac{d^2x}{dt^2}\left(\frac{dy}{dt}-\frac{dz}{dt}\right)+\frac{d^2y}{dt^2}\left(\frac{dz}{dt}-\frac{dx}{dt}\right)+\frac{d^2z}{dt^2}\left(\frac{dx}{dt}-\frac{dy}{dt}\right)\right\}^2}{\left(a^2\frac{dy}{dt}\frac{dz}{dt}+b^2\frac{dz}{dt}\frac{dx}{dt}+c^2\frac{dx}{dt}\frac{dy}{dt}\right)^3}=0,$$

where a, b, c are the sides and κ is double the area of the triangle of reference.

1783. A circle rolls on a fixed straight line, trace the curve which is enveloped by any tangent to the circle; proving that the whole arc enveloped corresponding to a complete revolution of the circle is $2a\left(2\sqrt{3}+\frac{\pi}{3}\right)$, and the area cut off the envelope by the fixed straight line is $a^2\left(\frac{5\pi}{4}+4\right)$, a being the radius.

1784. The curve $r=\frac{4a}{(\theta-a)^2}$ can be made to roll outside a parabola of latus rectum $4a$ so that its pole always lies on the tangent at the vertex, and the curvatures of the two curves at a point of contact P will be as $SP+a : SP-a$, where S is the focus. Also the curve $r=2a\theta$ can be made to roll inside the parabola so that its pole always lies on the axis, and its curvature bears to that of the parabola the ratio $SP+a : 2a$; so that at any point of contact the radius of curvature of the parabola is equal to the sum of the radii of curvature of the two rolling curves.

1785. The curve $r = b \sin \frac{b\theta}{a}$ rolls within an ellipse of axes $2a$, $2b$, starting with its pole at the end of the major axis: prove that the pole will remain always on the major axis and the curvatures of the two curves when touching at P will be as $b^2 : b^2 + SP.S'P$. Similarly with the curve $r = a \sin \frac{a\theta}{b}$ and the minor axis.

1786. The three curves of the last two questions touch at P and O, O_1, O_2 are the centres of curvature of the ellipse and the two roulettes: prove that

$$\{OO_1O_2P\} = \frac{a^2}{b^2}.$$

1787. The curves $2r = b\left(\epsilon^{\frac{b\theta}{a}} - \epsilon^{-\frac{b\theta}{a}}\right)$, $2r = a\left(\epsilon^{\frac{a\theta}{b}} + \epsilon^{-\frac{a\theta}{b}}\right)$ can be made to roll on an hyperbola whose transverse and conjugate axes are $2a$, $2b$, so that the poles trace out these axes respectively: the curvatures at any point of contact P will be as $a^2b^2 : a^2(b^2 + SP.S'P) : b^2(SP.S'P - a^2)$, and if O, O_1, O_2 be the centres of curvature $\{OO_1O_2P\} = -\frac{a^2}{b^2}$.

1788. A cardioid and cycloid whose axes are equal roll along the same straight line so as always to touch it at the same point, their vertices being simultaneously points of contact: prove that the cusp of the cardioid will always lie in the base of the cycloid and will be the point where the base touches its envelope. The curvatures of the two curves at their point of contact will be as 3 : 1.

1789. A curve rolls along a fixed straight line: prove that the curvature of a carried point is $\frac{d}{dp}\left(\frac{p}{r}\right)$, where r is the distance from the carried point to the point of contact and p the perpendicular from it on the directrix.

1790. The curve

$$\frac{a}{r} = 1 + \sec a \sin(\theta \sin a), \text{ or } \frac{a}{r} = 1 + \frac{\epsilon^{\theta \tan a} + \epsilon^{-\theta \tan a}}{2 \sec a},$$

rolls on a straight line: prove that the locus of its polé is a circle.

1791. A loop of a lemniscate rolls in contact with the axis of x: prove that the locus of the node is given by the equation

$$1 + \left(\frac{dy}{dx}\right)^2 = \left(\frac{a}{y}\right)^4,$$

and that, if ρ, ρ' be corresponding radii of curvature of this locus and of the lemniscate, $2\rho\rho' = a^2$.

1792. The curve $r^m = a^m \cos m\theta$ rolls along a straight line: prove that the radius of curvature of the path of the pole is $r\left(\frac{m+1}{m}\right)$.

1793. A plane curve rolls along a straight line: prove that the radius of curvature of the path of any point carried by the rolling curve is $\frac{r^2}{r-\rho\sin\phi}$, where r is the distance from the carried point to the point of contact, ϕ the angle which this distance makes with the directrix, and ρ the radius of curvature at the point of contact.

1794. A curve is generated by a point of a circle which rolls along a fixed curve: prove that the diameter of the circle through the generating point will envelope a curve generated as a roulette by a circle of half the dimensions on the same directrix.

1795. A parabola rolls along a straight line: prove that the envelope of its directrix is a catenary.

1796. Two circles, of radii b, $a-b$, respectively, roll within a circle of radius a, their points of contact with the fixed circle being originally coincident, and the circles rolling in opposite directions in such a manner that the velocities of points on the circles relative to their respective centres are equal: prove that they will always intersect in the point which was originally the point of contact.

1797. In a hypocycloid, the radii of the rolling and fixed circles are as $n : 2n+1$, where n is a whole number: prove that part of the locus of the common point of two tangents at right angles to each other is a circle.

1798. Prove that a graphical solution of the equation $\tan x = x$ can be found by drawing tangents to a cycloid from a cusp; the value of x which satisfies the equation being the whole angle through which the tangent has turned, the point of contact starting at the cusp.

1799. Tangents are drawn to a given cycloid inclined at a given angle 2α (the angle through which the tangent turns in passing from one point of contact to the other): prove that the straight line bisecting the external angle between them is tangent to an equal cycloid whose vertex is at a distance $2a\alpha\tan\alpha$ from the vertex of the given cycloid; and that the straight line bisecting the internal angle is normal to another equal cycloid whose vertex is at a distance $2a(1-\alpha\cot\alpha)$ from the vertex of the given cycloid, a being the radius of the generating circle.

1800. Find the envelopes of

$$(1) \quad x\cos^3\theta + y\sin^3\theta = a,$$

$$(2) \quad \frac{x^3}{a^3\cos\theta} + \frac{y^3}{b^3\sin\theta} = 1,$$

θ being the parameter in each case.

1801. A perpendicular OY is let fall from a fixed point O on any one of a series of straight lines drawn according to some fixed law: prove that, when OY is a maximum or minimum, Y is in general a point on the envelope; and that, if Y be not on the envelope, the line to which OY is the perpendicular is an asymptote to the envelope.

1802. Find the envelope of the system of circles

$$(x-a\lambda^2)^2+(y-2a\lambda)^2=a^2(1+\lambda^2)^2,$$

λ being the parameter.

1803. The envelope of the directrix of a parabola which has four-point contact with a given rectangular hyperbola is the curve

$$\left(\frac{a}{r}\right)^{\frac{2}{3}}=\cos\frac{2\theta}{3}.$$

1804. The envelope of the directrix of a parabola having four-point contact with a given curve is the locus of the point found by measuring along the normal outwards a length equal to half the radius of curvature.

1805. Prove that the envelope of the circle

$$x^2+y^2+a^2+b^2-2ax\cos\theta-2by\sin\theta=\left(\frac{x}{a}\cos\theta+\frac{y}{b}\sin\theta\right)(a^2\sin^2\theta+b^2\cos^2\theta)$$

is the ellipse $\frac{x^2}{a^2}+\frac{y^2}{b^2}=1$, and its inverse $\left(\frac{x^2+y^2}{a^2+b^2}\right)^2=\frac{x^2}{a^2}+\frac{y^2}{b^2}$.

1806. The envelope of the straight line

$$x\cos\phi+y\sin\phi=a(\cos n\phi)^{\frac{1}{n}}$$

is the curve whose polar equation is $r^{\frac{n}{1-n}}=a^{\frac{n}{1-n}}\cos\frac{n\theta}{1-n}$.

1807. On any radius vector of the curve $r=a\sec^n\frac{\theta}{n}$ is described a circle; the envelope is the curve $r=c\sec^{n-1}\frac{\theta}{n-1}$. Prove this geometrically when $n=2$, and when $n=3$.

1808. A parabola is described touching a given circle and having its focus at a given point on the circle: prove that the envelope of its directrix is a cardioid.

1809. A straight line is drawn through each point of the curve $r^m=a^m\cos m\theta$ at right angles to the radius vector: prove that the envelope of such lines is the curve $r^{\frac{m}{m-1}}=a^{\frac{m}{m-1}}\cos\frac{m}{m-1}\theta$.

1810. From the pole S is drawn SY perpendicular upon a tangent to the curve $r^m=a^m\cos m\theta$, and with S as pole and Y as vertex is drawn a curve similar to $r^n=a^n\cos n\theta$: prove that the envelope of such curves is

$$r^{\frac{mn}{m+n}}=a^{\frac{mn}{m+n}}\cos\frac{mn}{m+n}\theta.$$

1811. The negative pedal of the parabola $y^2=4ax$ with respect to the vertex is the curve $27ay^2=(x-4a)^3$.

1812. The envelope of the straight line $px+qy+rz=0$, subject to the condition

$$\frac{p}{q-r}+\frac{q}{r-p}+\frac{r}{p-q}=0,$$

is $$\left(\frac{x}{4}+y+z\right)^{-\frac{1}{2}}+\left(\frac{y}{4}+z+x\right)^{-\frac{1}{2}}+\left(\frac{z}{4}+x+y\right)^{-\frac{1}{2}}=0\,;$$

and, when the condition is $\dfrac{a^2p}{q-r}+\dfrac{b^2q}{r-p}+\dfrac{c^2r}{p-q}=0$, a, b, c being the sides of the triangle of reference ABC, the envelope is

$$u^{-\frac{1}{2}}+v^{-\frac{1}{2}}+w^{-\frac{1}{2}}=0,$$

where

$$u\equiv\tfrac{3}{2}(x+y+z)-x\cos\tfrac{2}{3}(B-C)-y\cos\{120^\circ+\tfrac{2}{3}(C-A)\}$$
$$-z\cos\{120^\circ+\tfrac{2}{3}(B-A)\},$$

and similarly for v, w.

1813. The contact of the curve $f(x, y, a)=0$ with its envelope will be of the second order if, at the point of contact,

$$\frac{d^2f}{da^2}=0,\quad\text{and}\quad\frac{d^2f}{da\,dx}\frac{df}{dy}=\frac{d^2f}{da\,dy}\frac{df}{dx}.$$

1814. Find the envelopes of the rectangular hyperbola

$$x^2-y^2-4ax\cos^3\alpha+4ay\sin^3\alpha+3a^2\cos 2\alpha=0,$$

and of the parabola

$$(x-a\cos^3\alpha)^2=2ay\sin^3\alpha+a^2\sin^4\alpha\,(2+\cos^2\alpha)\,;$$

proving that the conditions for osculation are satisfied in each case.

1815. Given a focus and the length and direction of the major axis of a conic, the envelope of the tangents at the ends of either latus rectum is two parabolas, and that of the normals at the same points two semi-cubical parabolas.

1816. The tangent at a point P of an ellipse meets the axes in T, t, and a parabola is described touching the axes in T, t: prove that the envelope of this parabola is an evolute of an ellipse, and if PM, PN be let fall perpendicular to the axes, MN will touch the parabola where it has contact with its envelope. The curvatures of the parabola and the envelope at the point of contact are as $2:3$.

1817. At each point $P(a\cos\theta,\ b\sin\theta)$ of an ellipse is described the parabola of four-point contact, and S is its focus: prove that the point where PS touches its envelope is (x, y) where

$$\frac{x}{a\cos^3\theta}=\frac{-y}{b\sin^3\theta}=\frac{2(a^2-b^2)}{a^2\cos^2\theta+b^2\sin^2\theta+(a^2-b^2)(\cos^2\theta-\sin^2\theta)}\,;$$

and, if P' be this point,

$$PP' = \frac{2CP \, . \, CD^2}{2CP^2 - CD^2},$$

where CP, CD are conjugate semi-diameters. Also prove that, when $a^2 = 2b^2$, the envelope is the curve

$$x^4 y^2 = a^4 b^2 \left(\frac{4}{9} - \frac{x^2}{a^2}\right)^3.$$

1818. At each point of a given ellipse is described another ellipse osculating the given ellipse at P and having one focus at the centre C: prove that its second focus will be the point P' found in the last question, and PP' will be the tangent at P' to the locus of P'.

1819. A given finite straight line of length $2c$ is a focal chord of an ellipse of given eccentricity e: prove that the envelope of the major axis is a four-cusped hypocycloid inscribed in a circle of radius ce; the envelope of the minor axis is that involute of a four-cusped hypocycloid, inscribed in a circle of radius $\frac{ce}{1-e^2}$, which passes through the centre and cuts the given segment at right angles; the envelope of the nearer latus rectum a similar involute touching the given segment; the envelopes of the farther latus rectum and farther directrix are also involutes of four-cusped hypocycloids; and the envelope of the nearer directrix is a circle of radius $2ce$.

HIGHER PLANE CURVES.

1820. Prove that a cubic which passes through the angular points, the mid points of the sides, and the centroid of a triangle, and also through the centre of a circumscribing conic, will also pass through the point of concourse of the straight lines each joining an angular point to the common point of the tangents to the conic at the ends of the opposite side.

[The equation of the cubic will be

$$lyz(y-z)+mzx(z-x)+nxy(x-y)=0,$$

and, if $(X:Y:Z)$, $(X':Y':Z')$ be the centre of the conic and the point of concourse,

$$YZ'+Y'Z=ZX'+Z'X=XY'+X'Y.$$

When the conic is a circle the cubic is the locus of a point such that, if with it as centre be described two conics, one circumscribing the triangle and the other touching its sides, their axes will be in the same directions.]

1821. Two cubics are drawn through four given points A, E, C, D, and through the three vertices of the quadrangle $ABCD$: prove that, if they touch at A, B, C, or D, the contact will be three-pointic.

[The equation of such a cubic may be taken to be

$$lx(y^2-z^2)+my(z^2-x^2)+nz(x^2-y^2)=0.]$$

1822. Two cubics are drawn through the four points (A, B, C, D) and the three vertices (E, F, G) of a quadrangle: prove that, if they touch at E, their remaining common point lies on FG and on the common tangent at E, and, if EG be the tangent at E, FG will be a tangent at F.

1823. Two cubics are drawn as in last question, and another common point lies on the axis of homology of the triangles ABC, EFG: prove that their remaining common point lies on the conic whose centre is D and which touches the sides of the triangle ABC.

1824. Prove that an infinite number of cubics can be drawn through the ends of the diagonals of a given quadrilateral, and through the three points where the straight lines joining a given point O to the intersection of two diagonals meets the third; also that the cubic which passes through O will have a node at O.

[The equation of such a cubic will be

$$\lambda xyz + (lx + my + nz)(x^2 + y^2 + z^2) - 2(lx^3 + my^3 + nz^3) = 0,$$

where $x \pm y \pm z = 0$ are the sides of the quadrilateral, and $lx = my = nz$ the point O; and the tangents at the node will be real if O lie within the convex quadrilateral or in one of the portions of space vertically opposite an angle of the convex quadrilateral.]

1825. A conic is drawn through four fixed points, and O, O' are two other fixed points which are conjugate with respect to every such conic: prove that the locus of the intersections of tangents drawn from O, O' to the conic is a sextic having six nodes and two cusps.

[The cusps are at O, O', three of the nodes are at the vertices A, B, C of the quadrangle, and the other three are points A', B', C' on BC, CA, AB such that the pencils

$$A'\{AOO'B\},\quad B'\{BOO'C\},\quad C'\{COO'A\},$$

are harmonic.

1826. The evolute of the parabola $y^2 = 4ax$ is its own polar reciprocal with respect to any conic whose equation is

$$\tfrac{1}{2}y^2 + \lambda(x - 2a)^2 = 27\lambda^3a^2:$$

the cissoid $x(x^2 + y^2) = ay^2$ is its own polar reciprocal with respect to any conic whose equation is $(x - a)^2 = 3\lambda^2x^2 - 2\lambda^3y^2$. Also the cubic $x(y^2 - x^2) = ay^2$ is its own reciprocal with respect to each of the latter family of conics.

1827. A cubic of the third class is its own reciprocal with respect to each of a family of conics, the triangle whose sides are the tangents to the cubic at the cusp and at the point of inflexion and the straight line joining the cusp and inflexion is self-conjugate to any one of these conics; and the cubic has double contact with each of the conics, the chord of contact passing through the inflexion · also each of the conics has double contact with another cubic having the same cusp and inflexion and the same tangents at those points.

[The cubics may be taken to be $x^3 = \pm y^2z$, and the conics are the family

$$\lambda^2z^2 - 3\lambda x^2 + 2y^2 = 0.]$$

1828. In any cuspidal cubic, A is the cusp, B the inflexion, C the common point of the tangents at A, B; any straight line through B meets the cubic again in P, Q: prove that the pencil $A\{BPQC\}$ is harmonic.

1829. The three asymptotes of the cubic (areal co-ordinates)

$$x^2(y+z)+y^2(z+x)+z^2(x+y)=0$$

meet in the point $(1:1:1)$, the cubic touches at each angular point the minimum ellipse circumscribing the triangle of reference, and its curvature at any point of contact is to that of the ellipse as $-2:1$. If from any point P on this curve AP, BP, CP be drawn to meet the opposite sides of the triangle of reference in A', B', C', the triangle $A'B'C'$ will be equal to the triangle ABC.

1830. The area of the loop of the cubic

$$4(y+z)(z+x)(x+y)=(x+y+z)^3$$

is

$$K\int_0^{\frac{\sqrt{5}-1}{2}} \sqrt{\frac{x}{1-x}(1-x-x^2)}\,dx,$$

where K is twice the area of the triangle of reference; and the radius of curvature of the loop at a point where the tangent is parallel to a side is to that at the point on the side as $5+\sqrt{5}:2$.

1831. The base BC of a triangle ABC being given, and the relation $\tan^2 A=\tan B\tan C$ between its angles; prove that the locus of its vertex is a lemniscate whose axis bisects BC at right angles, and whose foci are the ends of the second diagonal of a square on BC as diagonal. Investigate the nature of the singularity at B and C.

[Each is a triple point, two of the tangents being impossible.]

1832. A circle is described on a chord of a given ellipse, passing through a fixed point on the axis, as diameter: prove that the envelope is a bicircular quartic whose polar equation is

$$\left(\frac{r^2}{a^2}+2m\frac{r\cos\theta}{a}+m^2-1\right)\left(\frac{r^2}{b^2}+m^2-1\right)=m^2\frac{r^2}{a^2}\sin^2\theta,$$

ma being the distance from the centre of the fixed point.

[One focus of the envelope is always the fixed point, and the other axial foci are at distances from the fixed point given by the equation

$$m^2(a^2-b^2)z^3-2am(\overline{1-m^2}a^2-\overline{1-2m^2}b^2)z^2+$$
$$z(1-m^2)(\overline{1-m^2}a^4-\overline{2-5m^2}a^2b^2+b^4)-2am(1-m^2)b^2(\overline{1-m^2}a^2-b^2)=0.$$

Hence the origin is a double focus if $m=0$, e, or 1. When $m=0$ the envelope degenerates into the point circle at the centre and the circular points; when $m=e$ the equation for the remaining foci is $\{ez-2a(1-e^2)\}^2=0$, so that there are two pairs of coincident foci, and the envelope breaks up into two circles whose vector equations are $r_1:r_2=e:2\pm e$.]

1833. Given the circumscribed and inscribed circles of a triangle, the envelope of the polar circle is a Cartesian.

[The given centres being A, B and AC a straight line bisected in B, C will be the centre (or triple focus), B one of the single foci, and the distances of the others from C are given by the equation

$$cx^2 - 2(a^2 - 3ab + b^2)x + c(a-2b)^2 = 0,$$

where a, b are the radii and c the distance between the centres $(=\sqrt{a^2-2ab})$.]

1834. The equation of the nodal limaçon $r = 2(c\cos\theta - a)$ becomes, when the origin is moved along the initial line through a space $\frac{c^2-a^2}{c}$,

$$r^2 - 2r\left(c + \frac{a^2}{c}\cos\theta\right) + \left(\frac{c^2-a^2}{c}\right)^2 = 0,$$

(so that the curve is now its own inverse with respect to the pole); and, if OPQ be any radius vector from this pole, A, B the vertices,

$$\text{arc } BQ - \text{arc } AP = 8a \sin \tfrac{1}{2} AOP.$$

If S be the node and any circle be drawn touching the axis in S and meeting the curve again in P, Q, OPQ will be a straight line, the tangents to the curve at P, Q will intersect in a point R on the circle such that SR is parallel to the bisector of SOQ, the locus of R will be a cissoid, and that of R' (where RR' is a diameter of the circle) a circle.

1835. In the trisectrix $r = a(2\cos\theta \pm 1)$, S is the node, RSR' a chord to the outer loop, SpP, SqQ two chords inclined at angles of 60° to the former $\{RSP = PSQ = QSR' = 60°\}$, and A is the inner vertex: prove that P, A, q, R' are in one straight line and R, p, A, Q in another straight line at right angles to the former.

1836. A circle touches a given parabola at P and passes through the focus S, and the other two common tangents intersect in T: prove that SP is equally inclined to ST and to the axis of the parabola, the diameter of the circle through S bisects the angle PST, and the locus of T has for its equation

$$(y^2 + 28ax - 96a^2)^2 = 64a(3a - x)(7a - x)^2.$$

1837. The locus of the common points of circles of curvature of a parabola drawn at the ends of a focal chord is a nodal bicircular quartic which osculates the parabola in two points whose distance from the directrix is equal to the latus rectum.

[The node is an acnode, and the equation of the curve when the pole is at the node is

$$r = 2a(\cos\theta + \sqrt{3 + 4\cos^2\theta}).$$

Two foci are at infinity, and two are the points (vertex of parabola origin) $2x = 3a$, $2y = \pm 9a$.]

1838. The envelope of the radical axis of two circles of curvature of the ellipse $a^2y^2 + b^2x^2 = a^2b^2$ drawn at the ends of conjugate diameters is the sextic (of the class 6)

$$\left(\frac{x^2}{a^2} - \frac{y^2}{b^2}\right)^2 \left\{2\left(\frac{x^2}{a^2} + \frac{y^2}{b^2}\right) - 3\right\} = \left\{3\left(\frac{x^2}{a^2} + \frac{y^2}{b^2}\right) - 4\right\}^2,$$

having asymptotes $2\left(\frac{x}{a} \pm \frac{y}{b}\right) \pm 3 = 0$.

[The curve has four cusps, $x^2 = 2a^2$, $y = 0$; $y^2 = 2b^2$, $x = 0$; four acnodes, $\frac{x^2}{a^2} = \frac{y^2}{b^2} = \frac{2}{3}$, and two crunodes at infinity.]

1839. A circle is described with its centre on the arc of a given ellipse and radius $\sqrt{r^2 - c^2}$, where r is the focal distance and c a constant: prove that its envelope is a bicircular quartic which has a node at the nearer vertex when $c = a(1 - e)$, and four real axial foci when c is $< a(1 - e)$ or $> a(1 + e)$.

[The polar equation is, focus of ellipse being pole,

$$\left(r + \frac{c^2}{r}\right)^2 + 4ae\left(r + \frac{c^2}{r}\right)\cos\theta - 4b^2 = 0,$$

and any chord through the pole has two middle points on the auxiliary circle of the ellipse. The distances of the foci from the pole are given by the equation $r + \frac{c^2}{r} = 2(b^2 \pm ac)$, and the points of contact of the double tangent lie on the ellipse

$$\frac{a^2x^2}{b^2} + y^2 = b^2.]$$

1840. The straight line joining the points of contact of parallel tangents to the cardioid $r = 2a(1 - \cos\theta)$ always touches the curve $2r\cos\theta = a(1 - 4\cos^2\theta)$; and an infinite number of triangles can be inscribed in the cardioid whose sides touch the other curve.

[This envelope is a circular cubic having a double focus at the cusp of the cardioid and two single foci on the prime radius at the distances $-a$, $3a$ respectively from the cusp; and, if r_1, r_2, r_3 be the distances of any point on the curve from these three foci,

$$r_3 = 2r_1 + 3r_2;$$

r_1 being reckoned positive for the loop and negative for the sinuous branch. Another form of the equation is $2r\cos\frac{\theta}{3} = a$, the origin being at the centre of the fixed circle when the cardioid is generated as an epicycloid.]

1841. A point P moves so that OP is always a mean proportional between SP, HP; O, S, H being three fixed points in one straight line: prove that, if O lie between S and H, another system of three points O', S', H' can be found on the same straight line such that $O'P$ is always a mean proportional between $S'P$ and $H'P$; that O' will lie without $S'H'$, and the ratio $O'S' : O'H'$ will lie between $\sqrt{2}-1 : \sqrt{2}+1$ and $\sqrt{2}+1 : \sqrt{2}-1$.

[If $OS=a$, $OH=b$, $OO'=z$, $OS'=x$, $OH'=y$; $z=\dfrac{2ab}{a+b}$, and $\dfrac{x+y}{a+b}=\dfrac{xy}{ab}=\left(\dfrac{a-b}{a+b}\right)^2$. The locus of P is a circular cubic whose real foci are S, H, S', H', and a vector equation is

$$l \,.\, SP - m \,.\, HP + (m-l)\, S'P = 0,$$

where $$\frac{l}{b\left(a+b+\sqrt{a^2+b^2-6ab}\right)} = \frac{m}{a\left(a+b-\sqrt{a^2+b^2-6ab}\right)}.]$$

1842. With a point on the directrix of the parabola $y^2=4ax$ as centre is described a circle touching the parabola: prove that the locus of the common point of the other two common tangents to the circle and the parabola is the quartic

$$(y^2-2ax)^2+4a(x+2a)^2(4x+3a)=0;$$

also if on the normal at the point of contact of the circle and parabola be measured outwards a distance equal to one-sixth of the radius of curvature, the envelope of the polar of this point with respect to the circle is

$$y^4+8a(x-2a)^3=0.$$

1843. A circle drawn through the foci B, C of a rectangular hyperbola meets the curve in P, the tangent at P to the circle meets BC in O, and OQ is another tangent to the circle: prove that (1) the locus of Q is the lemniscate (Bernoulli's) whose foci are B, C and that OP is parallel to the bisector of the angle BQC; (2) if $OP'Q'$ be drawn at right angles to OP meeting the circle in P', Q', the locus of P' will be a circular cubic of which B, C are two foci, and the two other real foci coincide at A, a point dividing BC in the ratio $\sqrt{2}-1 : \sqrt{2}+1$, (B being the nearer point to O); (3) the vector equation is

$$(\sqrt{2}-1)\,CP' - (\sqrt{2}+1)\,BP' = 2AP',$$

or $AB \,.\, CP' + AC \,.\, BP' = \dfrac{BC}{\sqrt{2}} \,.\, AP'$; and (4) the angle $Q'QP$ exceeds the angle PQP' by a right angle.

1844. Any curve and its evolute have common foci, and touch each other in the (impossible) points of contact of tangents drawn from the foci.

1845. Trace the curve

$$4xy(x+y-a-b)+ab(x+y)=0;$$

and prove that if (x_1, y_1), (x_2, y_2) be the ends of a chord through the origin

$$x_1+x_2+y_1+y_2=a+b.$$

Also prove that the area of the loop is

$$2(a-b)^2\int_0^{\frac{\pi}{2}}\frac{\sin^2\theta\,d\theta}{\sqrt{1-\left(\frac{a-b}{a+b}\cos\theta\right)^2}}.$$

1846. A circle is described with its centre on the axis, and the points of contact of the common tangents to it and to the fixed circle $x^2+y^2=a^2$ lie on two straight lines: prove that the locus of the points of contact on the variable circle is the two curves

$$(x^2-y^2-2a^2)^2+(x\pm y)^4=2a^2(x\pm y)^2,$$

that these curves osculate in the points $\pm a$, 0, and that the area of each common loop is $a^2\left(\frac{\pi}{4}-\log 2\right)$.

1847. The fixed points S, H are foci of a lemniscate

$$(4n\,SP\,.\,HP=SH^2),$$

and the points U, V, V', U' its vertices, a circle through S, H meets the lemniscate in R, R' (on the same side of SH) and UR', VR', $V'R'$, $U'R'$ meet the circle again in Q, P, P', Q': prove that the straight lines QQ', PP' intersect SH in the same point as the tangent to the circle at R, and each is equally inclined to SR, HR. Also each of the points Q, P, P', Q' is such that its distance from R is a mean proportional between its distances from S and H, as also is its distance from the corresponding vertex. The locus of any one of the points for different circles is therefore an inverse of the lemniscate with respect to one of its vertices, the constant of inversion being the rectangle under the focal distances of the corresponding vertex, sign being regarded.

[The curves are axial circular cubics, similar to each other, and any one is the inverse of any other with respect to one of its vertices, the constant of inversion being always the rectangle under the distances of the centre of inversion from S, H. The loci of P, P' (and of Q, Q') are images of each other, or their centre of inversion is at ∞. Each of the four curves has one vertex peculiar to itself, and is its own inverse with respect to that vertex, the constant following the same rule as for two different curves. The linear dimensions of the loci of Q, P are as $\sqrt{n-1}:\sqrt{n+1}$. Each curve has S, H for two of its foci; and, for the locus of P, the two other real foci (F_3, F_4) divide SH in the ratios

$$SF_3:HF_3=\sqrt{n(n-1)}+\sqrt{n(n+1)}+1:-\sqrt{n(n-1)}+\sqrt{n(n+2)}+1,$$

$$HF_4:SF_4=\sqrt{n(n+1)}+\sqrt{n(n-1)}-1:\sqrt{n(n+1)}-\sqrt{n(n-1)}-1;$$

and, if r_1, r_2, r_3, r_4 denote SP, HP, F_3P, F_4P,

$$2\sqrt{n-1}\,r_3=(\sqrt{n-1}+\sqrt{n+1}+2\sqrt{n})\,r_1+(\sqrt{n-1}-\sqrt{n+1}-2\sqrt{n})\,r_2,$$

$$2\sqrt{n-1}\,r_4=(\sqrt{n-1}-\sqrt{n+1}+2\sqrt{n})\,r_1+(\sqrt{n-1}+\sqrt{n+1}-2\sqrt{n})\,r_2.]$$

1848. Tangents inclined at a given angle α are drawn to two given circles, whose radii are a, b and centres at a distance c: prove that the locus of their point of intersection is an epitrochoid, the fixed and rolling circles being each of radius $\sqrt{\dfrac{a^2+b^2-2ab\cos\alpha}{\sin\alpha}}$, and the distance of the generating point from the centre of the moving circle being c.

1849. In a three-cusped hypocycloid whose cusps are A, B, C, a chord APQ is drawn through A: prove that the tangents at P, Q will divide BC harmonically, and their point of intersection will lie on a conic passing through B, C; also the tangents to this conic at B, C pass through the centre of the hypocycloid.

1850. A tangent to a cardioid meets the curve again in P, Q: prove that the tangents at P, Q divide the double tangent harmonically, and the locus of their common point is a conic passing through the points of contact of the double tangent and having triple contact with the cardioid (two of the contacts impossible).

[The equation $X^{-\frac{1}{2}}+Y^{-\frac{1}{2}}+Z^{-\frac{1}{2}}=0$ will represent a cardioid when

$$X=x+iy,\quad Y=x-iy,\quad Z=a;$$

and a three-cusped hypocycloid when

$$X=x+y\sqrt{3},\quad Y=x-y\sqrt{3},\quad Z=9a-2x.]$$

1851. Chords of a Cartesian are drawn through the triple focus: prove that the locus of their middle points is

$$(r^2-bc)(r^2-ca)(r^2-ab)+a^2b^2c^2\sin^2\theta=0,$$

a, b, c being the distances of the single foci from the triple focus which is the origin.

1852. Two points describe the same circle of radius a with velocities which are to each other as $m:n$ (m, n being integers prime to each other and $n>m$); the envelope of the joining line is an epicycloid whose vertices lie on the given circle and the radius of whose fixed circle is $a\dfrac{n-m}{n+m}$. When $-m$ is put for m, the points must describe the circle in opposite senses, and the envelope is a hypocycloid. Hence may be deduced that the class number is $m+n$.

1853. An epicycloid is generated by circle of radius ma rolling upon one of radius $(n-m)a$, m, n being integers prime to each other, and in the moving circle is described a regular m-gon one of whose corners is the describing point; all the other corners will move in the same epicycloid, and the whole epicycloid will be completely generated by these m points in one revolution about the fixed circle. The same epicycloid may also be generated by the corners of a regular n-gon inscribed in a circle of radius na rolling on the same fixed circle with internal contact.

1854. In an epicycloid (or hypocycloid) whose order is $2p$ and class $p+q$, tangents are drawn to the curve from any point O on the circle through the vertices: their points of contact will be corners of two regular polygons of p and q sides respectively inscribed in the two moving circles by which the curve can be generated which touch the circle through the vertices in O.

1855. The locus of the common point of two tangents to an epicycloid inclined at a constant angle is an epitrochoid, for which the radius of the fixed and moving circles are respectively

$$\frac{a(a+2b)}{a+b}\,\frac{\sin\dfrac{a+b}{a+2b}\alpha}{\sin\alpha},\quad \frac{b(a+2b)}{a+b}\,\frac{\sin\dfrac{a+b}{a+2b}\alpha}{\sin\alpha},$$

and the distance of the generating point from the centre of the moving circle is

$$(a+2b)\,\frac{\sin\dfrac{b}{a+2b}\alpha}{\sin\alpha};$$

where a, b are the radii of the fixed and moving circles for the epicycloid, and α is the angle through which one tangent would turn in passing into the position of the other, always in contact with the curve.

1856. The pedal of a parabola with respect to any point O on the axis is a nodal circular cubic which is its own inverse with respect to the vertex A, the constant of inversion being the square on OA. If OO' be a straight line bisected in A, PQ a chord passing through O', OY, OZ perpendiculars on the tangents at P, Q, then A, Y, Z will be collinear and $AY \cdot AZ = AO^2$.

[If $OA = b$ and $4a$ be the latus rectum, the distances of the two single foci from O, the double focus, are given by the equation $x^2 + 4ax = 4ab$, and the vector equation is, for the loop,

$$\frac{r_2}{\sqrt{a+b}-\sqrt{a}} - \frac{r_3}{\sqrt{a+b}+\sqrt{a}} + \frac{2\sqrt{a}}{b}r_1 = 0,$$

r_1 being the distance from the internal focus. The difference of the arcs s_1, s_2 from the node O to corresponding points Y, Z on the loop and sinuous branch is determined by the equation

$$\frac{ds_2}{d\theta} - \frac{ds_1}{d\theta} = \frac{\sin\theta}{\cos^2\theta}\sqrt{(a+b)(a+b+\overline{3a-b}\cos^2\theta)}.]$$

1857. The area common to two ellipses which have the same centre and equal axes inclined at an angle α is

$$2ab\tan^{-1}\frac{2ab}{(a^2-b^2)\sin\alpha}.$$

1858. Perpendiculars are let fall upon the tangents to an ellipse from a point within it at a distance c from the centre: prove that the area of the curve traced out by the feet of these perpendiculars is

$$\frac{\pi}{2}(a^2+b^2+c^2).$$

1859. The areas of the curves

$$a^2y^2(x-b)^2=(a^2-x^2)(bx-a^2)^2,\ x^2+y^2=a^2,\ (b>a)$$

are A, A': prove that the limiting value of $\dfrac{A'-A}{b-a}$, as b decreases to a, is $6\pi a$.

1860. The sum of the products of each element of an elliptic lamina multiplied by its distance from the focus is $\frac{1}{3}Ma(2+e^2)$, M being the mass of the lamina, $2a$ the major axis, and e the excentricity; and the mean distance of all points within a prolate spheroid from one of the foci is $\frac{1}{4}a(3+e^2)$.

1861. Prove that the arc of the curve $y=\sqrt{a^2-b^2}\left(1-\cos\dfrac{x}{b}\right)$ between $x=0$, $x=2\pi b$, is equal to the perimeter of an ellipse of axes $2a$, $2b$: and determine the ratio of $a:b$ in order that the area included between the curve and the axis of x may be equal to the area of the ellipse. [$a:b=2:\sqrt{3}$.]

1862. Find the whole length of the arc enveloped by the directrix of an ellipse rolling along a straight line during a complete revolution; and prove that the curve will have two cusps if the excentricity of the ellipse exceed $\dfrac{\sqrt{5}-1}{2}$.

[The arc s is determined by the equation

$$\frac{ds}{d\psi} = \frac{a}{e}\left\{1 - \frac{e\cos\psi}{\sqrt{1 - e^2\sin^2\psi}} - \frac{(1-e^2)\, e\cos\psi}{(1 - e^2\sin^2\psi)^{\frac{3}{2}}}\right\},$$

ψ being the angle through which the directrix turns.]

1863. A sphere is described touching a given plane at a given point, and a segment of given curve surface is cut off by a plane parallel to the former: prove that the locus of the circular boundary of this segment is a sphere.

1864. Two catenaries touch each other at the vertex, and the linear dimensions of the outer are twice those of the inner; two common ordinates MPQ, mpq are drawn from the directrix of the outer: prove that the volume generated by the revolution of the arc Pp about the directrix is equal to $2\pi \times$ area $MQqm$.

1865. The area of the curve $r = a(\cos\theta + 3\sin\theta)^2 \div (\cos\theta + 2\sin\theta)^3$ included between the maximum and minimum radii is to the triangle formed by the radii and chord in the ratio 781 : 720 nearly.

1866. Prove the results stated below, A denoting in each case the whole area, $\bar{x}$ and $\bar{y}$ the co-ordinates of the centre of inertia of the area on the positive side of the axis of y;

(1) the curve $\quad (a^2 + x^2)\, y^2 - 4a^3 y + x^4 = 8a^4,$

$$A = 5\pi a^2,\ \bar{x} = \frac{8a}{5\pi}\log(2 + \sqrt{3}),\ \bar{y} = \frac{7a}{5};$$

(2) the curve $\quad y^2(3a^2 + x^2) - 4a^3 y + x^4 = 0,$

$$A = \frac{11}{9 + 4\sqrt{3}}\pi a^2,\ \bar{x} = \frac{4a}{\pi}\,\frac{4 - 3\log 3}{9 - 4\sqrt{3}},\ \bar{y} = \frac{a}{33}(21\sqrt{3} - 16);$$

(3) the curve $y^2(a^2 + x^2) - 4a^3 y + (x^2 - 2a^2)^2 = 0,$

$$A = \pi a^2,\ \bar{x} = \frac{8a}{\pi}\{\sqrt{3} - \log(2 + \sqrt{3})\},\ \bar{y} = a;$$

(4) the curve $y^2(a^2 + x^2) - 2max^2 y + x^4 = 0,\ (m > 1),$

$$A = \pi(m-1)^2 a^2,\ \bar{x} = \frac{4a}{3\pi}\,\frac{(m^2-1)^{\frac{3}{2}} + 3\{\sqrt{m^2-1} - m\log(m + \sqrt{m^2-1})\}}{(m-1)^2},$$

$$\bar{y} = (m-1)\,a.$$

1867. Prove that the curve whose equation is

$$y^2(a^2 + x^2) - 2m(m+1)\,a^3 y + 4x^4 - 4(m^2 + m - 1)\,a^2x^2 + (m+1)(4m^2 - 3m + 1)\,a^4 = 0$$

consists of three loops, the area of one of which is equal to the sum of the areas of the other two.

1868. For a loop of the curve $x^2y^2 - 4a^3y + (3a^2 - x^2)^2 = 0$,

$$A = a^2(2\pi - 3\sqrt{3}),\quad \bar{x} = 4a\frac{\{\sqrt{3} - \log(2 + \sqrt{3})\}}{2\pi - 3\sqrt{3}},\quad \bar{y} = \frac{a}{3}\frac{9\sqrt{3} - 4\pi}{2\pi - 3\sqrt{3}}.$$

1869. The area of a loop of the curve

$$y^2(4a^2 - x^2) - 4a^3y + (a^2 - x^2)^2 = 0 \text{ is } a^2(3 - 2\log 2).$$

[This curve breaks up into two hyperbolas.]

1870. In the curve

$$(ma^2 + x^2)y^2 - 2ay(a^2 - x^2) + \frac{(a^2 - x^2)^2}{1 + m} = 0,$$

$$A = \frac{\pi a^2}{\sqrt{m}(1 + m)}\{2(1 + m)^{\frac{3}{2}} - (2m + 3)\sqrt{m}\},$$

$$\bar{y} = \frac{a}{m}\frac{3(m + 1)^{\frac{3}{2}}\sqrt{m} - 3m^2 - 2m + 2}{3m + 4}.$$

1871. Trace the curve $x = \frac{a\cos 3\theta}{\sin 2\theta}$, $y = \frac{a\sin 3\theta}{\sin 2\theta}$, and prove that the internal area included by the four branches is $2\sqrt{3}\,a^2$.

1872. Trace the curve whose equation is $x = 2a\sin\frac{y}{x}$; and prove that each loop has the same area πa^2 and is bisected by the straight line joining the origin to the point where the tangent is parallel to the axis of y.

1873. The area of the loop of the curve $a^{2n-2}y^2 = n^2x^{2n}\frac{a - x}{a + x}$, when n is indefinitely increased, is $\sqrt{2\pi}\,a^2$; and the area between the curve $a^{2n-2}y^2 = nx^{2n}\frac{a - x}{a + x}$ and the asymptote, when n is indefinitely increased, is $2\sqrt{2\pi}\,a^2$.

1874. The areas of (1) the loop of the curve $y^{2n}(a + x) = x^{2n}(a - x)$, (2) the part between the curve and the asymptote differ by $\frac{\pi a^2}{n^2\sin\frac{\pi}{2n}}$.

1875. Prove that the whole arc of the curve $8a^2y^2 = x^2(a^2 - 2x^2)$ is πa; and for the part included in the positive quadrant, the centre of gravity of the area is $\left(\frac{3\pi a}{16\sqrt{2}}, \frac{a}{20}\right)$; the centre of gravity of the arc is $\left(\frac{5a}{3\sqrt{2}\pi}, \frac{a}{4\pi}\right)$; the centre of gravity of the volume generated by revolution about the axis of x is $\left(\frac{5a}{8\sqrt{2}}, 0\right)$; and that of the area of the surface generated is $\left(\frac{3a}{5\sqrt{2}}, 0\right)$.

1876. Prove that the curve $b^2y^2 = x^2(2a - x)$ is rectifiable if

$$b^2 = (9 \pm 6\sqrt{3})\,a^2.$$

1877. **The arc of** the curve $r(\epsilon^\theta + 1) = a(\epsilon^\theta - 1)$ measured from the **origin to a point** (r, θ) is $a\theta - r$; and the corresponding area **is** $\frac{1}{2}a^2\theta - ar$.

1878. The whole arc of the curve

$$x^{\frac{2}{3}} + y^{\frac{2}{3}} = a^{\frac{2}{3}} \text{ is } 5a\left\{1 + \frac{1}{2\sqrt{3}}\log(2 + \sqrt{3})\right\}.$$

1879. The arc of the curve

$$x = a(2 - 3\cos\theta + \cos 3\theta),\quad y = 3a\sqrt{2}\,(2\theta - \sin 2\theta),$$

from cusp to cusp, is $14a$.

1880. The curve $27(y^2 - 8ax - a^2)^2 = 8ax(9a + 8x)^2$ is rectifiable.

[**We may** put $y = at^3$, $8x = 3a(1 - t^2)^2$, and $8s = 3at^2(2 + t^2)$, measuring **from the** cusp.]

1881. **The** arc of the curve

$$x = a(6\sin\theta + \sin 3\theta),\quad y = 3a(2\theta + \sin 2\theta)$$

measured from the cusp (at **the origin) is** $a(12\sin\theta + \sin 3\theta)$.

1882. The curves whose intrinsic equations are

$$(1)\quad \frac{ds}{d\phi} = a\sec^4\frac{\phi}{2} \div \sqrt{\tan\frac{\phi}{2}},$$

$$(2)\quad \frac{ds}{d\phi} = 2a\frac{1 - \cos\phi}{(1 + \cos\phi)^3},$$

are both quintics: in (1), $s^2 = x^2 + \frac{9}{5}y^2$, and in (2), $s^2 = x^2 + \frac{16}{15}y^2$, x, y, s vanishing together and ϕ being measured from the axis of x; also the area of **the loop in (2) is** $\frac{8}{7}\frac{5^{\frac{5}{2}}}{3^{\frac{13}{2}}}a^2$ **and its centre** of gravity divides the axis in the ratio 63 : 80.

1883. **The curve** whose intrinsic equation is

$$s = \frac{3a}{4}\tan^2\frac{\phi}{2}\left(2 + \tan^2\frac{\phi}{2}\right)$$

is a quartic. $\left[y^4 + 8a(x - a)y^2 + \frac{64ax^3}{27} = 0.\right]$

1884. In the curve $y^4 - 6a^2xy + 3a^4 = 0$, the arc, measured from a point of contact of the double tangent through the origin, is $\frac{a^2}{y} - x$.

1885. The arc of the curve $ax = y^2 - 2a^2\log\frac{y}{a} - a^2$, measured from $(0, a)$ is $\frac{2y^2}{a} - x - 2a$.

1886. The whole area of the curve

$$x = a \sin\theta\,(15 - 5\sin^2\theta + 3\sin^4\theta),\quad y = 10a\cos^3\theta,$$

is $\frac{3225}{32}\pi a^2$; the arc in one quadrant is $17a$, and its centre of gravity lies outside the area.

1887. A hypocycloid is generated by a circle of radius na rolling within a circle of radius $(2n+1)a$, (n integral), and an involute is drawn passing through the cusps: prove that the area of this involute is to that of the fixed circle as

$$2n(n+1)(8n^2+8n-1) : (2n+1)^4;$$

and the arc of one to the arc of the other as $4n(n+1) : (2n+1)^2$.

1888. If u_m denote $\int_a^b x^m \sqrt{(x-a)(b-x)}\,dx$, and $m-1$ be positive,

$$2(m+3)u_{m+1} - (2m+3)(a+b)u_m + 2mabu_{m-1} = 0.$$

1889. Prove that the limiting values of

$$(1)\quad \left\{\sin\frac{\pi}{n}\sin\frac{2\pi}{n}\sin\frac{3\pi}{n}\quad \sin(n-1)\frac{\pi}{n}\right\}^{\frac{1}{n}},$$

$$(2)\quad \left\{\sin\frac{\pi}{n}\sin^2\frac{2\pi}{n}\sin^3\frac{3\pi}{n}\dots\sin^{n-1}(n-1)\frac{\pi}{n}\right\}^{\frac{2}{n^2}},$$

$$(3)\quad \left\{\left(1+\tan\frac{\pi}{4n}\right)\left(1+\tan\frac{2\pi}{4n}\right)\dots\left(1+\tan(n-1)\frac{\pi}{4n}\right)\right\}^{\frac{2}{n}},$$

when n is indefinitely increased are each equal to $\frac{1}{2}$.

1890. An arithmetical, a geometrical, and an harmonical progression have each the same number of terms, and the same first and last terms a and l; the sums of their terms are respectively s_1, s_2, s_3, and the continued products p_1, p_2, p_3: prove that, when the number of terms is indefinitely increased,

$$\frac{2s_1}{s_2} = \frac{l+a}{l-a}\log\frac{l}{a},\quad \frac{s_2^2}{s_1 s_3} = \frac{(a+l)^2}{4al},\quad \frac{p_1 p_3}{p_2^2} = 1.$$

[The last of these equations is true whatever be the number of terms.]

1891. Prove that, if $n+1$ be positive, the area included by the curve $x = a\cos^{2n+2}\theta$, $y = b\sin^{2n+2}\theta$ and the positive co-ordinate axes is

$$(n+1)\,ab\,\Gamma(n+1)\,\Gamma(n+2) \div \Gamma(2n+3),$$

and tends to $ab\sqrt{n\pi} \div 2^{2n+2}$ as n tends to infinity. Also, by considering the arc of this curve, prove that

$$\frac{(n+1)}{a+b}\int_{-1}^{1}\sqrt{a^2(1+x)^{2n} + b^2(1-x)^{2n}}\,dx > 2^{n+1} - 1 < 2^{n+1};$$

and that the limit, when n is infinite, of

$$n2^{-n}\int_{-1}^{1}\sqrt{a^2(1+x)^{2n} + b^2(1-x)^{2n}}\,dx \text{ is } 2(a+b).$$

1892. The lengths of two tangents to a parabola are a, b and the included angle ω: prove that the arc between the points of contact is

$$(a+b)\frac{a^2+b^2-ab(1-\cos\omega)}{a^2+b^2+2ab\cos\omega}$$
$$+\frac{a^2b^2\sin^2\omega}{(a^2+b^2+2ab\cos\omega)^{\frac{3}{2}}}\log\frac{b}{a}\frac{\sqrt{a^2+b^2+2ab\cos\omega}+b+a\cos\omega}{\sqrt{a^2+b^2+2ab\cos\omega}-a-b\cos\omega}.$$

1893. The general integral of the equation $\frac{d\theta}{\cos\theta}=\frac{d\phi}{\cos\phi}$ may be written $\sec^2\theta+\sec^2\phi+\sec^2\mu-2\sec\theta\sec\phi\sec\mu=1$; and that of the equation $\frac{dx}{\sqrt{x^2-1}}=\frac{dy}{\sqrt{y^2-1}}$ may be written $x^2+y^2-2\lambda xy+\lambda^2=1$.

[That is, the differential equation of all conics inscribed in the parallelogram whose sides are $x^2=a^2$, $y^2=b^2$, is $\frac{dx}{\sqrt{x^2-a^2}}=\frac{dy}{\sqrt{y^2-b^2}}$.]

1894. The complete integral of the equation

$$\frac{d\theta}{\sqrt{1-e^2\sin^2\theta}}=\frac{d\phi}{\sqrt{1-e^2\sin^2\phi}},$$

may be written in the form

$$\frac{a^2}{a^2+\lambda}\cos^2\frac{\theta+\phi}{2}+\frac{b^2}{b^2+\lambda}\sin^2\frac{\theta+\phi}{2}=\cos^2\frac{\theta-\phi}{2},$$

where $e^2=1-\frac{b^2}{a^2}$: prove that this is equivalent to the ordinary form

$\cos\theta\cos\phi+\sqrt{1-e^2\sin^2\mu}\sin\theta\sin\phi=\cos\mu$, where $\tan\frac{\mu}{2}=\sqrt{\frac{\lambda(b^2+\lambda)}{b^2(a^2+\lambda)}}$.

Also prove that a particular solution is

$$(1-e^2\sin^2\theta)(1-e^2\sin^2\phi)=1-e^2.$$

1895. The area of each curvilinear quadrangle formed by the four parabolas $y^2=mux-u^2$, when u has successively the values a, b, c, d, is

$$\frac{2}{3m}(\sqrt{b}-\sqrt{a})(\sqrt{d}-\sqrt{c})(c+\sqrt{cd}+d-a-\sqrt{ab}-b),$$

$(a<b<c<d)$, and this is equal to the area of the quadrangle included by the common chords, when $\sqrt{d}+\sqrt{a}=\sqrt{b}+\sqrt{c}$.

1896. In an elliptic annulus bounded by two confocal ellipses the density at any point varies as the square root of the rectangle under the focal distances: prove that the moment of inertia about an axis through the centre perpendicular to the plane is

$$\frac{M}{2}(a'b'-ab)\div\log\left(\frac{a'+b'}{a+b}\right);$$

M being the mass, $2a$, $2b$ the axes of one boundary, $2a'$, $2b'$ those of the other.

1897. In an ellipsoid $\frac{x^2}{a^2}+\frac{y^2}{b^2}+\frac{z^2}{c^2}=1$, the density at any point (x, y, z) is $F\left(\frac{x^2}{a^2}+\frac{y^2}{b^2}+\frac{z^2}{c^2}\right)$: prove that the moment of inertia about the axis of x is $M\frac{(b^2+c^2)}{3}\int_0^1 x^4 F(x^2)\,dx \div \int_0^1 x^2 F(x^2)\,dx$.

1898. The value of $\iiint \frac{dx\,dy\,dz}{xyz}$ taken over one of the continuous volumes bounded by the six spheres $U=ax$, $U=a'x$, $U=by$, $U=b'y$, $U=cz$, $U=c'z$, where $U\equiv x^2+y^2+z^2$ and aa', bb', cc' are positive, is

$$\pm\log\frac{a}{a'}\log\frac{b}{b'}\log\frac{c}{c'}.$$

1899. Prove that, if a, b, c, d be in descending order of magnitude,

$$\int_b^a \frac{dx}{\sqrt{(a-x)(x-b)(x-c)(x-d)}} = (1)\int_{-\infty}^{\infty}\frac{dx}{\sqrt{\{a-c+(b-c)x^2\}\{a-d+(b-d)x^2\}}}$$

$$=(2)\int_{-\infty}^{\infty}\frac{dx}{\sqrt{\{b-c+(a-c)x^2\}\{b-d+(a-d)x^2\}}}$$

$$=(3)\int_{-\infty}^{\infty}\frac{dx}{\sqrt{\{a-d+(a-c)x^2\}\{b-d+(b-c)x^2\}}}$$

$$=(4)\int_{-\infty}^{\infty}\frac{dx}{\sqrt{\{a-c+(a-d)x^2\}\{b-c+(b-d)x^2\}}}$$

$$=(5)\int_0^{\pi}\frac{d\theta}{\sqrt{(a-c)(b-d)-(a-b)(c-d)\sin^2\theta}}.$$

1900. Prove that, if a, b, c, d be in descending order of magnitude, and m any positive quantity,

$$\int_b^a \frac{\{(a-x)(x-b)(x-c)(x-d)\}^{m-1}}{\left\{\frac{(a-x)(x-d)}{a-d}+\frac{(x-b)(x-c)}{b-c}\right\}^{2m-1}}\,dx$$

$$=\int_d^c \frac{\{(a-x)(b-x)(c-x)(x-d)\}^{m-1}}{\left\{\frac{(a-x)(x-d)}{a-d}+\frac{(b-x)(c-x)}{b-c}\right\}^{2m-1}}\,dx;$$

$$\int_b^a \frac{(a-x)^{m-1}(x-c)^{m-1}\,dx}{\left\{\frac{(a-x)(x-d)}{a-d}+\frac{(x-b)(x-c)}{b-c}\right\}^{m}}$$

$$=\int_d^c \frac{(a-x)^{m-1}(c-x)^{m-1}\,dx}{\left\{\frac{(a-x)(x-d)}{a-d}+\frac{(b-x)(c-x)}{b-c}\right\}^{m}};$$

and
$$\int_b^a \frac{(x-b)^{m-1}(x-d)^{m-1}\,dx}{\left\{\frac{(a-x)(x-d)}{a-d}+\frac{(x-b)(x-c)}{b-c}\right\}^m}$$
$$=\int_d^c \frac{(b-x)^{m-1}(x-d)^{m-1}\,dx}{\left\{\frac{(a-x)(x-d)}{a-d}+\frac{(b-x)(c-x)}{b-c}\right\}^m}.$$

1901. Prove that the limiting values, when b increases to a, of

(1) $$\int_b^a \frac{(a-x)^{m-1}(x-c)^{m-1}\,dx}{\left\{\frac{(a-x)(x-d)}{a-d}+\frac{(x-b)(x-c)}{b-c}\right\}^{2m-1}},$$

(2) $$\int_b^a \frac{\{(a-x)(x-b)(x-c)(x-d)\}^{m-1}\,dx}{\left\{\frac{(a-x)(x-d)}{a-d}+\frac{(x-b)(x-c)}{b-c}\right\}^{2m-1}},$$

are respectively

(1) $\{\Gamma(m)\}^2 \div \Gamma(2m)$, (2) $(a-c)^{m-1}(a-d)^{m-1}\{\Gamma(m)\}^2 \div \Gamma(2m)$.

1902. The limiting values, when b increases to a, of

(1) $$\int_b^a \frac{(a-x)^{m-1}(x-c)^{m-1}\,dx}{\left\{\frac{(a-x)(x-d)}{a-d}+\frac{(x-b)(x-c)}{b-c}\right\}^{m}},$$

(2) $$\int_b^a \frac{(x-b)^{m-1}(x-d)^{m-1}\,dx}{\left\{\frac{(a-x)(x-d)}{a-d}+\frac{(x-b)(x-c)}{b-c}\right\}^{m-1}},$$

are respectively (1) $\frac{1}{m}(a-c)^{m-1}$, (2) $\frac{1}{m}(a-d)^{m-1}$.

1903. Prove that, if $a > b > c > d$,

$$\int_b^a \frac{dx}{\frac{(a-x)(x-d)}{a-d}+\frac{(x-b)(x-c)}{b-c}} = \sqrt{\frac{(a-d)(b-c)}{(a-b)(c-d)}}\tan^{-1}\sqrt{\frac{(a-b)(c-d)}{(a-d)(b-c)}},$$

$$\int_c^b \dots\, dx = \sqrt{\frac{(a-d)(b-c)}{(a-b)(c-d)}}\tan^{-1}\sqrt{\frac{(a-d)(b-c)}{(a-b)(c-d)}}.$$

1904. Having given $2x = r(\epsilon^\theta + \epsilon^{-\theta})$, $2y = r(\epsilon^\theta - \epsilon^{-\theta})$: prove that

$$\int_0^\infty\int_0^\infty V\,dx\,dy = \int_0^\infty\int_0^\infty V' r\,dr\,d\theta,$$

V being a function of x, y which becomes V' when their values are substituted.

1905. Having given $Xx^2 = Yy^2 = Zz^2 = \dots = xyz \dots$, prove that

$$\iiint \dots V\,dX\,dY\,dZ \dots = 2^{n-1}(n-2) \iiint \dots v\,(xyz\dots)^{n-2}\,dx\,dy\,dz \dots,$$

V being a function of $X, Y, Z, \dots$ which becomes v when their values are substituted, and n being the number of integrations.

1906. Having given $x + y + z = u$, $y + z = uv$, $z = uvw$, prove that

$$\int_0^\infty \int_0^\infty \int_0^\infty V\,dx\,dy\,dz = \int_0^\infty \int_0^1 \int_0^1 V'\,u^2 v\,du\,dv\,dw\,;$$

also, having given

$$x_1 + x_2 + x_3 + x_4 = u_1, \quad x_2 + x_3 + x_4 = u_1 u_2, \quad x_3 + x_4 = u_1 u_2 u_3, \quad x_4 = u_1 u_2 u_3 u_4,$$

prove that

$$\int_0^\infty \int_0^\infty \int_0^\infty \int_0^\infty V\,dx_1\,dx_2\,dx_3\,dx_4 = \int_0^\infty \int_0^1 \int_0^1 \int_0^1 V'\,u_1^3 u_2^2 u_3\,du_1\,du_2\,du_3\,du_4\,;$$

and the corresponding theorem with n variables.

1907. Having given

$$x_1 = r \sin\theta_1 \cos\theta_2, \quad x_2 = r\cos\theta_1 \cos\theta_3, \quad x_3 = r \sin\theta_1 \sin\theta_2 \cos\theta_4,$$
$$x_4 = r \sin\theta_1 \sin\theta_2 \sin\theta_4, \quad x_5 = r\cos\theta_1 \sin\theta_3 \cos\theta_5, \quad x_6 = r\cos\theta_1 \sin\theta_3 \sin\theta_5,$$

prove that

$$\int_0^\infty \int_0^\infty \dots V\,dx_1\,dx_2\,dx_3\,dx_4\,dx_5\,dx_6$$
$$= \int_0^\infty \int_0^{\frac{\pi}{2}} \int_0^{\frac{\pi}{2}} \dots V'\,r^5 \sin^3\theta_1 \cos^3\theta_1 \sin\theta_2 \sin\theta_3\,dr\,d\theta_1 \dots d\theta_5.$$

1908. Prove that $\iiint \dots dx_1\,dx_2 \dots dx_n$, taken over all real values of $x_1, x_2, \dots x_n$ for which

$$x_1^2 + x_2^2 + \dots + x_n^2 + 2m\,(x_1x_2 + x_2x_3 + x_3x_1 + \dots) \ngtr 1,$$

is equal to

$$\left(\frac{\pi}{1-m}\right)^{\frac{n}{2}} \sqrt{\frac{1-m}{1+m\,(n-1)}}\;\frac{1}{\Gamma\left(\frac{n}{2}+1\right)},$$

provided that m lies between $\dfrac{-1}{n-1}$ and 1.

1909. Prove that

$$\int_0^\infty \int_0^\infty \dots \frac{dx_1\,dx_2 + \dots + dx_n}{\{a^2 + x_1^2 + x_2^2 + \dots + x_{2n}^2\}^{\frac{2n+2r+1}{2}}} = \frac{\pi^n}{2a^{2r+1}}\;\frac{|\underline{n+r-1}\;|\underline{2r}}{|\underline{2n+2r-1}\;|\underline{r}}\,;$$

n, r being positive whole numbers.

1910. Prove that $\iiint \dots dx_1\, dx_2 \dots dx_n$ is equal to

$$\frac{1}{\sqrt{n+1}} \left\{\frac{\pi}{n(n+1)}\right\}^{\frac{n}{2}} \frac{1}{\Gamma\left(\frac{n}{2}+1\right)},$$

the limits of the integral being given by the equation

$$x_1^2 + x_2^2 + \dots + x_n^2 - x_1x_2 - x_2x_3 - \dots - x_{n-1}x_n = \frac{1}{2n(n+1)}.$$

1911. Prove that

$$\iiint \dots \frac{dx_1\, dx_2 \dots dx_n}{\sqrt{1-(x_1^2+x_2^2+\dots+x_n^2)^2}} = \frac{\pi^{\frac{n+1}{2}}}{2\Gamma\left(\frac{n}{2}\right)} \frac{\Gamma\left(\frac{n}{4}\right)}{\Gamma\left(\frac{n+2}{4}\right)},$$

and that $\qquad \iiint \dots \sqrt{\dfrac{1+x_1^2+x_2^2+\dots+x_n^2}{1-x_1^2-x_2^2-\dots-x_n^2}}\, dx_1\, dx_2 \dots dx_n$

$$= \frac{\pi^{\frac{n+1}{2}}}{2\Gamma\left(\frac{n}{2}\right)} \left\{\frac{\Gamma\left(\frac{n}{4}\right)}{\Gamma\left(\frac{n+2}{4}\right)} + \frac{\Gamma\left(\frac{n+2}{4}\right)}{\Gamma\left(\frac{n+4}{4}\right)}\right\};$$

the integral extending over all real values for which $x_1^2+x_2^2+\dots+x_n^2 \not> 1$.

1912. Prove that

$$\int_0^\infty \frac{x\,dx}{(1+x)^n-(1-x)^n} = \frac{\pi}{2n} \Sigma (-1)^{r-1} \sin\frac{r\pi}{n} \cos^{n-3}\frac{r\pi}{n},$$

r having all integral values from 1 to $\frac{n}{2}-1$ or $\frac{n-1}{2}$ according as n is an even or an odd integer.

1913. Having given the equation

$$(a^2-x^2)\frac{dy}{dx} - xy + a^2 = 0,$$

and that when $x = \frac{a}{2}$, $y = \frac{2\pi a}{3\sqrt{3}}$; prove that, when $x = 2a$,

$$y = \frac{a}{\sqrt{3}} \log(2+\sqrt{3});$$

and generally that $y = a^2 \cos^{-1}\left(\frac{x}{a}\right) + \sqrt{a^2-x^2}$ when $x^2 < a^2$, and

$$= a^2 \log\left(\frac{x+\sqrt{x^2-a^2}}{a}\right) \div \sqrt{x^2-a^2}$$

when $x > a$.

[It would seem that the only form of solution holding generally is

$$y = a^2 \int_0^\infty \frac{dz}{1+2xz+a^2z^2}.]$$

1914. Having defined X, X' by the equations

$$\sin x = x - \frac{x^3}{\underline{|3}} + \frac{x^5}{\underline{|5}} - \ldots + (-1)^{n-1}\frac{x^{2n-1}}{\underline{|2n-1}} + (-1)^n \frac{x^{2n}X}{\underline{|2n}},$$

$$\cos x = 1 - \frac{x^2}{\underline{|2}} + \frac{x^4}{\underline{|4}} - \ldots + (-1)^n \frac{x^{2n}}{\underline{|2n}} + (-1)^{n+1}\frac{x^{2n+1}X'}{\underline{|2n+1}},$$

prove that

$$\int_0^\infty \frac{X}{x}\,dx = \int_0^\infty \frac{X'}{x}\,dx = \frac{\pi}{2}.$$

1915. Prove that the limit, when n tends to ∞, of

$$\sqrt{n}\int_0^\pi \sin^n x\,dx \text{ is } \sqrt{2\pi};$$

and that this is also the limit of $\sqrt{n}\int_{\frac{\pi}{2}-a}^{\frac{\pi}{2}+a} \sin^n x\,dx$, where a has any value between 0 and $\frac{\pi}{2}$ excluding the former. Similarly the limit of

$$\sqrt{n}\int_{\frac{\pi}{2}-a}^{\frac{\pi}{2}+a} \sin^{2n+1} x\,dx \text{ or of } \sqrt{n}\int_{-a}^{a} (\cos x)^{2n}\,dx \text{ is } \sqrt{\pi}.$$

1916. Prove that the limits, when n tends to ∞, of

$$n^{\frac{3}{2}}\int_0^\pi \sin^{2n} x\,(1-\sin x)\,dx,$$

and of $n^{\frac{5}{2}}\int_0^\pi \sin^{2n} x\left(\frac{4n+2}{4n+1}\sin x - 1\right)dx$, are $\frac{1}{4}\sqrt{\pi}$ and $\frac{1}{32}\sqrt{\pi}$ respectively.

[In both these also the only portion of the integral which affects the result is that which arises from values of x differing very little from $\frac{\pi}{2}$.]

1917. The expansions of

$$\sqrt{n}\int_0^\pi \sin^{2n} x\,dx \text{ and } \sqrt{n}\int_0^\pi \sin^{2n+1} x\,dx,$$

when n is very large, are respectively

$$\sqrt{\pi}\left(1 - \frac{1}{8n} + \frac{1}{128n^2} + \frac{5}{1024n^3} + \ldots\right), \quad \sqrt{\pi}\left(1 - \frac{3}{8n} + \frac{25}{128n^2} - \frac{105}{1024n^3} + \ldots\right).$$

1918. Prove that

$$\int_0^1 \frac{(1+x^2)^n}{(1+x)^{n+1}}\,dx = 2\int_0^{\sqrt{2}-1} \frac{(1+x^2)^n}{(1+x)^{n+1}} = \int_1^2 \frac{(2-2x+x^2)^n}{x^{n+1}}\,dx$$

$$= 2\int_1^{\sqrt{2}} \frac{(2-2x+x^2)^n}{x^{n+1}}\,dx = 2\int_1^{\sqrt{2}} \frac{(2-2x^2+x^4)^n}{x^{2n+1}}\,dx\,;$$

and obtain in the same way the equation

$$\int_1^{\sqrt{2}} F\left(\frac{2-2x^2+x^4}{x^2}\right)\frac{dx}{x} = \int_1^{\sqrt{2}} F\left(\frac{2-2x+x^2}{x}\right)\frac{dx}{x}.$$

[The results are easily obtained by putting $x \equiv \tan z$. The more general theorem is

$$\int_1^a F\left(x^2+\frac{a^2}{x^2}\right)\frac{dx}{x} = \int_1^a F\left(x+\frac{a^2}{x}\right)\frac{dx}{x}.]$$

1919. Prove the following definite integrals:

(1) $\displaystyle\int_0^{\frac{\pi}{2}} \sin 2x \log\cot\frac{x}{2}\,dx = 1,\quad \int_0^\infty \frac{x^{2n-1}}{(1+x^2)^{n+1}}\log\frac{1+x^2}{x^2}\,dx = \frac{1}{n^2}\,;$

(2) $\displaystyle\int_0^\infty \frac{x\,dx}{1+x^2}\log\frac{1+x^2}{x^2} = \frac{\pi^2}{12},\quad \int_0^{\frac{\pi}{4}} \tan\left(\frac{\pi}{4}+x\right)\log\cot x\,dx = \frac{3\pi^2}{16}\,;$

(3) $\displaystyle\int_0^\infty \frac{\log x}{a^2+x^2}\,dx = \frac{\pi}{2a}\log a,\quad \int_0^\infty \log x \log\left(1+\frac{a^2}{x^2}\right)dx = \pi a(\log a - 1)\,;$

(4) $\displaystyle\int_{-1}^1 \frac{dx}{\sqrt{(1-2ax+a^2)(1+2bx+b^2)}} = \frac{2}{\sqrt{ab}}\tan^{-1}(\sqrt{ab}),$

$\displaystyle\frac{2}{\sqrt{ab}}\tan^{-1}\sqrt{\frac{b}{a}},\quad\text{or}\quad \frac{2}{\sqrt{ab}}\cot^{-1}(\sqrt{ab})\,;$

(1) $a<1,\ b<1$; (2) $a>1,\ b<1$; (3) $a>1,\ b>1$;

(5) $\displaystyle\int_0^1 \frac{\log x\,dx}{\sqrt{1-x}} = 4(\log 2 - 1),\quad \int_0^1 \frac{\log x\,dx}{(1+x^2)} = -\log 2\,;$

(6) $\displaystyle\int_0^\pi \frac{x\,dx}{1+\cos\alpha\sin x} = \frac{\pi\alpha}{\sin\alpha},\quad \int_0^\pi \frac{x\,dx}{(1+\cos\alpha\sin x)^2} = \pi\frac{(\alpha-\sin\alpha\cos\alpha)}{\sin^3\alpha},\quad \left(\alpha<\frac{\pi}{2}\right);$

(7) $\displaystyle\int_0^\infty x\sin 2rx\,\epsilon^{-x^2}dx = \frac{\sqrt{\pi}}{2}r\epsilon^{-r^2},\quad \int_0^\infty x^2\cos 2rx\,\epsilon^{-x^2}dx = \frac{\sqrt{\pi}}{4}\epsilon^{-r^2}(1-2r^2)\,;$

(8) $\int_0^\pi \frac{x \sin x\, dx}{\sqrt{1-c^2 \sin^2 x}} = \frac{\pi}{2c} \log \frac{1+c}{1-c}, \quad \int_0^{\frac{\pi}{2}} \frac{\log(1-c^2 \sin^2 x)}{\sin x}\, dx$

$= -(\sin^{-1} c)^2, \quad (c^2 < 1);$

(9) $\int_0^{\frac{\pi}{4}} \cos 2x \log \cot x\, dx = \frac{\pi}{4}, \quad \int_0^{\frac{\pi}{2}} \sin x \log \sin x\, dx = \log 2 - 1;$

(10) $\int_0^1 \frac{dx}{1-x^2} \tan^{-1}\left(\frac{1-x^2}{1+x^2} \cot a\right) = \frac{\pi}{4} \log \cot \frac{a}{2},$

$\int_0^1 \frac{dx}{1-x^2} \log\left(\frac{1-2x^2 \cos 2a + x^4}{4 \sin^2 a}\right) = \frac{1}{2}\left\{\left(\log \cot \frac{a}{2}\right)^2 - \frac{\pi^2}{4}\right\};$

(11) $\int_0^\pi \frac{x}{\sin x} \log(1 + \sin a \sin x)\, dx = \frac{\pi}{2} a(\pi - a), \quad a < \frac{\pi}{2};$

(12) $\int_0^\infty \frac{dx}{x} \log \frac{1 + 2x \sin a + x^2}{1 - 2x \sin a + x^2} = 2\pi a, \quad \left(a < \frac{\pi}{2}\right);$

(13) $\int_0^{\frac{\pi}{2}} \left(\frac{x}{\sin x}\right)^2 dx = 2\int_0^{\frac{\pi}{2}} x \cot x\, dx = \pi \log 2;$

(14) $\int_{-\infty}^{\infty} \frac{\sin^{2n} x}{x^2}\, dx = \int_0^\pi \sin^{2n-2} x\, dx = \frac{1.3.5 \ldots (2n-3)}{2.4.6 \ldots 2n-2}\pi;$

(15) $\int_{-\infty}^{\infty} \frac{\sin^{2n+1} x}{x}\, dx = \frac{1.3.5 \ldots 2n-1}{2.4.6 \ldots 2n}\pi;$

(16) $\int_{-\infty}^{\infty} \frac{\cos^n ax - \cos^n bx}{x^2}\, dx = \pi(b-a)\frac{3.5 \ldots (n-1)}{2.4 \ldots (n-2)}$

or $\pi(b-a)\frac{3.5 \ldots n}{2.4 \ldots (n-1)};$

(17) $\int_{-\infty}^{\infty} \frac{A \cos ax + B \cos bx + C \cos cx + \ldots}{x^2}\, dx$

$= -\pi(Aa + Bb + Cc + \ldots);$

(18) $\int_0^\infty \frac{A \cos ax + B \cos bx + C \cos cx + \ldots}{x}\, dx$

$= \log(a^{-A} b^{-B} c^{-C} \ldots)$

(the two last integrals are finite only when $A + B + C + \ldots = 0$);

(19) $\int_{-\infty}^{\infty} \frac{A_1 \cos a_1 x + A_2 \cos a_2 x + \ldots + A_s \cos a_s x}{x^{2n}}\, dx$

(when finite) $= \frac{(-1)^n}{\lfloor 2n-1} \pi \Sigma (A_r a_r^{2n-1});$

(20) $\int_{-\infty}^{\infty} \frac{A_1 \sin a_1 x + A_2 \sin a_2 x + \ldots + A_s \sin a_s x}{x^{2n+1}}\, dx$

(when finite) $= \frac{(-1)^n}{\lfloor 2n} \pi \Sigma (A_r a_r^{2n});$

$$(21)\quad \int_0^\infty \frac{A_1 \cos a_1 x + A_2 \cos a_2 x + \ldots + A_n \cos a_n x}{x^{2n+1}}\, dx$$

$$\text{(when finite)} = \frac{(-1)^{n-1}}{\lfloor 2n} \Sigma \left(A_r a_r^{2n} \log a_r\right);$$

$$(22)\quad \int_0^\infty \frac{A_1 \sin a_1 x + A_2 \sin a_2 x + \ldots + A_n \sin a_n x}{x^{2n}}\, dx$$

$$\text{(when finite)} = \frac{(-1)^{n}}{\lfloor 2n-1} \Sigma \left(A_r a_r^{2n-1} \log a_r\right);$$

$$(23)\quad \int_{-\infty}^\infty \frac{\sin^2 x}{x^2} F(\sin^2 x)\, dx = \int_0^\pi F(\sin^2 x)\, dx;$$

$$(24)\quad \int_{-\infty}^\infty \frac{\log(1 + m \sin^2 x)}{x^2}\, dx = 2\pi\left(\sqrt{1+m} - 1\right), \quad (m < 1);$$

$$(25)\quad \int_0^\infty \frac{dx}{x^2} \log\left(\frac{1 + 2n \cos ax + n^2}{1 + 2n \cos bx + n^2}\right) = \pi (b - a) \frac{n}{1+n}, \quad (n < 1);$$

$$(26)\quad \int_0^\infty \left(\frac{\log x}{x-1}\right)^3 dx = \pi^2, \quad \int_0^\infty \left(\frac{\log x}{x-1}\right)^4 dx = \tfrac{4}{3}\pi^2 + \tfrac{8}{45}\pi^4;$$

$$(27)\quad \int_0^\infty \left(\frac{\log x}{x-1}\right)^6 dx = 2\pi^2 + \tfrac{14}{3}\pi^4 + \tfrac{32}{105}\pi^6;$$

$$(28)\quad \int_0^\infty \left(\frac{\log x}{x-1}\right)^{2n+2} dx$$

$$= 2(n+1)\, \Sigma_1^\infty \frac{(r+1)(r+2)\ldots(r+2n) + (r-1)(r-2)\ldots(r-2n)}{r^{2n+2}};$$

$$(29)\quad \int_0^{\frac{\pi}{2}} \phi(\sin 2x) \sin x\, dx = \int_0^{\frac{\pi}{2}} \phi(\sin 2x) \cos x\, dx$$

$$= \sqrt{2} \int_0^{\frac{\pi}{4}} \phi(\cos 2x) \cos x\, dx;$$

$$(30)\quad \int_0^{\frac{\pi}{2}} (\log \sin x)^2\, dx = \int_0^{\frac{\pi}{2}} (\log \cos x)^2\, dx = \frac{\pi}{2}\left\{\frac{\pi^2}{12} + (\log 2)^2\right\};$$

$$(31)\quad \int_0^{\frac{\pi}{4}} \log \sin x \log \cos x\, dx = \frac{\pi}{4}\left\{(\log 2)^2 - \frac{\pi^2}{24}\right\},$$

$$\int_0^\pi \log(1 - \cos x) \log(1 + \cos x)\, dx = \pi\left\{(\log 2)^2 - \frac{\pi^2}{6}\right\};$$

$$(32)\quad \int_0^{\frac{\pi}{2}} \log \sin x \log \tan x\, dx = \frac{\pi^3}{16};$$

(33) $\int_0^{\frac{\pi}{2}} \frac{\log \sec x}{\sin x}\, dx = \frac{\pi^2}{8}$, $\int_0^{\frac{\pi}{2}} \frac{(\log \sec x)^3}{\sin x}\, dx = \frac{\pi^4}{16}$,

$$\int_0^{\frac{\pi}{2}} \frac{(\log \sec x)^5}{\sin x}\, dx = \frac{\pi^6}{8};$$

(34) $$\int_0^{\frac{\pi}{2}} \frac{(\log \sec x)^{2n-1}}{\sin x}\, dx = \lfloor 2n-1 \left\{\frac{1}{1^{2n}} + \frac{1}{3^{2n}} + \frac{1}{5^{2n}} + \dots\right\}$$
$$= \frac{(2^{2n}-1)\pi^{2n}}{4n} B_n;$$

(35) $$\int_0^\infty \frac{dx}{x} \log \frac{1 + 2n \sin x + n^2}{1 - 2n \sin x + n^2} = 2\pi \tan^{-1} n, \quad (n < 1)$$
$$\int_0^\infty \frac{dx}{x} \log \frac{1 + m \sin x}{1 - m \sin x} = \pi \sin^{-1}(m), \quad (m < 1);$$

(36) $$\int_0^\infty \frac{dx}{x \sin x} \log (1 + n \sin^2 x) = \pi (\sqrt{1+n} - 1), \quad (n < 1);$$

(37) $$\int_0^\infty \frac{dx}{x} \tan^{-1}(m \tan x) = \frac{\pi}{2} \log (1 + m),$$
$$\int_0^\infty \frac{dx}{x} \tan^{-1}(m \sin x) = \frac{\pi}{2} \log (\sqrt{1 + m^2} + m);$$

(38) $$\int_0^{\frac{\pi}{4}} \tan^{-1}\left(\frac{2\sqrt{\tan x}}{1 + \tan x}\right) dx = \frac{\pi^2}{24},$$
$$\int_0^\infty \frac{dx}{x} \tan^{-1}\left(\frac{m \sin x}{1 + m \cos x}\right) = \frac{\pi}{2} \log (1 + m), \quad (m < 1);$$

(39) $$\int_0^1 \frac{dx}{1-x} \log \frac{(1 + zx)^{\frac{1}{x}}}{1 + z} = \tfrac{1}{2} \{\log (1 + z)\}^2;$$

(40) $$\int_0^\infty \frac{\log x}{1 + x^n}\, dx = -\cos \frac{\pi}{n} \left(\frac{\pi}{n \sin \frac{\pi}{n}}\right)^2,$$
$$\int_0^\infty \frac{\log x}{x^n - 1}\, dx = \left(\frac{\pi}{n \sin \frac{\pi}{n}}\right)^2, \quad (n > 1);$$

(41) $$\int_0^\infty \frac{(\log x)^{2r+1}}{1 + x^n}\, dx = -\left(\frac{\pi}{n}\right)^{2r+2} \frac{d^{2r}}{d\theta^{2r}} \left(\frac{\cos \theta}{\sin^2 \theta}\right)_{\theta = \frac{\pi}{n}}, \quad (n > 1);$$

(42) $$\int_0^\infty \frac{(\log x)^{2r+1}}{x^n - 1}\, dx = \left(\frac{\pi}{n}\right)^{2r+2} \frac{d^{2r}}{d\theta^{2r}} \left(\frac{1}{\sin^2 \theta}\right)_{\theta = \frac{\pi}{n}}, \quad (n > 1);$$

(43) $$\int_0^\pi \frac{x}{\sin x} \log (1 + \tan^2 x)\, dx = \int_0^\pi \frac{x}{\sin x} \log \tan \left(\frac{\pi}{4} + \frac{x}{2}\right) dx;$$

(44) $\int_0^\pi \frac{x}{\sin x}\tan^{-1}(c\sin x)\,dx = \frac{\pi^2}{2}\log\left(c+\sqrt{1+c^2}\right);$

(45) $\int_0^{\frac{\pi}{4}} \log(\cot x - 1)\,dx = \frac{\pi}{8}\log 2,$

$$\int_0^\pi x\sin x\sqrt{1+c^2\sin^2 x}\,dx = \frac{\pi}{2}\left(1+\frac{1+c^2}{c}\tan^{-1}c\right);$$

(46) $\int_0^\infty \frac{dx}{x}\log\left\{\frac{1+2n\cos(x-a)+n^2}{1+2n\cos(x+a)+n^2}\right\}$

$$= 2\pi\tan^{-1}\left(\frac{m\sin a}{1+m\cos a}\right);$$

(47) $\int_0^\infty \frac{\epsilon^{-mx}\cos x}{\sqrt{x}}\,dx = \sqrt{\frac{\pi}{2}\frac{\sqrt{1+m^2}+m}{1+m^2}},$

$$\int_0^\infty \frac{\epsilon^{-mx}\sin x}{\sqrt{x}}\,dx = \sqrt{\frac{\pi}{2}\frac{\sqrt{1+m^2}-m}{1+m^2}};$$

(48) $\int_0^{\frac{\pi}{2}} \epsilon^{n\cos x}\cos(x+n\sin x)\,dx = \frac{\sin n}{n};$

(49) $\int_0^1 (1-x^n)^{\frac{1}{n}}\,dx = \int_0^1 (1+x^n)^{-\frac{2}{n}}\,dx = \frac{1}{2}\int_0^\infty (1+x^n)^{-\frac{2}{n}}\,dx$

$$= \frac{1}{2n}\left\{\Gamma\left(\frac{1}{n}\right)\right\}^2 \div \Gamma\left(\frac{2}{n}\right);$$

(50) $\int_0^\infty \frac{x^{m-1}dx}{x^{2n}+2x^n\cos na+1} = \frac{\sin(n-m)a}{\sin na}\,\frac{\pi}{n\sin\frac{m\pi}{n}},$

$$(m<2n,\quad na<\pi);$$

(51) $\int_0^\pi (\sin x)^{n-1}\,dx\int_0^\pi (\sin x)^n\,dx = \frac{2\pi}{n},$

$$\int_0^\pi (\sin x)^n\,dx\int_0^\pi (\sin x)^{-n}\,dx = \frac{2\pi}{n}\tan\frac{n\pi}{2},\quad (n<1).$$

1920. Prove that, if n be a positive integer,

$$\int_0^{\frac{\pi}{2}} (\log\tan x)^{2n}\,dx = \left(\frac{\pi}{2}\right)^{2n+1}\left(\frac{d^{2n}\sec z}{dz^{2n}}\right)_{z=0}.$$

1921. Prove that

$$\epsilon^{h\frac{d^2}{dx^2}}\{x\epsilon^{-kx^2}\} = \frac{x}{(1+4hk)^{\frac{3}{2}}}\,\epsilon^{-\frac{kx^2}{1+4hk}},$$

if $1+4hk$ be positive.

1922. Trace the curve

$$y = \frac{2}{\pi}\int_0^\infty \frac{\sin xz \sin^3 z}{z^3}\,dz.$$

[When x is between $-\infty$ and -2, $y=-1$; $x=-2$ to $x=0$, $y=x+\frac{x^2}{4}$; $x=0$ to $x=2$, $y=x-\frac{x^2}{4}$; $x>2$, $y=1$.]

1923. The limiting value of the infinite series

$$\left(\frac{1}{n}\right)^n + \left(\frac{2}{n}\right)^n + \left(\frac{3}{n}\right)^n + \ldots + \left(\frac{n-1}{n}\right)^n,$$

when n is ∞, is $\frac{1}{\epsilon - 1}$.

1924. Prove that

$$\int_0^\alpha \sqrt{1-e\cos x}\,dx - \int_0^\beta \sqrt{1-e\cos x}\,dx = \frac{2e\sin\alpha}{\sqrt{1+e\cos\alpha}} = \frac{2e\sin\beta}{\sqrt{1-e\cos\beta}},$$

if α, β be angles $<\pi$ such that $\tan\frac{\beta}{2} = \sqrt{\frac{1-e}{1+e}}\tan\frac{\alpha}{2}$.

1925. Prove that

$$2\int_0^\alpha \frac{dx}{(1-\sin\beta\cos x)^{\frac{3}{2}}} - \int_0^\pi \frac{dx}{(1-\sin\beta\cos x)^{\frac{3}{2}}} = \frac{4\sin\frac{\beta}{2}}{\cos^2\beta},$$

if β be any angle between $-\frac{\pi}{2}$ and $\frac{\pi}{2}$, and $\cos\alpha = \tan\frac{\beta}{2}$.

1926. Prove that, if $\phi(c) = \frac{1}{\pi}\int_0^\pi \log(1+c\cos x)\,dx$,

$$2\phi(c) - \phi\left(\frac{c^2}{2-c^2}\right) = \log\left(1-\frac{c^2}{2}\right).$$

1927. By means of the identity

$$\int_0^1 \frac{x^{a-1}\log x}{1-x}\,dx \equiv \int_0^1 (1-x)^{a-1}\log(1-x)\,\frac{dx}{x},$$

prove that

$$\frac{1}{a^2} + \frac{1}{(a+1)^2} + \frac{1}{(a+2)^2} + \ldots \text{to } \infty = \frac{1}{a} + \frac{1}{2a(a+1)} + \frac{|\underline{2}}{3a(a+1)(a+2)} + \ldots \text{to } \infty.$$

1928. Prove that, $p-q$ being a positive whole number,

$$\int_0^\infty \frac{\sin^p x}{x^q}\,dx = \frac{(-1)^{\frac{p-q}{2}}\pi}{|\underline{q-1}\,2^p}\left\{p^{q-1} - p(p-2)^{q-1} + \frac{p(p-1)}{|\underline{2}}(p-4)^{q-1} - \ldots\right.$$

$$\left.\text{to } \frac{p}{2} \text{ or } \frac{p+1}{2} \text{ terms}\right\},$$

$$\text{or} = \frac{(-1)^{\frac{p-q-1}{2}}}{\lfloor q-1 \, 2^{p-1}} \Big\{ p^{q-1} \log p - p\,(p-2)^{q-1} \log (p-2)$$

$$+ \frac{p\,(p-1)}{2} (p-4)^{q-1} \log (p-4) + \dots \text{ to } \frac{p}{2} \text{ or } \frac{p+1}{2} \text{ terms} \Big\},$$

according as $p-q$ is even or odd.

1929. From the identity

$$\int_0^{\frac{\pi}{2}} (a + \sin^2 x)^n \cos x \, dx = \int_0^{\frac{\pi}{2}} (1 + a - \sin^2 x)^n \sin x \, dx,$$

or otherwise, prove that

$$a^n + \frac{n}{3} a^{n-1} + \frac{n\,(n-1)}{2\,.\,5} a^{n-2} - \frac{n\,(n-1)\,(n-2)}{\lfloor 3\,.\,7} a^{n-3} + \dots$$

$$= (a+1)^n - \frac{2n}{3} (a+1)^{n-1} + \frac{2^2 n\,(n-1)}{3\,.\,5} (a+1)^{n-2}$$

$$- \frac{2^3 n\,(n-1)\,(n-2)}{3\,.\,5\,.\,7} (a+1)^{n-3} + \dots :$$

and prove in a similar manner that

$$a^n + na^{n-1} + \frac{3n\,(n-1)}{(\lfloor 2)^2} a^{n-2} + \frac{3\,.\,5n\,(n-1)\,(n-2)}{(\lfloor 3)^2} a^{n-3} + \dots$$

$$= (a+2)^n - n\,(a+2)^{n-1} + \frac{3n\,(n-1)}{(\lfloor 2)^2} (a+2)^{n-2}$$

$$- \frac{3\,.\,5n\,(n-1)\,(n-2)}{(\lfloor 3)^2} (a+2)^{n-3} + \dots.$$

1930. Prove that, c being < 1,

$$\pi \int_0^{\pi} \sin^{-1} (c \sin x)\, dx = \int_0^{\pi} \left\{ \tan^{-1} \left(\frac{2c \cos x}{1 - c^2} \right) \right\}^2 dx.$$

1931. On a straight line of length $a + b + c$ are measured at random two segments of lengths $a + c$, $b + c$ respectively: prove that the mean value of the common segment is $b + c - \frac{b^2}{3a}$, a being $> b$.

1932. A point is taken at random on a given finite straight line of length a: prove that the mean value of the sum of the squares on the two parts of the line is $\frac{2}{3} a^2$, and that the chance of the sum being less than this mean value is $\frac{1}{\sqrt{3}}$.

1933. A triangle is inscribed in a given circle whose radius is a: prove that, if all positions of the angular points be equally probable, the mean value of the perimeter is $\frac{12a}{\pi}$, and that the mean value of the radius of the inscribed circle is $a\left(\frac{12}{\pi^2}-1\right)$.

1934. The perimeter $(2a)$ of a triangle is given and all values of the sides for which the triangle is real are equally probable: prove that the mean value of the radius of the circumscribed circle is five times, and that the mean value of the radius of an escribed circle is seven times the mean value of the radius of the inscribed circle.

[The three mean values are $\frac{4\pi a}{105}$, $\frac{4\pi a}{21}$, $\frac{4\pi a}{15}$.]

1935. The whole perimeter $(2a)$ and one side (c) of a triangle are given, prove that the mean value of its area is $\frac{\pi}{8}c\sqrt{a(a-c)}$; and that the mean value of this mean value, c being equally likely to have any value from 0 to a, is $\frac{\pi}{30}a^2$.

1936. The mean value of the area of all acute-angled triangles inscribed in a given circle of radius a is $\frac{3a^2}{\pi}$, and the mean value of the area of all the obtuse-angled triangles is $\frac{a^2}{\pi}$.

1937. The mean value of the perimeter of all acute-angled triangles inscribed in a given circle of radius a is $\frac{48a}{\pi^2}$, and that of the perimeter of the obtuse-angled triangles is $\frac{16(\pi-1)a}{\pi^2}$.

1938. The mean value of the distance from one of the foci of all points within a given prolate spheroid is $a\frac{3+e^2}{4}$.

1939. The mean value of $\sqrt{xyz}$ where x, y, z are areal co-ordinates of a point within the triangle of reference is $\frac{4\pi}{105}$; and the mean value of $\sqrt{wxyz}$, where w, x, y, z are tetrahedral co-ordinates of a point within the tetrahedron of reference is $\frac{\pi^2}{320}$. Also the mean value of $(wxyz)^{n-1}$ is $6\,(\Gamma(n))^4 \div \Gamma(4n)$.

1940. Prove that the mean value of $\sqrt{x_1x_2x_3\ldots x_n}$ for all positive values of $x_1, x_2, \ldots$ such that $x_1+x_2+\ldots+x_n=1$ is $\Gamma(n)\left\{\Gamma\left(\frac{3}{2}\right)\right\}^n \div \Gamma\left(\frac{3n}{2}\right)$; and, more generally, that of $(x_1x_2\ldots x_n)^{r-1}$, r being positive, is

$$\Gamma(n)\{\Gamma(r)\}^n \div \Gamma(nr).$$

1941. In the equation $x^3 - qx + r = 0$ it is known that q and r both lie between -1 and $+1$; assuming all values between these limits to be equally probable, prove that the chance that all the roots of the equation shall be real is $2 \div 15\sqrt{3}$.

1942. A given finite straight line is divided at random in two points: prove that the chance that the three parts can be sides of an acute-angled triangle is $3 \log 2 - 2$.

1943. A rod is divided at random in two points, and it is an even chance that n times the sum of the squares on the parts is less than the square on the whole line: prove that

$$n(4\pi + 3\sqrt{3}) = 12\pi.$$

1944. On a given finite straight line are taken n points at random prove that the chance that one of the $n+1$ segments will be greater than half the line is $(n+1)\,2^{-n}$.

1945. A straight line is divided at random by two points: prove that the chance that the square on the middle segment shall be less than the rectangle under the other two is $(4\pi - 3\sqrt{3}) \div 9\sqrt{3}$; and the chance that the square on the mean segment of the three shall be less than the rectangle contained by the greatest and least is $\cdot 41841\ldots$

1946. A rod is divided at random in three points; the chance that one of the segments will be greater than half the rod is $\cdot 5$, and the chance that three times the sum of the squares on the segments will be less than the square on the whole is $\pi \div 6\sqrt{3}$. Also the chance that $4n$ times the sum of the squares on the segments will be less than $(n+1)$ times the square on the whole $(n > 3)$ is $\pi \div 2n^{\frac{3}{2}}$.

1947. A given finite straight line is divided at random in (1) four points, (2) n points; the chance that (1) four times, (2) n times the sum of the squares on the segments will be less than the square on the whole line is

$$(1)\quad \frac{3\pi^2}{100\sqrt{5}},\qquad (2)\quad \frac{1}{\sqrt{n+1}}\left\{\frac{\pi}{n(n+1)}\right\}^{\frac{n}{2}} \frac{\Gamma(n+1)}{\Gamma\left(\frac{n}{2}+1\right)}.$$

1948. A given finite straight line of length a is divided at random in two points; the chance that the product of the three segments will exceed $\frac{1}{108}a^3$ is

$$\frac{4}{9}\int_{\frac{4\pi}{9}}^{\frac{\pi}{9}} \sin\frac{x}{2}\sqrt{1 + 2\cos 3x}\,dx.$$

1949. The mean value of the distance between two points taken at random within a circle of radius a is $121a \div 45\pi$; the corresponding mean value for a sphere is $36a \div 35$. The mean distance of a random point within a given sphere from a fixed point, (1) without the sphere, (2) within the sphere, is

$$(1)\quad c+\frac{a^2}{5c},\quad (2)\quad \frac{3a}{4}+\frac{c^2}{2a}-\frac{c^4}{20a^3};$$

c being the distance of the fixed point from the centre of the sphere and a the radius of the sphere.

1950. The mean value of the distance of any point within a sphere of radius a from a point in a concentric shell of radius b is

$$\frac{3}{20}\,\frac{(a+b)(5a^2+7b^2)}{a^2+ab+b^2}.$$

1951. A rod is marked at random in three points; the chance that n times the sum of the squares on the segments will be less than the square on the whole is $\frac{\pi}{2}\left(\frac{4-n}{n}\right)^{\frac{3}{2}}$, or $\frac{\pi}{6\sqrt{3}}\left\{\frac{36}{n}-10-\left(\frac{12}{n}-3\right)^{\frac{3}{2}}\right\}$, according as n lies between 3 and 4 or between 2 and 3.

1952. A point in space is determined by taking at random its distances from three given points A, B, C: prove that the density of distribution at any point will vary directly as the distance from the plane and inversely as the product of the distances from A, B, C.

1953. Points P, Q, R are taken at random on the sides of a triangle ABC; the chance that the area of the triangle PQR will be greater than $(n+\frac{1}{4})$ of the triangle ABC, (n being positive and $< \frac{3}{4}$), is

$$\frac{3-4n}{4}-\frac{12n+1}{4}\log\frac{4}{4n+1}+8n^{\frac{3}{2}}\left(\frac{\pi}{3}-\tan^{-1}\sqrt{2n}\right).$$

1954. A rod is marked in four points at random, A bets B £50 even that no segment exceeds $\frac{7}{16}$ of the whole: prove that A's expectation is 3*s*. 11*d*. nearly.

1955. A given finite straight line is marked at random in three points; the chance that the square on the greatest of the four segments will not exceed the sum of the squares on the other three is

$$12\log 2-\pi-5.$$

1956. From each of n equal straight lines is cut off a piece at random; the chance that the greatest of the pieces cut off exceeds the sum of all the others is $1 : \underline{n-1}$; and the chance that the square on the greatest exceeds the sum of the squares on all the others is

$$\left(\frac{\pi}{4}\right)^{\frac{n-1}{2}} : \Gamma\left(\frac{n+1}{2}\right).$$

1957. A rod AB is marked at random in P, and points Q, R are then taken at random in AP, PB respectively: prove that the chance that the sum of the squares on AQ, RB will exceed the sum of those on QP, PR is ·5; but, when Q, R are first taken at random in AB and P then taken at random in QR, the chance of the same event is

$$\tfrac{5}{6}(3-2\log 2).$$

1958. Three points P, Q, R are taken at random on the perimeter of a given semicircle (including the diameter): prove that the mean value of the area of the triangle PQR is

$$\frac{3(\pi^2 - 8)}{2\pi^3} a^2,$$

a being the radius.

1959. A rod being marked at random in two points, the chance that **twice** the square on the mean segment will exceed the sum of the squares on the greatest and least segments is ·225 nearly.

1960. **The** curve $p(2a - r) = a^2$, p being the perpendicular from the pole on the tangent, consists **of an** oval and a sinuous branch: the oval being a circle, and the sinuous branch **the** curve

$$1 - \cos\theta = \left(1 - \frac{1}{\sqrt{2}}\right)\left(1 - \frac{a}{r}\frac{r+a}{r-a}\right).$$

1961. Trace the curve $r^2 = \dfrac{p^4}{p^2 - a^2}$, the prime radius passing through a point **of** the curve where $r = 2a$: discuss the nature of this point and prove that, if perpendiculars OY, OZ be let fall from the pole on the tangent and normal at **any** point, YZ will touch a fixed circle.

1962. Find the differential equation of a curve such that the foot of the perpendicular from a fixed **point on** the tangent lies **on** a fixed circle : and obtain the general integral **and** singular solution.

[Taking the fixed circle **to** be $x^2 + y^2 = a^2$, **and** the fixed point $(c, 0)$, the differential equation is

$$(y - px)^2 = a^2(1 + p^2) - c^2;$$

which is **of** Clairaut's form.]

1963. Reduce the equation $(x - py)\left(x - \dfrac{y}{p}\right) = c^2$ to Clairaut's form, by putting $x^2 = X$, $y^2 = Y$, **and** deduce the general integral **and singular** solution.

[The general integral is $\dfrac{x^2}{c^2 + \lambda} + \dfrac{y^2}{c^2 - \lambda} = \dfrac{1}{2}$, and the singular solution is $\pm x \pm y = c$.]

1964. Along the normal to a curve at P is measured a constant length PQ; O is a fixed point and the curve is such that the circle described about OPQ has a fixed tangent at O: find the differential equation of the curve, the general integral, and singular solution.

[Taking O for origin, and the fixed tangent at O for axis of x, the differential equation is $x^2 + 2xyp - y^2 = cy\sqrt{1 + p^2}$; the general integral is

$$x^2 + y^2 - 2ax + b^2 = 0, \text{ where } b^2(b^2 + c^2) = a^2 c^2,$$

and the singular solution is $x^2 + y^2 = \pm cy$. If the singular solution be deduced from the general integral, the student should account for the extraneous factors c and x.]

1965. The ordinate and normal from a point P of a curve to the axis of x are PM and PG: find a curve (1) in which PM^2 varies as PG; (2) in which the curvature varies as $PM^4 \div PG^3$, and prove that one species of curve satisfies both conditions.

[The curve (1) is the catenary $\frac{2y}{c} = m\epsilon^{\frac{x}{c}} + m^{-1}\epsilon^{-\frac{x}{c}}$; and the curve (2) is $\frac{2y}{c} = m\epsilon^{\frac{x}{c}} + n\epsilon^{-\frac{x}{c}}$, which coincides with the former when $mn = 1$; or $2y = A\cos\frac{x}{c} + B\sin\frac{x}{c}$.]

1966. Prove that the equation $2xy\frac{d^2y}{dx^2} - x\left(\frac{dy}{dx}\right)^2 + y\frac{dy}{dx} = 0$ is the general equation of a parabola touching the co-ordinate axes; and deduce (1) that, if in a series of such parabolas, the curvature has a given value when the tangent is in a certain given direction, the locus of the points where the tangent has this direction is an hyperbola with asymptotes parallel to the co-ordinate axes and passing through the origin where its tangent is in the given direction and its curvature is four times the given curvature, (2) that if a straight line from the origin meet one of the parabolas at right angles in the point (x, y) the radius of curvature at (x, y) will be

$$\frac{2xy\sqrt{x^2 + y^2 + 2xy\cos\omega}}{(x + y\cos\omega)(y + x\cos\omega)}\sin^2\omega,$$

where ω is the angle between the co-ordinate axes.

1967. Find the general solution of the equation

$$\left\{y\frac{d^2y}{dx^2} + 1 + \left(\frac{dy}{dx}\right)^2\right\}^2 = 4a\frac{d^2y}{dx^2}\left\{x\frac{d^2y}{dx^2} - \frac{dy}{dx}\left(1 + \frac{dy}{dx}^2\right)\right\};$$

and prove that a singular first integral is

$$x + y\frac{dy}{dx} + a\left(\frac{dy}{dx}\right)^2 = 0.$$

[The general solution is $x^2 + y^2 = 2a(\lambda y + \lambda^2 x + \mu)$, and one general first integral is $x + y\frac{dy}{dx} = a\left(\lambda^2 + 2\lambda\frac{dy}{dx}\right)$.]

1968. **Prove that** the equation

$$(x^2 + y^2 - 2xyp)^2 = 4a^2y^2(1 - p^2)$$

can be reduced to Clairaut's form by putting $x^2 - y^2 = 2z$; and obtain the general and singular solutions.

$[(x^2 - y^2 - \lambda x - a^2)^2 = a^2(a^2 - \lambda^2);\ x^2 - y^2 = \pm 2ay.]$

1969. Find the general and singular solutions of the equation

$$(x^3 - y^3)^2 (1 + p^2) = 2 (x^2 + py^2)^3.$$

[Reduce by putting $x^3 + y^3 = Y$, $3xy = X$: the solutions are

$$x^3 + y^3 - 3axy = a^3, \quad x^6 + 6x^3y^3 + y^6 = 0.]$$

1970. Find the general differential equation of a circle touching the parabola $y^2 = 4ax$ and passing through the focus; and deduce that the locus of the extremities of a diameter parallel, (1) to the axis of y, (2) to the axis of x, is

(1) $8y^2 (x + 2a)^3 = 27a (x^2 + y^2)^2$, (2) $(x^2 - y^2 + 4ax)^3 = 27ax (x^2 + y^2)^2$.

1971. The equation $x^2 \left(\frac{dy}{dx}\right)^2 + 2ab \frac{dy}{dx} + y^2 = 0$ has the singular solution $xy = ab$ and the general solution is formed by eliminating θ from the equations

$$x^2 = \lambda ab\, \epsilon^{-\theta} (1 - \sin\theta), \quad y^2 = \lambda^{-1} ab\, \epsilon^{\theta} (1 + \sin\theta).$$

1972. Solve the equation $\frac{x^2}{a} + \frac{y^2}{b} = \frac{a-b}{a+b} \frac{x - y\frac{dy}{dx}}{x + y\frac{dy}{dx}}$; and examine the nature of the solution $\frac{x^2}{a} + \frac{y^2}{b} = 1$.

[The general solution is $x^2 + y^2 = -\frac{2ab}{a+b} \log \left(\frac{\frac{x^2}{a} + \frac{y^2}{b} - 1}{C}\right)$, so that $\frac{x^2}{a} + \frac{y^2}{b} = 1$ is the particular integral corresponding to $C = 0$.]

1973. Find the general solution of the equation $y + x\frac{dy}{dx} = \frac{a}{\frac{dy}{dx}}$; and examine the relation of the curve $y^2 + 4ax = 0$ to the family of curves represented by the equation.

[The general solution is $(y^3 - 12axy - \lambda^3)^2 + (y^2 + 4ax)^3 = 0$, and each curve of the family has a cusp on the limiting curve $y^2 + 4ax = 0$.]

1974. The general solution of the equation $2y = x\frac{dy}{dx} + \frac{3a}{\frac{dy}{dx}}$ is

$$(y^2 - 4ax)(x^2 - 4\lambda y) + 2a\lambda xy = 27a^2\lambda^2;$$

and each such curve has a cusp lying on the curve $y^2 = 3ax$.

1975. The general solution of the equation $x\left(\frac{dy}{dx}\right)^2 - my\frac{dy}{dx} + a = 0$ is found by eliminating p between the equations

$$(2m-1)\, x = ap^{-2} + \lambda p^{\frac{1}{m-1}}, \quad (2m-1)\, y = 2ap^{-1} + \frac{\lambda}{m} p^{\frac{m}{m-1}};$$

except when $2m = 1$, when $p^2x = 2a\log p + \lambda$, $py = 4a\log p + 2a + 2\lambda$.

1976. Find the orthogonal trajectory of the circles

$$x^2 + y^2 - 2\lambda y + a^2 = 0,$$

λ being the parameter.

[$x^2 + y^2 - 2\mu y - a^2 = 0.$]

1977. Find the orthogonal trajectory of the rectangular hyperbolas $x^2 - y^2 - 2\lambda x + a^2 = 0$; and prove that one solution is a conic.

[The general solution is $y(3x^2 + y^2 - 3a^2) = 2\mu^3$, which has an oval lying within the conic $3x^2 + y^2 = 3a^2$, if $\mu^2 < a^2$, an acnode when $\mu^2 = a^2$ and lies altogether without the conic when $\mu^2 > a^2$.]

1978. Prove that the orthogonal trajectory of the family

$$r^n = \lambda^n \cos n\theta \text{ is } r^n = \mu^n \sin n\theta.$$

1979. The orthogonal trajectory of the system of ellipses

$$3x^2 + y^2 + 3a^2 + 2\lambda x = 0 \text{ is } y^3 = \mu(x^2 - y^2 - a^2).$$

1980. Integrate the equations:

(1) $x\dfrac{dy}{dx} = 2y + x^{n+1}y - x^n y\sqrt{y}$,

(2) $\dfrac{d^2y}{dx^2} - \cot x\dfrac{dy}{dx} + y\sin^2 x = 0$,

(3) $\dfrac{d^2x}{dt^2} + 2x + 2y = \dfrac{d^2y}{dt^2} + x + 3y = \cos nt$,

(4) $(1 - x^2)\dfrac{d^2y}{dx^2} - 2(n+1)x\dfrac{dy}{dx} = r(n+1)y$,

(5) $2xyz\dfrac{d^2z}{dxdy} + 2xy\dfrac{dz}{dx}\dfrac{dz}{dy} + 2x\dfrac{dz}{dx} + 2y\dfrac{dz}{dy} + z^2 = 0$,

(6) $\dfrac{d^2u}{dx^2} + \dfrac{d^2u}{dy^2} + \dfrac{d^2u}{dz^2} - \dfrac{d^2u}{dydz} - \dfrac{d^2u}{dzdx} - \dfrac{d^2u}{dxdy} = 0$,

(7) $\dfrac{d^2u}{dx^2} + \dfrac{d^2u}{dy^2} + \dfrac{d^2u}{dz^2} + 2\dfrac{d^2u}{dydz} = 0$,

(8) $\dfrac{d^3u}{dx^3} + \dfrac{d^3u}{dy^3} + \dfrac{d^3u}{dz^3} - 3\dfrac{d^3u}{dxdydz} = 0.$

1981. Having given the equation $x\dfrac{d^3y}{dx^3} + 2\dfrac{d^2y}{dx^2} + x\dfrac{dy}{dx} = 0$; and that, when $x = 0$, $y = 0$, $\dfrac{dy}{dx} = 1$, $\dfrac{d^2y}{dx^2} = 0$; prove that, when x is ∞, $y = \dfrac{\pi}{2}$.

1982. Integrate the equation

$$\frac{dy}{dx}(\sqrt{\cos y}\cos x + \sin y \sin x) = 2\cos y \sec x;$$

and examine the nature of the solution $y = \frac{\pi}{2}$.

[The general solution is $y + \lambda = 2\tan x\sqrt{\cos y}$.]

1983. Integrate the equation

$$\frac{dy}{dx}\cos x \cos(y - x) = \cos y;$$

and examine if the solution $y = 2x + 2r\pi - \frac{\pi}{2}$ is a general solution.

[The general solution is $\lambda \cos x = \sin(x - y)$.]

1984. Solve the equation

$$\left(\frac{d^2y}{dx^2} + y\right)\cot x + 2\left(\frac{dy}{dx} + y\tan x\right) = f(x),$$

by putting $y \equiv z \cos x$.

1985. The general solution of the equation

$$u_x(2x + 1 - u_{x+1}) = x^2$$

is

$$\frac{1}{u_x - x} = C + \Sigma\left(\frac{1}{x}\right).$$

1986. A complete primitive of the equation

$$(u_{x+1} - u_x)^2 = 2(u_{x+1} + u_x) + 3$$

is $u_x + 1 = (x + C)^2$; and another is $u_x + 1 = \{C(-1)^x - \frac{1}{2}\}^2$: also deduce one of these as the indirect solution corresponding to the other.

1987. Solve the equations

(1) $u_{x+1} + 3u_x - 4u_x^3 = 0$,

(2) $u_{x+1} = x(u_x + u_{x-1})$,

(3) $(u_{x+1} - u_x)^2 = u_{x+1} + u_x$,

(4) $2(u_{x+1} - u_x)^3 = (u_{x+1} + 2u_x)(u_x + 2u_{x+1})$.

1988. Prove that the limiting value of

$$2n + 1 - 2\log\frac{(2n+1)^n}{1 \cdot 3 \cdot 5 \ldots (2n-1)},$$

when n is an indefinitely great positive integer, is $\log 2$.

1989. Prove that, if r be a positive integer so depending upon x that $x - r\pi$ always lies between $-\frac{\pi}{2}$ and $\frac{\pi}{2}$, $\int_0^\infty \frac{x - r\pi}{x}\,dx = \frac{\pi}{2}\log 2$.

1990. Solve the equations:

(1) $y\dfrac{dy}{dx}(x^2+y^2+a^2)+x(x^2+y^2-a^2)=0,$

(2) $2y\dfrac{d^2y}{dx^2}-3\left(\dfrac{dy}{dx}\right)^2=4y^2,$

(3) $x^{n+2}\dfrac{d^{n+2}y}{dx^{n+2}}+x^{n+1}\dfrac{d^{n+1}y}{dx^{n+1}}=1,$

(4) $(x^2-yz)\dfrac{dz}{dx}+(y^2-zx)\dfrac{dz}{dy}=z^2-xy,$

(5) $\tan\dfrac{y-z}{2}\dfrac{dz}{dx}+\tan\dfrac{z-x}{2}\dfrac{dz}{dy}=\tan\dfrac{x-y}{2}.$

[(1) $(x^2+y^2+a^2)^2-4a^2x^2=\lambda,$ (2) $y=(A\cos x+B\sin x)^{-2},$

(3) $y=a_0+a_1x+a_2x^2+\ldots+a_nx^n+bx^n\log x+\dfrac{1}{2\,\underline{|n}}x^n(\log x)^2;$

(4) $y-z=(x-y)f(yz+zx+xy),$

(5) $\cos(y+z)+\cos(z+x)+\cos(x+y)$

$=f\{\sin(y+z)+\sin(z+x)+\sin(x+y)\}.$]

1991. Prove the following equation for Bernoulli's numbers:

$$\left(\frac{z}{e^z-1}=1-\frac{z}{2}+B_1\frac{z^2}{\underline{|2}}-B_2\frac{z^4}{\underline{|4}}+\ldots\right)$$

$$\tfrac{1}{2}=nB_1-\frac{n(n-1)(n-2)}{\underline{|3}}B_2+\frac{n(n-1)(n-2)(n-3)(n-4)}{\underline{|5}}B_3-\ldots$$

to $\dfrac{n-1}{2}$ terms, where n is an odd integer. The equation will still be true when n is an even integer if we multiply the last term by $\dfrac{n+1}{n}$, $\dfrac{n}{2}$ being then the number of terms.

1992. Two equal circles have radii $2a$, and the distance between their centres is $4c$, a series of circles is drawn, each touching the previous one of the series and touching the two given circles symmetrically: prove that the radius of the nth of such a series is

$$c\sin^2\alpha\div\sin(n\alpha+\beta)\sin(\overline{n-1}\,\alpha+\beta),$$

where $a=c\cos\alpha$. Deduce from this the result when $c=a$,

$$a\div(n+\lambda)(n+\lambda-1);$$

and the result when $c<a$.

[If $c=2a\div(p+p^{-1})$, the nth radius is $\dfrac{4(a^2-c^2)}{(bp^n-b'p^{-n})(bp^{n-1}-b'p^{-n+1})}$,

where $bb'=c^2$.]

1993. When $\psi(t) = (a + bt) \div (c + et)$, prove that

$$\psi^y(t) = \frac{b+c}{2e\cos\alpha} \frac{b\sin y\alpha - c\sin(y+2)\alpha - 2et\cos\alpha\sin(y+1)\alpha}{b\sin(y-1)\alpha - c\sin(y+1)\alpha - 2et\cos\alpha\sin y\alpha} - \frac{c}{e},$$

where $(b+c)^2 = 4(bc - ae)\cos^2\alpha$; and hence prove that the condition that ψ is a periodic function of the xth order is

$$\alpha x = i\pi, \text{ or } 4ae\cos^2\frac{i\pi}{x} + b^2 + c^2 - 2bc\cos\frac{2i\pi}{x} = 0,$$

i being an integer not a multiple of x.

Find $\psi^y(t)$ when $(b+c)^2 > 4(bc - ae)$, and discuss the special case when

$$(b+c)^2 = 4(bc - ae).$$

SOLID GEOMETRY.

I. *Straight Line and Plane.*

1994. The co-ordinates of four points are $a-b$, $a-c$, $a-d$; $b-c$, $b-d$, $b-a$; $c-d$, $c-a$, $c-b$; and $d-a$, $d-b$, $d-c$, respectively: prove that the straight line, joining the middle points of any two opposite edges of the tetrahedron of which they are the angular points, passes through the origin.

1995. Of the three acute angles which any straight line makes with three rectangular axes, any two are together greater than the third.

1996. The straight line joining the points (a, b, c), (a', b', c') will pass through the origin if $aa' + bb' + cc' = \rho\rho'$; ρ, ρ' being the distances of the points from the origin, and the axes rectangular. Obtain the corresponding equation when the axes are inclined respectively at angles whose cosines are l, m, n.

$$[aa' + bb' + cc' + (bc' + b'c)\,l + (ca' + c'a)\,m + (ab' + a'b)\,n$$
$$= \rho\rho' \sqrt{1 - l^2 - m^2 - n^2 + 2lmn}.]$$

1997. From any point P are drawn PM, PN perpendicular to the planes of zx, zy; O is the origin, and α, β, γ, θ the angles which OP makes with the co-ordinate planes and with the plane OMN: prove that

$$\operatorname{cosec}^2\theta = \operatorname{cosec}^2\alpha + \operatorname{cosec}^2\beta + \operatorname{cosec}^2\gamma.$$

1998. The equations of a straight line are given in the forms

$$(1) \quad \frac{a + mz - ny}{l} = \frac{b + nx - lz}{m} = \frac{c + ly - mx}{n};$$

$$(2) \quad \frac{a + mz - ny}{L} = \frac{b + nx - lz}{M} = \frac{c + ly - mx}{N};$$

obtain each in the standard form

$$\frac{x - x_0}{\lambda} = \frac{y - y_0}{\mu} = \frac{z - z_0}{\mu}.$$

$$[(1)\quad \frac{x-\dfrac{mc-nb}{l^2+m^2+n^2}}{l}=\frac{y-\dfrac{na-lc}{l^2+m^2+n^2}}{m}=\frac{z-\dfrac{lb-ma}{l^2+m^2+n^2}}{n};$$

$$(2)\quad \frac{x-\dfrac{Mc-Nb}{Ll+Mm+Nn}}{l}=\frac{y-\dfrac{Na-Lc}{Ll+Mm+Nn}}{m}=\frac{z-\dfrac{Lb-Ma}{Ll+Mm+Nn}}{n}.]$$

1999. A straight line moves parallel to a fixed plane and intersects two fixed straight lines (not in one plane): prove that the locus of a point which divides the intercepted segment in a given ratio is a straight line.

2000. Determine what straight line is represented by the equations

$$(1)\quad \frac{a+mz-ny}{m-n}=\frac{b+nx-lz}{n-l}=\frac{c+ly-mx}{l-m};$$

$$(2)\quad \frac{a+mz-ny}{mc'-nb'}=\frac{b+nx-lz}{na'-lc'}=\frac{c+ly-mx}{lb'-ma'}.$$

[(1) The straight line at infinity in the plane

$$x(m-n)+y(n-l)+z(l-m)=0;$$

unless $la+mb+nc=0$, in which exceptional case the line is indeterminate, and the locus of the equations is the plane

$$x(m-n)+y(n-l)+z(l-m)=a+b+c;$$

(2) the straight line at infinity in the plane

$$x(mc'-nb')+y(na'-lc')+z(lb'-ma')=0;$$

unless $la+mb+nc=0$, when the locus of the equations is the plane

$$x(mc'-nb')+y(na'-lc')+z(lb'-ma')=aa'+bb'+cc'.]$$

2001. The two straight lines

$$x+y+z=0,\quad \frac{yz}{b-c}+\frac{zx}{c-a}+\frac{xy}{a-b}=0,$$

are inclined to each other at an angle $\dfrac{\pi}{3}$.

2002. The cosine of the angle between the two straight lines determined by the equations

$$lx+my+nz=0,\quad ax^2+by^2+cz^2=0,$$

is
$$\frac{l^2(b+c)+m^2(c+a)+n^2(a+b)}{\sqrt{l^4(b-c)^2+\ldots+\ldots+2m^2n^2(a-b)(a-c)+\ldots+\ldots}}.$$

2003. A straight line moves parallel to the plane $y=z$ and intersects the curves

$$(1)\ \ y=0,\ z^2=cx;\quad (2)\ \ z=0,\ y^2=bx:$$

prove that the locus of its trace on the plane of yz is two straight lines.

[The locus of the moving straight line is $x=(y-z)\left(\dfrac{y}{b}-\dfrac{z}{c}\right)$.]

2004. The direction cosines of a number of fixed straight lines, referred to any system of rectangular axes, are (l_1, m_1, n_1), (l_2, m_2, n_2), &c.: prove that, if $\Sigma(l^2) = \Sigma(m^2) = \Sigma(n^2)$, and $\Sigma(mn) = \Sigma(nl) = \Sigma(lm) = 0$, when referred to one system of axes, the same equations will be true for any other system of rectangular axes. Also prove that, if these conditions be satisfied and a fixed plane be drawn perpendicular to each straight line, the locus of a point which moves so that the sum of the squares of its distances from the planes is constant will be a sphere having a fixed centre O which is the centre of inertia of equal particles at the feet of the perpendiculars drawn from O, and that the centre of inertia of equal particles at the feet of the perpendiculars drawn from any other point P lies on OP and divides OP in the ratio 2 : 1.

2005. A straight line always intersects at right angles the straight line $x + y = z = 0$, and also intersects the curve $y = 0$, $x^2 = az$: prove that the equation of its locus is

$$x^2 - y^2 = az.$$

2006. The equations

$$\frac{ax + hy + gz}{x} = \frac{hx + by + fz}{y} = \frac{gx + fy + cz}{z}$$

represent in general three straight lines, two and two at right angles to each other; but, if $a - \frac{gh}{f} = b - \frac{hf}{g} = c - \frac{fg}{h}$, they will represent a plane and a straight line normal to that plane.

2007. The two straight lines

$$\frac{x \pm a}{0} = \frac{\pm y}{\cos\alpha} = \frac{z}{\sin\alpha},$$

meet the axis of x in O, O'; and points P, P' are taken on the two respectively such that

(1) $OP = k \,.\, O'P'$; (2) $OP \,.\, O'P' = c^2$; (3) $OP + O'P' = 2c$:

prove that the equation of the locus of PP' is

(1) $(x + a)(y\sin\alpha + z\cos\alpha) = k(x - a)(y\sin\alpha - z\cos\alpha)$;

(2) $\dfrac{x^2}{a^2} - \dfrac{y^2}{c^2\cos^2\alpha} + \dfrac{z^2}{c^2\sin^2\alpha} = 1$;

(3) $\dfrac{xy}{\cos\alpha} - \dfrac{az}{\sin\alpha} = \dfrac{c}{a}(x^2 - a^2)$;

the points being taken on the same side of the plane xy.

[Denoting OP, $O'P'$ by 2λ, 2μ, the equations of PP' may be written

$$-\frac{x}{a} = \frac{y - (\lambda - \mu)\cos\alpha}{(\lambda + \mu)\cos\alpha} = \frac{z - (\lambda + \mu)\sin\alpha}{(\lambda - \mu)\sin\alpha},$$

so that, when any relation is given between λ, μ, the locus may be found immediately.]

2008. A triangle is projected orthogonally on each of three planes mutually at right angles: prove that the algebraical sum of the tetrahedrons which have these projections for bases and a common vertex in the plane of the triangle is equal to the tetrahedron which has the triangle for base and the common point of the planes for vertex.

[This follows at once from the equation $x\cos\alpha + y\cos\beta + z\cos\gamma = p$ on multiplying both members by the area of the triangle.]

2009. A plane is drawn through the straight line $\frac{x}{l} = \frac{y}{m} = \frac{z}{n}$: prove that the two other straight lines in which it meets the surface

$$(b-c)\,yz\,(mz-ny) + (c-a)\,zx\,(nx-lz) + (a-b)\,xy\,(ly-mx) = 0$$

are at right angles to each other.

2010. The direction cosines of three straight lines, which are two and two at right angles to each other, are (l_1, m_1, n_1), (l_2, m_2, n_2), (l_3, m_3, n_3), and

$$am_1n_1 + bn_1l_1 + cl_1m_1 = am_2n_2 + bn_2l_2 + cl_2m_2 = 0;$$

prove that $\quad am_3n_3 + bn_3l_3 + cl_3m_3 = 0$; and $\dfrac{a}{l_1l_2l_3} = \dfrac{b}{m_1m_2m_3} = \dfrac{c}{n_1n_2n_3}$.

2011. The equations of the two straight lines bisecting the angles between the two given by the equations

$$lx + my + nz = 0, \qquad ax^2 + by^2 + cz^2 = 0,$$

may be written

$$lx + my + nz = 0,\ l(b-c)\,yz + m(c-a)\,zx + n(a-b)\,xy = 0.$$

2012. The straight lines bisecting the angles between the two given by the equations

$$lx + my + nz = 0,\ ax^2 + by^2 + cz^2 + 2fyz + 2gzx + 2hxy = 0,$$

lie on the cone

$$x^2(nh-mg) + \ldots + \ldots + yz\,\{mh - ng + l(b-c)\} + \ldots + \ldots = 0.$$

2013. The lengths of two of the straight lines joining the middle points of opposite edges of a tetrahedron are x, y, ω is the angle between them, and a, a' the lengths of those edges of the tetrahedron which are not met by either x or y: prove that

$$4xy\cos\omega = a^2 \sim a'^2.$$

2014. The lengths of the three pairs of opposite edges of a tetrahedron are a, a'; b, b'; c, c': prove that, if θ be the acute angle between the directions of a and a',

$$2aa'\cos\theta = (b^2 + b'^2) \sim (c^2 + c'^2).$$

2015. The locus of a straight line which moves so as always to intersect the three fixed straight lines,

$$y = m(b-a),\ z = n(c-a);\ z = n(c-b),\ x = l(a-b);$$
$$x = l(a-c),\ y = m(b-c);$$

is

$$lyz(b-c) + mzx(c-a) + nxy(a-b) - mnx(b-c)^2 - \dots - \dots$$
$$= 2lmn(b-c)(c-a)(a-b):$$

and every such straight line also intersects the fixed line

$$\frac{a(x-al)}{l} = \frac{b(y-bm)}{m} = \frac{c(z-cn)}{n}.$$

2016. The straight line joining the centres of the two spheres, which touch the faces of the tetrahedron $ABCD$ opposite to A, B respectively and the other faces produced, will intersect the edges CD, AB (produced) in points P, Q respectively such that

$$CP : PD = \Delta ACB : \Delta ADB, \text{ and } AQ : BQ = \Delta CAD : \Delta CBD.$$

2017. On three straight lines meeting in a point O are taken points A, a; B, b; C, c respectively: prove that the intersections of the planes ABC, abc; aBC, Abc; AbC, aBc; and ABc, abC; all lie on one plane which divides each of the three segments harmonically to O.

2018. Through any one point are drawn three straight lines each intersecting two opposite edges of a tetrahedron $ABCD$; and a, f; b, g; c, h are the points where these straight lines meet the edges BC, AD; CA, BD; AB, CD: prove that

$$Ba \,.\, Ch \,.\, Dg = Bg \,.\, Ca \,.\, Dh,$$
$$Cb \,.\, Af \,.\, Dh = Ch \,.\, Ab \,.\, Df,$$
$$Ac \,.\, Bg \,.\, Df = Af \,.\, Bc \,.\, Dg,$$
$$Ab \,.\, Bc \,.\, Ca = Ac \,.\, Ba \,.\, Cb.$$

2019. Any point O is joined to the angular points of a tetrahedron $ABCD$, and the joining lines meet the opposite faces in a, b, c, d: prove that

$$\frac{Oa}{Aa} + \frac{Ob}{Bb} + \frac{Oc}{Cc} + \frac{Od}{Dd} = 1,$$

regard being had to the signs of the segments. Hence prove that the reciprocals of the radii of the eight spheres which can be drawn to touch the faces of a tetrahedron are the eight positive values of the expression

$$\pm\frac{1}{p_1} \pm \frac{1}{p_2} \pm \frac{1}{p_3} \pm \frac{1}{p_4};$$

p_1, p_2, p_3, p_4 being the perpendiculars from the corners on the opposite faces.

2020. The three diagonals of an octahedron intersect each other in one point, at right angles two and two, and perpendiculars are let fall from this point on the faces: prove that the feet of these perpendiculars lie on a sphere and will be corners of a hexahedron such that the perpendiculars on its faces each from the corresponding corners of the octahedron will all meet in a point.

[The faces of the octahedron will all touch a prolate conicoid of revolution of which the points of concourse are foci.]

2021. The areas of the faces of a tetrahedron $ABCD$ are denoted by A, B, C, D and the cosines of the dihedral angles

$$\hat{BC},\ \hat{CA},\ \hat{AB},\ \hat{DA},\ \hat{DB},\ \hat{DC}$$

respectively by a, b, c, f, g, h: prove that

$$\frac{A^2}{1-f^2-b^2-c^2-2fbc}=\frac{B^2}{1-a^2-g^2-c^2-2agc}=\frac{C^2}{1-a^2-b^2-h^2-2abh}$$

$$=\frac{D^2}{1-f^2-g^2-h^2-2fgh};$$

that $\qquad A^2=B^2+C^2+D^2-2BCa-2CDh-2DBg$, &c.; and that

$$\frac{A}{\sin\alpha}=\frac{B}{\sin\beta}=\frac{C}{\sin\gamma}=\frac{D}{\sin\delta};$$

where α, β, γ, δ are determined by equations of the type

$$\sin^2\alpha=1-\cos^2 BAC-\cos^2 CAD-\cos^2 DAB+2\cos BAC\cos CAD\cos DAB.$$

Also, if l, m, n, r be any real quantities, prove that

$$l^2+m^2+n^2+r^2>2mnf+2nlg+2lmh+2lra+2mrb+2nrc,$$

except when
$$\frac{l}{A}=\frac{m}{B}=\frac{n}{C}=\frac{r}{D}.$$

2022. Three straight lines are drawn, two and two at right angles, through a given point, and two of them lie respectively in two fixed planes: the locus of the third is a quadric cone with its circular sections parallel to the fixed planes.

2023. A point O is taken within a tetrahedron $ABCD$ so as to be the centre of inertia of four equal particles at the feet of the perpendiculars let fall from O on the faces: prove that the distances of O from the several faces are proportional respectively to the faces.

[If x, y, z, w be the distances of O from the faces A, B, C, D, the property stated gives us the equations

$$4x=(x+yh)+(x+zg)+(x+wa),\ \text{or}\ x=yh+zg+wa,\ \&c.$$

and, by projecting B, C, D on A, we have $A=Bh+Cg+Da$.]

2024. The point O is such that the sum of its distances from four fixed points A, B, C, D is the least possible: prove that any two opposite edges of the tetrahedron $ABCD$ subtend equal angles at O; and that, if AOA', BOB', COC', DOD' be drawn to meet the faces, the harmonic mean between AO, OA' will be one half the harmonic mean between AO, BO, CO, DO.

2025. The equation of a cone of revolution which can be drawn to touch the co-ordinate planes is

$$(lx)^{\frac{1}{2}} + (my)^{\frac{1}{2}} + (nz)^{\frac{1}{2}} = 0,$$

the ratios $l : m : n$ being given by the equations

$$\frac{m^2 + n^2 + 2mn\cos\alpha}{\sin^2\alpha} = \frac{n^2 + l^2 + 2nl\cos\beta}{\sin^2\beta} = \frac{l^2 + m^2 + 2lm\cos\gamma}{\sin^2\gamma};$$

where α, β, γ are the angles of inclination of the co-ordinate axes.

[For the solutions of these equations, see question 456.]

2026. The equations of the axes of the four cones of revolution which can be drawn to touch the co-ordinate planes are

$$\frac{x^2}{\sin^2\alpha} = \frac{y^2}{\sin^2\beta} = \frac{z^2}{\sin^2\gamma}.$$

2027. The inscribed sphere of a tetrahedron $ABCD$ touches the faces in A', B', C', D': prove that AA', BB', CC', DD' will meet in a point, if

$$\cos\frac{a}{2}\cos\frac{\alpha}{2} = \cos\frac{b}{2}\cos\frac{\beta}{2} = \cos\frac{c}{2}\cos\frac{\gamma}{2};$$

where a, α; b, β; c, γ are pairs of dihedral angles at opposite edges.

[For a sphere touching the face A and also the faces B, C, D produced; $\pi - a$, $\pi - \beta$, $\pi - \gamma$ must be written for a, β, γ; and for the sphere in the compartment vertically opposite the dihedral angle BC, $\pi - b$, $\pi - c$, $\pi - \beta$, $\pi - \gamma$ must be written for b, c, β, γ.]

2028. There can in general be drawn two quadric cones containing a given conic and three given points not in the plane of the conic.

[The general equation of a conicoid satisfying the conditions may be written

$$\frac{x^2}{a^2} + \frac{y^2}{b^2} + \frac{z^2}{c^2} - 1 = \lambda\left(p\frac{x}{a} + q\frac{y}{b} + r\frac{z}{c} - 1\right)\left(p'\frac{x}{a} + q'\frac{y}{b} + r'\frac{z}{c} - 1\right),$$

where λ is the only undetermined quantity.]

2029. The equations of the axes of the four cones of revolution which contain the co-ordinate axes are

$$\frac{x}{a(b+c-a)} = \frac{y}{b(c+a-b)} = \frac{z}{c(a+b-c)},$$

where
$$\frac{a}{\cos\alpha \pm 1} = \frac{b}{\cos\beta \pm 1} = \frac{c}{\cos\gamma \pm 1};$$

an odd number of negative signs being taken in the ambiguities, and α, β, γ being the angles of inclination of the co-ordinate axes.

2030. A point O is taken such that the three straight lines drawn through it, each intersecting two opposite edges of the tetrahedron $ABCD$, are two and two at right angles, and $\alpha, \beta, \gamma, \delta$ denote the perpendiculars let fall from O on the faces of the tetrahedron: prove that

$$\frac{1}{\beta^2} + \frac{1}{\gamma^2} + \frac{2\cos \widehat{AD}}{\beta\gamma} = \frac{1}{\alpha^2} + \frac{1}{\delta^2} + \frac{2\cos \widehat{BC}}{\alpha\delta}, \text{ \&c.}$$

and that

$$\frac{bc}{Ob \,.\, Oc} = \frac{ad}{Oa \,.\, Od}, \text{ \&c.}$$

II. *Linear Transformations. General Equation of the Second Degree.*

[The following simple method of obtaining the conditions for a surface of revolution is worthy of notice.

When the expression $ax^2 + by^2 + cz^2 + 2fyz + 2gzx + 2hxy$ is transformed into $AX^2 + BY^2 + CZ^2$, we obtain the coefficients A, B, C from the equivalence of the conditions that

$$\lambda(x^2 + y^2 + z^2) - ax^2 - \ldots$$

and

$$\lambda(X^2 + Y^2 + Z^2) - AX^2 - BY^2 - CZ^2$$

may break up into (real or impossible) linear factors: which is the case when $\lambda = A, B$, or C.

But, should two of the three coincide as $B = C$, then when $\lambda = B$ the corresponding factors coincide, or either expression must be a complete square. The conditions that the former expression may be a complete square when $\lambda = B$ give us

$$(B - a) f = -gh, \text{ \&c.},$$

or

$$B = a - \frac{gh}{f} = b - \frac{hf}{g} = c - \frac{fg}{h};$$

provided f, g, h be all finite.

Should we have $f = 0$, then gh must $= 0$; suppose then f and $g = 0$, then $B = c$, and we must have $(c - a)x^2 + (c - b)y^2 - 2hxy$ a square, whence

$$h^2 = (c - a)(c - b).$$

In the case of oblique axes, inclined two and two at angles α, β, γ, we must have

$$\lambda(x^2 + y^2 + z^2 + 2yz\cos\alpha + 2zx\cos\beta + 2xy\cos\gamma) - ax^2 - \ldots - 2fyz - \ldots$$

a complete square.

It follows that the three equations

$$(\lambda - a)(\lambda \cos \alpha - f) = (\lambda \cos \beta - g)(\lambda \cos \gamma - h),$$
$$(\lambda - b)(\lambda \cos \beta - g) = (\lambda \cos \gamma - h)(\lambda \cos \alpha - f),$$
$$(\lambda - c)(\lambda \cos \gamma - h) = (\lambda \cos \alpha - f)(\lambda \cos \beta - g),$$

must be simultaneously true; and the two necessary conditions may be found by eliminating λ.]

2031. Determine the nature of the curve traced by the point

$$x = a \cos\left(\theta + \frac{\pi}{3}\right), \quad y = a \cos \theta, \quad z = a \cos\left(\theta - \frac{\pi}{3}\right).$$

[A circle of radius $\sqrt{\frac{3}{2}}a$.]

2032. In two systems of rectangular co-ordinate axes, $\theta_1, \theta_2, \theta_3$ are the angles made by the axes of x', y', z' with the axis of z, and ϕ_1, ϕ_2, ϕ_3 the angles which the planes of zx', zy', zz' make with that of zx: prove that

$$\tan^2 \theta_1 + \frac{\cos(\phi_2 - \phi_3)}{\cos(\phi_1 - \phi_2)\cos(\phi_1 - \phi_3)} = 0,$$

with two similar equations.

2033. By direct transformation of co-ordinates, prove that the equation

$$x^2 + y^2 + z^2 + yz + zx + xy = a^2$$

represents an ellipsoid of revolution whose polar axis is one half of its equatoreal, and the equations of whose polar axis are $x = y = z$.

2034. Prove that the surface whose equation, referred to axes inclined each to each at an angle of $\frac{\pi}{3}$, is $yz + zx + xy + a^2 = 0$, is cut by the plane $x + y + z = 0$ in a circle whose radius is a and by the plane $x + y + z = 12a$ in a circle whose radius is $7a$.

2035. A quadric cone is described which touches the co-ordinate planes (rectangular): prove that an infinite number of systems of three planes, two and two at right angles, can be drawn to touch it; and that, if (l_1, m_1, n_1), (l_2, m_2, n_2), (l_3, m_3, n_3) be the direction cosines of any such system, the equation of the cone will be

$$(l_1 l_2 l_3 x)^{\frac{1}{2}} + (m_1 m_2 m_3 y)^{\frac{1}{2}} + (n_1 n_2 n_3 z)^{\frac{1}{2}} = 0.$$

2036. A point is taken and also the common point of its polar planes with respect to three given spheres: prove that the sphere in which these two points are ends of a diameter will contain a fixed circle and will cut each of the given spheres orthogonally.

2037. In the expression

$$ax^2 + by^2 + cz^2 + 2fyz + 2gzx + 2hxy + 2Ax + 2By + 2Cz + D,$$

prove that

$$A^2 (b + c) + \ldots + \ldots - fBC - gCA - hAB,$$

and

$$A^2 (bc - f^2) + \ldots + \ldots + 2BC (gh - af) + \ldots + \ldots$$

are invariants for all systems of rectangular co-ordinates having the same origin.

2038. Prove also that the coefficients in the following equation in λ

$$\begin{vmatrix} \lambda + a, & \lambda \cos \gamma + h, & \lambda \cos \beta + g, & A \\ \lambda \cos \gamma + h, & \lambda + b, & \lambda \cos \alpha + f, & B \\ \lambda \cos \beta + g, & \lambda \cos \alpha + f, & \lambda + c, & C \\ A, & B, & C, & D \end{vmatrix} = 0;$$

α, β, γ being the angles between the co-ordinate axes, are invariants for all systems of co-ordinates having the same origin.

2039. Assuming the formulæ for transforming from a system of co-ordinate axes inclined at angles α, β, γ to another inclined at angles α', β', γ' to be

$$x = l_1 X + m_1 Y + n_1 Z, \quad y = l_2 X + m_2 Y + n_2 Z, \quad z = l_3 X + m_3 Y + n_3 Z,$$

prove that

$$1 = l_1^2 + l_2^2 + l_3^2 + 2l_2 l_3 \cos \alpha + 2l_3 l_1 \cos \beta + 2l_1 l_2 \cos \gamma,$$

with similar equations in m and n; and that

$$\cos \alpha' = m_1 n_1 + m_2 n_2 + m_3 n_3 + (m_2 n_3 + m_3 n_2) \cos \alpha + \ldots + \ldots,$$

with similar equations in n, l, and l, m.

2040. Prove that, if $ax^2 + by^2 + cz^2$ become $AX^2 + BY^2 + CZ^2$ by any transformation of co-ordinates, the positive and negative coefficients will be in like number in the two expressions.

2041. The homogeneous equation

$$ax^2 + by^2 + cz^2 + 2fyz + 2gzx + 2hxy = 0$$

will represent a cone of revolution if

$$\frac{gh}{f} + \frac{g^2 - h^2}{b - c} = \frac{hf}{g} + \frac{h^2 - f^2}{c - a} = 0.$$

[These are of course equivalent to

$$a - \frac{gh}{f} = b - \frac{hf}{g} = c - \frac{fg}{h}.]$$

2042. The surface whose equation, referred to axes inclined at angles α, β, γ, is $ax^2+by^2+cz^2=1$, will be one of revolution, if

$$\frac{a\cos\alpha}{\cos\alpha-\cos\beta\cos\gamma}=\frac{b\cos\beta}{\cos\beta-\cos\gamma\cos\alpha}=\frac{c\cos\gamma}{\cos\gamma-\cos\alpha\cos\beta};$$

and the corresponding conditions for the surface $ayz+bzx+cxy=1$ are

$$\frac{a}{1\pm\cos\alpha}=\frac{b}{1\pm\cos\beta}=\frac{c}{1\pm\cos\gamma};$$

one, or three, of the ambiguities being taken negative.

2043. The equation

$$ax^2+by^2+cz^2+2fyz+2gzx+2hxy+2Ax+2By+2Cz+D=0$$

will in general represent a paraboloid of revolution, if

$$agh+f(g^2+h^2)=bhf+g(h^2+f^2)=cfg+h(f^2+g^2)=0;$$

and a cylinder of revolution if, in addition to these conditions,

$$Agh+Bhf+Cfg=0.$$

2044. The equation of a given hyperboloid of one sheet whose equation referred to its axes is

$$\frac{x^2}{a^2}+\frac{y^2}{b^2}-\frac{z^2}{c^2}=1$$

can be obtained in the form $x^2+y^2-z^2=d^2$ in an infinite number of ways, provided that a^2-c^2, b^2-c^2 are not both negative; and the new axes of x and y lie on the cone

$$\frac{x^2}{a^2}(b^2-c^2)+\frac{y^2}{b^2}(a^2-c^2)-\frac{z^2}{c^2}(a^2+b^2)=0,$$

and the new axis of z on the cone

$$\frac{x^2}{a^2}(b^2-c^2)+\frac{y^2}{b^2}(a^2-c^2)-\frac{z^2}{c^2}(a^2+b^2)+2(x^2+y^2+z^2)=0.$$

2045. The equation of a given hyperboloid may be obtained in the form

$$ayz+bzx+cxy=1$$

in an infinite number of ways; and, if α, β, γ be the angles between the co-ordinate axes in any such case, the expression

$$\frac{abc}{1-\cos^2\alpha-\cos^2\beta-\cos^2\gamma+2\cos\alpha\cos\beta\cos\gamma}$$

will be constant.

2046. Prove that the only conoid of the second degree is a hyperbolic paraboloid; and that it will be a right conoid if the two principal sections be equal parabolas.

[The equation of a conoid must be reducible to the form $z=f\left(\frac{y}{x}\right)$, and this will be of the second degree only when

$$f(m)=(A+Bm)\div(A'+B'm).]$$

2047. A cone is described having a plane section of a given sphere for base and vertex at a point O on the sphere; the subcontrary sections are parallel to the tangent plane at O.

2048. A cone whose vertex is the origin and base a plane section of the surface $ax^2 + by^2 + cz^2 = 1$ is a cone of revolution: prove that the plane of the base must touch one of the cylinders

$$(b-a)y^2 + (c-a)z^2 = 1,\quad (c-b)z^2 + (a-b)x^2 = 1,\quad (a-c)x^2 + (b-c)y^2 = 1.$$

2049. A cone is described whose base is a given conic and one of whose axes passes through a fixed point in the plane of the conic: prove that the locus of the vertex is a circle.

2050. The locus of the feet of the perpendiculars let fall from a fixed point on the tangent planes to the cone $ax^2 + by^2 + cz^2 = 0$ is a plane curve: prove that it must be a circle, and that the point must lie on one of the three systems of straight lines

$$x = 0,\quad b(c-a)y^2 = c(a-b)z^2, \text{ \&c.}$$

[One only of the three systems is real.]

2051. Prove also that, when the point lies on one of these straight lines, the plane of the circle is perpendicular to the other; and that a plane section of the cone perpendicular to one of the straight lines will have a focus where it meets that straight line, and the excentricity will be equal to $\frac{1}{a}\sqrt{(b-a)(c-a)}$.

2052. A plane cuts the cone $ayz + bzx + cxy = 0$ in two straight lines at right angles to each other: prove that the normal to the plane at the origin also lies on the cone.

2053. The centre of the surface

$$a(x^2 + 2yz) + b(y^2 + 2zx) + c(z^2 + 2xy) - 2Ax - 2By - 2Cz + 1 = 0$$

is (X, Y, Z): prove that

$$A^3 + B^3 + C^3 - 3ABC = (a^3 + b^3 + c^3 - 3abc)(X^3 + Y^3 + Z^3 - 3XYZ);$$

and that the surface will be a cylinder whose principal sections are rectangular hyperbolas, if $a + b + c = 0$, $A + B + C = 0$.

[In this case the axis of the cylinder will be

$$x - \frac{Aa}{a^2+b^2+c^2} = y - \frac{Bb}{a^2+b^2+c^2} = z - \frac{Cc}{a^2+b^2+c^2}.]$$

2054. The radius r of the central circular sections of the surface $ayz + bzx + cxy = 1$ is given by the equation

$$abcr^6 + (a^2 + b^2 + c^2)r^4 = 4;$$

and the direction cosines $(l : m : n)$ of the sections by the equations

$$\frac{l}{a}(m^2 + n^2) = \frac{m}{b}(n^2 + l^2) = \frac{n}{c}(l^2 + m^2) = -lmnr^2.$$

2055. The semi-axes of a central section of the surface

$$ayz + bzx + cxy + abc = 0,$$

made by a plane whose direction cosines are l, m, n, are given by the equation

$$r^4 (2bcmn + \ldots - a^2l^2 - \ldots) - 4abcr^2 (amn + \ldots) + 4a^2b^2c^2 = 0.$$

2056. The section of the surface $yz + zx + xy = a^2$ by the plane $lx + my + nz = p$ will be a parabola if $l^{\frac{1}{2}} + m^{\frac{1}{2}} + n^{\frac{1}{2}} = 0$; and that of the surface $x^2 + y^2 + z^2 - 2yz - 2zx - 2xy = a^2$ will be a parabola if

$$mn + nl + lm = 0.$$

2057. Prove that the section of the surface $u = 0$ by the plane $lx + my + nz = 0$ will be a rectangular hyperbola, if

$$l^2 (b + c) + m^2 (c + a) + n^2 (a + b) = 2mnf + 2nlg + 2lmh;$$

and a parabola, if

$$l^2 (bc - f^2) + \ldots + \ldots + 2mn (gh - af) + \ldots + \ldots = 0;$$

and explain why this last equation becomes identical when

$$gh = af, \ hf = bg, \ fg = ch.$$

[The surface when these conditions are satisfied is a parabolic cylinder, and every plane section will obviously be a parabola, reckoning two parallel straight lines as a limiting case.]

2058. Prove that, when $bg = hf$ and $ch = fg$, the equation

$$u \equiv ax^2 + by^2 + cz^2 + 2fyz + 2gzx + 2hxy + 2Ax + 2By + 2Cz + D = 0$$

represents in general a paraboloid, the direction cosines of whose axis are as $(0 : g : -h)$.

2059. Prove that the tangent lines drawn from the origin to the surface $u = 0$ lie on the cone

$$Du - (Ax + By + Cz + D)^2 = 0;$$

and investigate the condition that the surface u may be a cone from the consideration that this locus will then become two planes.

2060. The generators drawn through the point (X, Y, Z) of the surface $ayz + bzx + cxy + abc = 0$ will be at right angles, if

$$X^2 + Y^2 + Z^2 = a^2 + b^2 + c^2.$$

2061. The generators of the surface $u = 0$ drawn through a point (X, Y, Z) will be at right angles, if

$$\left(\frac{dU}{dX}\right)^2 \left(\frac{d^2U}{dY^2} + \frac{d^2U}{dZ^2}\right) + \ldots + \ldots = 2\frac{dU}{dY}\frac{dU}{dZ}\frac{d^2U}{dYdZ} + \ldots + \ldots$$

2062. Normals are drawn to a conicoid at points lying along a generator: prove that they will generate a hyperbolic paraboloid whose principal sections are equal parabolas.

[It is obvious that the surface generated is a right conoid.]

2063. The axes of the two surfaces

$$Ax^2 + By^2 + Cz^2 - (ax + by + cz)^2 = \epsilon^4,$$

$$\left(\frac{x^2}{A} + \frac{y^2}{B} + \frac{z^2}{C}\right)\left(\frac{a^2}{A} + \frac{b^2}{B} + \frac{c^2}{C} - 1\right) - \left(\frac{ax}{A} + \frac{by}{B} + \frac{cz}{C}\right)^2 = 1,$$

are coincident in direction.

2064. The two conicoids

$$ax^2 + by^2 + cz^2 + 2fyz + 2gzx + 2hxy = 1, \quad Ax^2 + By^2 + Cz^2 = 1,$$

have one, and in general only one, system of conjugate diameters coincident in direction; but, if

$$\frac{1}{A}\left(a - \frac{gh}{f}\right) = \frac{1}{B}\left(b - \frac{hf}{g}\right) = \frac{1}{C}\left(c - \frac{fg}{h}\right),$$

there will be an infinite number of such systems, the direction of one diameter being the same in all.

[If l, m, n be the direction cosines of any one of such a system, we have the equations

$$al + hm + gn = \lambda Al, \quad hl + bm + fn = \lambda Bm, \quad gl + fm + cn = \lambda Cn,$$

giving for λ the cubic

$$(a - \lambda A)(b - \lambda B)(c - \lambda C) - f^2(a - \lambda A) - g^2(b - \lambda B) - h^2(c - \lambda C) + 2fgh = 0;$$

which may be written in the form

$$1 = \frac{gh}{af - gh - \lambda Af} + \frac{hf}{bg - hf - \lambda B} + \frac{fg}{ch - fg - \lambda C}.]$$

2065. Prove that eight conicoids can in general be drawn containing a given conic and touching four given planes.

2066. The equation of the polar reciprocal of the surface

$$ax^2 + by^2 + cz^2 + 2fyz + 2gzx + 2hxy = 1$$

with respect to a sphere, centre (X, Y, Z) and radius k, is

$$\Delta\{X(x - X) + Y(y - Y) + Z(z - Z)\}^2 = (bc - f^2)(x - X)^2 + \ldots$$
$$+ 2(gh - af)(y - Y)(z - Z) + \ldots,$$

where Δ is the discriminant.

2067. Prove that, if $l_1, l_2, \ldots l_7$ be constants so determined that the expression

$$l_1u_1^2 + l_2u_2^2 + \ldots + l_7u_7^2,$$

where $u_1, u_2, \ldots u_7$ are given linear functions, is the product of two factors, the two planes corresponding to these factors will be conjugate to each other with respect to any conicoid which touches the seven planes $u=0$; and that, when the expression is a complete square, the corresponding plane is the eighth plane which touches every conicoid drawn to touch the other seven.

2068. Seven points of a conicoid being given, an eighth is thereby determined; eight points $A_1, A_2, \ldots A_8$ being given, from every seven is determined an eighth accordingly, giving the points $B_1, B_2 \ldots B_8$: prove that the relation between the A points and the B points is reciprocal, and that the straight lines $A_1B_1, A_2B_2, \ldots$ all meet in one point.

2069. The straight line, on which lies the shortest distance between two generators of the same system of a conicoid, meets the two in A, B, and any generator of the opposite system meets them in P, Q respectively: prove that the lengths x, y of AP, BQ are connected by a constant relation of the form

$$axy + bx + cy + d = 0.$$

2070. Two fixed generators of one system of a conicoid are met by two of the opposite system in the points A, B; P, Q; respectively, and A, B are fixed: prove that the lengths x, y of AP, BQ are connected by a constant relation of the form

$$axy + bx + cy = 0.$$

2071. An hyperboloid of revolution is drawn containing two given straight lines which do not intersect: prove that the locus of its axis is a hyperbolic paraboloid, and that its centre lies on one of the generating lines through the vertex of the paraboloid.

III. *Conicoids referred to their axes.*

2072. The curve traced out on the surface $\dfrac{y^2}{b} + \dfrac{z^2}{c} = x$ by the extremities of the latus rectum of any section made by a plane through the axis of x lies on the cone $y^2 + z^2 = 4x^2$.

2073. The locus of the middle points of all straight lines passing through a fixed point and terminated by two fixed planes is a hyperbolic cylinder, unless the fixed planes are parallels.

2074. An ellipsoid and a hyperboloid are concentric and confocal: prove that a tangent plane to the asymptotic cone of the hyperboloid will cut the ellipsoid in a section of constant area.

2075. The locus of the centres of all plane sections of a given conicoid drawn through a given point is a similar and similarly situated conicoid, on which the given point and the centre of the given surface are ends of a diameter.

2076. An ellipse and a circle have a common diameter, and on any chord of the ellipse parallel to this as diameter is described a circle whose plane is parallel to that of the given circle: prove that the locus of these circles is an ellipsoid.

2077. Of two equal circles one is fixed and the other moves parallel to a given plane and intersects the former in two points: prove that the locus of the moving circle is an elliptic cylinder. If instead of circles we take any two conics of which one is fixed and the other moves parallel to a given plane without rotation in its own plane, and always intersects the fixed one in two points, the locus of the moving conic is a cylinder.

[There is no need to use co-ordinates of any kind.]

2078. A given ellipsoid is generated by the motion of a point fixed in a certain straight line, which straight line moves so that three other points fixed in it lie always one in each of the principal planes: prove that there are four such systems of points; and that, if the corresponding four straight lines be drawn through any point on the ellipsoid, the angle between any two is equal to the angle between the remaining two.

[If x, y, z be the point on the ellipsoid $\frac{x^2}{a^2}+\frac{y^2}{b^2}+\frac{z^2}{c^2}=1$, the direction cosines of the four straight lines will be $\frac{x}{a}, \pm\frac{y}{b}, \pm\frac{z}{c}$.]

2079. Prove that, when a straight line moves so that three fixed points in it always lie in three rectangular planes, the normals drawn at different points of the straight line to the ellipsoids which are traced out by those points will in any one position of the straight line all lie on an hyperboloid.

[When l, m, n are the direction cosines of the straight line, the locus of the normals is

$$\frac{yz}{mn}(b-c)+\ldots+(b-c)(b+c-2a)\frac{x}{l}+\ldots+2(b-c)(a-b)(a-c)=0,$$

$2a$, $2b$, $2c$ being the axes of the ellipsoid.

2080. From a fixed point O on an ellipsoid are let fall perpendiculars, (1) on any three conjugate diameters, (2) on any three conjugate diametral planes of the ellipsoid: prove that in each case the plane passing through the feet of the perpendiculars passes through a fixed point, and that this point in (2) lies on the normal to the ellipsoid at O.

[If (X, Y, Z) be the point O, the fixed point in (1) is given by

$$\frac{x}{a^2X} = \frac{y}{b^2Y} = \frac{z}{c^2Z} = \frac{1}{a^2+b^2+c^2},$$

and in (2) by

$$\frac{x-X}{\frac{X}{a^2}} = \frac{y-Y}{\frac{Y}{b^2}} = \frac{z-Z}{\frac{Z}{c^2}} = -\frac{1}{\frac{1}{a^2}+\frac{1}{b^2}+\frac{1}{c^2}}.]$$

2081. At each point of a generating line of a conicoid is drawn a straight line in the tangent plane at right angles to the generator: prove that the locus of such straight lines is a hyperbolic paraboloid whose principal sections are equal parabolas.

2082. The three acute angles made by any system of equal conjugate diameters of an ellipsoid will be always together equal to two right angles, if

$$2(b^2+c^2-2a^2)(c^2+a^2-2b^2)(a^2+b^2-2c^2)+27a^2b^2c^2=0;$$

$2a$, $2b$, $2c$ being the axes. Deduce the condition that an infinite number of systems of three generators can be found on the cone

$$Ax^2+By^2+Cz^2=0,$$

such that the sum of the acute angles in any such system is equal to two right angles.

[The condition is found by eliminating λ from the equations

$$A(A-\lambda)^{-1}+B(B-\lambda)^{-1}+C(C-\lambda)^{-1}=0, \quad \lambda^2=2ABC;$$

and, if A, B, C be roots of the equation

$$z^3-3p_1z^2+3p_2z-p_3=0,$$

the result is $$p_3^2+12p_1p_2p_3+p_1^3p_3=16p_2^3.]$$

2083. The locus of the axes of sections of the surface

$$ax^2+by^2+cz^2=1,$$

made by planes containing the straight line $\frac{x}{l}=\frac{y}{m}=\frac{z}{n}$, is the cubic cone

$$(b-c)yz(mz-ny)+(c-a)zx(nx-lz)+(a-b)xy(ly-mx)=0.$$

2084. Two generators of the hyperboloid $\frac{x^2}{a^2}+\frac{y^2}{b^2}-\frac{z^2}{c^2}=1$ drawn through a point O intersect the principal elliptic section in points P, P' at the ends of conjugate diameters: prove that

$$OP^2+OP'^2=a^2+b^2+2c^2.$$

2085. The generators of a given conicoid are orthogonally projected upon a plane perpendicular to one of the generators: prove that their projections all pass through a fixed point.

2086. The orthogonal projections of the generators of the conicoid $ax^2 + by^2 + cz^2 = 1$ on the plane $lx + my + nz = 0$ in general envelope a conic which degenerates if $al^2 + bm^2 + cn^2 = 0$; and which is similar to the section of the reciprocal surface $\frac{x^2}{a} + \frac{y^2}{b} + \frac{z^2}{c} = m^4$ by the plane.

2087. From different points of the straight line $\frac{x}{l} = \frac{y}{m} = \frac{z}{n}$, asymptotic straight lines are drawn to the hyperboloid $\frac{x^2}{a^2} + \frac{y^2}{b^2} - \frac{z^2}{c^2} = 1$: prove that they will lie on the two planes

$$\left(\frac{x^2}{a^2} + \frac{y^2}{b^2} - \frac{z^2}{c^2}\right)\left(\frac{l^2}{a^2} + \frac{m^2}{b^2} - \frac{n^2}{c^2}\right) = \left(\frac{lx}{a^2} + \frac{my}{b^2} - \frac{nz}{c^2}\right)^2.$$

2088. The asymptotes of sections of the conicoid $ax^2 + by^2 + cz^2 = 1$ made by planes parallel to $lx + my + nz = 0$ lie on the two planes

$$(l^2bc + m^2ca + n^2ab)(ax^2 + by^2 + cz^2) = abc\,(lx + my + nz)^2.$$

2089. The locus of points from which rectilinear asymptotes at right angles to each other can be drawn to the conicoid $ax^2 + by^2 + cz^2 = 1$ is the cone

$$a^2(b + c)x^2 + b^2(c + a)y^2 + c^2(a + b)z^2 = 0.$$

2090. The locus of the asymptotes drawn from a point (X, Y, Z) to the system of confocal conicoids

$$\frac{x^2}{a^2 + \lambda} + \frac{y^2}{b^2 + \lambda} + \frac{z^2}{c^2 + \lambda} = 1$$

is the cone

$$\frac{(x - X)(b^2 - c^2)}{Yz - Zy} + \frac{(y - Y)(c^2 - a^2)}{Zx - Xz} + \frac{(z - Z)(a^2 - b^2)}{Xy - Yx} = 0.$$

2091. A plane which contains two parallel generators of a given conicoid must pass through the centre, and touch the asymptotic cone.

2092. A sphere is described having for a great circle a plane section of a given conicoid: prove that the plane in which it again meets the conicoid intersects the plane of the former circle in a straight line which lies in one of two fixed planes.

[With the usual notation, the two planes are

$$\frac{x}{a^2}\sqrt{a^2 - b^2} \pm \frac{z}{c^2}\sqrt{b^2 - c^2} = 0.]$$

2093. **In the hyperboloid** $\frac{x^2}{a^2} + \frac{y^2 - z^2}{b^2} = 1$, $(a > b)$, the spheres, of which one series of circular sections of the hyperboloid are great circles, will have a common radical plane.

[If the sections be parallel to $y\sqrt{a^2 - b^2} + z\sqrt{a^2 + b^2} = 0$, the common radical plane will be $y\sqrt{a^2 - b^2} - z\sqrt{a^2 + b^2} = 0$.]

2094. Two generators of the paraboloid $\frac{x^2}{a} - \frac{y^2}{b} = 4z$ are drawn through the point $(X, 0, Z)$: prove that the angle between them is

$$\cos^{-1}\left(\frac{a - b + Z}{a + b + Z}\right).$$

2095. The perpendiculars let fall from the vertex of a hyperbolic paraboloid on the generators lie on two quadric cones whose circular sections are parallel to the principal parabolic sections of the paraboloid.

[The equation of the paraboloid being $\frac{x^2}{a^2} - \frac{y^2}{b^2} = \frac{2z}{c}$, those of the two cones are $x^2 + y^2 + 2z^2 \pm xy\left(\frac{a}{b} + \frac{b}{a}\right) = 0$.]

2096. Through A, A' the ends of the real principal axis of an hyperboloid of one sheet are drawn two generators of the same system, and any generator of the opposite system meets them in P, P' respectively: prove that the rectangle contained by AP, $A'P'$ is constant.

[If the equation of the hyperboloid be $\frac{x^2}{a^2} + \frac{y^2}{b^2} - \frac{z^2}{c^2} = 1$, and $AA' = 2a$, the constant rectangle is equal to $b^2 + c^2$.]

2097. The least distance between two generators of the same system in an hyperboloid of revolution of one sheet cannot exceed the diameter of the principal circular section.

2098. The equation of the cone generated by straight lines drawn through the origin parallel to normals to the ellipsoid $\frac{x^2}{a^2} + \frac{y^2}{b^2} + \frac{z^2}{c^2} = 1$ at points where it is met by the confocal $\frac{x^2}{a^2 - \lambda} + \frac{y^2}{b^2 - \lambda} + \frac{z^2}{c^2 - \lambda} = 1$ is

$$\frac{a^2x^2}{a^2 - \lambda} + \frac{b^2y^2}{b^2 - \lambda} + \frac{c^2z^2}{c^2 - \lambda} = 0.$$

2099. The points on a given conicoid, the normals at which intersect the normal at a given point, lie on a quadric cone whose vertex is the given point.

[With the usual notation, (X, Y, Z) being the given point, the equation of the cone is

$$\frac{X(b^2 - c^2)}{x - X} + \frac{Y(c^2 - a^2)}{y - Y} + \frac{Z(a^2 - b^2)}{z - Z} = 0.]$$

2100. Normals are drawn to a central conicoid at the ends of three conjugate diameters: prove that their orthogonal projections on the plane through the three ends will meet in a point.

2101. The six normals drawn to the ellipsoid $\frac{x^2}{a^2}+\frac{y^2}{b^2}+\frac{z^2}{c^2}=1$ drawn from the point (x_0, y_0, z_0) all lie on the cone

$$(b^2-c^2)\frac{x_0}{x-x_0}+(c^2-a^2)\frac{y_0}{y-y_0}+(a^2-b^2)\frac{z_0}{z-z_0}=0;$$

and the normals drawn from the same point to any confocal will also lie on the same cone.

2102. The normals at the ends of a chord of a given conicoid intersect each other: prove that the chord will be normal to some one confocal conicoid.

2103. The six normals drawn to the conicoid $ax^2+by^2+cz^2=1$, from any point on one of the lines

$$a(b-c)x=\pm b(c-a)y=\pm c(a-b)z,$$

will lie on a cone of revolution.

2104. The normals to the ellipsoid $\frac{x^2}{a^2}+\frac{y^2}{b^2}+\frac{z^2}{c^2}=1$ at points on the plane $l\frac{x}{a}+m\frac{y}{b}+n\frac{z}{c}=1$ all intersect one straight line: prove that normals at all points lying on the plane $\frac{x}{al}+\frac{y}{bm}+\frac{z}{cn}+1=0$ also intersect the same straight line; and that the necessary condition is

$$(m^2n^2-l^2)(b^2-c^2)^2+(n^2l^2-m^2)(c^2-a^2)^2+(l^2m^2-n^2)(a^2-b^2)^2=0.$$

Also prove that, when $l=m=n=1$, the normals all intersect the straight line

$$ax(b^2-c^2)=by(c^2-a^2)=cz(a^2-b^2).$$

2105. The normals to the paraboloid $\frac{y^2}{b}+\frac{z^2}{c}=2x$, at points on the plane $px+qy+rz=1$, will all intersect one straight line if

$$p^2(b-c)^2+2p(q^2b-r^2c)(b-c)=2(q^2b+r^2c).$$

2106. Prove that a tangent plane to the cone $\frac{2x^2}{b-c}+\frac{y^2}{b}-\frac{z^2}{c}=0$ will meet the paraboloid $\frac{y^2}{b}+\frac{z^2}{c}=2x$ in points the normals at which all intersect the same straight line; and the surface generated by this straight line has for its equation

$$2(b-c)\{x(by^2-cz^2)-bc(y^2-z^2)\}^2=(by^2-cz^2)(by^2+cz^2)^2.$$

2107. A section of the conicoid $ax^2+by^2+cz^2=1$ is made by a plane parallel to the axis of z, and the trace of the plane on xy is normal to the ellipse

$$\frac{ax^2}{(a-c)^2}+\frac{by^2}{(b-c)^2}=\frac{c^2}{(c^2-ab)^2}:$$

prove that the normals to the conicoid at points in this plane all intersect one straight line.

2108. Through a fixed point (x_0, y_0, z_0) are drawn straight lines each of which is an axis of some plane section of the conicoid

$$ax^2+by^2+cz^2=1:$$

prove that the locus of these lines is the cone

$$a(b-c)\frac{x_0}{x-x_0}+b(c-a)\frac{y_0}{y-y_0}+c(a-b)\frac{z_0}{z-z_0}=0.$$

2109. In a fixed plane are drawn straight lines each of which is an axis of some plane section of a given conicoid: prove that the envelope of these lines is a parabola.

2110. Straight lines are drawn in a given direction, and the tangent planes drawn through each straight line to a given conicoid are at right angles to each other: prove that the locus of such straight lines is a cylinder of revolution or a plane.

[With a central conicoid $\frac{x^2}{a^2}+\frac{y^2}{b^2}+\frac{z^2}{c^2}=1$, the locus is

$$x^2+y^2+z^2-(lx+my+nz)^2=a^2+b^2+c^2-p^2,$$

where l, m, n are the direction cosines of the given direction, and p the central distance of a tangent plane perpendicular to the given direction. With the paraboloid $\frac{y^2}{b}+\frac{z^2}{c}=x$, the locus is

$$m(ly-mx)+n(lz-nx)=b(l^2+n^2)+c(l^2+m^2).]$$

2111. A cone is described having for base the section of the conicoid $ax^2+by^2+cz^2=1$ made by the plane $lx+my+nz=0$, and intersects the conicoid in a second plane perpendicular to the former: prove that the vertex must lie on the surface

$$(l^2+m^2+n^2)(ax^2+by^2+cz^2-1)=2(lx+my+nz)(alx+bmy+cnz).$$

2112. The cone described with vertex (X, Y, Z) and base the curve determined by the equations $ax^2+by^2+cz^2=1$, $lx+my+nz=p$, will meet the conicoid again in the plane

$$(aX^2+bY^2+cZ^2-1)(lx+my+nz-p)=2(lX+mY+nZ-p)(axX+byY+czZ-1).$$

2113. A chord AB of a conicoid is drawn normal at A and the central plane conjugate to AB meets the tangent plane at A in a straight line, through which is drawn a plane intersecting the conicoid in a conic U: prove that the cone whose vertex is A and base the conic U has for its axes the normals at A to the conicoid and to the two confocals through A.

2114. Through the vertex of an enveloping cone of a given conicoid $ax^2 + by^2 + cz^2 = 1$ is drawn a similar concentric and similarly situated conicoid: prove that this conicoid will meet the cone in a plane curve which will touch the given conicoid if the vertex lie on the conicoid

$$ax^2 + by^2 + cz^2 = 4.$$

2115. A tangent plane is drawn to an ellipsoid and another plane drawn parallel to it so that the centre of the ellipsoid divides the distance between them in the ratio 1 : 4: prove that, if a cone be drawn enveloping the ellipsoid and have its vertex on the latter plane, the C.G. of the volume cut off this cone by the former plane will be a fixed point.

[The equation of the ellipsoid referred to conjugate diameters being $\frac{x^2}{a^2} + \frac{y^2}{b^2} + \frac{z^2}{c^2} = 1$, $x + a = 0$ the tangent plane, $x = 4a$ the parallel plane, the C. G. is $\left(\frac{a}{4}, 0, 0\right)$.]

2116. Straight lines are drawn through the point (x_0, y_0, z_0) such that their conjugates with respect to the paraboloid $\frac{y^2}{b} + \frac{z^2}{c} = 2x$ are perpendicular to them respectively: prove that the locus of these straight lines is the cone

$$\frac{y_0}{y - y_0} - \frac{z_0}{z - z_0} + \frac{b - c}{x - x_0} = 0;$$

and that their conjugates envelope the parabola

$$\frac{yy_0}{b} + \frac{zz_0}{c} = x + x_0, \quad \left(-\frac{yy_0}{b}\right)^{\frac{1}{2}} + \left(\frac{zz_0}{c}\right)^{\frac{1}{2}} + (b - c)^{\frac{1}{2}} = 0.$$

2117. A straight line is perpendicular to its conjugate with respect to a certain conicoid: prove that it is also perpendicular to its conjugate with respect to any conicoid confocal with the former.

2118. Any generator of the surface $y^2 + z^2 - x^2 = m$ will be perpendicular to its conjugate with respect to the surface

$$ax^2 + by^2 + cz^2 + 2fyz + 2gzx + 2hxy = 1,$$

if $$bc - f^2 = ca - g^2 = ab - h^2 \text{ and } af = gh.$$

2119. An hyperboloid of one sheet and an ellipsoid are concentric and every generator of the hyperboloid is perpendicular to its conjugate with respect to the ellipsoid: prove that their equations, referred to rectangular axes, may be obtained in the forms

$$x^2 - 2yz = m^2, \quad \frac{y^2}{2b} + \frac{z^2}{2c} + \frac{x^2}{b + c} = 1;$$

and that the locus of the conjugate straight lines is

$$\frac{x^2}{(b + c)^2} - \frac{yz}{2bc} = \frac{1}{m^2}.$$

[If $2b = 2c = m^2$ this locus is the hyperboloid itself, the ellipsoid being a sphere.]

2120. In the two conicoids

$$ax^2 + by^2 + cz^2 = 1, \quad Ax^2 + By^2 + Cz^2 = 1,$$

eight generators of the first are respectively perpendicular to their conjugates with respect to the second.

2121. A fixed point O being taken, P is any point such that the polar planes of O, P with respect to a given conicoid are perpendicular to each other: prove that the locus of P is the plane bisecting chords which are perpendicular to the polar plane of O.

2122. A hyperbolic paraboloid whose principal sections are equal is drawn through two given straight lines not in one plane: prove that the locus of its vertex is a straight line.

2123. Prove that, when two conicoids have in common two generators of one system, they have also common two generators of the opposite system.

2124. Two given straight lines not in one plane are generators of a conicoid: prove that the polar plane of any given point with respect to the conicoid passes through a fixed point.

2125. Two conicoids touch each other in three points: prove that they either touch in an infinite number of points or have four common generators.

2126. Generators of the same system of the hyperboloid

$$x^2 + y^2 - m^2z^2 = a^2$$

are drawn at the ends of a chord of the principal circle which subtends a given angle 2α at the centre: prove that the locus of the straight line which intersects both at right angles is the hyperboloid of revolution

$$x^2 + y^2 - \frac{z^2}{m^2 \cos^2 \alpha} = a^2 \cos^2 \alpha \left(\frac{1 + m^2}{1 + m^2 \cos^2 \alpha}\right)^2.$$

2127. A cone is described with vertex (X, Y, Z) and base the curve

$$S \equiv ax^2 + by^2 + cz^2 = 1, \quad px + qy + rz = 1:$$

prove that the equation of the plane in which the cone again meets the conicoid $S = 1$ is

$$2(aXx + bYy + cZz - 1)(pX + qY + rZ - 1)$$
$$= (aX^2 + bY^2 + cZ^2 - 1)(px + qy + rz - 1).$$

[The cone will intersect the conicoid in two planes at right angles to each other if

$$(p^2 + q^2 + r^2)(aX^2 + bY^2 + cZ^2 - 1) = 2(apX + bqY + crZ)(pX + qY + rZ - 1);$$

and in two parallel planes if $\dfrac{aX}{p} = \dfrac{bY}{q} = \dfrac{cZ}{r}$, that is if the vertex lie on the diameter conjugate to either plane section.]

2128. Tangent planes are drawn to a series of confocal conicoids parallel to a given plane: prove that the locus of the points of contact is a rectangular hyperbola which intersects both focal curves.

[The equations of the locus will be, with the usual notation,

$$lx + my + nz = \frac{b^2 - c^2}{\frac{y}{m} - \frac{z}{n}} = \frac{c^2 - a^2}{\frac{z}{n} - \frac{x}{l}} \left(= \frac{a^2 - b^2}{\frac{x}{l} - \frac{y}{m}} \right).]$$

2129. Two circular sections of the ellipsoid $\frac{x^2}{a^2} + \frac{y^2}{b^2} + \frac{z^2}{c^2} = 1$ are such that the sphere on which both lie is of constant radius mb: prove that the locus of the centre of this sphere is the hyperbola

$$y = 0,\ \frac{x^2}{a^2 - b^2} - \frac{z^2}{b^2 - c^2} = 1 - m^2;\ (a^2 > b^2 > c^2).$$

2130. A sphere of radius r has real double contact with the ellipsoid $\frac{x^2}{a^2} + \frac{y^2}{b^2} + \frac{z^2}{c^2} = 1$, and lies altogether within the ellipsoid: prove that the locus of its centre is the ellipse $\frac{x^2}{a^2 - c^2} + \frac{y^2}{b^2 - c^2} = 1 - \frac{r^2}{c^2}$, $z = 0$; and, if there be real double contact and the sphere lie altogether without the ellipsoid, the locus of the centre is the ellipse

$$\frac{y^2}{a^2 - b^2} + \frac{z^2}{a^2 - c^2} = \frac{r^2}{c^2} - 1,\ x = 0;\ (a^2 > b^2 > c^2).$$

[In the first case, r must lie between $\frac{c^2}{a}$, $\frac{c^2}{b}$; in the second, r must lie between $\frac{a^2}{b}$, $\frac{a^2}{c}$; and, in both cases, only a part of the ellipse can be traced out by the centre.]

2131. In an hyperboloid of revolution in which the excentricity of the generating hyperbola is $\sqrt{\frac{3}{2}}$, a cube can be placed with one diagonal along the axis of the hyperboloid and six edges lying along generators of the hyperboloid.

2132. A cone whose vertex is O meets a conicoid in two plane sections A, B; two other conicoids are described touching the former along A, B respectively and passing through O: prove that these two conicoids will touch at O, and will have a common plane section in the polar plane of O with respect to the first conicoid.

2133. The axes of sections of the conicoid $\frac{x^2}{a} + \frac{y^2}{b} + \frac{z^2}{c} = 1$ made by planes parallel to $lx + my + nz = 0$ lie on the two planes

$$\frac{x^2}{a}(b - c) + \ldots + \ldots + 2al^2\left(\frac{1}{c} - \frac{1}{b}\right)\frac{yz}{mn} + \ldots + \ldots$$

$$+ \left\{l^2\left(\frac{1}{b} - \frac{1}{c}\right) + \ldots + \ldots\right\}\left(\frac{ayz}{mn} + \frac{bzx}{nl} + \frac{cxy}{lm}\right) = 0.$$

2134. Two points are taken in the surface of a polished hollow ellipsoidal shell and a ray proceeding from one after one reflexion passes through the other: prove that the number of possible points of incidence is in general 8; but if the two points be ends of a diameter the number is 4, and these four points are the ends of two diameters which lie on a quadric cone containing the axes of the ellipsoid and of the central section perpendicular to the given diameter.

IV. *Tetrahedral Co-ordinates.*

2135. A plane meets the edges of a tetrahedron in six points and six other points are taken, one on each edge, so that each edge is divided harmonically: prove that the six planes, each passing through one of these six latter points and the edge opposite to it, will meet in a point.

2136. The opposite edges of a tetrahedron $ABCD$ are, two and two, at right angles: prove that the three shortest distances between opposite edges meet in the point

$$x(AB^2 + AC^2 + AD^2 - k) = y(BC^2 + BD^2 + BA^2 - k) = \dots = \dots,$$

k being the sum of the squares on any pair of opposite edges.

2137. Prove that any conicoid which touches seven of the planes

$$\pm lx \pm my \pm nz + rw = 0$$

will touch the eighth; and that its centre will lie on the plane

$$l^2x + m^2y + n^2z + r^2w = 0.$$

Prove that this plane bisects the part of each edge of the tetrahedron of reference which is intercepted by the given planes.

2138. Determine the condition that the straight line $\frac{x}{p} = \frac{y}{q} = \frac{z}{r}$ may touch the conicoid

$$lyz + mzx + nxy + l'xw + m'yw + n'zw = 0;$$

and thence prove that the equation of the tangent plane at the point (0, 0, 0, 1) is

$$l'x + m'y + n'z = 0.$$

2139. The general equation of a conicoid touching the faces of the tetrahedron of reference may be written

$$lqrx^2 + mrpy^2 + npqz^2 + lmnw^2 + (lp - mq - nr)(lxw + pyz)$$
$$+ (mq - nr - lp)(myw + qzx) + (nr - lp - mq)(nzw + rxy) = 0.$$

Prove that this will be a ruled surface if

$$l^2p^2 + m^2q^2 + n^2r^2 > 2mnqr + 2nlrp + 2lmpq;$$

and that, when $lp = mq = nr$, the straight lines joining the points of contact each to the opposite corner of the tetrahedron will meet in a point.

2140. A hyperbolic paraboloid is drawn containing the sides AB, BC, CD, DA of a quadrangle not in one plane: prove that, if P be any point on this paraboloid,

$$\text{vol. } PBCD : \text{vol. } PABC = \text{vol. } PCDA : \text{vol. } PDAB :$$

and that, if any tangent plane to the paraboloid meet AB, CD in P, Q respectively,

$$AP : BP = DQ : CQ.$$

2141. The locus of the centres of all conicoids which have in common four generators, two of each system, is a straight line.

2142. Perpendiculars are let fall from the point (x, y, z, w) on the faces of the tetrahedron of reference, and the feet of these perpendiculars lie in one plane: prove that

$$\frac{A^2}{x} + \frac{B^2}{y} + \frac{C^2}{z} + \frac{D^2}{w} = 0,$$

A, B, C, D being the areas of the faces of the tetrahedron.

2143. The volume of the ellipsoid which has its centre at the point $(X : Y : Z : W)$, and to which the tetrahedron of reference is self-conjugate, is $8\pi V\sqrt{XYZW} \div (X + Y + Z + W)^2$, where V is the volume of the tetrahedron.

2144. A tetrahedron is self-conjugate with respect to a given sphere: prove that each edge is perpendicular to the direction of the opposite edge, and that all the plane angles at one of the solid angles are obtuse.

2145. The opposite edges of a tetrahedron are two and two at right angles to each other, and in each face is described a circle of which the centroid and the centre of perpendiculars of that face are ends of a diameter: prove that the four circles so described lie on one sphere; and that this sphere (4), the circumscribed sphere (1), the polar sphere or sphere to which the tetrahedron is self-conjugate (2), the sphere bisecting the edges (3), and the sphere of which the centroid and the centre of perpendiculars of the tetrahedron are ends of a diameter (5), have all a common radical plane. Taking R, ρ to represent the radii of the circumscribed and polar spheres and δ the distance between their centres, $\delta^2 = R^2 + 3\rho^2$; and the distances from the common radical plane of the centres of these five spheres are

$$(1)\ \frac{R^2+\rho^2}{\delta},\quad (2)\ -\frac{2\rho^2}{\delta},\quad (3)\ \frac{R^2-\rho^2}{2\delta},\quad (4)\ \frac{R^2-3\rho^2}{3\delta},\quad (5)\ \frac{R^2-5\rho^2}{4\delta},$$

the radii of the five are

$$(1)\ R,\quad (2)\ \rho,\quad (3)\ \tfrac{1}{2}\sqrt{R^2-\rho^2},\quad (4)\ \frac{R}{3},\quad (5)\ \frac{\delta}{4};$$

and the centres of the spheres (3), (4), (5) divide the distance between the centres of (1) and (2) in the respective ratios 1 : 1, 2 : 1, 3 : 1.

2146. A tetrahedron is such that a sphere can be drawn touching its six edges: prove that any two of the four tangent cones drawn to this sphere from the corners of the tetrahedron have a common tangent plane and a common plane section; and that the planes of the common sections will all six meet in a point.

2147. A tetrahedron is such that the straight lines joining its angular points to the points of contact of the inscribed sphere with the respectively opposite faces meet in a point: prove that, at any point of contact, the edges of the tetrahedron which bound the corresponding face subtend equal angles.

2148. The tangent planes at A, B, C, D, to the sphere circumscribing the tetrahedron $ABCD$, form a tetrahedron $abcd$: prove that Aa, Bb, Cc, Dd will meet in a point if

$$BC \,.\, AD = CA \,.\, BD = AB \,.\, CD.$$

2149. Each edge of a tetrahedron is equal to the opposite edge: prove that the diameter of the circumscribed sphere is $\sqrt{\dfrac{a^2+b^2+c^2}{2}}$, where a, b, c are the edges bounding any one face.

2150. A conicoid circumscribes a tetrahedron $ABCD$ and the tangent planes at A, B, C, D form the tetrahedron $abcd$: prove that, if Aa, Bb intersect, Cc, Dd will also intersect.

2151. Four points are taken on a conicoid and the straight line joining one of the points to the pole of the plane containing the other three passes through the centre: prove that the tangent plane at that point is parallel to the plane of the other three.

2152. The equation of a conicoid being

$$mnyz + nlzx + lmxy + lrxw + mryw + nrzw = 0;$$

prove that it cannot be a ruled surface, and that it will be an elliptic paraboloid if

$$\frac{1}{l^2} + \frac{1}{m^2} + \frac{1}{n^2} + \frac{1}{r^2} = \tfrac{1}{3}\left(\frac{1}{l} + \frac{1}{m} + \frac{1}{n} + \frac{1}{r}\right)^2.$$

2153. The surface

$$lyz + mzx + nxy + l'xw + m'yw + n'zw = 0$$

will be a cylinder, if

$$ll'(m+n-l) + mm'(n+l-m) + nn'(l+m-n) = 2lmn,$$

and

$$ll'(m'+n'-l) + mm'(n'+l-m') + nn'(l+m'-n') = 2lm'n'.$$

[The relations

$$ll'(m+n'-l') + mm'(n'+l'-m) + nn'(l'+m-n') = 2l'mn',$$

$$ll'(m'+n-l') + mm'(n+l'-m') + nn'(l'+m'-n) = 2l'm'n,$$

will of course also be satisfied, the system being equivalent to the two-fold relation

$$(ll')^{\frac{1}{2}} + (mm')^{\frac{1}{2}} + (nn')^{\frac{1}{2}} = 0,$$

$$\left(\frac{m'n'}{mn}\right)^{\frac{1}{2}} + \left(\frac{n'l'}{nl}\right)^{\frac{1}{2}} + \left(\frac{l'm'}{lm}\right)^{\frac{1}{2}} = 1,$$

the first of which is the single condition for the surface to be a cone. See question (141).]

2154. The rectangles under the segments of chords of a certain sphere drawn through the four points A, B, C, D (not in one plane), are l, m, n, r, and the radius of the sphere is ρ: prove that

$$\begin{vmatrix} 0, & 1, & 1, & 1, & 1, & 1 \\ 1, & 0, & AB^2, & AC^2, & AD^2, & l+\rho^2 \\ 1, & BA^2, & 0, & BC^2, & BD^2, & m+\rho^2 \\ 1, & CA^2, & CB^2, & 0, & CD^2, & n+\rho^2 \\ 1, & DA^2, & DB^2, & DC^2, & 0, & r+\rho^2 \\ 1, & l+\rho^2, & m+\rho^2, & n+\rho^2, & r+\rho^2, & 0 \end{vmatrix} = 0.$$

2155. The enveloping developable of the two conicoids

$$lx^2 + my^2 + nz^2 + rw^2 = 0, \qquad l'x^2 + m'y^2 + n'z^2 + r'w^2 = 0,$$

will meet the planes of the faces of the tetrahedron of reference in the conics

$$w = 0, \quad \frac{ll'x^2}{lr' - l'r} + \frac{mm'y^2}{mr' - m'r} + \frac{nn'z^2}{nr' - n'r} = 0, \quad \&c.$$

2156. The perpendiculars p, q, r, s let fall from the corners of a finite tetrahedron on a moving plane are connected by the equation

$$Ap^2 + Bq^2 + Cr^2 + Ds^2 + 2Fqr + 2Grp + 2Hpq + 2F'ps + 2G'qs + 2H'rs = 0;$$

prove that the envelope is in general a conicoid, which degenerates to a plane curve if

$$\begin{vmatrix} A, & H, & G, & F' \\ H, & B, & F, & G' \\ G, & F, & C, & H' \\ F', & G', & H', & D \end{vmatrix} = 0.$$

V. *Focal Curves: Reciprocal Polars.*

2157. The equations of the focal lines of the cone $ayz + bzx + cxy = 0$ are

$$\frac{(cy+bz)^2}{y^2+z^2} = \frac{(az+cx)^2}{z^2+x^2} = \frac{(bx+ay)^2}{x^2+y^2}.$$

2158. A parallelogram of minimum area is circumscribed about the focal ellipse of a given ellipsoid, and from its angular points taken in order are let fall perpendiculars p_1, p_2, p_3, p_4 on any tangent plane to the ellipsoid: prove that

$$p_1p_3 + p_2p_4 = 2c^2,$$

$2c$ being the length of that axis of the ellipsoid which is normal to the plane of the focal ellipse.

2159. The perpendiculars from the ends of two conjugate diameters of the focal ellipse on any tangent plane to the ellipsoid are $\varpi_1, \varpi_2, \varpi_3, \varpi_4$, and the perpendicular from the centre is p: prove that

$$\varpi_1\varpi_3 + \varpi_2\varpi_4 = p^2 + c^2.$$

2160. With any two points of the focal ellipse as foci can be described a prolate spheroid touching an ellipsoid along a plane curve, and the contact will be real when the common point of the tangents to the focal at the two foci lies without the ellipsoid.

[The plane of contact is the polar with respect to the ellipsoid of this common point.]

2161. Four straight lines can be drawn in a given direction so as to intersect both focal curves of an ellipsoid, and they will lie on a cylinder of revolution whose radius is $\sqrt{a^2 - p^2}$; a being the semi major axis and p the perpendicular from the centre on a tangent plane normal to the given direction.

2162. The cones whose common vertex is (X, Y, Z) and whose bases are the real focal curves of the ellipsoid $\frac{x^2}{a^2} + \frac{y^2}{b^2} + \frac{z^2}{c^2} = 1$ being denoted by U_1 and U_2 whose discriminants are respectively

$$\frac{Z^4}{(a^2 - c^2)(b^2 - c^2)}, \quad \frac{Y^4}{(a^2 - b^2)(c^2 - b^2)},$$

the cone $\lambda U_1 + U_2 = 0$ will be a cone of revolution if

$$\frac{X^2(b^2 - c^2)}{1 - \lambda} + \frac{Y^2(c^2 - a^2)}{\lambda} - Z^2(a^2 - b^2) + \frac{(b^2 - c^2)(c^2 - a^2)(a^2 - b^2)}{a^2 - c^2 - \lambda(a^2 - b^2)} = 0.$$

2163. With a given point as vertex is described a cone of revolution whose base is a plane section of a given conicoid: prove that the plane of this section will envelope a fixed cone whose vertex lies on one of the axes of the enveloping cone drawn from the given point to the given conicoid.

2164. This straight line joining the points of contact of a common tangent plane to the two conicoids

$$ax^2 + by^2 + cz^2 = 1, \quad (a - \lambda)x^2 + (b - \lambda)y^2 + (c - \lambda)z^2 = 1,$$

subtends a right angle at the centre.

2165. Through a given point can in general be drawn two straight lines either of which is a focal line of any cone having its vertex on the straight line and enveloping a given conicoid: and, if two such cones be drawn with their vertices one on each straight line, a prolate conicoid of revolution can be inscribed in them having its focus at the given point.

2166. A point O is taken on the umbilical focal conic of a conicoid: prove that there exist two points L such that, if any plane A be drawn through L and a be its pole, Oa will be normal to the plane through O containing the intersection of A with the polar of L.

2167. With a given point as vertex there can in general be drawn one tetrahedron self-conjugate to a given conicoid and such that the edges meeting in the point are two and two at right angles; but when the given point lies on a focal curve the number of such tetrahedrons is infinite.

2168. A tetrahedron circumscribes a prolate ellipsoid of revolution whose foci are S, S', so that the focal distance (from S) of each angular point is normal to the opposite face: prove that the diameter of the sphere circumscribing the tetrahedron is three times the major axis of the ellipsoid, and that the centroid of the tetrahedron and the centre O of the circumscribed sphere divide SS' in the ratios $1:3$, $3:-1$ respectively.

2169. The vertical angles of two principal sections of a quadric cone are α, β: prove that the ratio of the axes of any section normal to a focal line is $\cos\alpha : \cos\beta$.

2170. A sphere is described with centre $(X, 0, Z)$ intersecting the ellipsoid $\frac{x^2}{a^2}+\frac{y^2}{b^2}+\frac{z^2}{c^2}=1$ in two circles: prove that the points of contact with the sphere of common tangents to the sphere and ellipsoid lie on the two planes

$$\left(1-\frac{X^2}{a^2-b^2}-\frac{Z^2}{c^2-b^2}\right)\left\{\frac{\left(x-\frac{a^2X}{a^2-b^2}\right)^2}{a^2-b^2}+\frac{\left(z-\frac{c^2Z}{c^2-b^2}\right)^2}{c^2-b^2}\right\}$$

$$=\left\{\frac{X(z-Z)}{a^2-b^2}-\frac{Z(x-X)}{c^2-b^2}\right\}^2.$$

2171. The circumscribing developable of two conicoids, which have not common plane sections, will in general contain four plane conics, which are double lines on the developable.

2172. In a given tetrahedron are inscribed a series of closed surfaces each similar to a given closed surface without singular points: prove that the one of maximum volume will be such that the normals at the points of contact will be generators of the same system of an hyperboloid.

2173. Two conicoids having for their equations $U=0$, $U'=0$, the discriminant of $\lambda U+U'$ is $\lambda^4\Delta+\lambda^3\Theta+\lambda^2\Phi+\lambda\Theta'+\Delta'$: prove that the condition that hexahedra can be described whose six faces touch U and whose eight corners lie upon U' is

$$\Theta^4-4\Theta^2\Phi\Delta+8\Theta\Theta'\Delta^2-16\Delta^3\Delta'=0,$$

and the condition that hexahedra can be described whose twelve edges are tangent lines to U and whose eight corners lie upon U' is

$$2\Theta^4-9\Theta^2\Phi\Delta+27\Theta\Theta'\Delta^2-81\Delta^3\Delta'=0.$$

VI. *General Functional and Differential Equations.*

2174. A surface is generated by a straight line which always intersects the two fixed straight lines

$$x = a, \quad y = mz; \quad x = -a, \quad y = -mz:$$

prove that the equation of the surface generated is of the form

$$\frac{maz - xy}{a^2 - x^2} = f\left(\frac{mxz - ay}{a^2 - x^2}\right).$$

2175. The general functional equation of surfaces generated by a straight line which intersects the axis of z and the circle $z = 0$, $x^2 + y^2 = a^2$, is

$$x^2 + y^2 = \left\{a + zf\left(\frac{y}{x}\right)\right\}^2;$$

and the general differential equation is

$$(x^2 + y^2)(px + qy - z) = a^2 (px + qy)^2.$$

2176. The general functional equation of surfaces generated by a straight line which always intersects the axis of z is

$$z = f\left(\frac{y}{x}\right) + x\phi\left(\frac{y}{x}\right);$$

and the differential equation is

$$rx^2 + 2sxy + ty^2 = 0.$$

2177. The differential equation of a family of surfaces, such that the perpendicular from the origin on the normal always lies in the plane of xy, is

$$z(p^2 + q^2) + px + qy = 0.$$

2178. The differential equation of a family of surfaces, generated by a straight line which is always parallel to the plane of xy and whose intercept between the planes of yz, zx is always equal to a, is

$$(px + qy)^2 (p^2 + q^2) = a^2p^2q^2.$$

2179. The general differential equation of surfaces, generated by a straight line, (1) always parallel to the plane $lx + my + nz = 0$, (2) always intersecting the straight line $\frac{x}{l} = \frac{y}{m} = \frac{z}{n}$, is

(1) $(m + nq)^2 r - 2(m + nq)(l + np)s + (l + np)^2 t = 0,$

(2) $(ly - mx)^2 (q^2r - 2pqs + p^2t) + 2(ly - mx)(nx - lz)(qr - ps)$
$\quad + 2(ly - mx)(ny - mz)(qs - pt)$
$\quad + (nx - lz)^2 r + 2(nx - lz)(ny - mz)s + (ny - mz)^2 t = 0.$

VII. *Envelopes.*

2180. The envelope of the plane $lx+my+nz=a$; l, m, n being parameters connected by the equations

$$l+m+n=0,\ l^2+m^2+n^2=1,$$

is the cylinder

$$(y-z)^2+(z-x)^2+(x-y)^2=3a^2.$$

2181. Find the envelope of the planes

$$(1)\quad \frac{x}{a}\cos(\theta-\phi)+\frac{y}{b}\cos(\theta-\phi)+\frac{z}{c}\sin(\theta+\phi)=\sin(\theta-\phi),$$

$$(2)\quad \frac{x}{a}\cos(\theta-\phi)+\frac{y}{b}(\cos\theta+\cos\phi)+\frac{z}{c}(\sin\theta+\sin\phi)=1;$$

both when θ, ϕ are parameters, and when θ only is a parameter.

[The envelope of (1) when both θ, ϕ are parameters is the hyperboloid

$$\frac{x^2}{a^2}-\frac{y^2}{b^2}+\frac{z^2}{c^2}=1,$$

and when θ only is a parameter, the plane (1) always passes through a fixed generator of this hyperboloid; the envelope of (2) when θ, ϕ are parameters is the ellipsoid

$$\frac{2x^2}{a^2}+\frac{2x}{a}+\frac{y^2}{b^2}+\frac{z^2}{c^2}=0,$$

when θ alone is a parameter the envelope is a cone whose vertex is the point $(-a,\ b\cos\phi,\ c\sin\phi)$.]

2182. The envelope of the plane

$$\frac{x}{\sin\theta\cos\phi}+\frac{y}{\sin\theta\sin\phi}+\frac{z}{\cos\theta}=a$$

is the surface

$$x^{\frac{2}{3}}+y^{\frac{2}{3}}+z^{\frac{2}{3}}=a^{\frac{2}{3}}.$$

2183. The envelope of all paraboloids to which a given tetrahedron is self-conjugate is the planes each of which bisects three edges of the tetrahedron.

[More generally, if a conicoid be drawn touching a given plane and such that a given tetrahedron is self-conjugate to it, there will be seven other fixed planes which it always touches, the equations of the eight planes referred to the given tetrahedron being

$$\pm px\pm qy\pm rz+w=0.]$$

2184. A prolate ellipsoid of revolution can be described having two opposite umbilics of a given ellipsoid as foci and touching the given ellipsoid along a plane curve: and this will be the envelope of one system of spheres, each of which has a circular section of the ellipsoid for a great circle.

2185. Spheres are described on a series of parallel chords of a given ellipsoid as diameter: prove that they will have double contact with another ellipsoid, and that the focal ellipse of this envelope will be the diametral section of the given ellipsoid which is conjugate to the chords. Also, if a, b, c be the axes of the given ellipsoid, and α, β, γ of the envelope,

$$\alpha^2+\beta^2-\gamma^2=a^2+b^2+c^2;$$

γ being that axis which is perpendicular to the focal ellipse.

2186. A series of parallel plane sections of a given ellipsoid being taken, on each as a principal section is described another ellipsoid of given form; the envelope is an ellipsoid touching the given one along a central section at any point of which the tangent plane is perpendicular to the planes of the parallel sections.

2187. The envelope of a sphere, intersecting a given conicoid in two planes and passing through the centre, is a quartic which touches the given conicoid along a sphero-conic.

VIII. *Curvature.*

2188. From any point of a curve equal small lengths s are measured in the same sense along the curve, and along the circle of absolute curvature at the point, respectively: prove that the distance between the ends of these lengths is ultimately

$$\frac{s^3}{6\rho}\sqrt{\frac{1}{\sigma^2}+\frac{1}{\rho^2}\left(\frac{d\rho}{ds}\right)^2},$$

ρ, σ being the radii of curvature and torsion respectively at the point.

2189. Find the radius of absolute curvature and of torsion at any point of the curves

(1) $x=a(3t-t^3),\quad y=3at^2,\quad z=a(3t+t^3);$

(2) $x=2at^3(1+t),\quad y=at^3(t+2),\quad z=at^2(t^2+2t+2).$

2190. The radius of absolute curvature (ρ) at any point of a rhumb line is $a\cos\theta\div\sqrt{1-\sin^2\theta\cos^2\alpha}$, where θ is the latitude, and α the angle at which the line crosses the meridians; and the radius of torsion is

$$\frac{a}{\sin\alpha\cos\alpha}(1-\sin^2\theta\cos^2\alpha)\ \text{ or }\ \frac{a^2\tan\alpha}{a^2-\rho^2\cos^2\alpha}.$$

2191. Two surfaces have complete contact of the nth order at a point: prove that there are $n+1$ directions of normal section for which the curves of section have contact of the $n+1^{\text{th}}$ order; and hence prove that two conicoids which have double contact with each other intersect in plane curves.

2192. Prove that it is in general possible to determine a paraboloid, whose principal sections are equal parabolas, and which has a complete contact of the second order with a given surface at a given point.

2193. A paraboloid can in general be drawn having a complete contact of the second order with a given surface at a given point, and such that all normal sections through the point have four-point contact.

2194. A skew surface is capable of generation in two ways by the motion of a straight line, and at any point of it the absolute magnitudes of the principal radii of curvature are a, b: prove that the angle between the generators which intersect in the point is $\cos^{-1}\left(\frac{a-b}{a+b}\right)$.

2195. The points on the surfaces

$$(1) \quad xyz = a\,(yz + zx + xy),$$
$$(2) \quad xyz = a^2\,(x + y + z),$$
$$(3) \quad x^3 + y^3 + z^3 - 3xyz = a^3,$$

at which the indicatrix is a rectangular hyperbola lie on the cones

$$(1) \quad x^4\,(y + z) + y^4\,(z + x) + z^4\,(x + y) = 0,$$
$$(2) \quad x^3 + y^3 + z^3 + xyz = 0,$$
$$(3) \quad yz + zx + xy = 0,$$

respectively; and in (3) these points lie on the circle

$$x + y + z = a, \quad x^2 + y^2 + z^2 = a^2.$$

2196. A surface is generated by a straight line moving so as always to intersect the two straight lines

$$x = \frac{a}{2}, \quad y = z\tan\frac{\alpha}{2}; \quad x = -\frac{a}{2}, \quad y = -z\tan\frac{\alpha}{2};$$

and λ, μ are the distances of the points where the generator meets these straight lines from the points where the axis of x meets them; prove that the principal radii of curvature at any point on the first straight line are given by the equation

$$a^2\rho^2\sin^2\alpha - 2a\rho\sin\alpha\,\frac{d\lambda}{d\mu}\,(\lambda - \mu\cos\alpha)\,\sqrt{a^2 + \mu^2\sin^2\alpha}$$
$$= \left(\frac{d\lambda}{d\mu}\right)^2 (a^2 + \mu^2\sin^2\alpha)^2.$$

2197. A surface is generated by the motion of a variable circle, which always intersects the axis of x, and is parallel to the plane of yz. At a point on the axis of x, r is the radius of the circle, and θ the angle which the diameter through the point makes with the axis of z: prove that the principal radii of curvature at this point are given by the equation

$$\rho^2 r + \left(\frac{dx}{d\theta}\right)^2 (\rho - r) = 0.$$

2198. A surface is generated by a straight line which always intersects a given circle and the normal to the plane of the circle drawn through its centre; θ is the angle which the generator makes with this normal, and ϕ the angle which the projection of the generator on the plane of the circle makes with a fixed radius: prove that the principal radii of curvature at the point where the generator meets the normal are

$$a\frac{d\theta}{d\phi} \div \sin\theta\,(\cos\theta \pm 1);$$

and that at the point where it meets the circle, the principal radii are given by the equation

$$\rho^2\left(\frac{d\theta}{d\phi}\right)^2 + a\rho\cos\theta = a^2.$$

2199. A surface is generated by a straight line, which is always parallel to the plane of xy, and touches the cylinder $x^2+y^2=a^2$: prove that, if ρ be a principal radius of curvature at the point whose co-ordinates are $(a\cos\theta + r\sin\theta,\ a\sin\theta - r\cos\theta,\ z)$

$$\rho^2\left(\frac{dz}{d\theta}\right)^2 + \rho\left(r\frac{d^2z}{d\theta^2} - a\frac{dz}{d\theta}\right)\sqrt{r^2+\left(\frac{dz}{d\theta}\right)^2} = \left\{r^2+\left(\frac{dz}{d\theta}\right)^2\right\}^2.$$

2200. A straight line moves so as always to intersect the circle $x^2+y^2=a^2$, $z=0$, and be parallel to the plane of zx; prove that the measure of specific curvature at the point $(a\cos\phi,\ a\sin\phi,\ 0)$ is

$$-\frac{1}{a^2}\frac{\cos^2\phi}{(1-\sin^2\theta\sin^2\phi)^2}\left(\frac{d\theta}{d\phi}\right)^2;$$

θ being the angle which the generator through the point makes with the axis of z.

2201. A circle of constant radius a moves so as to intersect the axis of x, its plane being parallel to the plane of yz: prove that, at the point

$$(x,\ a\sin\phi + a\sin\overline{\phi-\theta},\ a\cos\phi + a\cos\overline{\phi-\theta}),$$

the measure of specific curvature of the surface generated is

$$\frac{dx}{d\phi}\left(\frac{dx}{d\phi}\cos\theta - \frac{d^2x}{d\phi^2}\sin\theta\right) \div \left\{\left(\frac{dx}{d\phi}\right)^2 + a^2\sin^2\theta\right\}^2.$$

2202. In a right conoid whose axis is the axis of z, prove that the radius of curvature of any normal section at a point $(r\cos\theta,\ r\sin\theta,\ z)$ is

$$\frac{\left\{r^2+\left(\frac{dr}{d\theta}\right)^2+\left(\frac{dz}{d\theta}\right)^2\right\}\sqrt{r^2+\left(\frac{dz}{d\theta}\right)^2}}{r\frac{d^2z}{d\theta^2} - 2\frac{dr}{d\theta}\frac{dz}{d\theta}},$$

and deduce the equation

$$\rho^2\left(\frac{dz}{d\theta}\right)^2 + \rho r\frac{d^2z}{d\theta^2}\sqrt{r^2+\left(\frac{dz}{d\theta}\right)^2} = \left\{r^2+\left(\frac{dz}{d\theta}\right)^2\right\}^2$$

for the principal radii of curvature at the point.

2203. A straight line moves so as always to intersect the axis of z and make a constant angle a with it: prove that, if ρ be a principal radius of curvature of the surface generated at the point whose coordinates are $(r \sin a \cos \phi,\ r \sin a \sin \phi,\ z + r \cos a)$,

$$\rho^2 \sin^2 a \left(\frac{dz}{d\phi}\right)^2 + \rho \sin a \sqrt{r^2 + \left(\frac{dz}{d\phi}\right)^2} \left(r^2 \cos^2 a + 2 \cos a \left(\frac{dz}{d\phi}\right)^2 + r \frac{d^2 z}{d\phi^2}\right)$$
$$= \left\{r^2 + \left(\frac{dz}{d\phi}\right)^2\right\} \left\{r^2 \sin^2 a + \left(\frac{dz}{d\phi}\right)^2\right\}.$$

2204. Investigate the nature of the contact of the surfaces

$$xyz = a^2 (x + y + z), \quad x (y - z)^2 + 4a^2 (x + y + z) = 0,$$

at any point on the line $x = 0$, $y + z = 0$; proving that the principal radii of curvature of either surface are $\dfrac{(a^2 + y^2)^2 + 2a^4}{\pm 2a^2 y}$.

2205. Prove that, in the surface

$$(y^2 + z^2)\left(\overline{2x - y}^2 + z^2\right) = 4a^2 z^2,$$

(1) the points where the indicatrix is parabolic lie on the cylinder $x^2 + z^2 = a^2$; (2) the lines $y = 0$, $z = 0$; $y = 2x$, $z = 0$, are nodal lines, the tangent planes at any point being respectively

$$z^2 (a^2 - X^2) = X^2 y^2, \quad z^2 (a^2 - X^2) = X^2 (y - 2x)^2.$$

2206. An ellipsoid is described with its axes along the co-ordinate axes and touching the fixed plane $px + qy + rz = 1$: prove that the locus of the centres of principal curvature at the point of contact is the surface whose equation is

$$(px + qy + rz - 1)(p^3 yz + q^3 zx + r^3 xy) = xyz (p^2 + q^2 + r^2)^2.$$

2207. The direction cosines of the normal to the conicoid

$$\frac{x^2}{a} + \frac{y^2}{b} + \frac{z^2}{c} = 1$$

at a certain point are l, m, n, and the angle between the geodesics joining the point to the umbilics is ϕ: prove that

$$\cos^2 \phi = \frac{\{l^2 a (c - b) + m^2 b (c + a - 2b) + n^2 c (a - b)\}^2}{\{-l^2 a (b - c) + m^2 b (c - a) + n^2 c (a - b)\}^2 + 4m^2 n^2 bc (a - b)(a - c)}.$$

STATICS.

I. *Composition and Resolution of Forces.*

2208. A point O is taken in the plane of a triangle ABC and a, b, c are the mid points of the sides: prove that the system of forces Oa, Ob, Oc is equivalent to the system OA, OB, OC.

[The result is true when O is not in the plane ABC.]

2209. Forces P, Q, R act along the sides of a triangle ABC and their resultant passes through the centres of the inscribed and circumscribed circles: prove that

$$\frac{P}{\cos B - \cos C} = \frac{Q}{\cos C - \cos A} = \frac{R}{\cos A - \cos B}.$$

2210. Four points A, B, C, D lie on a circle and forces act along the chords AB, BC, CD, DA in the senses indicated by the order of the letters, each force being inversely proportional to the chord along which it acts: prove that the resultant passes through the common points of (1) AD, BC; (2) AB, DC; (3) the tangents at B, D; (4) the tangents at A, C.

[Of course this proves that these four points are collinear.]

2211. In a triangular lamina ABC, AD, BE, CF are the perpendiculars, and forces BD, CD, CE, AE, AF, BF are applied to the lamina: prove that their resultant passes through the centre of the circumscribed circle and through the point of concourse of the straight lines each joining an angular point to the intersection of tangents to the circle ABC at the ends of the opposite side.

[The equation of the line of action of the resultant in trilinear co-ordinates is

$$x \sin(B - C) + y \sin(C - A) + z \sin(A - B) = 0,$$

which passes through the points

$$(\cos A : \cos B : \cos C), (\sin A : \sin B : \sin C).]$$

2212. Three equal forces act at the corners of a triangle ABC, each perpendicular to the opposite side: prove that, if the magnitude of each force be represented by the radius of the circle ABC, the magnitude of the resultant will be represented by the distance between the centres of the inscribed and circumscribed circles.

2213. The resultant R of any number of forces $P_1, P_2, P_3, \ldots$ is determined in magnitude by the equation

$$R^2 = \Sigma(P^2) + 2\Sigma P_r P_s \cos(\widehat{P_r, P_s}),$$

where $\widehat{P_r, P_s}$ denotes the angle between the directions of P_r, P_s.

2214. The centre of the circumscribed circle of a triangle ABC is O, and the centre of perpendiculars is L: prove that the resultant of forces LA, LB, LC will act along LO and be equal to $2LO$.

2215. Three parallel forces act at the points A, B, C and are to each other as $b+c : c+a : a+b$, where a, b, c are the lengths of the sides of the triangle ABC: prove that their resultant passes through the centre of the circle inscribed in the triangle whose corners bisect the sides of the triangle ABC.

2216. The position of a point P such that forces acting along PA, PB, PC, and equal to $l \,.\, PA$, $m \,.\, PB$, $n \,.\, PC$ may be in equilibrium is determined by the areal co-ordinates $(l : m : n)$.

2217. Forces act along the sides of a triangle ABC and are proportional to the sides; AA', BB', CC' bisect the angles of the triangle: prove that, if the forces be turned in the same sense about the points A', B', C' respectively, each through the angle

$$\tan^{-1}\left(-\cot\frac{B-C}{2}\cot\frac{C-A}{2}\cot\frac{A-B}{2}\right),$$

there will be equilibrium.

2218. Forces in equilibrium act along the sides AD, BD, CD, BC, CA, AB of a frame $ABCD$, prove the following construction for a force diagram: take any one of the points (D) as focus and inscribe a conic in the triangle ABC; let d be the second focus and let fall da, db, dc perpendiculars on the sides of the triangle ABC, then $abcd$ will form the force diagram; that is, bc will be perpendicular to AD and proportional to the force along AD, and so for the other sides.

2219. Four points A, B, C, D are taken in a plane, perpendiculars are drawn from D on BC, CA, AB and a circle drawn through the feet of these perpendiculars, and another circle is drawn with centre D and radius equal to the diameter of the former circle; other circles are similarly determined with their centres at A, B, C. Prove that these four circles will intersect by threes in four points a, b, c, d, and that the diagrams $abcd$, $ABCD$ will be reciprocal force diagrams.

2220. A triangular frame ABC is kept in equilibrium by three forces at right angles to the sides, and S is the point of concourse of their lines of action, O the centre of the circle ABC; SS' is a straight line bisected in O: prove that the stresses at A, B, C are perpendicular and proportional to $S'A$, $S'B$, $S'C$.

2221. A number of light rigid rods are freely jointed at their extremities so as to form a polygon, and are in equilibrium under a system of forces perpendicular and proportional to the respective sides of the polygon and all meeting in one point: prove that the polygon is inscribable in a circle, and, if O be the centre of the circle, S the point of concourse of the lines of action, SS' a straight line bisected in O, that the stress at any angular point P of the polygon is perpendicular and proportional to $S'P$.

[The points S, S' will be foci of a conic which can be drawn to touch the lines of action of all the stresses at the angular points, and the circle circumscribing the polygon is the auxiliary circle of this conic. If $ABCD$... be the corners of the polygon, and $A'B'C'D'$... those of the polygon formed by the lines of action of the stresses at A, B, C,..., the diagrams $SA'B'C'D'$..., $S'ABCD$... will be reciprocal force diagrams.]

2222. Two systems of three forces (P, Q, R), (P', Q', R') act along the sides of a triangle ABC: prove that the two resultants will be parallel if

$$\begin{vmatrix} P, & Q, & R \\ P', & Q', & R' \\ BC, & CA, & AB \end{vmatrix} = 0.$$

2223. A lamina rests in a vertical plane with one corner A against a smooth inclined plane and another point B is attached to a fixed point C in the plane by a fine string, G is the C. G., and the distances of A, G, C from B are all equal: prove that, when the inclination of the inclined plane to the horizon is half the angle ABG, every position is one of equilibrium.

2224. Perpendiculars SK, SK' are drawn from a focus on the asymptotes of an hyperbola, and P is a point such that the rectangle KP, $K'P$ is constant: prove, from Statical considerations, that the tangent to the locus of P at a point where it meets the auxiliary circle of the hyperbola will touch the hyperbola, and that the normal will pass through S.

2225. Forces proportional to the sides a_1, a_2,... of a closed polygon act at points dividing the sides taken in order in the ratios $m_1 : n_1$, $m_2 : n_2$,... and each force makes the same angle θ in the same sense with the corresponding side: prove that there will be equilibrium if

$$\Sigma\left(\frac{n-m}{n+m}a^2\right) = 4 \cot\theta \times \text{area of the polygon}.$$

2226. The lines of action of a system of forces are generators of the same system of a hyperboloid: prove that the least distance of any generator of the opposite system from the central axis of the forces is proportional to the cotangent of the angle between the directions of the two straight lines.

2227. A system of co-planar forces whose components are (X_1, Y_1), (X_2, Y_2), ... act at the points (x_1, y_1), (x_2, y_2), ... and are equivalent to a single couple: prove that there will be equilibrium if each force be turned about its point of application in the same sense through the angle θ, where

$$\tan\theta = \frac{\Sigma(Xy - Yx)}{\Sigma(Xx + Yy)}.$$

2228. The sums of the moments of a given system of forces about three rectangular axes are respectively L, M, N; and the sums of the components in the directions of these axes are X, Y, Z: prove that

$$LX + MY + NZ$$

is independent of the particular system of axes.

[It is equal to RG, where R is the resultant force and G the minimum resultant couple.]

2229. Forces P, Q, R, P', Q', R' act along the edges BC, CA, AB, DA, DB, DC of a tetrahedron respectively: prove that there will be a single resultant if

$$\frac{P}{BC}\frac{P'}{AD} + \frac{Q}{CA}\frac{Q'}{BD} + \frac{R}{AB}\frac{R'}{CD} = 0;$$

and that the forces will be equivalent to a single couple if

$$\frac{P'}{AD} = \frac{R}{AB} - \frac{Q}{CA}, \quad \frac{Q'}{BD} = \frac{P}{BC} - \frac{R}{AB}, \quad \frac{R'}{CD} = \frac{Q}{CA} - \frac{P}{BC}.$$

2230. The necessary and sufficient conditions for the equilibrium of four equal forces acting at a point (not necessarily in one plane) are that the angle between the lines of action of any two is equal to that between the lines of action of the remaining two.

2231. Necessary and sufficient conditions of equilibrium for a system of forces acting on a rigid body are that the sum of the moments of all the forces about each edge of any one finite tetrahedron shall severally be equal to zero.

2232. Forces acting on a rigid body are represented by the edges of a given tetrahedron, three acting from one angular point towards the opposite face and the other three along the sides taken in order of the opposite face: prove that the product of the resultant force and of the minimum resultant couple will be the same whichever angular point be taken.

[The product will be represented on the same scale by $18V$, V being the volume of the tetrahedron.]

2233. A portion of a curve surface of continuous curvature is cut off by a plane, and at a point in each element of the portion a force proportional to the element is applied in direction of the normal: prove that, if all the forces act inwards or all outwards, they will in the limit have a single resultant.

2234. A system of forces acting on a rigid body is reducible to a single couple: prove that it is possible, by rotation about any proposed point, to bring the body into such a position that the forces, acting at the same points of the body in the same directions in space, shall be in equilibrium.

2235. A given system of forces is to be reduced to a force acting through a proposed point and a couple: prove that if the proposed point lie on a fixed straight line and through it be drawn always the axis of the couple, the extremity of this axis will lie on another fixed straight line.

2236. A given system of forces is to be reduced to two, both parallel to a fixed plane; straight lines representing these forces are drawn from the points where their lines of action are met by a fixed straight line which intersects both at right angles: prove that the locus of the other extremities of these straight lines is a hyperbolic paraboloid.

2237. Prove that the central axis of two forces P, Q intersects the shortest distance c between their lines of action, and divides it in the ratio

$$Q(Q + P\cos\theta) : P(P + Q\cos\theta),$$

θ being the angle between their directions. Also prove that the moment of the principal couple is

$$\frac{cPQ\sin\theta}{\sqrt{P^2 + Q^2 + 2PQ\cos\theta}}.$$

2238. A given system of forces is reduced to two, one of which F acts along a given straight line: prove that

$$\frac{1}{F} = \frac{\cos\theta}{R} + \frac{c\sin\theta}{G};$$

θ being the angle which the given straight line makes with the central axis, c the shortest distance between them, R the resultant force, and G the principal couple.

2239. A given system of forces is to be reduced to two at right angles to each other: prove that the shortest distance between their lines of action cannot be less than $2G \div R$. More generally, when the two are inclined at an angle 2α, the shortest distance cannot be less than $2G \div R\tan\alpha$.

2240. A given system of forces is reduced to two P, Q, and the shortest distances of their lines of action from the central axis are x, y respectively: prove that

$$P^2(R^2x^2 + G^2) = Q^2(R^2y^2 + G^2).$$

2241. Two forces act along the straight lines

$$x = a,\ y = z\tan\alpha;\quad x = -a,\ y = -z\tan\alpha:$$

prove that their central axis lies on the surface

$$x(y^2 + z^2)\sin 2\alpha = 2ayz,$$

the co-ordinates being rectangular.

2242. Two forces given in magnitude act along two straight lines not in one plane, a third force given in magnitude acts through a given point, and the three have a single resultant: prove that the line of action of the third force must lie on a certain cone of revolution.

[If R be the resultant force and G the principal couple which are together equivalent to the two given forces, P the third force, and a the distance of its point of application from the central axis of the two, the semi-vertical angle of the cone is

$$\cos^{-1}\left(\frac{GR}{P\sqrt{G^2+R^2a^2}}\right);$$

from which the conditions necessary for the possibility are obvious.]

2243. Forces X, Y, Z act along the three straight lines

$$y=b,\ z=-c;\quad z=c,\ x=-a;\quad x=a,\ y=-b;$$

respectively: prove that they will have a single resultant if

$$aYZ+bZX+cXY=0;$$

and that the equations of the line of action will be any two of the three

$$\frac{y}{Y}-\frac{z}{Z}+\frac{a}{X}=0,\quad \frac{z}{Z}-\frac{x}{X}+\frac{b}{Y}=0,\quad \frac{x}{X}-\frac{y}{Y}+\frac{c}{Z}=0.$$

II. *Centre of Gravity* (or *Inertia*).

2244. A rectangular board of weight W is supported in a horizontal position by vertical strings at three of its angular points; a weight $5W$ being placed on the board, the tensions of the strings become W, $2W$, $3W$: prove that the weight must be at one of the angular points of a hexagon whose opposite sides are equal and parallel, and whose area is to that of the board as 3 : 25.

2245. Particles are placed at the corners of a tetrahedron respectively proportional to the opposite faces: prove that their centre of gravity is at the centre of the sphere inscribed in the tetrahedron.

2246. A uniform wire is bent into the form of three sides of a polygon AB, BC, CD, and the centre of gravity of the whole wire is at the intersection of AC, BD: prove that, if E be the common point of AB, DC produced,

$$EB : BC : CE = AB^2 : BC^2 : CD^2.$$

2247. A thin uniform wire is bent into the form of a triangle ABC, and particles of weights P, Q, R are placed at the angular points: prove that, if the centre of gravity of the particles coincide with that of the wire,

$$P : Q : R = b+c : c+a : a+b.$$

2248. The straight lines, each joining an angular point of the triangle ABC to the common point of the tangents to the circle ABC at the ends of the opposite side, all meet in O: prove that, if perpendiculars be let fall from O on the sides, O will be the centroid of the triangle formed by joining the feet of these perpendiculars.

2249. Prove that a point O can always be found within a tetrahedron $ABCD$ such that, if Oa, Ob, Oc, Od be perpendiculars from O on the respective faces, O will be the centroid of the tetrahedron $abcd$; and that the distances of O from the faces will be respectively proportional to the faces.

[The point O, for either the triangle or the tetrahedron, is the point for which the sum of the squares of the distances from the sides or faces is the least possible.]

2250. Two uniform similar rods AB, BC, rigidly united at B and suspended freely from A, rest inclined at angles α, β to the vertical: prove that

$$\frac{AB}{BC} = \sqrt{1 + \frac{\sin\beta}{\sin\alpha}} - 1.$$

2251. Two uniform rods AB, BC are freely jointed at B and moveable about A, which is fixed; find at what point in BC a smooth prop should be applied so as to enable the rods to rest in one straight line inclined at a given angle to the horizon.

[If the weights of the rods be W, W', the point required must divide BC in the ratio $W' : W + W'$.]

2252. Four weights are placed at four fixed points in space, the sum of two of the weights being given and also the sum of the other two: prove that their centre of gravity lies on a fixed plane, and within a certain parallelogram in that plane.

2253. A polygon is such that the angles α_1, α_2, α_3, ... which its sides make with any fixed straight line satisfy the equations

$$\Sigma(\cos 2\alpha) = 0, \quad \Sigma(\sin 2\alpha) = 0:$$

prove that if O be the point which is the centre of gravity of equal particles placed at the feet of the perpendiculars from O on the sides, then the centre of gravity of equal particles, placed at the feet of the perpendiculars from any other point P, will bisect OP.

[Such a polygon has the property that the locus of a point, which moves so that the sum of the squares on its distances from the sides is constant, is a circle.]

2254. The limiting position of the centre of gravity of the area included between the area of a quadrant of an ellipse bounded by the axes and the corresponding quadrant of the auxiliary circle, when the ellipse approaches the circle as its limit, will be a point whose distance from the major axis is twice its distance from the minor.

2255. A curve is divided symmetrically by the axis of x and is such that the centre of gravity of the area included between the ordinates $x = 0$, $x = h$, is at a distance $\frac{2n-1}{3n-1}h$ from the origin: prove that the equation of the curve is

$$y^n = cx^{n-1}.$$

2256. The circle is the only curve in which the centre of gravity of the area included between any two radii drawn from a fixed point and the curve lies on the straight line bisecting the angle between the radii.

2257. Obtain the differential equation of a curve such that the centre of gravity of any arc measured from a fixed point lies on the straight line bisecting the angle between the radii drawn to the ends of the arc; and prove that the curve is a lemniscate of Bernoulli, with its radii drawn from the node, or a circle.

[The equation is $r\sqrt{r^2 + \left(\frac{dr}{d\theta}\right)^2} = a^2$, the general solution of which is $r^2 = a^2 \sin 2(\theta + \alpha)$, and a singular solution is $r = a$.]

III. *Equilibrium of Smooth Bodies.*

2258. A rectangular board is supported with its plane vertical by two smooth pegs and rests with one diagonal parallel to the straight line joining the pegs: prove that the other diagonal will be vertical.

2259. A rectangular board whose sides are a, b, is supported with its plane vertical on two smooth pegs in the same horizontal line at a distance c: prove that the angle θ made by the side a with the vertical when in equilibrium is given by the equation

$$2c \cos 2\theta = b \cos\theta - a \sin\theta.$$

2260. A uniform rod, of length c, rests with one end on a smooth elliptic arc whose major axis is horizontal and with the other on a smooth vertical plane at a distance h from the centre of the ellipse: prove that, if θ be the angle which the rod makes with the horizon and $2a$, $2b$ the axes of the ellipse,

$$2b \tan\theta = a \tan\phi, \text{ where } a\cos\phi + h = c\cos\theta;$$

and explain the result when $a = 2b = c$, $h = 0$.

2261. A rod of length a, whose centre of gravity is at a distance b from its lower extremity, rests in neutral equilibrium with the upper extremity on a fixed vertical plane and the lower extremity on an elliptic arc whose axes are $2a$, $2b$: prove that the moments about the centre of the ellipse of the three forces which keep the rod in equilibrium are in the constant ratios $-a : -b : a+b$.

2262. A lamina in the form of a rhombus made up of two equilateral triangles rests with its plane vertical between two smooth pegs in the same horizontal plane at a distance equal to a quarter of the longer diagonal: prove that either a side or a diagonal of the rhombus must be vertical, and that the stable position is that in which a diagonal is vertical.

2263. A straight uniform rod has smooth small rings attached to its extremities, one of which slides on a fixed vertical straight wire and the other on a fixed wire in the form of a parabolic arc whose axis coincides with the straight wire and whose latus rectum is twice the length of the rod: prove that in the position of equilibrium (stable when the vertex is upwards), the rod will be inclined at an angle of 30° to the horizon. Which is the position of stable equilibrium when the vertex is downwards?

2264. An elliptic lamina of axes $2a$, $2b$, rests with its plane vertical on two smooth pegs in the same horizontal line at a distance c: prove that, when $c < b\sqrt{2}$ or $> a\sqrt{2}$, the only positions of equilibrium are when one axis is vertical; and that, when $c > b\sqrt{2}$ and $< a\sqrt{2}$, the positions in which an axis is vertical are both stable and there are positions of unstable equilibrium in which the pegs are ends of conjugate diameters.

2265. A rectangular lamina rests in a vertical plane with one corner against a smooth vertical wall and an opposite side against a smooth peg: the position of equilibrium is given by the equation

$$c\sqrt{a^2+b^2} = 2b\,(b\sin\theta - a\cos\theta) + \sin\theta\,(b\cos\theta + a\sin\theta)^2;$$

where $2a$, $2b$ are the sides (the latter in contact with the peg), θ the angle which the diagonal through the point of contact makes with the vertical, and c the distance of the peg from the wall.

2266. Two similar uniform straight rods of lengths $2a$, $2b$, rigidly united at their ends at an angle α, rest over two smooth pegs in the same horizontal plane: prove that the angle which the rod $2a$ makes with the vertical is given by the equation

$$c\,(a+b)\sin(2\theta-\alpha) = a^2\sin\alpha\sin\theta - b^2\sin\alpha\sin(\alpha-\theta),$$

c being the distance between the pegs.

2267. A uniform lamina in the form of a parallelogram rests with two adjacent sides on two smooth pegs in the same horizontal plane at a distance c, $2h$ is the length of the diagonal through the intersection of the two sides, α, β, θ the angles which this diagonal makes with the sides and with the vertical: prove that

$$h\sin\theta\sin(\alpha+\beta) = c\sin(\beta-\alpha+2\theta).$$

2268. A uniform triangular lamina ABC, rough enough to prevent sliding, is attached to a fixed point O by three fine strings OA, OB, OC, and on the lamina is placed a weight w: prove that the tensions of the strings are as $OA\,(W+3xw) : OB\,(W+3yw) : OC\,(W+3zw)$, where W is the weight of the lamina and x, y, z the areal co-ordinates, measured on the triangle ABC, of the point where w is placed. Also prove that the least possible value of w for which the tensions can be equal is

$$\tfrac{1}{3}W\,.\,OA\left(\frac{1}{OB}+\frac{1}{OC}-\frac{2}{OA}\right),$$

where OA is the longest string.

2269. A lamina in the form of an isosceles triangle rests with its plane vertical and its two equal sides each in contact with a smooth peg, the pegs being in the same horizontal plane: prove that the axis of the triangle makes with the vertical the angle 0 or $\cos^{-1}\left(\frac{h\sin\alpha}{3c}\right)$; h being the length of the axis, α the vertical angle, and c the distance between the pegs.

2270. A uniform rod AB of length $2a$ is freely moveable about A; a smooth ring of weight P slides on the rod and has attached to it a fine string which passes over a pulley at a height a vertically above A and supports a weight Q hanging freely: find the position of equilibrium of the system; and prove that, if in this position the rod and string are equally inclined to the vertical,

$$2Q(Qb - Wa)^2 = P^2Wab.$$

2271. A portion of a parabolic lamina, cut off by a focal chord inclined at an angle α to the axis, rests with its chord horizontal on two smooth pegs in the same horizontal line at a distance c: prove that the latus rectum of the parabola is $c\sqrt{5}\sin^2\alpha$, that the distance between the pegs is $\frac{1}{\sqrt{5}} \times$ length of the bounding chord, and that the centre of gravity of the lamina bisects the distance between the mid points of the bounding chord and of the straight line joining the pegs.

2272. A portion of a parabolic lamina cut off by a focal chord inclined at an angle α to the axis rests on two smooth pegs at a distance b, with its chord c parallel to the distance between the pegs and inclined at an angle β to the vertical: prove that

$$\frac{5b^2}{c^2} = \frac{3\cos(2\alpha + \beta) + 17\cos\beta}{3\cos(2\alpha + \beta) + \cos\beta}.$$

2273. A small smooth heavy ring is capable of sliding on a fine elliptic wire whose major axis is vertical; two strings attached to the ring pass through small smooth rings at the foci and sustain given weights: prove that, if there be equilibrium in any position in which the whole string is not vertical, there will be equilibrium in every position. Prove also that, when this is the case, the pressure on the wire will be a maximum when the sliding ring is in the highest or lowest positions, and a minimum when its distances from the foci are respectively as the weights sustained.

[The maximum pressures are

$$w_1(1+e) + w_2(1-e), \quad w_1(1-e) + w_2(1+e),$$

and the minimum is $2\sqrt{1-e^2}\sqrt{w_1w_2}$; where w_1, w_2 are the weights sustained at the upper and lower foci, and e the excentricity of the ellipse. When $w_1(1-e) > w_2(1+e)$, the pressure will be a maximum in the highest position, and a minimum in the lowest, and there will be no other maximum or minimum pressures.]

2274. A uniform regular tetrahedron has three corners in contact with the interior of a fixed smooth hemispherical bowl of such magnitude that the completed sphere would circumscribe the tetrahedron: prove that every position is one of equilibrium; and that, if P, Q, R be the pressures at the corners and W the weight of the tetrahedron,

$$2(QR + RP + PQ) = 3(P^2 + Q^2 + R^2 - W^2).$$

2275. A heavy uniform tetrahedron rests with three of its faces against three fixed smooth pegs and the fourth face horizontal: prove that the pressures on the pegs are as the areas of the faces respectively in contact.

2276. A heavy uniform ellipsoid is placed on three smooth pegs in the same horizontal plane so that the pegs are at the extremities of a system of conjugate diameters: prove that there will be equilibrium, and that the pressures on the pegs will be one to another as the areas of the corresponding conjugate central sections.

2277. Seven equal and similar uniform rods AB, BC, CD, DE, EF, FG, GA are freely jointed at their extremities and rest in a vertical plane supported by rings at A and C, which are capable of sliding on a smooth horizontal rod: prove that, θ, ϕ, ψ being the angles which BA, AG, GF make with the vertical,

$$\tan\theta = 5\tan\phi = 3\tan\psi.$$

2278. Two spheres of densities ρ, σ and radii a, b, rest in a paraboloid whose axis is vertical and touch each other at the focus: prove that $\rho^3 a^{10} = \sigma^3 b^{10}$; also that, if W, W' be their weights and R, R' the pressures at the points of contact with the paraboloid,

$$\frac{R}{W} - \frac{R'}{W'} = \frac{1}{2}\left(\frac{R}{W'} - \frac{R'}{W}\right).$$

2279. Four uniform similar rods freely jointed at their extremities form a parallelogram, and at the middle points of the rods are small smooth rings joined by light rigid bars. The parallelogram is suspended freely from an angular point; find the stresses along the bars and the pressures of the rings on the rods, and prove that (1) if the parallelogram be a rectangle the stresses will be equal, (2) if a rhombus the pressures will be equal.

IV. *Friction.*

2280. Find the least coefficient of friction between a given elliptic cylinder and a particle, in order that for all positions of the cylinder in which the axis is horizontal, the particle may be capable of resting vertically above the axis.

[If the axes of the transverse section be $2a$, $2b$, the least coefficient of friction is $\tan^{-1}\left(\frac{a^2-b^2}{2ab}\right)$.]

2281. Two given weights of different material are laid on a given inclined plane and connected by a string in a state of tension inclined at a given angle to the intersection of the plane with the horizon, and the lower weight is on the point of motion: determine the coefficient of friction of the lower weight and the magnitude and direction of the force of friction on the upper weight.

2282. A weight w rests on a rough inclined plane ($\mu < 1$) supported by a string which, passing over a smooth pulley at the highest point of the plane, sustains a weight $> \mu w$ and $< w$ hanging vertically: prove that the angle between the two positions of the plane in which w is in limiting equilibrium is $2 \tan^{-1} \mu$.

2283. Two weights of similar material connected by a fine string rest on a rough circular arc on which the string lies: prove that the angle subtended at the centre by the distance between the limiting positions of either weight is $2 \tan^{-1} \mu$.

2284. A uniform rod rests with one extremity against a rough vertical wall, the other being supported by a string of equal length fastened to a point in the wall: prove that the least angle which the string can make with the wall is $\tan^{-1}(3\mu^{-1})$.

2285. A uniform rod of weight W rests with one end against a rough vertical plane and with the other end attached to a string which passes over a smooth pulley vertically above the former end and supports a weight P: find the limiting positions of equilibrium, and prove that equilibrium will be impossible if $P < W \cos \epsilon$, ϵ being the angle of friction.

2286. Two weights support each other on a rough double inclined plane by means of a fine string passing over the vertex, and both weights are on the point of motion: prove that, if the plane be tilted until both weights are again on the point of motion, the angle through which the plane will be turned is twice the angle of friction.

2287. A uniform heavy rod rests, with one extremity against a rough vertical wall, supported by a smooth horizontal bar parallel to the wall, and the angles between the rod and wall in the limiting positions of equilibrium are a, β: prove that the coefficient of friction is

$$\frac{\sin^3 \beta - \sin^3 a}{\sin^2 a \cos a + \sin^2 \beta \cos \beta}.$$

2288. A heavy uniform rod of weight W rests inclined at an angle θ to the vertical in contact with a rough cylinder of revolution whose axis is horizontal and whose diameter is equal in length to the rod: the rod is maintained in its position by a fine string in a state of tension which passes from one end of the rod to the other round the cylinder: prove that the tension of the string cannot be less than

$$W \cos(\theta + \epsilon) \div 2 \sin \epsilon,$$

where ϵ is the angle of friction.

2289. A square lamina has a string of length equal to that of a side attached at one of the corners; the string is also attached to a fixed point in a rough vertical wall, and the lamina rests with its plane vertical and perpendicular to that of the wall: prove that, if the coefficient of friction be 1, the angle which the string makes with the wall lies between $\frac{\pi}{4}$ and $\frac{1}{2} \tan^{-1} \frac{1}{2}$.

[More generally, if the lamina be rectangular, of sides a, b, and the length of the string be a, and the wall be inclined at an angle α to the horizon, the angle θ which the string makes with the wall in a position of limiting equilibrium is given by the equation

$$\sin(2\theta - \alpha + \beta \pm \epsilon) - 2\sin 2\theta \cos(\alpha + \beta \mp \epsilon) = \sin(\alpha - \beta \pm \epsilon);$$

where $a = b \tan\beta$, and ϵ is the angle of friction.]

2290. Two weights P, Q, of similar material, resting on a rough double inclined plane, are connected by a fine string passing over the common vertex, and Q is on the point of motion down the plane: prove that the greatest weight which can be added to P without disturbing the equilibrium is

$$\frac{P \sin 2\epsilon \sin(\alpha + \beta)}{\sin(\alpha - \epsilon)\sin(\beta - \epsilon)},$$

α, β being the angles of inclination of the planes and ϵ the angle of friction.

2291. A uniform rod rests with one extremity against a rough vertical wall ($3\mu = 7$), the other extremity being supported by a string three times the length of the rod attached to a point in the wall: prove that the tangent of the angle which the string makes with the wall in the limiting position of equilibrium is $\frac{5}{27}$ or $\frac{1}{3}$.

2292. A given weight resting upon a rough inclined plane is connected with a weight P by means of a string passing over a rough peg, P hanging freely; the angles of friction for the peg and plane are λ, λ' respectively, ($\lambda > \lambda'$): prove that the inclination of the string to the plane in limiting equilibrium, when P is a maximum or minimum, is $\lambda - \lambda'$.

2293. A weight W is supported on a rough inclined plane of inclination α by a force P, whose line of action makes an angle ι with the plane and whose component in the plane makes an angle β with the line of greatest inclination in the plane: prove that equilibrium will be impossible if

$$\mu^2(1 + \cos 2\alpha \cos 2\iota - \sin 2\alpha \sin 2\iota \cos\beta) > 2\sin^2\alpha \sin^2\beta \cos^2\iota.$$

2294. A heavy particle is attached to a point in a rough inclined plane by a fine weightless rigid wire and rests on the plane with the wire inclined at an angle θ to the line of greatest inclination in the plane; determine the limits of θ, the angle of inclination of the plane being $\tan^{-1}(\mu \operatorname{cosec}\beta)$.

[The limiting values of θ are β, $\pi - \beta$, and θ must not lie between these limits.]

2295. Two weights A, B connected by a fine string are lying on a rough horizontal plane; a given force P ($> \mu\sqrt{A^2 + B^2}$ and $< \mu A + \mu B$) is continually applied to A so as just to move A and B very slowly in the plane: prove that A and B will describe concentric circles whose radii are $a \operatorname{cosec}\beta$, $a \cot\beta$, where $\cos\beta = \dfrac{P^2 - \mu^2 A^2 - \mu^2 B^2}{2\mu^2 AB}$.

V. *Elastic Strings.*

2296. A string whose extensibility varies as the distance from one end is stretched by any force: prove that its extension is equal to that of a string of equal length, of uniform extensibility equal to that at the centre of the former, when stretched by an equal force.

2297. An elastic string rests on a rough inclined plane with the upper end fixed to the plane: prove that its extension will lie between the limits $\frac{l^2}{2\lambda}\frac{\sin(\alpha \pm \epsilon)}{\cos\epsilon}$; α being the inclination of the plane, ϵ the angle of friction, and l, λ the lengths of the whole string and of a portion of it whose weight is equal to the modulus.

2298. Two weights P, Q are connected by an elastic string without weight which passes over two small rough pegs A, B in the same horizontal line at a distance a, Q is just sustained by P, and $AP = b$, $BQ = c$: P and Q are then interchanged, and $AQ = b'$, $BP = c'$: obtain equations for determining the natural length of the string, its modulus, and the coefficients of friction at A and B.

2299. A weight P just supports another weight Q by means of a fine elastic string passing over a rough cylinder of revolution whose axis is horizontal; W is the modulus and a the radius of the cylinder: prove that the extension of the part of the string in contact with the cylinder is $\frac{a}{\mu}\log\left(\frac{Q+W}{P+W}\right)$.

2300. A heavy extensible string, uniform when unextended, hangs symmetrically over a cylinder of revolution whose axis is horizontal, a portion whose length in the position of rest is $a - h$ hanging vertically on each side: prove that the natural length of the part of the string in contact with the cylinder is $2\sqrt{2ah}\log(\sqrt{2}+1)$; a being the radius of the cylinder, and $2h$ the length of a portion of the string whose weight (when unextended) is equal to the modulus: also prove that the extension of either of the vertical portions of the string is $(\sqrt{a}-\sqrt{h})^2$.

2301. An extensible string is laid on a cycloidal arc whose plane is vertical and vertex upwards, and when stretched by its own weight is just in contact with the whole of the cycloid, the natural length of the string being equal to the perimeter of the generating circle: prove that the modulus is the weight of a portion of the string whose natural length is twice the diameter of the generating circle.

2302. A heavy elastic string whose natural length is $2l$ is placed symmetrically on the arc of a smooth cycloid whose axis is vertical and vertex upwards, and a portion of string whose natural length is x hangs vertically at each cusp: prove that

$$2\sqrt{a\lambda} = (x+\lambda)\tan\frac{l-x}{2\sqrt{a\lambda}};$$

$2a$ being the length of the axis of the cycloid, and λ the natural length of a portion of the string whose weight is equal to the modulus.

2303. A smooth right cylinder whose base is a cardioid is placed with the axis of the cardioid vertical and vertex upwards, and a heavy extensible string rests symmetrically upon the upper part in contact with a portion of the cylinder whose length is twice the axis of the cardioid, and the length of string whose weight is equal to the modulus is equal to the length of the axis: prove that the natural length of the string is to its length when resting on the cylinder as $\log(2+\sqrt{3}) : \sqrt{3}$.

2304. An extensible string of natural length $2l$ just surrounds a smooth lamina in the form of a cardioid, its free extremities being at the cusp, and remains in equilibrium under the action of an attractive force varying as the distance and tending to the centre of the fixed circle (when the cardioid is described as an epicycloid): prove that

$$\sqrt{\frac{\lambda}{2k\mu}} = 2a \cot\left(\frac{l}{2}\sqrt{\frac{k\mu}{2\lambda}}\right);$$

a being the radius of the fixed circle, $2kl$ the mass of the string, λ the modulus, and μr the acceleration of the force on unit mass at a distance r.

VI. *Catenaries, Attractions, &c.*

2305. An endless heavy chain of length $2l$ is passed over a smooth cylinder of revolution whose axis is horizontal; c is the length of a portion of the chain whose weight is equal to the tension at the lowest point, and 2ϕ the angle between the radii drawn to the points where the chain leaves the cylinder: prove that

$$\tan\phi + \frac{\pi-\phi}{\sin\phi}\log\tan\left(\frac{\pi}{4}+\frac{\phi}{2}\right) = \frac{l}{c}.$$

2306. In a common catenary A is the vertex, P, Q two points at which the tangents make angles ϕ, 2ϕ respectively with the horizon, and the tangents at A, Q meet in O: prove that the arc AP is equal to the horizontal distance between O and Q.

2307. Four pegs A, B, C, D are placed at the corners of a square, BC being vertically downwards, and an endless uniform inextensible string passes round the four hanging in two festoons: prove that

$$\frac{1}{\sin\alpha\log\cot\frac{\alpha}{2}} - \frac{1}{\sin\beta\log\cot\frac{\beta}{2}} = 2,$$

$$\frac{1}{\tan\frac{\alpha}{2}\log\cot\frac{\alpha}{2}} - \frac{1}{\tan\frac{\beta}{2}\log\cot\frac{\beta}{2}} = \frac{l}{a};$$

α, β being the angles which the tangents at B, C make with the vertical, l the length of the string, and a the length of a side of the square.

2308. A heavy uniform chain rests in limiting equilibrium on a rough circular arc whose plane is vertical, in contact with a quadrant of the circle one end of which is the highest point of the circle: prove that

$$(1-\mu^2)\,\epsilon^{\mu\frac{\pi}{2}} = 2\mu.$$

2309. A heavy uniform chain rests in limiting equilibrium on a rough cycloidal arc whose axis is vertical and vertex upwards, one extremity being at the vertex and the other at a cusp: prove that

$$(1+\mu^2)\,\epsilon^{\mu\frac{\pi}{2}} = 3.$$

2310. A uniform inextensible string hangs in the form of a common catenary, the forces at any point being X, Y perpendicular and parallel to the axis: prove that

$$\sin\phi\,\frac{dX}{d\phi} + \cos\phi\,\frac{dY}{d\phi} + 2X\sec\phi = 0\,;$$

where ϕ is the angle which the normal at the point makes with the axis.

2311. To each point of a chain hanging under gravity only in the form of the catenary $s = c\tan\phi$ is applied a horizontal force proportional to s^{-3}: prove that the form will be unaltered.

2312. A uniform inextensible string can rest in the form of a two-cusped epicycloid under the action of a constant force always tending from the centre of the moving circle.

2313. A uniform inextensible string rests in the form of a circle under a force which is always proportional to the square root of the tension: prove that the force is proportional to the distance from a fixed point on the circle and that its line of action always touches a certain cardioid.

[More generally, if the force vary as the nth power of the tension, its line of action will always touch an epicycloid generated by a circle of radius $a \div 2\,(2-n)$ rolling on a circle of radius

$$a\,(1-n) \div (2-n).]$$

2314. A uniform chain is kept in equilibrium in the form of an ellipse by repulsive forces F_1, F_2 in the foci: prove that

$$\frac{1}{r_1}\,\frac{d}{dr_1}\left(r_1^{\frac{3}{2}}\,r_2^{\frac{1}{2}}\,F_1\right) = \frac{1}{r_2}\,\frac{d}{dr_2}\left(r_2^{\frac{3}{2}}\,r_1^{\frac{1}{2}}\,F_2\right);$$

where r_1, r_2 are the focal distances.

2315. A uniform chain is in equilibrium in the form of an equiangular spiral and the tension is proportional to the radius vector: prove that the force is constant and makes a constant angle with the radius vector. When the chain is in equilibrium in the form of an equiangular spiral under a constant force which at any point makes an angle θ with the normal,

$$\sin\theta\frac{d\theta}{d\phi}=\sin\theta+\cot a\cos\theta;$$

where ϕ is the angle which the tangent makes with a fixed direction.

2316. A uniform chain can rest in the form of a common catenary under the action of a constant force if the force at any point make with the axis an angle θ determined by the equation

$$2\frac{d}{d\theta}(\tan\phi\sqrt{\sin\theta})-\sqrt{\sin\theta}=0;$$

ϕ being the angle which the normal at the point makes with the axis.

2317. A heavy uniform chain fastened at two points rests in the form of a parabola under the action of two forces, one (A) parallel to the axis and constant, and the other (F) tending from the focus: prove that $3F=A+B\cos^2\phi$, ϕ being the angle through which the tangent has turned since leaving the vertex and B a constant.

2318. Find the law of repulsive force tending from a focus under which an endless uniform chain can be kept in equilibrium in the form of an ellipse; and, if there be two such forces, one in each focus and equal at equal distances, prove that the tension at any point varies inversely as the conjugate diameter.

2319. A uniform chain rests in the form of a cycloid whose axis is vertical under the action of gravity and of a certain normal force, the tension at the vertex vanishing: prove that the tension at any point is proportional to the vertical height above the vertex, and that the normal force at any point bears to the force of gravity the ratio

$$(3\cos^2\theta-1) : 2\cos\theta;$$

where θ is the angle which the normal makes with the vertical.

2320. A heavy chain of variable density suspended from two points hangs in the form of a curve whose intrinsic equation is $s=f(\phi)$, the lowest point being origin: prove that the density at any point will vary inversely as $\cos^2\phi f'(\phi)$.

2321. A string is kept in equilibrium in the form of a closed curve by the action of a repulsive force tending from a fixed point, and the density at each point is proportional to the tension: prove that the force at any point is inversely proportional to the chord of curvature through the centre of force.

2322. A uniform chain is in equilibrium under the action of certain forces; from a fixed point O is drawn a straight line Op parallel to the tangent at any point P of the chain and proportional to the tension at P: prove that, (1) the tangent at p to the locus of p is parallel to the resultant force at P, (2) the ultimate ratio of small corresponding arcs at p, P is proportional to the resultant force at P.

2323. A uniform heavy chain rests in contact with a smooth arc in a vertical plane of such a form that the pressure at any point per unit of length is equal to m times the weight of a unit of length: prove that the intrinsic equation of the curve will be

$$\frac{ds}{d\phi} = \frac{a}{(m + \cos\phi)^2};$$

that, when $m > 1$, the horizontal distance between two consecutive vertices is $\dfrac{2\pi a}{(m^2-1)^{\frac{3}{2}}}$, the arc between the same points $\dfrac{2m\pi a}{(m^2-1)^{\frac{3}{2}}}$, and the vertical distance between the lines of highest and lowest points $2a \div (m^2 - 1)$. When $m = 1$, the curve is the first negative pedal of a parabola from the focus.

2324. A uniform heavy string is attached to two points in the surface of a smooth cone of revolution whose axis is vertical and rests with every point of its length in contact with the cone: prove that the curve of equilibrium is such that its differential equation, when the cone is developed into a plane, is $p(r+c) = a^2$, the vertex of the cone being pole.

2325. A uniform chain is laid upon the arc of a smooth curve which is the evolute of a common catenary so that a portion hangs vertically below the cusp of a length equal to the diameter of the catenary at the vertex: prove that the resolved vertical tension at any point of the arc is constant, and that the resolved vertical pressure per unit of length is equal to the weight of a unit of length of the chain. Also, in the curve whose intrinsic equation is

$$s = a\sin\phi \div \sqrt{1 + \cos^2\phi},$$

where ϕ is measured from the horizontal tangent, if a uniform chain be bound tightly on any portion of it so that the tension at a vertex is equal to the weight of a length $a\sqrt{2}$ of the chain, the resolved vertical pressure per unit will be equal to the weight of a unit of length and the resolved vertical tension at any point will be twice the weight of the chain intercepted between that point and the vertex.

[The height above the directrix of the C. G. of the portion of chain included between two cusps is

$$\frac{a}{\sqrt{2}}\log(\sqrt{2}+1) + a,$$

and the area included between the directrix, the curve, and the tangents at two consecutive cusps is $\pi a^2 \div \sqrt{2}$.]

2326. A heavy uniform chain just rests upon a rough curve in the form of the arc of a four-cusped hypocycloid, occupying the space between two consecutive cusps at which the tangents are horizontal and vertical respectively: prove that

$$2\mu\epsilon^{\mu\frac{\pi}{2}} = \mu^2 + 3.$$

2327. Find a curve such that the c. g. of any arc lies in a straight line drawn in a given direction through the intersection of the tangents at the ends of the arc.

[It is obvious that the common catenary satisfies the condition, and it will be found that, when the arc is uniform, no other curve does so. When the density is variable and the curve such that $\frac{ds}{d\phi}\cos^2\phi$ varies inversely as the density, the condition will be satisfied if the direction from which ϕ is measured be at right angles to the given direction.]

2328. A uniform chain rests in a vertical plane on a rough curve in the form of an equiangular spiral whose constant angle between the normal and radius vector is equal to the angle of friction, one end being at a point where the tangent is horizontal: prove that, for limiting equilibrium, the chain will subtend at the pole an angle equal to twice the angle of friction. (The chain makes an obtuse angle with the radius vector to the highest point.)

2329. A uniform wire in the form of a lemniscate of Bernoulli attracts a particle at the node, the force varying as the distance: prove that the attraction of any arc is the same as that of a circular arc of the same material touching the lemniscate at its vertices and intercepted between the same radii from the node. The same property will hold for an equiangular spiral when the force varies inversely as the distance, and for a rectangular hyperbola when the force varies inversely as the cube of the distance, and generally for any curve

$$r^n = a^n \sin n\theta,$$

if the force vary as r^{n-1}.

DYNAMICS, ELEMENTARY.

I. *Rectilinear Motion: Impulses.*

2330. A ball A impinges on another ball B, and after impact the directions of motion of A and B make equal angles θ with the previous direction of A: determine θ, and prove that, when $A = B$, $\tan\theta = \sqrt{e}$, where e is the coefficient of restitution.

[In general $B(1-e)\cos 2\theta = A + eB$.]

2331. A smooth inelastic ball, mass m, is lying on a horizontal table in contact with a vertical wall and is struck by another ball, mass m', moving in a direction normal to the wall and inclined at an angle α to the common normal at the point of impact: prove that the angle θ, through which the direction of motion of the striking ball is turned, is given by the equation $m\cot\theta\cot\alpha = m + m'$.

2332. Two equal balls A, B are lying very nearly in contact on a smooth horizontal table; a third equal ball impinges directly on A, the three centres being in one straight line: prove that if $e > 3 - 2\sqrt{2}$, the final velocity of B will bear to the initial velocity of the striking ball the ratio $(1+e)^2 : 4$.

2333. Equal particles $A_1, A_2, \ldots A_n$ are fastened at equal intervals a on a fine string of length $(n-1)a$ and are then laid on a horizontal table at n consecutive angular points of a regular polygon of p sides $(p > n)$, each equal to a; a blow P is applied to A_1 in direction A_1A_p: prove that the impulsive tension of the string A_rA_{r+1} is

$$P\cos^r\alpha\,\frac{(1+\sin\alpha)^{n-r} - (1-\sin\alpha)^{n-r}}{(1+\sin\alpha)^n - (1-\sin\alpha)^n};$$

where $p\alpha$ is equal to 2π.

2334. A circle has a vertical diameter AB, and two particles fall down two chords AP, PB respectively, starting simultaneously from A, P: prove that the least distance between them during the motion is equal to the distance of P from AB.

2335. A number of heavy particles start at once from the vertex of an oblique circular cone whose base is horizontal and fall down generating lines of the cone: prove that at any subsequent instant they will all lie in a subcontrary section.

2336. The locus of a point P such that the times of falling down PA, PB to two fixed points A, B may be equal is a rectangular hyperbola in which AB is a diameter and the normals at A, B are vertical.

2337. The locus of a point P such that the time of falling down PA to a fixed point A is equal to the time of falling vertically from A to a fixed straight line is one branch of an hyperbola in which one asymptote is vertical and the other perpendicular to the fixed straight line.

[The other branch of the hyperbola is the locus of a point P such that the time down AP is equal to the time from the straight line vertically to P.]

2338. A parabola is placed with its axis vertical and vertex downwards: prove that the time of falling down any chord to the vertex is equal to the time of falling vertically through a space equal to the parallel focal chord.

2339. An ellipse is placed with its major axis vertical: prove that the time of descent down any chord to the lower vertex, or from the higher vertex, is proportional to the length of the parallel diameter.

2340. The radii of two circles in one vertical plane, whose centres are at the same height, are a, b and the distance between their centres is c (which is greater than $a+b$): prove that the shortest time of descent from one circle to the other down a straight line is $\sqrt{\dfrac{2}{g}\dfrac{c^2-(a+b)^2}{a+b}}$.

2341. The radii of two circles in one vertical plane are a, b, the distance between their centres c, and the inclination of this distance to the vertical is α: prove that, when $c > (a+b)$, the time of shortest descent down a straight line from one circle to the other is equal to the time of falling vertically through a space $\dfrac{c^2-(a+b)^2}{a+b+c\cos\alpha}$; and, when $a > (b+c)$, the shortest time from the outer to the inner is $\sqrt{\dfrac{2}{g}\dfrac{(a-b)^2-c^2}{a-b+c\cos\alpha}}$, and from the inner to the outer $\sqrt{\dfrac{2}{g}\dfrac{(a-b)^2-c^2}{a-b-c\cos\alpha}}$, α being the angle which the line of centres makes with the vertical measured upwards from the centre of the outer circle. Also prove that, when $c\cos\alpha > (a+b)$, there will be a maximum time of descent from one circle to the other down a straight line, and this time will be $\sqrt{\dfrac{2}{g}\dfrac{c^2-(a+b)^2}{c\cos\alpha-(a+b)}}$.

[In the first case, the length of the line of shortest descent is

$$\frac{c^2 - (a+b)^2}{\sqrt{c^2 + (a+b)^2 + 2(a+b)c\cos a}},$$

and the angle which it makes with the vertical is

$$\cos^{-1}\frac{a+b+c\cos a}{\sqrt{c^2 + (a+b)^2 + 2(a+b)c\cos a}};$$

and similarly in the other cases.]

2342. A parabola is placed with its axis horizontal: prove that the length of the straight line of shortest descent from the curve to the focus is one third of the latus rectum.

2343. A parabola is placed with its plane vertical and its axis inclined at an angle $3a$ to the vertical: prove that the straight line of shortest descent from the curve to the focus is inclined at an angle a to the vertical.

2344. An ellipse is placed with its major axis vertical: prove that the straight line of quickest descent from the curve to the lower focus (or from the higher focus to the curve) is equal in length to the latus rectum, provided the excentricity exceed $\frac{1}{2}$.

2345. Two straight lines OP, OQ from a given point O to a given circle in the same vertical plane are such that the times of falling down them are equal: prove that PQ passes through a fixed point.

2346. Two circles A, B in the same vertical plane are such that the centre of A is the lowest point of B; through each point P on B are drawn two straight lines to A such that the times down them are equal to the time from P to the centre of A: prove that the chord of A joining the ends of these lines will touch a fixed circle concentric with A.

2347. There are two given circles in one vertical plane and from each point of one are drawn the two straight lines of given time of descent (t) to the other: prove that the chord joining the ends of these lines envelopes a conic, whose focus is vertically below the centre of the former circle at a depth $\frac{1}{4}gt^2$.

2348. Two weights W, W' move on two inclined planes, and are connected by a fine string passing over the common vertex, the whole motion being in one plane: prove that the centre of gravity of the weights describes a straight line with uniform acceleration equal to

$$g\,\frac{W'\sin\beta - W\sin\alpha}{(W+W')^2}\sqrt{W^2 + W'^2 + 2WW'\cos(\alpha+\beta)};$$

where α, β are the inclinations of the planes.

2349. When there is equilibrium in the single moveable pulley, the weight is suddenly doubled and the power is halved: prove that, in the ensuing motion, the tensions of the strings are the same as in equilibrium.

2350. In the system of pullies in which each hangs by a separate string, P just supports W: prove that, if P be removed and another weight Q be substituted, the centre of gravity of Q and W will descend with uniform acceleration

$$g\frac{W(Q-P)^2}{(Q+W)(P^2+QW)}.$$

2351. In any machine without friction and inertia a weight P supports a weight W, both hanging by vertical strings; these weights are removed and weights P', W', respectively substituted: prove that, if in the subsequent motion P' and W' always move vertically, their centre of gravity will descend with acceleration

$$g\frac{(WP'-W'P)^2}{(P'+W')(W^2P'+P^2W')}.$$

2352. Two weights each of 1 lb. support each other by means of a fine string passing over a moveable pulley to which is attached another string passing over another pulley and supporting a weight of 2 lbs.; to this pulley is similarly attached another string supporting a weight of 4 lbs., and so on, the last string passing over a fixed pulley and supporting a weight of 2^n lbs.: prove that, if the rth weight, reckoning from the top, be gently raised through a space of 2^r-1 inches, all the other weights will each fall one inch; and, if the rth weight be in any way gradually brought to rest, all the weights will come to rest at the same instant. (The pulleys are of insensible mass.)

2353. A fine uniform string of length $2a$ is in equilibrium, passing over a small smooth pulley, and is just displaced: prove that the velocity of the string when just leaving the pulley is $\sqrt{ag}$.

2354. A large number of equal particles are fastened at unequal intervals to a fine string and then collected into a heap at the edge of a smooth horizontal table with the extreme one just hanging over the edge; the intervals are such that the times between successive particles being carried over the edge are equal: prove that, if c_n be the length of string between the nth and $n+1^{\text{th}}$ particle, and v_n the velocity just after the $n+1^{\text{th}}$ particle has been carried over,

$$c_n=nc_1,\quad v_n=nv_1.$$

Deduce the law of density of a string collected into a heap at the edge of the table with the end just over the edge, in order that equal masses may always pass over in equal times.

[The density must vary inversely as the square root of the distance from the end.]

2355. A large number of equal particles are attached at equal intervals to a string and the whole is heaped up close to the edge of a smooth horizontal table with the extreme particle just over the edge: prove that, if v_n denote the velocity just before the $n+1^{\text{th}}$ particle is set in motion,

$$v_n^2 = \frac{ag}{3}\frac{(n+1)(2n+1)}{n},$$

where a denotes the length between two consecutive particles. Calculate the dissipated energy, and prove that, when a is indefinitely diminished, the end of the string, in the limit, descends with uniform acceleration $\frac{1}{3}g$.

[The whole energy dissipated, just before the $n+1^{\text{th}}$ particle is set in motion, is $\frac{1}{6}aw(n^2-1)$, where w is the weight of each particle.]

2356. A large number of equal particles are attached at equal intervals a to a fine string which passes through a very short fine tube in the form of a semicircle, and initially there are $2r$ particles on one side of the tube, the highest being at the tube, and r particles on the other side, the lowest being in contact with a horizontal table where the remaining particles are gathered together in a heap: prove that, if v_n denote the velocity just before the nth additional particle is set in motion,

$$v_n^2 = \frac{nag}{3}\left\{2 + \frac{n-1}{(n+3r-1)^2}\right\};$$

and deduce the corresponding result for a uniform chain hanging over a small pulley.

II. *Parabolic Motion.*

2357. A heavy particle is projected from a given point A in a given direction: determine its velocity in order that it may pass through another given point B.

[If the polar co-ordinates of B referred to A be (a, α), and β be the angle which the given direction makes with the horizontal initial line, the space due to the velocity of projection will be

$$a\cos^2\alpha \div 4\cos\beta\sin(\beta-\alpha).]$$

2358. A particle moving under gravity passes through two given points: prove that the locus of the focus of its path is an hyperbola whose foci are the two given points.

2359. The distances of three points in the path of a projectile from the point of projection are r_1, r_2, r_3, and the angular elevations of the three points above the point of projection are α_1, α_2, α_3: prove that

$$r_1\cos^2\alpha_1\sin(\alpha_2-\alpha_3) + r_2\cos^2\alpha_2\sin(\alpha_3-\alpha_1) + r_3\cos^2\alpha_3\sin(\alpha_1-\alpha_2) = 0.$$

2360. A number of heavy particles are projected from the same point at the same instant: prove that their lines of instantaneous motion at any subsequent instant will meet in a point, and that this point will ascend with uniform acceleration g.

2361. A number of heavy particles are projected in a vertical plane from one point at the same instant with equal velocities: prove that at any subsequent instant they will all lie on a circle whose centre descends with acceleration g and whose radius increases uniformly with the time. Also if, instead of having equal velocities, the velocity of any particle whose angle of projection is θ be that due to a height $a \sin^2\theta$, the particles will at any subsequent instant all lie on a circle.

2362. Two points A, B in the path of a projectile are such that the direction of motion at B is parallel to the bisector of the angle between the direction of motion at A and the direction of gravity: prove that the time from A to B is equal to that in which the velocity at A would be generated in a particle falling from rest under gravity.

2363. A number of particles are projected from the same point with velocities such that their components in a given direction are all equal: prove that the locus of the foci of their parabolic paths is another parabola whose focus is the point of projection, semi latus rectum the space due to the given component velocity, and the direction of whose axis makes with the vertical an angle which is bisected by the given direction.

2364. A particle is projected from a given point so as just to pass over a vertical wall whose height is b and distance from the point of projection a: prove that, when the area of the parabolic path described before reaching the horizontal plane through the point of projection is a maximum, the range is $\frac{3}{2}a$ and the height of the vertex of the path $\frac{9}{8}b$.

2365. A particle is projected from a point at the foot of one of two parallel vertical smooth walls so as after three reflexions at the walls to return to the point of projection, and the last impact is direct: prove that $e^3 + e^2 + e = 1$, and that the vertical heights of the three points of impact above the point of projection are as $e^2 : 1 - e^2 : 1$.

2366. A heavy particle, mass m, is projected from a point A so as after a time t to be at a point B: prove that the action in passing from A to B is $m\left\{\frac{AB^2}{t} + \frac{g^2t^3}{12}\right\}$, and is a minimum when the focus of the path lies in AB.

2367. In the parabolic path of a projectile, AB is a focal chord: prove that the time from A to B is always equal to the time of falling vertically from rest through a space equal to AB; and that the action in passing from A to B is also equal to the action in falling vertically from rest through a space equal to AB.

2368. A heavy particle is projected from a point in a horizontal plane in such a manner that at its highest point it impinges directly on a vertical plane from which it rebounds, and after another rebound from the horizontal plane returns to the point of projection: prove that the coefficient of restitution is $\frac{1}{2}$.

[The equation for e is $2e^2 + e - 1 = 0$; the student should account for the root -1.]

2369. A heavy particle, for which $e=1$, falls down a chord from the highest point of a vertical circle, and after reflexion at the arc describes a parabolic path passing through the lowest point: prove that the inclination of the chord to the vertical is $\frac{1}{2}\cos^{-1}\left(\frac{-1+\sqrt{3}}{4}\right)$. If the particle fall from the centre down a radius and after reflexion pass through the lowest point, the inclination to the vertical will be $\cos^{-1}\frac{1}{4}$.

2370. A particle is projected from a given point with given velocity up an inclined plane of given inclination so as after leaving the plane to describe a parabola: prove that the loci of the focus and vertex of the parabola for different lengths of the plane are both straight lines.

2371. A particle, for which $e=1$, is projected from the middle point of the base of a vertical square towards one of the angles, and after being reflected at the sides containing that angle falls to the opposite angle: prove that the space due to the velocity of projection bears to the length of a side of the square the ratio 45 : 32.

[More generally, when the particle is projected from the same point at an angle α to the horizon, the space due to the velocity of projection must be to the length of a side as $9 : 16\cos\alpha(3\sin\alpha - 4\cos\alpha)$; and $3\tan\alpha$ must lie between 4 and 9.]

2372. A particle ($e=1$) is projected with a given velocity from a given point in one of two planes equally inclined to the horizon and intersecting in a horizontal line, and after reflexion at the other plane returns to its starting point and is again reflected on the original path; determine the direction of projection and prove that the inclination of each plane must be 45°. Also, if the planes be not equally inclined to the horizon, prove that they must be at right angles and that the inclination of projection to the horizon (θ) is given by the equation

$$\cos(\theta+2\alpha)\cos\theta + \frac{a}{h}\sin\alpha\cos^2\alpha = 0,$$

where h is the space due to the velocity of projection, a the distance from the line of intersection, and α the inclination of the plane from which the particle starts.

[This equation has two roots θ_1, θ_2, and the times of flight in the two paths will be as $\cos(\theta_1+2\alpha) : \cos(\theta_2+2\alpha)$.]

2373. A particle being let fall on a fixed inclined plane bounds on to another fixed inclined plane, the line of intersection being horizontal, and the time between the planes is given: prove that the locus of the point from which the particle is let fall is in general a parabolic cylinder, but will be a plane if $\tan\alpha\tan(\alpha+\beta)=e$, where α, β are the angles of inclination of the planes.

2374. A heavy particle projected at an angle α to an inclined plane whose inclination to the vertical is ι, rebounds from the plane: prove that, if $2\tan\alpha=(1-e)\tan\iota$, the successive parabolic paths will be similar arcs of parabolas, and will all touch two fixed straight lines, one of which is normal to the plane and the other inclined to it at an angle

$$\tan^{-1}\left(\frac{1-e}{2e}\tan\alpha\right).$$

2375. A particle projected from a point in an inclined plane at the rth impact strikes the plane normally and at the nth impact is at the point of projection: prove that $e^n - 2e^r + 1 = 0$.

2376. A particle is projected from a given point in a horizontal plane at an angle a to the horizon, and after one rebound at a vertical plane returns to the point of projection: prove that the point of impact must lie on the straight line

$$y(1+e) = x \tan a,$$

x, y being measured horizontally and vertically from the point of projection. When the velocity of projection and not the direction is given, the locus of the point of impact is the ellipse

$$x^2 + y^2(1+e)^2 = 4ehy,$$

where h is the space due to the velocity of projection.

2377. A particle is projected from a given point with given velocity so as, after one reflexion at an inclined plane passing through the point, to return to the point of projection: prove that the locus of the point of impact is also the ellipse

$$x^2 + (1+e)^2 y^2 = 4ehy,$$

with the notation of the last question.

2378. A heavy particle is projected from a point in a plane whose inclination to the horizon is 30° in a vertical plane perpendicular to the inclined plane: prove that, if all directions of projection in that vertical plane are equally probable, the chance of the range on the inclined plane being at least one-third of the greatest possible range is ·5.

2379. A particle is projected from a point midway between two smooth parallel vertical walls, and after one impact at each wall returns to the point of projection: prove that the heights of the points of impact above the point of projection will be as $e(2e+1) : 2+e$, their depths below the highest point reached by the particle as

$$(1+2e-e^2)^2 : (1-2e-e^2)^2;$$

and that this highest point lies in a fixed vertical straight line whose distance from the point of projection is the less of the two lengths

$$\tfrac{1}{4}a\frac{(1-e^2)}{e}, \quad \tfrac{1}{4}a(1+e)^2,$$

a being the distance between the walls. Also, if the three parabolic paths be completed, each will meet the horizontal plane through the point of projection in fixed points.

III. *Motion on a smooth Curve under the action of Gravity.*

2380. A heavy particle is projected up a smooth parabolic arc whose axis is vertical and vertex upwards with a velocity due to the depth below the tangent at the vertex: prove that, whatever be the length of the arc, the parabola described by the particle after leaving the arc, will pass through a fixed point.

2381. A heavy particle falls down a smooth curve in a vertical plane of such a form that the resultant force on the particle in every position is equal to its weight: prove that the radius of curvature at any point is twice the intercept of the normal cut off by the horizontal line of zero velocity.

2382. A heavy particle is projected so as to move on a smooth parabolic arc whose axis is vertical and vertex upwards: prove that the pressure on the curve is always proportional to the curvature.

2383. A heavy particle is projected from the vertex of a smooth parabolic arc whose axis is vertical and vertex downwards with a velocity due to a height h, and after passing the extremity of the arc proceeds to describe an equal parabola freely: prove that, if c be the vertical height of the extremity of the arc, the latus rectum is $4(h-2c)$.

2384. A parabola is placed with its axis horizontal and plane vertical and a heavy smooth particle is projected from the vertex so as to move on the concave side of the arc: prove that the vertical height attained before leaving the arc is two-thirds of the greatest height attained; and that, if 2θ be the angle described about the focus before leaving the curve

$$h = a(\tan^3\theta + 3\tan\theta),$$

and the latus rectum of the free path will be $4a\tan^2\theta$; h being the space due to the initial velocity and $4a$ the latus rectum of the parabolic arc.

2385. Two heavy particles, connected by a fine string passing through a small fixed ring, describe horizontal circles in equal times: prove that the circles must lie in the same horizontal plane.

2386. A heavy particle P is attached by two strings to fixed points A, B in the same horizontal plane and is projected so as just to describe a vertical circle; the string PB is cut when P is in its lowest position, and P then proceeds to describe a horizontal circle: prove that $3\cos 2PAB = 2$; and that, in order that the tension of the string PA may be unaltered, the angle APB must be a right angle.

2387. Two given weights are attached at given points of a fine string which is attached to a fixed point, and the system revolves with uniform angular velocity about the vertical through the fixed point in a state of relative equilibrium: prove the equations

$$\tan\theta' = \frac{\Omega^2}{g}(a\sin\theta + a'\sin\theta') = \tan\theta + \frac{m}{m+m'}\,\frac{a'\Omega^2}{g}\sin\theta';$$

where a, a' are the lengths of the upper and lower strings, m, m' the masses of the particles, θ, θ' the angles which the strings make with the vertical, and Ω the common angular velocity.

2388. A heavy particle is projected so as to move on a smooth circular arc whose plane is vertical and afterwards to describe a parabola freely: prove that the locus of the focus of the parabolic path is an epicycloid formed by a circle of radius a rolling on a circle of radius $2a$; $4a$ being the radius of the given circle.

2389. A cycloidal arc is placed with its axis vertical and vertex upwards and a heavy particle is projected from the cusp up the concave side of the curve with the velocity due to a height h: prove that the latus rectum of the parabola described after leaving the arc is $h^2 \div 4a$, where a is the radius of the generating circle; also that the locus of the focus of the parabola is the cycloid which is enveloped by that diameter of the generating circle which passes through the generating point.

2390. In a certain curve the vertical ordinate of any point bears to the vertical chord of curvature at that point the constant ratio $1 : m$, and a particle is projected from the point where the tangent is vertical along the curve with any velocity: prove that the vertical height attained before leaving the curve bears to the space due to the velocity of projection the constant ratio $4 : 4 + m$.

2391. A smooth heavy particle is projected from the lowest point of a vertical circular arc with a velocity due to a space equal in length to the diameter $2a$, and the length of the arc is such that the range of the particle on the horizontal plane through the point of projection is the greatest possible: prove that this range is equal to $a\sqrt{9 + 6\sqrt{3}}$.

2392. Two triangles CAB, cAb have a common angle A and the sum of the sides containing that angle is the same in each, BC, bc intersect in D: prove that in the limit when b moves up to B,

$$CD : DB = AB : AC.$$

2393. Two equal parabolas have the same axis and the focus of the outer is the vertex of the inner one, MPp, NQq are common ordinates: prove that the area of the surface generated by the revolution of the arc PQ about the axis bears to the area $MpqN$ a constant ratio.

2394. Common ordinates from the major axis are drawn to two ellipses which have a common minor axis and the outer of which touches the directrices of the inner: prove that the area of the surface generated by the intercepted arc of the inner ellipse revolving about the major axis will bear a constant ratio to the corresponding intercepted area of the outer.

[In general if PM be the ordinate and PG the normal to any given curve at P both terminated by the same fixed straight line, and MP be produced to p so that $Mp = PG$ in length, the area of the surface generated by an elementary arc PP' will bear the constant ratio $2\pi : 1$ to the corresponding area $Mpp'M'$.]

2395. A diameter AB of a circle being taken, P is a point on the circle near to A and the tangent at P meets BA produced in T: prove that ultimately the difference of BA, BP bears to AT the ratio $1 : 2$.

2396. The tangent to a curve at a point B meets the normal at a point A in T; C is the centre of curvature at A and O a point on AC: prove that, in the limit when B moves up to A, the difference of OA and OB bears to AT the ratio $OC : OA$.

2397. In an arc PQ of continued curvature R is a point at which the tangent is parallel to PQ: prove that the ultimate ratio $PR : RQ$ when PQ is diminished indefinitely is one of equality.

2398. The tangents at the ends of an arc PQ of continued curvature meet in O: prove that the ultimate ratio of

$$OP + OQ - \text{arc } PQ : \text{arc } PQ - \text{chord } PQ,$$

as PQ is indefinitely diminished, is $2 : 1$.

2399. Three contiguous points being taken on a curve, the tangents form a triangle and the normals a similar triangle: prove that the ultimate ratios of these triangles when the points tend to coincidence at P is $1 : \left(\frac{d\rho}{ds}\right)^2$; ρ being the radius of curvature at P and s the arc to P from some fixed point of the curve.

2400. A point O is taken in the plane of a given closed oval, P is any point on the curve, and QPQ' a straight line drawn in a given direction so that $QP = PQ'$ and that each bears a constant ratio $n : 1$ to OP: prove that, as P moves round the curve, Q, Q' will trace out two closed loops the sum of whose areas is double the area of the given oval.

[When O is within the oval, the loops will intersect if $n > 1$, and touch if $n = 1$; when O is without the curve, the loops will intersect if n be less than a certain value (always < 1) which depends on the position of O.]

2401. Two contiguous points O, O' are taken on the outer of two confocal ellipses and tangents OP, OQ, $O'P'$, $O'Q'$ drawn to the inner, P' coinciding with P when O' moves up to O: prove that in the limit

$$PP' : QQ' = OP^2 : OQ^2.$$

2402. At a point P of a curve is drawn the circle of curvature, and small arcs PQ, Pq are taken such that the tangents at Q, q are parallel: prove that Qq generally varies as PQ^2, but, if P be a point of maximum or minimum curvature, Qq will vary as PQ^3; also that the angle which Qq makes with the tangent at P is, in the former case two-thirds and in the latter three-fourths of the angle which the tangent at Q or q makes with that at P.

2403. Three equal particles A, B, C move on the arc of a given circle in such a way that their centre of gravity remains fixed: prove that, in any position, their velocities are as $\sin 2A : \sin 2B : \sin 2C$.

2404. The velocities at three points of a central orbit are inversely as the sides of the triangle formed by the tangents at these points: prove that the centre of force is the point of concourse of the straight lines joining each an angular point of this triangle to the common point of the tangents to its circumscribed circle at the ends of its opposite side.

2405. A parabola is described under a force in the focus S, and along the focal distance SP is measured a given length SQ; QR drawn parallel to the normal at P meets the axis in R: prove that the velocity at P bears to the velocity at the vertex the ratio $QR : 2SQ$.

2406. Prove that the equation $2V^2 = F . PV$ is true when a body is moving in a resisting medium, F being the extraneous force and PV the chord of curvature in the direction of F.

2407. Two points P, Q move as follows; P describes an ellipse under acceleration to the centre, and Q describes relatively to P an ellipse of which P is the centre under acceleration to P, and the periodic times in these ellipses are equal: prove that the absolute path of Q is an ellipse concentric with the path of P.

2408. Two bodies are describing concentric ellipses under a centre of force in the common centre: prove that the relative orbit of either with respect to the other is an ellipse, and examine under what circumstances it can be a circle.

[The bodies must be at apses simultaneously, and either the sums of the axes of their two paths equal, or the differences.]

2409. In a central orbit the velocity of the foot of the perpendicular from the centre of force on the tangent varies inversely as the length of the chord of curvature through the centre of force.

2410. Different points describe different circles uniformly, the acceleration in each varying as the radius of the circle: prove that the periodic times will be equal.

[Kinematic similarity.]

2411. A particle describes an hyperbola under a force tending to a focus: prove that the rate at which areas are described by the central radius vector is inversely proportional to the length of that radius.

2412. A rectangular hyperbola is described by a point under acceleration parallel to one of the asymptotes: prove that at a point P the acceleration is $2U^2 . MP \div CM^2$, MP being drawn, in direction of the acceleration, from the other asymptote, C the centre, and U the constant component velocity parallel to the other asymptote.

2413. A point describes a cycloid under acceleration tending always to the centre of the generating circle: prove that the acceleration is constant and that the velocity varies as the radius of curvature at the point.

2414. A particle constrained to move on an equiangular spiral is attracted to the pole by a force proportional to the distance: prove that, in whatever position the particle be placed at starting (at rest), the time of describing a given angle about the centre of force will be the same.

[This follows at once from properties of similar figures.]

2415. An endless string, on which runs a small smooth bead, encloses a fixed elliptic lamina whose perimeter is less than the length of the string; the bead is projected so as to keep the string in a state of tension: prove that it will move with constant velocity, and that the tension of the string will vary inversely as the rectangle under the focal distances.

2416. A small smooth bead runs on an endless thread enclosing a lamina in the form of an oval curve, and the bead is projected so as to describe a curve of continuous curvature in the plane of the lamina under no forces but the tensions of the thread: prove that the tension will vary inversely as the harmonic mean between the lengths of the two parts of the string not in contact with the lamina; and apply this result to prove that the chord of curvature of an ellipse at a point P in a given direction is twice the harmonic mean between the tangents from P to the confocal which touches a straight line drawn through P in the given direction; any tangent which is drawn from P outwards being reckoned negative.

2417. A parabola is described with constant velocity under the action of two equal forces one of which tends to the focus: prove that either force varies inversely as the focal distance.

2418. A particle is describing an ellipse about a centre of force μr^{-2}; at a certain point μ receives a small increment $\Delta\mu$ and the excentricity is unaltered: prove that the point is an extremity of the minor axis and that the major axis $2a$ is diminished by $\frac{a}{\mu}\Delta\mu$.

2419. A particle is describing an ellipse about a centre of force μr^{-2} and at a certain point μ receives a small increment $\Delta\mu$: prove the following equations for determining the corresponding alterations in the major axis $2a$, the excentricity e, and the longitude of the apse ϖ,

$$\frac{r\Delta a}{a(2a-r)} = \frac{e\Delta e}{1-e^2}\frac{r}{a-r} = \frac{e\Delta\varpi}{\sin(\theta-\varpi)} = -\frac{1}{\mu}\Delta\mu;$$

r, θ being polar co-ordinates (from the centre of force) of the point at which the change takes place.

2420. In an elliptic orbit about the focus, when the particle is at a distance r from the centre of force the direction of motion is suddenly turned through a small angle $\Delta\beta$: prove that the consequent alteration in the longitude of the apse is $\frac{1}{e^2}\left(1+e^2-\frac{r}{a}\right)\Delta\beta$, $2a$ being the length of the major axis and e the excentricity.

2421. At any point in an elliptic orbit about the focus, the velocity v receives a small increment Δv: prove that the consequent alterations in the excentrity e and the longitude of the apse ϖ are given by the equations

$$\frac{\Delta e}{b^2(2a-r)} = \frac{e\Delta\varpi}{ab\sqrt{a^2e^2-(r-a)^2}} = \frac{2v\Delta v}{\mu ae(2a-r)}.$$

2422. In an elliptic orbit about the centre the resolved part of the velocity at any point perpendicular to one of the focal distances is constant; and if the whole velocity be resolved into two, one perpendicular to each focal distance, each will vary as the rectangle under the focal distances.

2423. A particle moves along AP a rough chord of a circle under the action of a force to B varying as the distance and AB is a diameter; the particle starts from rest at A and comes to rest again at P: prove that the co-efficient of friction is $\frac{1}{2}\tan PAB$.

2424. A number of particles start from the same point with the same velocity and are acted on by a central force varying as the distance: prove that the ellipses described are enveloped by an ellipse having its centre at the centre of force and a focus at the point of projection.

2425. An ellipse is described by a particle under the action of two forces tending to the foci and each varying inversely as the square of the distance: prove that

$$\frac{2a^5}{b^4} = \frac{(\mu\omega^2 + \mu'\omega'^2)(\omega + \omega')^4}{\omega^4\omega'^4};$$

a, b being the axes of the ellipse, and ω, ω' the angular velocities at any point about the foci.

2426. Two fixed points of a lamina slide along two straight lines fixed in space (in the plane of the lamina) so that the angular velocity of the lamina is constant: prove that (1) every fixed point of the lamina describes an ellipse under acceleration tending to the common point of the two fixed straight lines and proportional to the distance; (2) every straight line fixed in the lamina envelopes during its motion an involute of a four-cusped hypocycloid; (3) the motion of the lamina is completely represented by supposing a circle fixed in the lamina to roll uniformly with internal contact on a circle of double the radius fixed in space; (4) for a series of points in the lamina lying in one straight line the foci of the ellipses described lie on a rectangular hyperbola.

2427. A lamina moves in its own plane so that two fixed points of it describe straight lines with accelerations f, f': prove that the acceleration of the centre of instantaneous rotation is

$$\sqrt{f^2 + f'^2 - 2ff'\cos\alpha} \div \sin\alpha,$$

where α is the angle between the straight lines.

[The accelerations f, f' must satisfy the equations

$$\cos^2\theta \frac{d}{d\theta}(f\sec\theta) = \cos^2\theta' \frac{d}{d\theta'}(f'\sec\theta'),$$

$$f\cos\theta + f'\cos\theta' = c\omega^2,$$

where θ, θ' are the angles which the straight line joining the two points makes with the fixed straight lines, and ω is the angular velocity of the lamina.]

2428. Two points A, B of a lamina describe the two straight lines Ox, Oy fixed in space (in the plane of the lamina), P is any other point of the lamina, and QQ' any diameter of the circle AOB; PQ, PQ' meet the circle again in R, R': prove that OR, OR' will be the directions of two conjugate diameters of the locus of P.

2429. Two points fixed in a lamina move upon two straight lines fixed in space and the velocity of one of the points is uniform: prove that every other point in the lamina moves so that its acceleration is constant in direction and varies inversely as the cube of the distance from a fixed straight line.

[If A describe Ox with uniform velocity U and B describe Oy at right angles to Ox, then if P be any other point fixed in the lamina and PA, PB meet the circle on AB in a, b, the acceleration of P will be always parallel to Oa and vary inversely as the cube of the distance from Ob; and, if PM be drawn parallel to Oa to meet Ob, the acceleration of P will be $$U^2 . AP^2 \div AB^2 . PM^3.]$$

2430. A lamina moves in its own plane so that two points fixed in the lamina describe straight lines with equal accelerations: prove that the acceleration of the centre of instantaneous rotation is constant in direction, and that the acceleration of any point fixed in the lamina is constant in direction.

2431. Two ellipses are described about a common attractive force in their centre; the axes of the two are coincident in direction and the sum of the axes of one is equal to the difference of the axes of the other: prove that, if the describing particles be at corresponding extremities of the major axes at the same instant and be moving in opposite senses, the straight line joining them will be of constant length and of uniform angular velocity during the motion.

2432. A lamina moves in such a manner that two straight lines fixed in the lamina pass through two points fixed in space: prove that the motion of the lamina is completely represented by supposing a circle fixed in the lamina to roll with internal contact on a circle of half the radius fixed in space.

2433. A lamina moves in its own plane with uniform angular velocity so that two straight lines fixed in the lamina pass each through one of two points fixed in space: prove that the acceleration of any point fixed in the lamina is compounded of two constant accelerations, one tending to a fixed point, and the other in a direction which revolves with double the angular velocity of the lamina.

2434. A triangular lamina ABC moves so that the point A lies on a straight line bc fixed in space, and the side BC passes through a point a fixed in space, and the triangles ABC, abc are equal and similar: prove that the motion of the lamina is completely represented by supposing a parabola fixed in the lamina to roll upon an equal parabola fixed in space, similar points being in contact.

2435. A particle describes a parabola under a repulsive force from the focus, varying as the distance, and another force parallel to the axis which at the vertex is three times the former; find the law of this latter force; and prove that, if two particles describe the same parabola under the action of these forces, their lines of instantaneous motion will intersect in a point which lies on a fixed confocal parabola.

[The second force is always three times the first.]

2436. Two particles describe curves under the action of central attractive forces, and the radius vector of either is always parallel and proportional to the velocity of the other: prove that the curves will be similar ellipses described about their centres.

DYNAMICS OF A PARTICLE.

I. *Rectilinear Motion, Kinematics.*

2437. A heavy particle is attached by an extensible string to a fixed point, from which the particle is allowed to fall freely; when the particle is in its lowest position the string is of twice its natural length: prove that the modulus is four times the weight of the particle, and find the time during which the string is extended beyond its natural length.

[The time is $2\sqrt{\frac{a}{g}}\tan^{-1}\sqrt{2}$.]

2438. A particle at B is attached by an elastic string at its natural length to a point A and attracted by a force varying as the distance to a point C in BA produced, A dividing BC in the ratio $1:3$, and the particle just reaches the centre of force: prove that the velocity will be greatest at a point which divides CA in the ratio $8:7$.

2439. A particle is attracted to a fixed point by a force $\mu\,(\text{dist.})^{-2}$, and repelled from the same point by a constant force f; the particle is placed at a distance a from the centre, at which point the attractive force is four times the magnitude of the repulsive, and projected directly from the centre with velocity V: prove that (1) the particle will move to infinity or not according as $V^2 >$ or $< 2af$; (2) that, if x, $x+c$ be the distances from the centre of force of two positions of the particle, the time of describing the given distance c between them will be greatest when $x(x+c) = 4a^2$. Also, when $V = \sqrt{2af}$ or $3\sqrt{2af}$, determine the time of describing any distance.

[When $V = \sqrt{2af}$, the time of reaching a distance x from the centre of force is

$$\frac{2}{\sqrt{f}}\left\{\sqrt{2x} - \sqrt{2a} - \sqrt{a}\log\left(\frac{\sqrt{2a}+\sqrt{x}}{\sqrt{2a}-\sqrt{x}}\right) + 2\sqrt{a}\log\left(\sqrt{2}+1\right)\right\};$$

and, when $V = 3\sqrt{2af}$, the time is

$$\frac{2}{\sqrt{f}}\left\{\sqrt{2x} - \sqrt{2a} + 2\tan^{-1}\frac{1}{\sqrt{2}} - 2\tan^{-1}\sqrt{\frac{x}{2a}}\right\}.]$$

2440. The accelerations of a point describing a curve are resolved into two, along the radius vector and parallel to the prime radius: prove that these accelerations are respectively

$$\frac{\cot\theta}{r}\frac{d}{dt}\left(r^2\frac{d\theta}{dt}\right)+\frac{d^2r}{dt^2}-r\left(\frac{d\theta}{dt}\right)^2 \text{ and } -\frac{1}{r\sin\theta}\frac{d}{dt}\left(r^2\frac{d\theta}{dt}\right).$$

2441. The motion of a point is referred to two axes Ox, Oy, of which Ox is fixed and Oy revolves about the origin: prove that the accelerations in these directions at any time t are

$$\frac{d^2x}{dt^2}-\frac{1}{y\sin\theta}\frac{d}{dt}\left(y^2\frac{d\theta}{dt}\right);\quad \frac{d^2y}{dt^2}+\frac{\cot\theta}{y}\frac{d}{dt}\left(y^2\frac{d\theta}{dt}\right)-y\left(\frac{d\theta}{dt}\right)^2;$$

where θ denotes the angle between the axes.

2442. A point P is taken on the tangent to a given curve at a point Q, and O is a fixed point on the curve, the arc $OQ=s$, $QP=r$, and ϕ is the angle through which the tangent revolves as the point of contact passes from O to Q: prove that the accelerations of P in direction QP and in the direction at right angles to this, in the sense in which ϕ increases, are respectively

$$\frac{d^2s}{dt^2}+\frac{d^2r}{dt^2}-r\left(\frac{d\phi}{dt}\right)^2,\quad \frac{1}{r}\frac{d}{dt}\left(r^2\frac{d\phi}{dt}\right)+\frac{d\phi}{dt}\frac{ds}{dt}.$$

2443. A point describes a curve of double curvature, and its polar co-ordinates at the time t are (r, θ, ϕ): prove that its accelerations (1) along the radius vector, (2) perpendicular to the radius vector in the plane of θ and in the sense in which θ increases, and (3) perpendicular to the plane of θ in the sense in which ϕ increases, are respectively

$$(1)\quad \frac{d^2r}{dt^2}-r\left(\frac{d\theta}{dt}\right)^2-r\sin^2\theta\left(\frac{d\phi}{dt}\right)^2,$$

$$(2)\quad \frac{1}{r}\frac{d}{dt}\left(r^2\frac{d\theta}{dt}\right)-r\sin\theta\cos\theta\left(\frac{d\phi}{dt}\right)^2,$$

$$(3)\quad \frac{1}{r\sin\theta}\frac{d}{dt}\left(r^2\sin^2\theta\frac{d\phi}{dt}\right).$$

2444. A point describes a parabola in such a manner that its velocity, at a distance r from the focus, is $\sqrt{\frac{2f}{r}(r^2-c^2)}$, where f, c are constant: prove that its acceleration is compounded of f parallel to the axis and $f\frac{c^2}{r^2}$ along the radius vector from the focus.

2445. A point describes a semi-ellipse bounded by the minor axis, and its velocity at a distance r from the focus is $a\sqrt{\frac{f(a-r)}{r(2a-r)}}$, where $2a$ is the length of the major axis and f a constant acceleration: prove that the acceleration of the point is compounded of two, each varying inversely as the square of the distance, one tending to the nearer focus and the other from the farther focus.

2446. A point is describing a circle, and its velocity at an angular distance θ from a fixed point on the circle varies as $\sqrt{1+\cos^2\theta+\sin^2\theta}$: prove that its acceleration is compounded of two tending to fixed points at the extremities of a diameter, each varying inversely as the fifth power of the distance and equal at equal distances.

2447. A point describes a circle under acceleration, constant, but not tending to the centre: prove that the point oscillates through a quadrant and that the line of action of the acceleration always touches a certain epicycloid.

[The radius of the fixed circle of the epicycloid is $\frac{2a}{3}$ and of the moving circle $\frac{a}{6}$, a being the radius of the circle described by the point.]

2448. A parabola is described with accelerations F, A, tending to the focus and parallel to the axis respectively: prove that

$$\frac{1}{r^3}\frac{d}{dr}(Fr^2)+\frac{dA}{dr}=0,$$

r being the focal distance.

2449. A point describes an ellipse under accelerations F_1, F_2 tending to the foci, and r_1, r_2 are the focal distances of the point: prove that

$$\frac{1}{r_1^2}\frac{d}{dr_1}(F_1r_1^2)=\frac{1}{r_2^2}\frac{d}{dr_2}(F_2r_2^2).$$

2450. The parabola $y^2=4ax$ is described under accelerations X, Y parallel to the axes: prove that

$$2x\frac{dY}{dx}-y\frac{dX}{dx}+3Y=0.$$

2451. A point describes a parabola under acceleration which makes a constant angle α with the normal, and θ is the angle described from the vertex about the focus in a time t: prove that

$$\epsilon^{\theta\tan\alpha}\left(\frac{d\theta}{dt}\right)^2 \propto (1+\cos\theta)^2;$$

and find the law of acceleration.

[The acceleration varies as $\cos^3\frac{\theta}{2}\,\epsilon^{-\theta\tan\alpha}$, which is easily expressed as a function of the focal distance.]

2452. A point P describes a circle of radius $4a$ with uniform angular velocity ω about the centre, and another point Q describes a circle of radius a with angular velocity 2ω about P: prove that the acceleration of Q varies as the distance of P from a certain fixed point.

2453. The only curve which can be described under constant acceleration in a direction making a constant angle with the normal is an equiangular spiral.

2454. An equiangular spiral is described by a point with constant acceleration in a direction making an angle ϕ with the normal: prove that

$$\sin\phi\frac{d\phi}{d\theta}=2\sin\phi+\cot a\cos\phi,$$

a being the constant angle of the spiral and θ the angle through which the radius vector has turned from a given position.

2455. The parabola $y^2=2cx$ is described by a point under acceleration making a constant angle a with the axis and the velocity when the acceleration is normal is V: prove that, at any point (x, y) of the parabola, the acceleration is $V^2c^2\div(c\cos a-y\sin a)^3$; and that, if when the acceleration is normal the particle is moving towards the vertex, the time in which the direction of motion will turn through a right angle will be $c\div V\sin 2a\cos a$.

2456. A lamina moves so that two straight lines fixed in it pass through two points fixed in space and the angular velocity is uniform: prove that any point fixed in the lamina, whose distance from the point of intersection of the two straight lines is twice the diameter of the circle described by that point, will move under acceleration whose line of action always touches a three-cusped hypocycloid.

2457. The catenary $s=c\tan\phi$ is described under acceleration which at any point makes an angle ϕ with the normal on the side towards the vertex: prove that the acceleration varies inversely as the cube of the distance from the directrix.

2458. A point describes a parabola, starting from rest at the vertex, under acceleration which makes with the tangent an angle $\tan^{-1}(2\tan\theta)$, where θ is the angle through which the tangent has turned: prove that the acceleration varies as $\sqrt{4r-3a}\,.\,r^{-2}$, where r is the focal distance and a the initial value of r.

2459. The curve whose intrinsic equation is $s=a\tan 2\phi$ is described by a point under constant acceleration: prove that the direction of the acceleration makes with the tangent an angle $\theta-2\phi$, where θ is given by the equation

$$2\frac{d}{d\theta}(\tan 2\phi\sqrt{\cos\theta})=\sqrt{\cos\theta}.$$

2460. A point describes an epicycloid under acceleration tending to the centre of the fixed circle: prove that the pedal of the epicycloid with respect to the centre will also be described under acceleration tending to the same point.

2461. The intrinsic equation of a curve is $s = f(\phi)$ and the curve is described under accelerations X, Y parallel to the tangent and normal at the origin (where $\phi = 0$): prove that

$$\cos\phi\left(\frac{dY}{d\phi} - 3X\right) - \sin\phi\left(\frac{dX}{d\phi} + 3Y\right) + \frac{f''(\phi)}{f'(\phi)}(Y\cos\phi - X\sin\phi) = 0.$$

2462. The curve $s = f(\phi)$ is described by a point with constant acceleration which is at the origin in direction of the normal: prove that its inclination θ to this direction at any other point is given by the equation

$$\left(3 - \frac{d\theta}{d\phi}\right)\tan(\phi - \theta) f'(\phi) = f''(\phi).$$

2463. A catenary is described by a point under acceleration whose vertical component is constant (f): prove that the horizontal component when the tangent makes an angle ϕ with the horizon is

$$f\cos\phi\,\text{cosec}^3\phi\,(1 + m\cos\phi + \cos^2\phi).$$

2464. A curve is described under constant acceleration parallel to a straight line which revolves uniformly: prove that the curve is a prolate, common, or curtate cycloid; or a circle.

2465. A point describes a certain curve and initially the acceleration is normal; when the direction of motion has turned through an angle ϕ the direction of acceleration has turned through an angle 2ϕ in the same sense: prove that the acceleration varies as $\cos\phi\,\frac{d\phi}{ds}$, as does the angular velocity of the tangent, and that the velocity varies as $\cos\phi$.

2466. A parabola is described under constant acceleration and θ, ϕ are the angles which the direction of the acceleration at any point and the tangent at that point make respectively with the directrix: prove that

$$3\frac{d}{d\theta}\left(\tan\phi\cos^{\frac{4}{3}}\theta\right) = \cos^{\frac{1}{3}}\theta.$$

2467. A point moves under constant acceleration which is initially normal, and when the direction of motion has turned through an angle ϕ the direction of acceleration has turned through an angle $m\phi$ (m constant) in the same sense: prove that the intrinsic equation of the curve described is

$$\frac{ds}{d\phi} = c\left(\cos\overline{m-1}\,\phi\right)^{\frac{2-m}{m-1}};$$

and determine the curve when $m = 1$, 2, or 3.

2468. A cycloid is described under constant acceleration and θ, ϕ are the angles which the directions of motion and of acceleration at any point make with the tangent and normal at the vertex respectively: prove that

$$\sin\theta = \cos\theta \sin(\phi - 2\theta) \log\left\{n \tan\left(\frac{\phi}{2} - \theta\right)\right\};$$

or that $\phi = 2\theta$.

2469. A point describes an ellipse under accelerations to the foci which are, one to another at any point, inversely as the focal distances; find the law of either acceleration, prove that the velocity of the point varies inversely as the conjugate diameter, and that the periodic time is $\frac{\pi}{\omega}\left(\frac{a}{b} + \frac{b}{a}\right)$, where ω is the angular velocity about the centre at the end of either axis.

[This path so described is also a brachystochrone between any two points for a certain force in the centre.]

2470. The cusp of a cardioid is S and the centre of the fixed circle (by which it can be generated as an epicycloid) is C, and the cardioid is described under accelerations F, F' tending to S, C respectively: prove that

$$\frac{4}{r^3}\frac{d}{dr}(Fr^4) + \frac{3}{r'}\frac{d}{dr'}\left\{\frac{F'(r'^2 - a^2)^2}{r'}\right\} = 0,$$

where r, r' are the distances from S, C, and $a = SC$. Also prove that, if the angular velocity about the cusp be constant, F will be constant, F' will vary as r', and at the apse $2F + F' = 0$.

2471. A point P starts from A and moves along a straight line with uniform velocity V; a point Q starts from B and moves always towards P with uniform velocity v: prove that, if $V > v$, the least distance c between P and Q is

$$a(\sin\alpha)^{1-m}(1 - \cos\alpha)^m \div (1+m)^{\frac{1+m}{2}}(1-m)^{\frac{1-m}{2}},$$

and, if t be the time after which they are at this distance,

$$Vt = \frac{2mc - a(m - \cos\alpha)}{1 - m^2};$$

where $m = \frac{v}{V}$, $a = AB$, and α is the angle which AB makes with the path of P.

2472. A point P is describing a parabola whose focus is S under acceleration always at right angles to SP, the plane in which the motion takes place having a constant velocity parallel to the axis, and equal to the velocity of P parallel to the axis in the parabola at the end of the latus rectum: prove that the path of P in space is a "curve of pursuit" to S described with a constant velocity equal to that of S.

2473. A point describes a curve which lies on a cone of revolution and crosses all the generating lines at a constant angle, under acceleration whose direction always intersects the axis: prove that the acceleration makes a constant angle with the axis and varies inversely as the cube of the distance from the vertex.

2474. The straight lines AP, BP joining a moving point P to two fixed points A, B have constant angular velocities 2ω, 3ω: prove that the acceleration of P is compounded of a constant acceleration along AP and an acceleration varying as BP along PB.

[These accelerations are $12\omega^2 AB$, and $16\omega^2 PB$ respectively.]

2475. A point describes a rhumb line on a sphere so that the longitude increases uniformly: prove that the whole acceleration varies as the cosine of the latitude and at any point makes with the normal an angle equal to the latitude.

[If α be the constant angle at which the curve crosses the meridians, and ω be the rate at which the longitude increases, the three accelerations resolved as in (2443) will be

$$-f\sin^2\theta,\ f\cos 2\alpha \sin\theta\cos\theta,\ f\sin 2\alpha\sin\theta\cos\theta,$$

where $f\sin^2\alpha = \omega^2 \times$ radius of the sphere.]

2476. A point P describes a circle under acceleration tending to a point S and varying as SP, S being a point which moves on a fixed diameter initially passing through P: prove that, if θ be the angle described about the centre in a time t, $\sqrt{m}\sin\theta = \sqrt{m+1}\sin(t\sqrt{\mu})$, and the distance of S from the centre $=\dfrac{a}{m}\sec^3\theta$; where a is the radius of the circle and m constant.

2477. A point describes an arc of a circle so that its acceleration is always proportional to the nth power of its velocity: prove that the direction of the acceleration of the point always touches a certain epicycloid generated by a circle of radius $a \div 2(3-n)$ rolling on a circle of radius $a(2-n)\div(3-n)$; where a is the radius of the described circle.

II. *Central Forces.*

2478. Prove that the parabola $y^2 = 4ax$ can be described under a constant force parallel to the axis of y and a force proportional to y parallel to the axis of x; also, under two forces $4\mu(c+x)$, μy parallel to the axes of x and y respectively.

2479. A particle is acted on by a force parallel to the axis of y whose acceleration is μy, and is initially projected with a velocity $a\sqrt{\mu}$ parallel to the axis of x at a point where $y = a$: prove that it will describe a catenary.

2480. A particle is acted on by a force parallel to the axis of y whose acceleration (always towards the axis of x) is μy^{-2}, and, when $y = a$, is projected parallel to the axis of x with velocity $\sqrt{\dfrac{2\mu}{a}}$: prove that it will describe a cycloid.

2481. Two equal particles attract each other with a force varying inversely as the square of the distance and are projected simultaneously with equal velocities at right angles to the joining line: prove that, if each velocity be equal to that in a circle at the same distance, each particle will describe a semi-cycloid.

2482. A cardioid is described with constant angular velocity about the cusp under a constant force to the cusp and another constant force: prove that the magnitude of the latter is double that of the former and that its line of action always touches an epicycloid generated by a circle of radius a rolling upon one of radius $2a$; $8a$ being the length of the axis of the cardioid.

2483. The force to the origin under which the hyperbola

$$r \cos 2\theta = 2\sqrt{2}\, a \cos \theta$$

can be described will vary as $(\sqrt{a^2+r^2}+a)^3 \div r^5$.

2484. The perpendicular SY is let fall from the origin upon the tangent at any point P of the curve $r^2 = a^2 \sin 2\theta$, and the locus of Y is described under a force to S: prove that this force will vary as SP^{-13}.

2485. In a central orbit the resolved velocity at any point perpendicular to the radius vector is equal to the velocity in a circle at that distance: prove that the orbit is a reciprocal spiral.

2486. A particle moves under a constant repulsive force from a fixed point, and is projected with a velocity which is to that in a circle at the same distance under an equal attractive force as $\sqrt{2} : 1$: prove that the orbit is the curve whose equation is of the form

$$r^{\frac{3}{2}} = a^{\frac{3}{2}} \sec \tfrac{3}{2}\theta.$$

2487. The force to the pole under which the pedal of a given curve $r = f(p)$ can be described will vary as $rp^{-5}\left(2r - p\dfrac{dr}{dp}\right)$; and, if the given curve be $r^{\frac{3}{2}} \sin \frac{3}{2}\theta = a^{\frac{3}{2}}$, this force will be constant.

2488. An orbit described under a constant force tending to a fixed point will be the pedal of one of the curves represented by the equation $a^2r^2 = p^5 + bp^4$, where a and b are constants.

2489. A parabola is described about a centre of force in C, the centre of curvature at the vertex A: prove that the force at any point P of the parabola varies as $CP\,(AS+SP)^{-3}$, where S is the focus.

2490. The force tending to the pole under which the evolute of the curve $r = f(p)$ can be described will vary inversely as

$$(r^2 - p^2)^2 \frac{d}{dp}\left(r\frac{dr}{dp}\right) \div \sqrt{r^2 + r^2\left(\frac{dr}{dp}\right)^2 - 2rp\frac{dr}{dp}}.$$

2491. A particle P is projected from a point A at right angles to a straight line SA and attracted to the fixed point S by a force varying as cosec PSA: prove that the rate of describing areas about A will be uniformly accelerated.

2492. A particle is projected at a distance a with velocity equal to that in a circle at the same distance and at an angle of 45° with the distance, and attracted to a fixed point by a force which at a distance r is equal to $\mu r^{-5}(a^2 + 3r^2)$: prove that the equation of the path is $r = a \tan\left(\frac{\pi}{4} - \frac{\theta}{2}\right)$, and that the time to the centre of force is

$$\frac{a^2}{\sqrt{2\mu}}\left(2 - \frac{\pi}{2}\right).$$

2493. A particle is attracted to a fixed point by a force which at a distance r is equal to

$$\mu r^{-7}(3a^8 + 3a^4r^4 - r^8),$$

and is projected from a point at a distance a from the centre with a velocity equal to that in a circle at the same distance and in a direction making an angle $\cot^{-1} 2$ with the distance: prove that the equation of the orbit is

$$r^2 = a^2 \tan\left(\frac{\pi}{4} - 2\theta\right),$$

and that the time to the centre of force is

$$\frac{1}{4\sqrt{\mu}} \log 2.$$

2494. A particle is describing a circle under the action of a constant force in the centre and the force is suddenly increased to ten times its former magnitude: prove that the next apsidal distance will be equal to one fourth the radius of the circle.

2495. A particle is describing a central orbit in such a manner that the velocity at any point is to the velocity in a circle at that distance as $1 : \sqrt{n}$: prove that $p \propto r^n$, p being the perpendicular from the centre of force on the tangent at a point whose distance is r, and that the force will vary inversely as r^{2n+1}. If the force be repulsive and the velocity at any point be to that in a circle at that distance under an equal attractive force as $1 : \sqrt{n}$, the particle will describe a path having two asymptotes inclined at an angle $\frac{\pi}{n+1}$.

2496. A particle acted on by a central force $\mu r^{-3}(4r - 3a)$ is projected at a distance a, an angle 45°, and with a velocity which is to the velocity from infinity as $\sqrt{2} : \sqrt{5}$: prove that the equation of the path is $a = r(1 + \sin\theta\cos\theta)$, and that the time from projection to an apse is $\frac{a^{\frac{3}{2}}}{9\sqrt{3\mu}}(4\pi - 3\sqrt{3})$.

2497. A portion of an epicycloid is described under a force tending to the centre of the fixed circle: prove that, if a straight line be drawn from any fixed point always parallel and proportional to the radius of curvature in the epicycloid, the extremity of this line will describe a central orbit.

2498. The curve whose intrinsic equation is $s = a(\epsilon^{m\phi} - \epsilon^{-m\phi})$ is described under a central attractive force, the describing point being initially at an apse at a distance $c = 2ma \div (1 + m^2)$ from the centre of force: prove that the force varies as $r(r^2 + c^2)^{-2}$, and that the end of a straight line drawn from a fixed point always parallel and proportional to the radius of curvature in the path will also describe a central orbit.

2499. A particle describing a parabola about a force in the focus comes to the apse at which point the law of force changes, and the force varies inversely as the distance until the particle next comes to an apse when the former law is restored; there are no instantaneous changes in magnitude: prove that the major axis of the new elliptic orbit will be $m^2 a \div (m^2 - 1)$, where $4a$ is the latus rectum of the parabola and m is that root of the equation $x^2(1 - \log x) = 1$ which lies between $\sqrt{\epsilon}$ and ϵ, and that the excentricity will be $1 - \dfrac{2}{m^2}$.

2500. In an orbit described under a central force a straight line is drawn from a fixed point perpendicular to the tangent and proportional to the force, and this straight line describes equal areas in equal times: prove that the differential equation of the orbit is of the form

$$\left(\frac{1}{p}\right)^{\frac{4}{3}} = \left(\frac{1}{c}\right)^{\frac{4}{3}} \pm \left(\frac{r}{a^2}\right)^{\frac{4}{3}},$$

and that the rectangular hyperbola described about the centre is a particular case.

2501. A uniform chain rests under normal and tangential forces which at any point of the chain are $-n$, t per unit of length of the chain: prove that a particle whose mass is equal to that of a unit of length of the chain can describe the same curve under the action of normal and tangential forces $2n$, t at the same point.

2502. A centre of force varying inversely as the nth power of the distance moves in the circumference of a circle and a particle describes an arc of the same circle under the action of the force: prove that the velocity of the centre of force must bear to the velocity of the particle the constant ratio $5 - n : 1 - n$, and that, when the acceleration of the force at a distance r is μr^{-3}, the time of describing a semicircle is $4a^2\sqrt{\dfrac{2}{\mu}}$.

2503. A particle P is repelled from a fixed point S by a force varying as $(\text{distance})^{-2}$ and attracts another particle Q with a force varying as $(\text{distance})^{-3}$; initially P and Q are equidistant from S in opposite directions, P is at rest, and the accelerations of the two forces are equal: prove that Q, if projected at right angles to SQ with proper velocity, will describe a parabola with S for focus.

2504. A particle P is repelled from two fixed points S, S' by forces, varying each as (distance)$^{-2}$ and equal at equal distances, and attracts another particle Q with a force varying as (distance)$^{-3}$; initially P, Q divide SS' internally and externally in the same ratio, P is at rest, and the accelerations of the forces on the two particles are equal: prove that, if Q be projected at right angles to SS' with velocity equal to that in a circle at the same distance from P, it will describe an ellipse of which S, S' are foci.

2505. A particle P is describing a parabola under the action of gravity, S is the focus and the straight line drawn through P at right angles to SP touches its envelope in Q: prove that the velocity of Q varies as SP.

2506. A particle P describes a central orbit, centre of force S, and through P is drawn a straight line at right angles to PS, which line touches its envelope in Q: prove that the velocity of Q varies as

$$\frac{1}{r^2}\left(\frac{d^2r}{d\theta^2}+r\right),$$

and is constant when the orbit is a parabola with its focus at S.

2507. A particle acted on by an attractive central force

$$\mu r\,(r^2-a^2)^{-2}$$

is projected from an apse at a distance na with a velocity which is to the velocity in a circle at the same distance as $\sqrt{n^2-1} : n$: prove that the path will be an arc of an epicycloid and that the time before reaching the cusp is

$$\frac{\pi}{4\sqrt{\mu}}\,a^2\,(n^2-1)^{\frac{3}{2}}.$$

2508. A particle describes an involute of a circle under a force in the centre of the circle: prove that the force, at any point at a distance r, will vary as $r\,(r^2-a^2)^{-2}$, where a is the radius of the circle.

2509. A smooth horizontal disc revolves with angular velocity ω about a vertical axis at which is placed a material particle acted on by an attractive force, of acceleration equal to ω^2. distance, to a certain point of the disc: prove that the path of the particle on the disc will be a cycloid, and that its co-ordinates in space, when the disc has turned through an angle θ, are $a\theta\sin\theta$, $a(\sin\theta-\theta\cos\theta)$, the former being measured along the straight line which initially joins the particle to the centre of force.

2510. An orbit is described under a force P tending to a fixed point S and a normal force N: prove that

$$p^2\frac{d}{dr}\left(Nr\frac{dr}{dp}\right)+\frac{d}{dr}\left(Pp^3\frac{dr}{dp}\right)=0.$$

2511. A particle is projected from a point at a distance a from a fixed centre of force whose acceleration at distance $r = 2\mu r (r^2 + a^2)^{-2}$, with a velocity $\sqrt{\mu} \div 2a^2$, and in a direction making an angle α with the distance: prove that the orbit is a circle whose radius is $a \div \sin \alpha$.

2512. A particle describes a conic under the action of a centre of force at a point O on the transverse axis: prove that the time of passing from one extremity of the ordinate through O to the other will be

$$2\sqrt{\frac{2}{\mu}}\,\frac{2\alpha - \sin 2\alpha}{\sin^3 \alpha},$$

the acceleration of the force at any point P being

$$\mu OP^{-2}\left(\frac{1}{OP} + \frac{1}{Op}\right)^{-3},$$

where POp is a chord through O; and $\cos \alpha : 1$ being the ratio which the distance of O from the centre bears to the semi-major axis.

III. *Constrained Motion on Curves or Surfaces: Particles joined by Strings.*

2513. A particle, mass m, is constrained to move on a curve under the action of forces such that the particle, if projected from a certain point of the curve with velocity v, would describe the curve freely: prove that, when projected from that point with velocity V, the pressure on the curve at any point will be $m (V^2 - v^2) \div \rho$, where ρ is the radius of curvature.

2514. A particle is acted on by two forces, one parallel to a fixed straight line and constant, the other tending from a fixed point and varying as (distance)$^{-2}$, and is constrained to move on a parabola whose focus is the fixed point and axis parallel to the fixed line: prove that the pressure is always proportional to the curvature, and that, if the velocity vanish at a point where the magnitudes of the forces are equal, the pressure will also vanish.

2515. A particle is attracted to two fixed points by two forces, the acceleration of either force at a distance r being $\mu r^{-2} (a^2 + r^2)$, and is placed at rest at a point at which the two forces are equal and the distances of the particle from the centres of force unequal: prove that it will proceed to oscillate in a hyperbolic arc of which the centres of force are foci.

2516. A particle is acted on by a repulsive force tending from a fixed point and by another force in a fixed direction; when at a distance r from the fixed point the accelerations of these forces are

$$\frac{\mu}{r^2}\left(1 - \frac{r}{a}\right),\ \frac{\mu}{r^2}\left(\frac{r^2}{c^2} + \frac{r}{a}\right)$$

respectively: prove that the particle, abandoned motionless to the action of these forces at a point where they are equal in magnitude, will proceed to describe a parabola with its focus at the fixed point and its axis in the fixed direction.

2517. A particle is placed in a smooth parabolic groove which revolves in its own plane about the focus with uniform angular velocity ω, and the particle describes in space an equal confocal parabola under an attractive force in the focus: prove that this force at any point is measured by $\omega^2 r (3r - 4c) \div 4c$, r being the focal distance and $2c$ the latus rectum.

2518. A bead moves on a smooth elliptic wire and is attached to the foci by two similar elastic strings, of equal natural lengths, which remain extended throughout the motion: prove that, if projected with proper velocity, the velocity will always vary as the conjugate diameter.

2519. Two particles A, B are together in a smooth circular tube; A attracts B with a force whose acceleration is ω^2 distance and moves along the tube with uniform angular velocity 2ω; B is initially at rest: prove that the angle ϕ subtended by AB at the centre after a time t is given by the equation

$$\log \tan \frac{\pi - \phi}{4} = \omega t.$$

2520. A force f resides at the centre of a rough circular arc and from a point of the circle a particle is projected with a velocity $V (> \sqrt{af})$ along the interior of the circle: prove that the normal pressure on the curve will be diminished one half after the time

$$\frac{1}{\mu}\sqrt{\frac{a}{f}} \log \left(\frac{\sqrt{2af} + \sqrt{V^2 + af}}{V + \sqrt{af}}\right);$$

where a denotes the radius of the circle and μ the coefficient of friction.

2521. A heavy particle is projected horizontally so as to move on the interior of a smooth hollow sphere of radius a and the velocity of projection is $\sqrt{2ga}$: prove that, when the particle again moves horizontally, its vertical depth below the highest point of the sphere is equal to its initial distance from the lowest point.

2522. A heavy particle is attached to a fixed point by a fine inextensible string of length a, and, when the string is horizontal and at its full length, the particle is projected horizontally at right angles to the string with the velocity due to a height $2a \cot 2\alpha$: prove that the greatest depth to which it will fall is $a \tan \alpha$.

2523. A particle slides in a vertical plane down a rough cycloidal arc whose axis is vertical, starting from the cusp and coming to rest at the vertex: prove that the coefficient of friction is given by the equation

$$\mu^2 \epsilon^{\mu\pi} = 1.$$

[More generally, if the particle come to rest at the lowest point and θ be the angle which the tangent at the starting point makes with the horizon,

$$\mu\epsilon^{\mu\theta} = \sin\theta - \mu\cos\theta.]$$

2524. A rough wire in the form of an arc of an equiangular spiral whose constant angle is $\cot^{-1}(2\mu)$ is placed with its plane vertical and a heavy particle falls down it, coming to rest at the first point where the tangent is horizontal: prove that at the starting point the tangent makes with the horizon an angle double the angle of friction, and that during the motion the velocity will be greatest when the angle ϕ which the tangent makes with the horizon is given by the equation

$$(2\mu^2-1)\sin\phi+3\mu\cos\phi=2\mu.$$

2525. A heavy particle falls down the arc of a four-cusped hypocycloid, starting at a cusp where the tangent is vertical and coming to rest at the next cusp: prove that, if μ be the coefficient of friction,

$$\mu\epsilon^{\mu\frac{\pi}{2}}(4\mu^2-1)=8\mu^2+3.$$

2526. Find the equation for the curve on which, if a smooth particle be constrained to move under a force varying as $(\text{distance})^{-7}$, the pressure will be constant; and prove that Bernoulli's lemniscate is a particular case.

2527. Three equal and similar particles, repelling each other with forces varying as the distance, are connected by equal inextensible strings and are at rest; one of the strings is cut: prove that the subsequent angular velocity of either of the uncut strings will vary as

$$\sqrt{\frac{1-2\cos\theta}{2+\cos\theta}},$$

where θ is the angle between them.

2528. Two heavy particles are placed on a smooth cycloidal arc whose axis is vertical and are connected by a fine string passing along the arc; the distance of either particle from its position of equilibrium measured along the arc is initially c: prove that the time of reaching to a distance s from the position of equilibrium will be

$$2\sqrt{\frac{a}{g}}\log\left(\frac{s}{c}+\sqrt{\frac{s^2}{c^2}-1}\right),$$

where a is the radius of the generating circle.

2529. An elliptic wire is placed with its minor axis vertical and on it slides a smooth ring to which are attached strings which pass through smooth fixed rings at the foci and sustain each a particle of weight equal to the weight of the ring: determine the velocity which the particle must have at the highest point in order that the velocity at the lowest point may be equal to that at the end of the major axis.

[The required velocity is that due to a height

$$b(1-2e^2)\div 2e^2(1+2e^2).]$$

2530. Two particles of masses p, q are connected by a fine inextensible string which passes through a small fixed ring; p hangs vertically and q is held so that the adjacent string is horizontal: prove that when q is let go the initial tension of the string is $pqg \div (p+q)$, and the initial radius of curvature of the path of q bears to the initial distance of q from the ring the ratio $3\{p^2+(p+q)^2\}^{\frac{3}{2}} : p(p+q)(4p+3q)$.

2531. A particle in motion on the surface $z = \phi(x, y)$ under the action of gravity describes a curve in a horizontal plane with velocity u: prove that, at every point of the path,

$$\frac{u^2}{g}\left\{\left(\frac{dz}{dy}\right)^2\frac{d^2z}{dx^2} - 2\frac{dz}{dx}\frac{dz}{dy}\frac{d^2z}{dxdy} + \left(\frac{dz}{dx}\right)^2\frac{d^2z}{dy^2}\right\} + \left\{\left(\frac{dz}{dx}\right)^2 + \left(\frac{dz}{dy}\right)^2\right\}^2 = 0;$$

the axis of z being vertical.

2532. In a smooth surface of revolution whose axis is vertical a heavy particle is projected so as to move on the surface and describe a path which differs very little from a horizontal circle: prove that the time of a vertical oscillation is $\pi\sqrt{\dfrac{kr\cos\alpha}{g(k+3r\sin\alpha\cos^2\alpha)}}$, where k is the distance from the axis, r the radius of curvature of the meridian curve, and α the inclination of the normal to the vertical in the mean position of the particle.

2533. A heavy particle is projected inside a smooth paraboloid of revolution whose vertex is its lowest point and the greatest and least vertical heights of the particle above the vertex are h_1, h_2, the velocities at these points being V_1, V_2: prove that $V_1^2 = 2gh_2$, $V_2^2 = 2gh_1$, and that throughout the motion the pressure of the particle on the paraboloid will vary as the curvature of the generating parabola.

[The pressure $= 2W(h_1+a)(h_2+a) \div a\rho$, where W is the weight of the particle, $4a$ the latus rectum, and ρ the radius of curvature of the generating parabola. Also ρ', the radius of absolute curvature of the path of the particle, is given by the equation

$$4\left(\frac{h_1+h_2-z}{\rho'}\right)^2 = \frac{\{h_1h_2 + a(h_1+h_2-z)\}^2}{a(a+z)^3} + \frac{h_1h_2}{(a+z)(h_1+h_2-z)},$$

where z denotes the height of the particle.]

2534. A heavy particle is moving upon a given smooth surface of revolution under the action of a force P parallel to the axis: prove that the equation of the projection of the path on a plane perpendicular to the axis is

$$Pf(u) = h^2u^2\left\{u + \frac{d^2u}{d\theta^2} + f(u)\frac{d}{d\theta}\left(f(u)\frac{du}{d\theta}\right)\right\},$$

where $\left(\dfrac{1}{u}, \theta\right)$ are polar co-ordinates measured from the trace of the axis and the equation of the surface is $u^2\dfrac{dz}{du} = f(u)$, the axis of the surface being the axis of z.

2535. Two particles of masses m, m' lying on a smooth horizontal table are connected by an inextensible string at its full length and passing through a small fixed ring in the table; the particles are at distances a, a' from the ring and are projected with velocities V, V' at right angles to the string so that the parts of the string revolve in the same sense: prove that either particle will describe a circle uniformly if $mV^2a' = m'V'^2a$; and that the second apsidal distances will be a', a respectively if
$$mV^2a^2 = m'V'^2a'^2.$$

2536. Two particles m, m', connected by a string which passes through a small fixed ring, are held so that the string is horizontal and the distances from the ring are a, a'; the particles are simultaneously set free and proceed to describe paths whose initial radii of curvature are ρ, ρ': prove that
$$\frac{m}{\rho} = \frac{m'}{\rho'}, \quad \frac{1}{\rho} + \frac{1}{\rho'} = \frac{1}{a} + \frac{1}{a'}.$$

2537. Two particles m, m' are connected by a string, m' lies on a smooth horizontal table and m is held so that the part of the string (of length a) which is not in contact with the table makes an angle α with the horizon: prove that, when m is set free, the initial radius of curvature of its path is
$$\frac{3a\{m^2 + m'(2m + m')\cos^2\alpha\}^{\frac{3}{2}}}{m'(m + m')\{m\cos\alpha + (2m + 3m')\cos^3\alpha\}}.$$

2538. Two particles A and B are connected by a fine string; A rests on a rough horizontal table and B hangs vertically at a distance a below the edge of the table, A being in limiting equilibrium; B is now projected horizontally with a velocity V in the plane normal to the edge of the table: prove that A will begin to move with acceleration
$$\mu V^2 \div (\mu + 1)\, a,$$
and that the initial radius of curvature of the path of B will be $a(\mu + 1)$, where μ is the coefficient of friction.

2539. A smooth surface of revolution is generated by the curve $x^2y = a^3$ revolving about the axis of y, which is vertically downwards, and a heavy particle is projected with a velocity due to its depth below the horizontal plane through the origin so as to move on the surface: prove that it will cross all the meridians at a constant angle.

2540. A heavy particle is projected so as to move on a smooth curve in a vertical plane starting from a point where the tangent is vertical; the form of the curve is such that for any velocity of projection the particle will abandon the curve when it is at a vertical height above the point of projection which bears a constant ratio $2 : m + 1$ to the greatest height subsequently attained: prove that the equation of the curve is $y^m = cx^{m-1}$, where c is constant.

2541. A smooth wire in the form of a circle is made to revolve uniformly in a horizontal plane about a point A in its circumference with angular velocity ω; a small ring P slides on the wire and is initially at rest at its greatest distance c from A: prove that its distance from A after any time t will be $2c \div (\epsilon^{\omega t} + \epsilon^{-\omega t})$ and that the tangent to the path in space of P bisects the angle between PA and the radius to P.

2542. Two equal particles are connected by a fine string and one lies on a smooth horizontal table, the string passing through a small fixed ring in the table to the other particle, which is vertically below the ring; the first particle is projected on the table at right angles to the string with a velocity due to a height $c \div n(n+1)$, where c is the distance from the ring: prove that the next apsidal distance will be equal to cn^{-1} and the velocity will then bear to the initial velocity the ratio $n : 1$; also that the radius of curvature of the initial path of the projected particle is $4c \div (n^2 + n + 2)$.

2543. A heavy particle of weight W is attached to a fixed point by a fine extensible string of natural length a and modulus λ, and the particle is projected so as to make complete revolutions in a vertical plane: prove that if properly set in motion the angular velocity of the string will be uniform, provided that λ be not less than six times W, and that the equation of the curve described by the particle is

$$\frac{3r}{2a} = 2 - \frac{3W}{\lambda}\cos\theta,$$

the straight line from which θ is measured being drawn from the fixed point vertically upwards.

2544. Two particles whose masses are p, q are connected by a fine inextensible string passing through a small fixed ring, and p hangs vertically while q describes a path deviating very little from a horizontal circle: prove that the distance of p at any time from its mean position is $A_1 \sin(n_1 t + B_1) + A_2 \sin(n_2 t + B_2)$; where n_1, n_2 are the positive roots of the equation in x

$$\left\{x^2 - \frac{3g}{c}(1 - \cos\alpha)\right\}\left\{x^2 - \frac{g}{c\cos\alpha}(1 + 3\cos^2\alpha)\right\} = \frac{9g^2}{c^2}\cos\alpha(1 - \cos\alpha);$$

where c is the mean distance from the ring and $q\cos\alpha \equiv p$.

2545. A heavy particle is projected so as to move on a rough inclined plane, the coefficient of friction being $n\tan\alpha$ and the inclination of the plane α: prove that the intrinsic equation of the path will be

$$\frac{ds}{d\phi} = \frac{V^2}{g\sin\alpha}\,\frac{1}{\cos\phi\,(1+\sin\phi)^{1+n}(1-\sin\phi)^{1-n}},$$

where V is the velocity at the highest point. Also prove that, if two points be taken at which the directions of motion make equal angles ϕ with the direction at the highest point, and V_1, V_2 be the velocities, ρ_1, ρ_2, r the radii of curvature, at these points and at the highest point,

$$V_1 V_2 \cos^2\phi = V^2, \quad \rho_1\rho_2\cos^6\phi = r^2.$$

2546. In the last question, in the particular case when $n=1$, prove that the time of moving from one of the two points to the other is

$$\frac{V}{g\sin\alpha}\left\{\log\tan\left(\frac{\pi}{4}+\frac{\phi}{2}\right)+\frac{\sin\phi}{\cos^2\phi}\right\};$$

the arc described is

$$\frac{V^2}{2g\sin\alpha}\left\{\log\tan\left(\frac{\pi}{4}+\frac{\phi}{2}\right)+\frac{\sin\phi}{\cos^2\phi}+\frac{2\sin\phi}{\cos^4\phi}\right\};$$

the horizontal space described is

$$\frac{2V^2}{3g\sin\alpha}\,\frac{\sin\phi\,(3-\sin^2\phi)}{\cos^3\phi};$$

and that

$$\frac{1}{\rho_1}-\frac{1}{\rho_2}=\frac{4\sin\phi\cos\phi}{r}.$$

2547. A heavy particle moves on a smooth curve in a vertical plane of such a form that the pressure on the curve is constant and equal to m times the weight of the particle: prove that the intrinsic equation of the path is

$$\frac{ds}{d\phi}=\frac{a}{(m+\cos\phi)^2},$$

ϕ being measured downwards from a fixed horizontal line; that the difference of the greatest and least vertical depths of the particle is $2ma\div(m^2-1)^2$; the time from one vertex to the next in the same horizontal plane is $2\pi m\frac{a}{g}\div(m^2-1)^{\frac{3}{2}}$; and the arc between these points $\pi a\,(1+2m^2)\div(m^2-1)^{\frac{5}{2}}$. Also the greatest breadth of a loop is

$$a\left\{(1+2m^2)\sqrt{m^2-1}-3m^2\cos^{-1}\left(\frac{1}{m}\right)\right\}\div m\,(m^2-1)^{\frac{5}{2}}.$$

2548. A particle is placed at rest in a rough tube $(4\mu=3)$ which revolves uniformly in one plane about one extremity and is acted on by no force but the pressures of the tube: prove that the equation of the path of the particle is

$$5r=a\,(4\epsilon^{\frac{1}{2}\theta}+\epsilon^{-2\theta}).$$

2549. A rectilinear tube inclined at an angle α to the vertical revolves with uniform angular velocity ω about a vertical axis which intersects the tube, and a heavy particle is projected from the stationary point of the tube with a velocity $g\cos\alpha\div\omega\sqrt{\sin\alpha}$; find the position of the particle at any given time before it attains relative equilibrium; and prove that the equilibrium is unstable.

[The particle will describe a space s along the tube in the time

$$\frac{1}{\omega\sqrt{\sin\alpha}}\log\left(\frac{a}{a-s}\right),$$

where $a\omega^2 \sin\alpha = g\cos\alpha$; and the equation of motion is

$$\frac{d^2s}{dt^2} = \omega^2 (s - a)\sin\alpha.]$$

2550. A smooth parabolic tube of latus rectum l is made to revolve about its axis, which is vertical, with angular velocity $\sqrt{\frac{g}{l}}$, and a heavy particle is projected up the tube: prove that the velocity of the particle is constant and that the greatest height to which the particle rises in the tube is double that due to the velocity of projection.

2551. A smooth parabolic tube revolves with uniform angular velocity about its axis, which is vertical, and a heavy particle is placed within the tube very near the lowest point; find the least angular velocity which the tube can have in order that the particle may rise; and prove that, if it rise, its velocity will be proportional to its distance from the axis; also that, if one position be one of relative equilibrium, every position will be such.

2552. A curved tube is revolving uniformly about a vertical axis in its plane and is symmetrical about that axis; the angular velocity is $\sqrt{\frac{g}{a}}$, where a is the radius of curvature at the vertex: prove that the equilibrium of a particle placed at the vertex will be stable or unstable according as the conic of closest contact is an ellipse or hyperbola.

2553. A circular tube of radius a revolves uniformly about a vertical diameter with angular velocity $\sqrt{\frac{ng}{a}}$, and a particle is projected from its lowest point with such velocity that it can just reach the highest point: prove that the time of describing the first quadrant is

$$\sqrt{\frac{a}{(n+1)g}}\log\left(\sqrt{n+2} + \sqrt{n+1}\right).$$

2554. A circular tube containing a smooth particle revolves about a vertical diameter with uniform angular velocity ω, find the position of relative equilibrium; and prove that the particle will oscillate about this position in a time $2\pi \div \omega\sin\alpha$, α being the angle which the normal at the point makes with the vertical.

2555. A heavy particle is placed in a tube in the form of a plane curve which revolves with uniform angular velocity ω about a vertical axis in its plane, and the particle oscillates about a position of relative equilibrium: prove that the time of oscillation is

$$\frac{2\pi}{\omega}\sqrt{\frac{r\sin\alpha}{k - r\sin\alpha\cos^2\alpha}},$$

k being the distance from the axis, r the radius of curvature, and α the inclination of the normal to the vertical, at the point of equilibrium.

2556. A straight tube inclined to the vertical at an angle α revolves with uniform angular velocity ω about a vertical axis whose shortest distance from the tube is a and contains a smooth heavy particle which is initially placed at its shortest distance from the axis: prove that the space s which the particle describes along the tube in a time t is given by the equation

$$2s = \frac{g \cos \alpha}{\omega^2 \sin^2 \alpha} (\epsilon^{\omega t \sin \alpha} + \epsilon^{-\omega t \sin \alpha} - 2) + a (\epsilon^{\omega t \sin \alpha} - \epsilon^{-\omega t \sin \alpha}).$$

2557. A heavy particle is attached to two points in the same horizontal plane at a distance a by two extensible strings each of natural length a, and is set free when each string is at its natural length: prove that the radius of curvature of the initial path of the particle is

$$2\sqrt{3}a \div (m \sim n),$$

the moduli of the strings being respectively m and n times the weight of the particle.

2558. Three equal particles P, Q, Q', for any two of which $e = 1$, move in a smooth fine circular tube of which AB is a vertical diameter; P starts from A, and Q, Q' at the same instant in opposite senses from B, the velocities being such that at the first impact all three have equal velocities: prove that throughout the whole motion the straight line joining any two particles is either horizontal or passes through one of two fixed points (images of each other with respect to the circle); and that the intervals of time between successive impacts are all equal.

2559. A point P describes the curve $y = a \log \sec \frac{x}{a}$ with a velocity which varies as the cube of the radius of curvature and has attached to it a particle Q by means of a string of length a; when P is at the origin, Q is at the corresponding centre of curvature and its velocity is equal and opposite to that of P: prove that throughout the motion the velocity of Q will be equal in magnitude to that of P, and that Q is always the pole of the equiangular spiral of closest contact with the given curve at P.

IV. *Motion of Strings on Curves or Surfaces.*

2560. A uniform heavy chain is placed on the arc of a smooth vertical circle, its length being equal to that of a quadrant and one extremity being at the highest point of the circle: prove that in the beginning of the motion the resultant vertical pressure on the circle bears to the resultant horizontal pressure the ratio $\pi^2 - 4 : 4$.

2561. A string of variable density is laid on a smooth horizontal table in the form of a curve such that the curvature is everywhere proportional to the density and tangential impulses are applied at the ends: prove that the equation for determining the impulsive tension T at any point is $T = A\epsilon^{\phi} + B\epsilon^{-\phi}$, where ϕ is the angle which the tangent makes with a fixed direction; and that, if the curve be an equiangular spiral, the initial direction of motion of any point will be at right angles to the radius vector.

2562. A number of material particles P_1, P_2, ... of masses m_1, m_2, ... connected by inextensible strings are placed on a horizontal plane so that the strings are sides of an unclosed polygon each of whose angles is $\pi - \alpha$, and an impulse is applied to P_1 in the direction P_2P_1: prove that

$$m_r(T_{r+1}\cos\alpha - T_r) = m_{r+1}(T_r - T_{r-1}\cos\alpha),$$

where T_r is the impulsive tension of the rth string; and deduce the equation

$$\frac{d^2T}{ds^2} - \frac{1}{\mu}\frac{d\mu}{ds}\frac{dT}{ds} - \frac{T}{\rho^2} = 0$$

for the impulsive tension in the case of a fine chain. From either equation deduce the result of the last question.

2563. A fine chain of variable density is placed on a smooth horizontal table in the form of a curve in which it would hang under the action of gravity and two impulsive tensions applied to its ends, which are to each other in the same ratio as the tensions at the same points in the hanging chain: prove that the whole will move without change of form parallel to the straight line which was vertical in the hanging chain.

2564. A heavy uniform string PQ, of which P is the lower extremity, is in motion on a smooth circular arc in a vertical plane, O being the centre and OA the horizontal radius: prove that the tension at any point R of the string is

$$W\frac{\gamma}{\alpha}\left\{\frac{\sin\gamma}{\gamma}\cos(\gamma+\theta) - \frac{\sin\alpha}{\alpha}\cos(\alpha+\theta)\right\},$$

where θ, 2α, 2γ are the angles AOP, POQ, POR respectively, and W the weight of the string.

2565. A portion of a heavy uniform string is placed on the arc of a four-cusped hypocycloid, occupying the space between two adjacent cusps, and runs off the curve at the lower cusp where the tangent is vertical: prove that the velocity which the string will have when just leaving the arc will be that due to a space of nine-tenths the length of the string.

2566. A uniform string is placed on the arc of a smooth curve in a vertical plane and moves under the action of gravity: prove the equation of motion

$$\frac{d^2s}{dt^2} = \frac{g}{l}(y_2 - y_1),$$

l being the length of the string, s the arc described by any point of it at a time t, and y_1, y_2 the depths of its ends below a fixed horizontal straight line.

2567. A uniform heavy string APB is in motion on a smooth curve in a vertical plane, and on the horizontal ordinate from a fixed vertical line to A, P, B are taken lengths equal to the arcs measured from a fixed point of the curve to A, P, B respectively: prove that the ends of these lengths are the corners of a triangle whose area is always proportional to the tension at P.

2568. A uniform heavy string is placed on the arc of a smooth cycloid whose axis is vertical and vertex upwards: determine the motion, and prove that, so long as the whole of the string is in contact with the cycloid, the tension at any given point of the string is constant throughout the motion and greatest at the middle point (measured on the arc).

2569. A uniform heavy chain is in motion on the arc of a smooth curve in a vertical plane and the tangent at the point of greatest tension makes an angle ϕ with the vertical: prove that the difference between the depths of the extremities is $l \cos \phi$.

2570. A uniform inextensible string is at rest in a smooth groove, which it just fits, and a tangential impulse P is applied at one end: prove that the normal impulse per unit of length at a distance s (along the arc) from the other end is $Ps \div a\rho$, where a is the whole length of the string and ρ the radius of curvature at the point considered.

2571. A straight tube of uniform bore is revolving uniformly in a horizontal plane about a vertical axis at a distance c from the tube, and within the tube is a smooth uniform chain of length $2a$ which is initially at rest with its middle point at the distance c from the axis of revolution: prove that the chain in a time t will describe a space $\frac{1}{2}c\,(\epsilon^{\omega t} - \epsilon^{-\omega t})$ along the tube, and that the tension of the chain at a point distant x from its middle point is

$$\frac{m}{4a}(a^2 - x^2)\,\omega^2,$$

where m is the mass of the chain and ω the angular velocity.

2572. A circular tube of radius a revolving with uniform angular velocity ω about a vertical diameter contains a heavy uniform rigid wire which just fits the tube and subtends an angle 2α at the centre: prove that the wire will be in relative equilibrium if the radius to its middle point make with the vertical an angle whose cosine is

$$g \div a\omega^2 \cos \alpha,$$

and that the stress along the wire is a minimum at the lowest point of the tube (provided the wire pass through that point) and a maximum at the point whose projection on the axis bisects the distance between the projections of the ends of the wire. Discuss which position of equilibrium is stable, proving the equation of motion

$$a\alpha \frac{d^2\theta}{dt^2} + \sin \alpha \sin \theta \,(g - a\omega^2 \cos \alpha \cos \theta) = 0,$$

where θ is the angle which the radius to the middle point of the wire makes with the vertical.

[The highest position of equilibrium is always unstable; the oblique position is stable if it is possible, the time of a small oscillation being

$$\frac{2\pi}{\omega \sin\beta}\sqrt{\frac{2a}{\sin 2a}}, \quad \text{where } a\omega^2 \cos a \cos\beta = g;$$

and the lowest position is stable when $a\omega^2 \cos a < g$, the time of a small oscillation being

$$2\pi\sqrt{\frac{aa}{\sin a}\,\frac{1}{g - a\omega^2 \cos a}}.]$$

2573. A pulley is fixed above a horizontal plane; over the pulley passes a fine inextensible string which has two equal uniform chains fixed to its ends; in the position of equilibrium a length a of each chain is vertical, and the rest is coiled up on the table. One chain is now drawn up through a space na: form the equation of motion, and prove that the system will next come to instantaneous rest when the upper end of the other chain is at a depth ma below its mean position, where

$$(1 - m)\,\epsilon^m = (1 + n)\,\epsilon^{-n}.$$

Also, when $n = 1$, prove that $m = \cdot 5623$ nearly.

V. *Resisting Media. Hodographs.*

2574. A heavy particle is projected vertically upwards, the resistance of the air being mass × (velocity)^2 ÷ c; the particle in its ascent and descent has equal velocities at two points whose respective heights above the point of projection are x, y: prove that

$$\epsilon^{2\frac{a-x}{c}} + \epsilon^{-2\frac{a-y}{c}} = 2.$$

2575. A heavy particle moves in a medium in which the resistance varies as the square of the velocity, v, v', u are its velocities at the two points where its direction of motion makes angles $-\phi$, ϕ with the horizon and at the highest point, and ρ, ρ', r are the radii of curvature at the same two points and the highest point respectively: prove that

$$\frac{1}{v^2} + \frac{1}{v'^2} = \frac{2\cos^2\phi}{u^2}, \quad \frac{1}{\rho} + \frac{1}{\rho'} = \frac{2\cos^3\phi}{r}.$$

2576. A heavy particle moves in a medium whose resistance varies as the $2n^{\text{th}}$ power of the velocity; v, v', u are the velocities of the particle when its direction of motion makes angles $-\phi$, ϕ with the horizon and at the highest point, and ρ, ρ', r are the radii of curvature at the same two points and the highest point respectively: prove that

$$\frac{1}{v^{2n}} + \frac{1}{v'^{2n}} = \frac{2\cos^{2n}\phi}{u^{2n}}, \quad \frac{1}{\rho^n} + \frac{1}{\rho'^n} = 2\left(\frac{\cos^3\phi}{r}\right)^n.$$

2577. A small smooth bead slides on a fine wire whose plane is vertical and the height of any point of which is $a\sin\frac{2s}{c}$, s being the arc measured from the lowest point, in a medium whose resistance is mass × (velocity)^2 ÷ c, and starts from the point where $8s = \pi c$: prove that the velocity acquired in falling to the lowest point is $\sqrt{ag}$.

2578. A heavy particle slides on a smooth curve whose plane is vertical in a medium whose resistance varies as the square of the velocity, and in any time describes a space which is to the space described in the same time by a particle falling freely in vacuo as $1 : 2n$: prove that the curve must be a cycloid whose vertex is its highest point, and that the starting point of the particle must divide the arc between two cusps in the ratio $2n-1 : 2n+1$.

2579. A heavy particle falls down the arc of a smooth cycloid whose axis is vertical and vertex upwards in a medium whose resistance is mass $\times$ (velocity)$^2 \div 2c$, its distance along the arc from the vertex being initially c: prove that the time to the cusp will be

$$\sqrt{\frac{8a}{g}\left(\frac{4a}{c}-1\right)},$$

where $2a$ is the length of the axis.

2580. A particle is projected from a fixed point A in a medium whose resistance is measured by $3\omega \times$ velocity and attracted by a fixed point S by a force whose acceleration is $2\omega^2 \times$ distance: prove that the particle will describe a parabola tending in the limit to come to rest at S.

[Taking $SA = a$, and u, v to be the component velocities at A along and perpendicular to SA, the equation of the path is

$$\{(u + a\omega)\, y - vx\}^2 = av\,\{vx - (u + 2a\omega)\, y\},$$

and the length of the latus rectum is

$$a^2v^2\omega \div (\overline{u + a\omega}^2 + v^2)^{\frac{3}{2}}.]$$

2581. A heavy particle moves in a circular tube whose plane is vertical in a medium whose resistance is mass $\times$ (velocity)$^2 \div 2c$, starting from a point in the upper semicircle where the normal makes the angle $\tan^{-1}\frac{c}{a}$ with the vertical: prove that the kinetic energy at any time while the particle moves through the semicircle which begins at this point is proportional to the distance of the particle from the bounding diameter.

2582. A point describes a straight line under acceleration tending to a fixed point and varying as the distance: prove that the corresponding point of the hodograph will move under the same law of acceleration.

2583. The curves $r^m = a^m \sin m\theta$, $r^n = a^n \sin n\theta$ will be each similar to the hodograph of the other when described about a centre of force in the pole, provided that $mn + m + n = 0$. Prove this property geometrically for both curves when $m = 1$, $2n = -1$.

2584. A point describes a certain curve in such a manner that its hodograph is described as if under a central force in its pole, and T, N are the tangential and normal accelerations of the point: prove that

$$N = 3T\frac{ds}{d\rho} = \left(\frac{c^4}{\rho}\right)^{\frac{1}{3}};$$

where ρ is the radius of curvature, s the arc measured from a fixed point, and c a constant: also prove that the acceleration at any point of the hodograph will vary as

$$\frac{1}{\rho}\left(9 + \overline{\frac{d\rho}{ds}}\Big|^2 - 3\rho\frac{d^2\rho}{ds^2}\right),$$

and that if the intrinsic equation of the curve be

$$\frac{ds}{d\phi} = a \sec^3 m\phi,$$

the equation of the hodograph will be $r = a \sec m\theta$.

2585. A point describes half the arc of a cardioid, oscillating symmetrically about the vertex, in such a way that the hodograph is a circle with the pole in the circumference: prove that the acceleration of the point describing the cardioid varies as $2r - 3a$, r being the distance from the cusp and $2a$ the length of the axis: also prove that the direction of acceleration changes at double the rate of the direction of motion.

2586. A heavy particle of weight W is moving in a medium in which the resistance varies as the nth power of the velocity, and F is the resistance when the direction of motion makes an angle ϕ with the horizon: prove that

$$W = nF\cos^n\phi \int \sec^{n+1}\phi \, d\phi.$$

2587. A heavy particle is projected so as to move on a rough plane inclined to the horizon at the angle of friction: prove that the hodograph of the path is a parabola and that the intrinsic equation of the path is

$$s = \frac{V^2}{4g\sin\alpha}\left\{\frac{3\sin\phi + 2\sin^2\phi}{(1+\sin\phi)^2} + \log\tan\left(\frac{\pi}{4} + \frac{\phi}{2}\right)\right\};$$

where V is the velocity at the highest point and α the angle of friction.

2588. Two points P, Q describe two curves with equal velocities, and the radius vector of Q is always parallel to the direction of motion of P: shew how to find P's path when Q's path is given; and prove that (1) when Q describes a straight line P describes a catenary, (2) when Q describes the circle $r = a\cos\theta$, P describes a circle of radius a, (3) when Q describes the cardioid $r = a(1 + \cos\theta)$, P describes a two-cusped epicycloid.

2589. A circle is described by a point in a given time under the action of a force tending to a fixed point within the circle: prove that, for different positions of the centre of force, the action during a whole revolution varies inversely as the minimum chord which can be drawn through the point.

[In any closed oval under a central force to a point within it the action during a whole revolution $= \frac{2A}{P}\int_0^{2\pi}\frac{r^3}{p^2}\,d\phi$, where r is the radius vector from the centre of force, p the perpendicular from the centre of force on the tangent, ϕ the angle which the tangent makes with some fixed straight line, A the area of the oval, and P the periodic time.]

2590. A point describes a parabola under a central force in the vertex: prove that the hodograph is a parabola whose axis is at right angles to the axis of the described parabola.

[In general if any conic be described under any central force the hodograph is another conic which will be a parabola when the described conic passes through the centre of force.]

2591. A point P describes a catenary in such a manner that a straight line drawn from a fixed point parallel and proportional to the velocity of P sweeps out equal areas in equal times: prove that the direction of P's acceleration makes with the normal at P an angle $\tan^{-1}(\frac{2}{3}\tan\phi)$, where ϕ is the angle through which the direction of motion has turned in passing from the vertex.

2592. A circle is described under a constant force not tending to the centre: prove that the hodograph is Bernoulli's lemniscate.

2593. A curve is described with constant acceleration and its hodograph is a parabola with its pole at the focus: prove that the intrinsic equation of the described curve is

$$\frac{ds}{d\phi} = a\sec^3\frac{\phi}{2}.$$

2594. A point describes a curve so that the hodograph is a circle described with constant velocity and with the pole on its circumference: prove that the described curve is a cycloid described as if by a heavy particle falling from cusp to cusp.

2595. A point describes a certain curve with acceleration initially along the normal, and the direction of acceleration changes at double the rate of the direction of motion and in the same sense: prove that the hodograph will be a circle with the pole on its circumference.

2596. A particle is constrained to move in an elliptic tube under two forces to the foci, each varying inversely as the square of the distance and equal at equal distances, and is just displaced from the position of unstable equilibrium: prove that the hodograph is a circle with the pole on its circumference.

[The particle will oscillate over a semi-ellipse bounded by the minor axis, and the hodograph corresponding to this will be a complete circle with the diameter through its pole parallel to the minor axis.]

DYNAMICS OF A RIGID BODY.

I. *Moments of Inertia, Principal Axes.*

2597. The density of an ellipsoid at any point is proportional to the product of the distances of the point from the principal planes: prove that the moments of inertia about the principal axes are

$$\tfrac{1}{4}m\,(b^2+c^2),\ \tfrac{1}{4}m\,(c^2+a^2),\ \tfrac{1}{4}m\,(a^2+b^2),$$

where m is the mass and a, b, c the semi-axes.

2598. Prove the following construction for the principal axes at O, the centroid of a triangular lamina ABC: draw the circle OBC, and in it the chords Ob, Oc parallel to AC, AB respectively, and let Bb, Cc meet in L; then, if aa' be the diameter of the circle drawn through L, Oa, Oa' will be the directions of the principal axes at O.

2599. Prove the following construction for the principal axes at the centre O of a lamina bounded by a parallelogram $ABCD$: draw the circle OBC and in it chords Ob, Oc parallel to AB, BC, and let BC, bc meet in L; then, if aa' be the diameter of this circle drawn through L, Oa, Oa' will be the directions of the principal axes at O.

2600. Prove that any lamina is kinetically equivalent to three particles, each of one third the mass of the triangle, placed at the corners of a maximum triangle inscribed in the ellipse whose equation, referred to the principal axes at the centre of inertia, is $Ax^2+By^2=2AB$, where mA, mB are the principal moments of inertia and m the mass.

2601. Prove that any rigid body is kinetically equivalent to three equal uniform spheres, each of one third the mass of the body, whose centres are corners of a maximum triangle inscribed in the ellipse

$$\frac{x^2}{C-A}+\frac{y^2}{C-B}=2,\ z=0,$$

and whose common radius is $\sqrt{\tfrac{5}{2}(A+B-C)}$; the equation of the ellipsoid of gyration being $\dfrac{x^2}{A}+\dfrac{y^2}{B}+\dfrac{z^2}{C}=1$, and $A<B<C$.

[Since $A+B$ can never be less than C the radius will always be real, but for the spheres not to intersect in any of their positions it will be necessary that $23(C-B)>20A$, which could not be satisfied by a body of form approaching spherical. As the spheres need only be ideal for simplification of calculation, this condition is of no importance.]

2602. A straight line is at every point of its course a principal axis of a given rigid body: prove that it passes through the centre of inertia.

2603. A tetrahedron is kinetically equivalent to six particles at the middle points of the edges, each $\frac{1}{10}$ the mass of the tetrahedron, and one at the centroid of mass $\frac{2}{5}$ the mass of the tetrahedron.

2604. The principal moments of inertia of a rigid body, whose mass is unity, at the centre of inertia are A, B, C, and $a^2+b^2+c^2+r^2$ is a principal moment of inertia at the point (a, b, c), the principal axes at the centre of inertia being axes of co-ordinates: prove that

$$\frac{a^2}{A-r^2}+\frac{b^2}{B-r^2}+\frac{c^2}{C-r^2}=1.$$

2605. The locus of the points at which two principal moments of inertia of a given rigid body are equal is the focal curves of the ellipsoid of gyration at the centre of inertia.

2606. The locus of the points at which one of the principal axes passes through a given point, which lies in one of the principal planes at the centre of inertia, is a circle.

2607. The locus of the points at which one of the principal axes of a given rigid body is in a given direction is a rectangular hyperbola with one asymptote in the given direction.

2608. In a triangular lamina any one of the sides is a principal axis at the point bisecting the distance between its mid-point and the foot of the perpendicular from the opposite corner.

2609. In any uniform tetrahedron, if one edge be at any point a principal axis so also will the opposite edge; the necessary condition is that the directions of the two edges shall be perpendicular; and the point at which an edge is a principal axis divides the distance between the mid-point and the foot of the shortest distance between it and the opposite edge in the ratio 1 : 2.

2610. Straight lines are drawn in the plane of a given lamina through a given point; the locus of the points at which they are principal axes of the lamina is a circular cubic.

2611. The locus of the straight lines drawn through a given point, each of which is at some point of its course a principal axis of a given rigid body, is the cone

$$a(B-C)yz+b(C-A)zx+c(A-B)xy=0,$$

A, B, C being the principal moments of inertia at the given point, a, b, c the co-ordinates of the centre of inertia and the principal axes at the given point the axes of reference. Also prove that the locus of the points at which these straight lines are principal axes is the curve

$$x^2+y^2+z^2=\frac{(B-C)yz}{cy-bz}=\frac{(C-A)zx}{az-cx}\left\{=\frac{(A-B)xy}{by-ax}\right\}.$$

[The equation of the cone on which these straight lines lie retains the same form when A, B, C denote the principal moments of inertia at the centre of inertia, and the co-ordinate axes are parallel to the principal axes at the centre of inertia.]

2612. The principal axes at a certain point are parallel to the principal axes at the centre of inertia: prove that the point must lie on one of the principal axes at the centre of inertia.

2613. The different straight lines which can be drawn through the point (x, y, z), each of which is at some point of its course a principal axis of a given rigid body, will lie on a cone of revolution if

$$x(B-C) = y(C-A) = z(A-B),$$

the principal axes at the centre of inertia being co-ordinate axes and A, B, C the principal moments of inertia.

II. *Motion about a fixed Axis.*

2614. A circular disc rolls in one plane on a fixed plane, its centre describing a straight line with uniform acceleration f; find the magnitude and position of the resultant of the impressed forces.

[The resultant is a force Mf acting parallel to the plane at a distance from the centre of the disc of one half the radius on the side opposite to the plane.]

2615. A piece of uniform fine wire of given length is bent into the form of an isosceles triangle and revolves about an axis through its vertex perpendicular to its plane: prove that the centre of oscillation will be at the least possible distance from the axis of revolution when the triangle is right-angled.

2616. A heavy sphere of radius a and a heavy rod of length $2a$ swing, the one about a horizontal tangent, the other about a horizontal axis perpendicular to its length through one end, each through a right angle to its lowest position, and the pressures on the axis in the lowest positions are equal: prove that the weights are as 35 : 34.

2617. The centre of percussion of a triangular lamina one of whose sides is the fixed axis bisects the straight line joining the opposite corner with the mid-point of the side.

2618. A lamina $ABCD$ is moveable about AB which is parallel to CD: prove that its centre of percussion will be at the common point of AC and BD if $AB^2 = 3CD^2$.

2619. In the motion of a rigid body about a horizontal axis under the action of gravity, prove that the pressure on the axis is reducible to a single force at every instant of the motion only when the axis of revolution is a principal axis at the point M which is nearest to the centre of inertia: and, if the axis be a principal axis at another point M and the forces be reduced to two acting at M, N respectively, the former will be equal and opposite to the weight of the body.

2620. A rough uniform rod of length $2a$ is placed with a length $c\,(>a)$ projecting over the edge of a horizontal table, the rod being initially in contact with the table and perpendicular to the edge: prove that the rod will begin to slide over the edge when it has turned through an angle whose tangent is $\dfrac{\mu a^2}{a^2+9(c-a)^2}$, μ being the coefficient of friction.

2621. A uniform beam capable of motion about one end is in equilibrium; find at what point a blow must be applied perpendicular to the rod in order that the impulse on the fixed end may be $\dfrac{1}{n}$th of the blow.

[The distance of the point from the fixed end must be to the length of the rod in the ratio $\sqrt{n-1} : \sqrt{3n}$.]

2622. A uniform beam moveable about its middle point is in equilibrium in a horizontal position, a particle whose mass is one-fourth that of the beam and such that the coefficient of restitution is 1 is let fall upon one end and is afterwards grazed by the other end of the beam: prove that the height from which the particle is let fall bears to the circumference of the circle described by an end of the beam the ratio $49(2n+1) : 48$, where n is a positive integer.

2623. A smooth uniform rod is revolving about its middle point, which is fixed on a horizontal table, when it strikes an inelastic particle at rest whose mass is one-sixth of its own, and the angular velocity of the rod is immediately reduced one-ninth: find the point of impact, and prove that, when the particle leaves the rod, the direction of motion of the particle will make with the rod an angle of 45°.

[The point of impact must bisect one of the halves of the rod, and during the subsequent motion

$$\left(\frac{dr}{dt}\right)^2+(2a^2+r^2)\,\omega^2=\frac{16}{9}a^2\Omega^2,\quad (2a^2+r^2)\,\omega=2a^2\Omega,$$

where r is the distance of the particle from the centre of the rod, and ω the angular velocity of the rod at any time, Ω the angular velocity before impact.]

2624. A smooth uniform rod is moving on a horizontal table uniformly about one end and impinges on a particle of mass equal to its own, the distance of the particle from the fixed end being $\dfrac{1}{n}$th of the length of the rod: prove that the final velocity of the particle will be to its initial velocity in the ratio

$$\sqrt{(5n^2-1)(n^2+3)} : 4n.$$

(In this case also $e=0$.)

2625. A uniform rod (mass m) is moving on a horizontal table about one end and driving before it a smooth particle (mass p) which starts from rest close to the axis of revolution: prove that, when the particle is at a distance r from the axis, its direction of motion will make with the rod the angle $\cot^{-1}\sqrt{1+\frac{pr^2}{mk^2}}$, where mk^2 is the moment of inertia of the rod about the axis of revolution.

2626. A uniform circular disc of mass m is capable of motion in a vertical plane about its centre and a rough particle of mass p is placed on it close to the highest point: prove that the angle θ through which the disc will turn before the particle begins to slide is given by the equation

$$\cos\theta\left(1+\frac{3pa^2}{mk^2}\right)-\frac{1}{\mu}\sin\theta=\frac{2pa^2}{mk^2},$$

where a is the radius and mk^2 the moment of inertia of the disc.

2627. A uniform rod, capable of motion in a vertical plane about its middle point, has attached to its ends by fine strings two particles which hang freely; when the rod is in equilibrium inclined at an angle α to the vertical one of the strings is cut: prove that the initial tension of the other string is

$$mpg \div (m + 3p\sin^2\alpha),$$

and that the radius of curvature of the initial path of the particle is

$$9lp\sin^3\alpha \div m\cos\alpha,$$

m, p being the masses of the rod and of a particle, and l the length of the string.

2628. A uniform rod moveable about one end is held in a horizontal position, and to a point of the rod is attached a heavy particle by means of a string: prove that the initial tension of the string when the rod is allowed to fall freely is

$$mpga\,(4a-3c) \div (4ma^2+3pc^2),$$

where m, p are the masses of the rod and particle, $2a$ the length of the rod, and c the distance of the string from the fixed end: also prove that the initial path of the particle referred to horizontal and vertical axes will be the curve

$$ma\,(4a-3c)\,y^2+90c^2l\,(ma+pc)\,x=0,$$

where l denotes the length of the string.

2629. A uniform rod moveable about one end has attached to the other end a heavy particle by a fine string; initially the rod and string are in one horizontal straight line without motion: prove that the radius of curvature of the initial path of the particle will be

$$4ab \div (a+9b),$$

where a, b denote the lengths of the rod and string; and explain why the result does not depend on the masses of the two.

2630. A uniform rod, of length $2a$ and mass m, capable of motion about one end, is held in a horizontal position and on the rod slides a small smooth ring of mass p: prove that, when the rod is set free, the radius of curvature of the initial path of the ring will be

$$\frac{9c^2}{4a-3c}\left(1+\frac{p}{m}\frac{c}{a}\right),$$

where c is the initial distance of the ring from the fixed end.

2631. A uniform rod capable of motion about one end has attached at the other end a particle by means of a fine string, and the system is abandoned freely to the action of gravity when the rod makes an angle α with the string which is vertical: prove that the radius of curvature of the initial path of the particle is

$$9l\left(1+\frac{2p}{m}\right)\sin^2\alpha \div \cos\alpha\,(2-3\sin^2\alpha);$$

where m, p are the masses, and l the length of the string.

2632. A uniform rod is moveable about one end on a smooth horizontal table and to the other end is attached a particle by a fine string; at starting the rod and string are in one straight line, the particle is at rest, but the rod in motion: prove that when the rod and string are next in a straight line the angular velocities of the rod and string will be as $b : a$, or as

$$b\,\{3p\,(a-b)^2 - ma^2\} \;:\; a\,\{3p\,(a-b)^2 + ma\,(a-2b)\},$$

where m, p are the masses, and a, b the lengths of the rod and string.

III. *Motion in Two Dimensions.*

2633. Two equal uniform rods AB, BC, freely jointed at B and moveable about A, start from rest in a horizontal position, BC passing over a smooth peg whose distance from A is $4a\sin\alpha$ (where $3\sin\alpha < 2$): prove that, when BC leaves the peg, the angular velocity of AB is

$$\sqrt{\frac{3g}{2a}\,\frac{\cos\alpha}{1+\sin^2 2\alpha}},$$

where $2a$ is the length of either rod.

2634. A uniform rod of length $2a$ rests with its lower end at the vertex of a smooth surface of revolution whose axis is vertical and passes through a smooth fixed ring in the axis at a distance b from the vertex: the time of a small oscillation will be

$$2\pi\sqrt{\frac{c}{3g}\,\frac{a^2+3(b-a)^2}{b^2-ac}},$$

where c is the radius of curvature at the vertex.

2635. Two heavy particles are fixed to the ends of a fine wire in the form of a circular arc, which rests with its plane vertical on a rough horizontal plane, and α, β are the angles which the radii through the particles make with the vertical: prove that the time of a small oscillation will be

$$4\pi\sqrt{\frac{a}{g}\frac{\sin\frac{\alpha}{2}\sin\frac{\beta}{2}}{\cos\frac{\alpha+\beta}{2}}}.$$

2636. Two equal and similar uniform rods, freely jointed at a common extremity, rest symmetrically over two smooth pegs in the same horizontal plane so that each rod makes an angle α with the vertical: prove that the time of a small oscillation will be

$$2\pi\sqrt{\frac{9a}{g}\frac{\cos\alpha}{1+3\cos^2\alpha}},$$

where $2a$ is the length of either rod.

2637. A lamina with its centre of inertia fixed is at rest, and is struck by a blow at the point (a, b) normally to its plane: prove that the equation of the instantaneous axis is $Aax + Bby = 0$, the axes of co-ordinates being the principal axes at the centre of inertia and A, B being the principal moments of inertia; also that, if (a, b) lie on a certain straight line, there will be no impulse at the fixed point.

2638. A uniform heavy rod revolves uniformly about one end in such a manner as to describe a cone of revolution: determine the pressure on the fixed point and the relation between the angle of the cone and the time of revolution; and prove that, if θ, ϕ be the angles which the vertical makes with the rod and with the direction of pressure,

$$4\tan\phi = 3\tan\theta.$$

2639. A fine string of length $2b$ is attached to two points in the same horizontal plane at a distance $2a$ and carries a particle p at its middle point; a uniform rod of length $2c$ and mass m has at each end a ring through which the string passes and is let fall from a symmetrical position in the same straight line as the two points: prove that the rod will not reach the particle if

$$(a+b-2c)(m+2p)m < 2(2c-a)p^2.$$

2640. A heavy uniform chain is collected into a heap and laid on a horizontal table and to one end is attached a fine string which, passing over a smooth fixed pulley vertically above the heap, is attached to a weight equal to the weight of a length a of the chain: prove that the length of the chain raised before the weight first comes to instantaneous rest is $a\sqrt{3}$, and that when the weight next comes to rest the length of chain which is vertical is ax, where x is given by the equation

$$\left(\frac{1+\sqrt{3}}{1+x}\right)^2 = \epsilon^{\sqrt{3}-x},$$

and that x is nearly equal to $\frac{1}{\sqrt{3}}$.

2641. A uniform rod of length c has at its ends small smooth rings which slide on two fixed elliptic arcs whose planes are vertical and semi-axes are a, b; $a+c$, $b+c$ respectively, and are inclined at angles α, $\frac{\pi}{2}+\alpha$ to the horizon: determine the motion of the rod and the pressures on the arcs, the rod being initially vertical.

2642. A circular disc rolls on a rough cycloidal arc whose axis is vertical and vertex downwards, the length of the arc being such that the curvature at either end of the arc is equal to that of the circle: prove that, if the contact be initially at one end of the arc, the point on the auxiliary circle of the cycloid which corresponds to the point of contact will move with uniform velocity which is independent of the radius of the disc; and that the normal pressure R and the force of friction F in any position of the disc are given by the equations

$$3R = W(5\cos\theta - 2\cos\alpha),\quad 3F = W\sin\theta,$$

where W is the weight of the disc, θ the angle which the common normal makes with the vertical, and α the initial value of θ.

2643. A uniform sphere rolls from rest down a given length l of a rough inclined plane and then traverses a smooth portion of the plane of length ml; find the impulse which takes place when perfect rolling again begins, and prove that the subsequent velocity is less than would have been the case if the whole plane had been rough; if $m=120$, in the ratio 67 : 77.

[The ratio in general is $2+\sqrt{25+35m} : 7\sqrt{m+1}$.

2644. A straight tube AB of small bore, containing a smooth uniform rod of the same length, is closed at the end B and in motion about the fixed end A with angular velocity ω: prove that, if the end B be opened, the initial stress at a point P of the rod is equal to

$$M\omega^2 AP \cdot PB \div 2AB,$$

M being the mass of the rod.

2645. The ends of a uniform heavy rod are fixed by smooth rings to the arc of a circle which is made to revolve uniformly about a fixed vertical diameter; find the positions of relative equilibrium, and prove that any such position in which the rod is not horizontal will be stable.

[If a be the radius of the circle, ω its angular velocity, 2α the angle which the rod subtends at the centre, there will be no inclined positions of equilibrium unless $a\omega^2 \cos a > g$: if $a\omega^2 \cos a \cos \beta = g$, the time of a small oscillation about the inclined position will be $\frac{2\pi}{\omega \sin \beta} \sqrt{1 + \frac{1}{3} \tan^2 a}$; the time of oscillation about the lowest position will be

$$2\pi \sqrt{1 + \tfrac{1}{3} \tan^2 a} \div \sqrt{\frac{g}{a} - \omega^2 \cos a};$$

and, when $g = a\omega^2 \cos a$, the equation of motion will be

$$(1 + \tfrac{1}{3} \tan^2 a) \frac{d^2\theta}{dt^2} + \omega^2 \sin \theta (1 - \cos \theta) = 0.]$$

2646. A smooth semicircular disc rests with its plane vertical and vertex upwards on a smooth horizontal table and on it rest two equal uniform rods, each of which passes through two smooth fixed rings in a vertical line; the disc is slightly displaced, and in the ensuing motion one rod leaves the disc when the other is at the vertex: prove that

$$\frac{m}{p} = \frac{4 \sin a - (1 + \sin \beta)^2 (2 - \sin \beta)}{\sin^2 \beta};$$

where m, p are the masses of the disc and of either rod, a the angle which the radius to either point of contact initially makes with the horizon, and $\beta \equiv \cos^{-1} (2 \cos a)$.

[When the one rod leaves the disc, the pressure of the other on the disc is $pg (1 - \sin^2 \beta)$.]

2647. A uniform rod moves with one end on a smooth horizontal plane and the other end attached to a string which is fixed to a point above the plane; when the rod and string are in one straight line the rod is let go: prove that the angular velocity of the string when vertical will be $\sqrt{\frac{g}{l}\left(1 - \frac{h}{a + l}\right)}$ and its angular acceleration

$$\frac{g}{a + l} \sqrt{\frac{a + l - h}{a - l + h}},$$

a, l, h being the lengths of the rod and string and the height of the fixed point above the plane respectively.

2648. A uniform beam rests with one end on a smooth horizontal table and has the other attached to a fixed point by means of a string of length l: prove that the time of a small oscillation in a vertical plane will be

$$2\pi \sqrt{\frac{2l}{g}}.$$

2649. A sphere rests on a rough horizontal plane with half its weight supported by an extensible string attached to the highest point, whose extended length is equal to the diameter of the sphere: prove that the time of small oscillations of the sphere parallel to a vertical plane is $2\pi \sqrt{\frac{14a}{15g}}$.

2650. Two equal uniform rods AB, BC, freely jointed at B, are placed on a smooth horizontal table at right angles to each other and a blow is applied to A at right angles to AB: prove that the initial velocities of A, C are in the ratio 8 : 1.

2651. Two equal uniform rods AB, BC, freely jointed at B, are laid on a smooth horizontal table so as to include an angle α and a blow is applied at A at right angles to AB; determine the initial velocity of C, and prove that it will begin to move parallel to AB if $9 \cos 2\alpha = 1$.

2652. Five equal uniform rods, freely jointed at their extremities, are laid in one straight line on a horizontal table and a blow applied at the centre at right angles to the line: prove that, initially,

$$\frac{v}{14a} = \frac{v_1}{5a} = \frac{v_2}{-a} = \frac{\omega_1}{9} = \frac{\omega_2}{-3};$$

where v, v_1, v_2 are the velocities of the three rods, ω_1, ω_2 the angular velocities of the two pairs of rods, and $2a$ the length of each rod.

2653. Four equal uniform rods AB, BC, CD, DE, freely jointed at B, C, D, are laid on a horizontal table in the form of a square and a blow is applied at A at right angles to AB from the inside of the square: prove that the initial velocity of A is 79 times that of E.

2654. Two equal uniform rods AB, BC, freely jointed at B and moveable about A, are lying on a smooth horizontal table inclined to each other, at an angle α; a blow is applied to C at right angles to BC in a direction tending to decrease the angle ABC: prove that the initial angular velocities of AB, BC will be in the ratio $\cos\alpha : 8 - 3\cos^2\alpha$; that θ, the least value of the angle ABC during the motion is given by the equation

$$8(5 - 3\cos\theta)(2 - \cos^2\alpha) = (1 - \cos\alpha)^2 (16 - 9\cos^2\alpha):$$

also prove that, when $\alpha = \frac{\pi}{2}$, the angular velocities of the rods when in a straight line will have one of the ratios $-1 : 3$, or $3 : -5$.

2655. A heavy uniform rod resting in stable equilibrium within a smooth ellipsoid of revolution about its major axis, which is vertical, is slightly displaced in a vertical plane: prove that the length of the equivalent simple pendulum is $ace(3e^2 + 1) \div 6(a - c)$, where $2a$ is the length of the rod, $2c$ the latus rectum, and e the excentricity of the generating ellipse.

2656. A uniform rod of length $2a$ rests in a horizontal position with its ends on a smooth curve which is symmetrical about a vertical axis: prove that the time of a small oscillation will be

$$2\pi \sqrt{\frac{ar\cos\alpha(1 + 2\cos^2\alpha)}{3g(a - r\sin^3\alpha)}},$$

r being the radius of curvature of the curve and α the angle which the normal makes with the vertical at either end of the rod.

2657. Four equal rods of length a and mass m are freely jointed so as to form a rhombus one of whose diagonals is vertical; the ends of the other diagonal are joined by an extensible string at its natural length and the system falls through a height h on to a fixed horizontal plane: prove that, if θ be the angle which any rod makes with the vertical at a time t after the impact,

$$(1+3\sin^2\theta)\left(\frac{d\theta}{dt}\right)^2 = \frac{18gh}{a^2}\,\frac{\sin^2\alpha}{1+3\sin^2\alpha} + \frac{6g}{a}(\cos\alpha-\cos\theta) - \frac{3\lambda}{2ma\sin\alpha}(\sin\theta-\sin\alpha)^2;$$

where α is the initial value of θ and λ the modulus of the string.

IV. *Miscellaneous.*

2658. A square is moving freely about a diagonal with angular velocity Ω, when one of the corners not in that diagonal becomes fixed; determine the impulse on the fixed point, and prove that the instantaneous angular velocity is $\frac{1}{7}\Omega$.

[If V be the previous velocity of the point which becomes fixed the impulse will be $\frac{4}{7}MV$.]

2659. A uniform heavy rod of length a, freely moveable about one end, is initially projected in a horizontal plane with angular velocity Ω: prove that the equations of motion are

$$\sin^2\theta\frac{d\phi}{dt}=\Omega,\quad a\left(\frac{d\theta}{dt}\right)^2 = 3g\cos\theta - a\Omega^2\cot^2\theta;$$

where θ, ϕ are respectively the angles which the rod makes with the vertical (downwards from the fixed end) and which the projection of the rod on the horizontal plane makes with its initial position: also, if the least value of θ be $\frac{\pi}{3}$, prove that the resolved vertical pressure on the fixed point when $\theta=\frac{\pi}{3}$ will be $\frac{31}{16}W$, where W is the weight of the rod.

[The vertical pressure on the fixed point in any position is

$$\frac{W}{8}\left(11+9\cos 2\theta+\frac{4a\Omega^2}{g}\cos\theta\right);$$

and the horizontal pressure is $\frac{W}{8}\left(9\sin 2\theta+\frac{4a\Omega^2}{g}\sin\theta\right)$, in the vertical plane through the rod.]

2660. A uniform heavy rod moveable about one end moves in such a manner that the angle which it makes with the vertical never differs much from α: prove that the time of its small oscillations will be

$$2\pi\sqrt{\frac{2a}{3g}\frac{\cos\alpha}{1+3\cos^2\alpha}};$$

where a is the length of the rod.

2661. A centre of force whose acceleration is μ (distance) is at a point O, and from another point A at a distance a are projected simultaneously an infinite number of particles in a direction at right angles to OA and with velocities in arithmetical progression from $\frac{3}{7}a\sqrt{\mu}$ to $\frac{11}{7}a\sqrt{\mu}$: prove that, when after any lapse of time all the particles become suddenly rigidly connected together, the system will revolve with angular velocity $\frac{13}{14}\sqrt{\mu}$.

[If the limits of the velocity be $n_1 a\sqrt{\mu}$, $n_2 a\sqrt{\mu}$, and the time elapsed $\theta\div\sqrt{\mu}$, the common angular velocity of the rigidly connected particles will be $3(n_1+n_2)\sqrt{\mu}\div 6\cos^2\theta + 2(n_1^2+n_1n_2+n_2^2)\sin^2\theta$.]

2662. A uniform heavy rod is suspended by two inextensible strings of equal lengths attached to its ends and to two fixed points whose distance is equal and parallel to the length of the rod; an angular velocity about a vertical axis through its centre is suddenly communicated to the rod such that it just rises to the level of the fixed points: find the impulsive couple, and prove that the tension of either string is suddenly increased sevenfold.

2663. Two equal uniform heavy rods AB, BC, freely jointed at B, rotate uniformly about a vertical axis through A, which is fixed, with angular velocity Ω: prove that the angles α, β which the rods make with the vertical are given by the equations

$$(8\sin\alpha + 3\sin\beta)\cot\alpha = (9\sin\alpha + 6\sin\beta)\cot\beta = \frac{9g}{a\Omega^2};$$

where a is the length of each rod.

2664. A perfectly rough horizontal plane is made to revolve with uniform angular velocity about a vertical axis which meets the plane in O; a heavy sphere is projected on the plane at a point P so that its centre is initially in the same state of motion as if the sphere had been placed freely on the plane at a point Q and set in motion by the impulsive friction only: prove that the centre of the sphere will describe uniformly a circle of radius OQ, and whose centre R is such that OR is equal and parallel to QP.

2665. A perfectly rough plane inclined at an angle α to the horizon is made to revolve with uniform angular velocity Ω about a normal and a heavy motionless sphere is placed upon it and set in motion by the tangential impulse: prove that the ensuing path of the centre will be a prolate, a common, or a curtate cycloid, according as the initial point of contact is without, upon, or within the circle whose equation is

$$2\Omega^2(x^2+y^2) = 35gx\sin\alpha,$$

the axis of y being horizontal and the point where the axis of revolution meets the plane the origin. Also prove that, if the initial point of contact be the centre of this circle, the path will be a straight line.

2666. A rough hollow cylinder of revolution whose axis is vertical is made to revolve with uniform angular velocity Ω about a fixed generator and a heavy uniform sphere is rolling on the concave surface: prove that the equation of motion is

$$\left(\frac{d\phi}{dt}\right)^2 = C + \tfrac{10}{7}\frac{a+b}{b}\Omega^2 \cos\phi;$$

where ϕ is the angle which the common normal to the sphere and cylinder makes at a time t with the plane containing the fixed generator and the axis of the cylinder, and $a+b$, a are the radii of the cylinder and sphere respectively.

2667. A rough plane is made to revolve at a uniform rate Ω about a horizontal line in itself and a sphere is set in motion upon it: determine the motion, and prove that, if when the plane is horizontal the centre of the sphere is vertically above the axis of revolution and moving parallel to it, the contact will cease when the plane has turned through an angle θ given by the equation

$$\frac{11\cos\theta}{6} = \frac{a\Omega^2}{g} + \left(\tfrac{5}{12} - \frac{a\Omega^2}{g}\right)\left(\epsilon^{\theta\sqrt{\frac{5}{7}}} + \epsilon^{-\theta\sqrt{\frac{5}{7}}}\right);$$

where a is the radius of the sphere.

2668. A uniform heavy rod is free to move about one end in a vertical plane which is itself constrained to revolve about a vertical axis through the fixed end at a uniform rate Ω, and the greatest and least angles which the rod makes with the vertical during the motion are α, β: prove that

$$a\Omega^2(\cos\alpha + \cos\beta) = 3g,$$

where a is the length of the rod: also prove that, when $3g = 2a\Omega^2\cos\alpha$, the time of a small oscillation will be $\dfrac{2\pi}{\Omega\sin\alpha}$.

2669. Two heavy uniform rods of lengths $2a$, $2b$ and masses A, B are freely jointed at a common end and are moveable about the other end of A, and the rods fall from a horizontal position of instantaneous rest: prove that the radius of curvature of the initial path of the free end of B will be $2ab(A+B)^2 \div \{aA^2 + b(2A+B)^2\}$.

2670. A rigid body is in motion about its centre of inertia under no forces, and at a certain instant, when the instantaneous axis is the straight line whose equations are

$$x\sqrt{A(B-C)} = z\sqrt{C(A-B)},\ y=0,$$

a point on the cylinder

$$x^2(A-B) + z^2(B-C) + zx\sqrt{\frac{(A-B)(B-C)}{AC}}(C+A) = B(C-A)$$

is suddenly fixed: prove that the new instantaneous axis will be perpendicular to the direction of the former. (The axes of co-ordinates are, as usual, the principal axes at the centre of inertia, and A, B, C the squares of the semi-axes of the principal ellipsoid of gyration.)

2671. A number of concentric spherical shells of equal indefinitely small thickness revolve about a common axis through the centre, each at a uniform rate proportional to the nth power of its radius; the shells become suddenly rigidly united; prove that the subsequent angular velocity bears to the previous angular velocity of the outermost shell the ratio $5 : n + 5$.

2672. An infinite number of concentric spherical shells of equal small thickness are revolving about diameters all in one plane with equal angular velocities, and the axis of revolution of the shell whose radius is r is inclined at an angle $\cos^{-1}\dfrac{r}{a}$ to the axis of the outermost shell: prove that, when united into a solid sphere, the axis of revolution will make an angle $\tan^{-1}\dfrac{3\pi}{16}$ with the former axis of the outer shell.

2673. Prove that any possible given state of motion of a rigid straight rod may be represented by a single rotation about any one of an infinite number of axes lying in a certain plane.

2674. A free rigid body is in motion about its centre of inertia when another point of the rigid body is suddenly fixed and the body then assumes a state of permanent rotation about an axis through that point: prove that the point must lie on a certain rectangular hyperbola.

[With the notation of (2669) the point to be fixed must satisfy the equations

$$(B-C)\frac{x}{A\omega_1} + (C-A)\frac{y}{B\omega_2} + (A-B)\frac{z}{C\omega_3} = 0,$$

$$(A\omega_1 x + B\omega_2 y + C\omega_3 z)\left(\overline{B-C}\,\frac{x}{\omega_1} + \overline{C-A}\,\frac{y}{\omega_2} + \overline{A-B}\,\frac{z}{\omega_3}\right) + (B-C)(C-A)(A-B) = 0,$$

where ω_1, ω_2, ω_3 are the previous component angular velocities; also the new axis of revolution must be parallel to the normal to the invariable plane of the previous motion.]

2675. A rigid body is in motion under the action of no forces and its centre of inertia is at rest; when the instantaneous axis is a certain given line of the body a point rigidly connected with the body is suddenly fixed, and the new instantaneous axis is parallel to one of the principal axes at the centre of inertia: prove that the point to be fixed must lie on a certain hyperbola one asymptote of which is the given principal axis.

2676. A free rigid body is at a certain instant in a state of rotation about an axis through its centre of inertia when a given point of the body becomes suddenly fixed: determine the new instantaneous axis, and prove that there are three directions of the former instantaneous axis for which the new axis will be in the same direction; and these three directions are along conjugate diameters of the principal ellipsoid of inertia.

2677. A rigid body is in motion under the action of no forces with its centre of inertia at rest and the instantaneous axis is describing a plane in the body: prove that, if a point in that diameter of the principal ellipsoid of inertia which is conjugate to this plane be suddenly fixed, the new instantaneous axis will be parallel to the former.

2678. Two equal uniform rods AB, BC, freely jointed at B and in one straight line, are moving uniformly in a direction normal to their length on a smooth horizontal table when the point A becomes suddenly fixed: prove that the initial angular velocities of the rods will be in the ratio $3 : -1$, that the least subsequent obtuse angle between them will be $\cos^{-1}(-\frac{2}{3})$, and that when next in one straight line their angular velocities will be as $1 : 9$.

2679. Three equal uniform rods AB, BC, CD, freely jointed at B and C, are lying in one straight line on a smooth horizontal table when a blow is applied at their centre in a direction normal to the line of the rods: prove that

$$\frac{d\theta}{dt}\sqrt{1+\sin^2\theta} = \Omega,$$

where θ is the angle through which the outer rods have turned in a time t and Ω their initial angular velocity. Prove also that the velocity of BC will be $\frac{2a\Omega}{3}\left\{1+\frac{\cos\theta}{\sqrt{1+\sin^2\theta}}\right\}$, and that the direction of the stress at B or C will make with BC the angle $\tan^{-1}(\frac{2}{3}\tan\theta)$.

2680. Two equal uniform rods AB, BC, freely jointed at B, are in motion on a smooth horizontal table and their angular velocities are ω_1, ω_2 when the angle between them is θ: prove that $(\omega_1+\omega_2)(5-3\cos\theta)$ and $5(\omega_1^2+\omega_2^2)-6\omega_1\omega_2\cos\theta$ are both constant throughout the motion.

2681. Three equal uniform rods (for all of which $e=0$), freely jointed at common ends, are laid in one straight line on a smooth horizontal table and the two outer are set in motion about the ends of the middle rod with equal angular velocities, (1) in the same sense, (2) in opposite senses: prove that, (1) when the outer rods make the greatest angle with the direction of the middle rod produced on each side the common angular velocity of the three will be $\frac{4}{7}\omega$, and (2) that after the impact of the two outer rods the triangle formed by the three will move with velocity $\frac{1}{3}a\omega$, where a is the length of a rod.

2682. A uniform rod of length $2a$ has attached to one end a particle by a string of length b and the rod and string placed in one straight line on a smooth horizontal table; the particle is then projected at right angles to the string: prove that the greatest angle which the string can make with the rod (produced) will be

$$2\sin^{-1}\sqrt{\frac{a}{12b}\left(1+\frac{m}{p}\right)},$$

where m, p are the masses; also that, if after a time t the rod and string make angles θ, ϕ with their initial directions,

$$(k^2 + ab\cos\overline{\phi-\theta})\frac{d\theta}{dt} + (b^2 + ab\cos\overline{\phi-\theta})\frac{d\phi}{dt} = (a+b)V,$$

$$k^2\left(\frac{d\theta}{dt}\right)^2 + b^2\left(\frac{d\phi}{dt}\right)^2 + 2ab\cos\overline{\phi-\theta}\frac{d\theta}{dt}\frac{d\phi}{dt} = V^2;$$

where $k^2 \equiv \frac{1}{3}a^2\left(4+\frac{m}{p}\right)$ and V is the initial velocity of the particle.

2683. A circular disc capable of motion about a vertical axis through its centre normal to its plane is set in motion with angular velocity Ω, and at a given point of it is placed freely a rough uniform sphere: prove the equations of motion

$$\tfrac{7}{2}r^2\frac{d\theta}{dt} + k^2\omega = k^2\Omega,$$

$$\omega^2(r^2+k^2)(b^2+k^2) = k^4\Omega^2,$$

$$\frac{d^2r}{dt^2} - r\left(\frac{d\theta}{dt}\right)^2 + \tfrac{2}{7}r\omega\frac{d\theta}{dt} = 0,$$

r, θ being the polar co-ordinates of the point of contact at the time t, measured from the centre of the disc, ω the angular velocity of the disc, b the initial value of r and $4mk^2 = 7pc^2$, where m, p are the masses of the sphere and disc and c the radius of the disc.

[These equations are all satisfied by

$$r = b,\ \frac{d\theta}{dt} = \tfrac{2}{7}\omega = \tfrac{2}{7}\frac{k^2\Omega}{k^2+b^2}.]$$

2684. A circular disc lies flat on a smooth horizontal table, on which it can move freely, and has wound round it a fine string carrying a particle which is projected with a velocity V from a point of the disc in a direction normal to the perimeter of the disc: prove that

$$\frac{Vt}{ak^2} + 1 = \sec\frac{\theta}{k},\quad \phi = k\tan\frac{\theta}{k} - \theta;$$

where θ, ϕ are the angles through which the string and the disc have turned at a time t, a is the radius, and $2k^2 \equiv 3 + \frac{m}{p}$, m, p being the masses of the disc and particle.

2685. Two equal circular discs lying flat on a smooth horizontal table are connected by a fine string coiled round each, which is wound up until the discs are in contact with each other and are both on the same side of the tangent string: one of the discs has its centre fixed and can move freely about it, and the other is projected with a velocity V at right angles to the tangent string: prove that after a time t either disc will have turned through an angle $\sqrt{1+\frac{V^2t^2}{5a^2}}-1$ and the string will have turned through an angle $\frac{\sqrt{5}}{2}\tan^{-1}\frac{Vt}{a\sqrt{5}}$, where a is the radius of either disc.

2686. A smooth straight tube of length $2a$ and mass m, lying on a horizontal table, contains a particle, mass p, which just fits it; the system is set in motion by a blow at right angles to the tube: prove that

$$\left(\frac{dr}{d\theta}\right)^2 (b^2+c^2) = (r^2-c^2)(b^2+r^2);$$

r being the distance of the particle from the mid-point of the tube when the tube has turned through an angle θ, c the initial value of r, and

$$3b^2 \equiv a^2\left(1+\frac{m}{p}\right).$$

2687. A circular disc of mass m and diameter d can move on a smooth horizontal plane about a fixed point A in its perimeter, and a fine string is wound round it carrying a particle of mass p; the particle is initially projected from the disc at the other end of the diameter through A with velocity V normal to the perimeter and the disc is then at rest: prove that the angular velocity of the string will vanish when the length unwound is that which initially subtended at A an angle θ such that

$$8p(\theta\tan\theta+1)\cos^2\theta+3m=0;$$

and that the angular velocity of the disc is then

$$\frac{V}{d}(-\theta\sin\theta\cos\theta)^{-\frac{1}{2}}.$$

2688. A rough sphere of radius a moves on the concave surface of a vertical cylinder of revolution of radius $a+b$, and the centre of the sphere is initially moving horizontally with a velocity V: prove that the depth of the centre below its initial position after a time t is

$$\frac{5gb^2}{2V^2}(1-\cos nt), \text{ where } 7b^2n^2 \equiv 2V^2;$$

also prove that, in order that perfect rolling may be maintained, the coefficient of friction must not be less than

$$12bg \div 7V^2.$$

2689. A cylinder of revolution is fixed with its axis horizontal and a rough sphere is projected so as to move in contact with the cylinder, being initially in its lowest position with its centre moving horizontally in a direction which makes an angle α with the axis of the cylinder: prove that, for the sphere to reach the highest point, the initial velocity must not be less than

$$\sqrt{\frac{67}{7}\frac{bg}{\sin^2\alpha}};$$

a, $a+b$ being the radii of the sphere and cylinder.

[The equations of motion are

$$\left(\frac{d\phi}{dt}\right)^2 = \frac{V^2}{b^2}\sin^2\alpha - \frac{10g}{7b}(1-\cos\phi),$$

$$\frac{dz}{dt} = V\cos\alpha\cos\left(\phi\sqrt{\tfrac{2}{7}}\right),$$

$$a\omega = V\sqrt{\tfrac{7}{2}}\cos\alpha\sin\left(\phi\sqrt{\tfrac{2}{7}}\right);$$

where z is the distance described by the centre of the sphere parallel to the axis, ϕ the angle through which the common normal has turned, and ω the angular velocity of the sphere about that normal, after a time t.]

2690. A sphere, radius a, is in motion on the surface of a cylinder of revolution of radius $a+b$ whose axis makes an angle α with the vertical and is initially in contact with the lowest generator, its centre moving in a direction perpendicular to the generator with such a velocity that the sphere just makes complete revolutions: prove the equations of motion

$$7\left(\frac{d\phi}{dt}\right)^2 = \frac{g}{b}\sin\alpha\,(17+10\cos\phi),$$

$$a\frac{d\omega}{dt} = \frac{dz}{dt}\frac{d\phi}{dt},$$

$$7\left(\frac{dz}{dt}\right)^2 + 2a^2\omega^2 = 10gz\cos\alpha;$$

z being the distance described by the centre of the sphere parallel to the axis of the cylinder, ϕ the angle through which the common normal has turned, and ω the angular velocity about the normal, after a time t.

2691. A rough sphere of radius a rolls in a spherical bowl of radius $a+b$, the centre of the sphere being initially at the same height as the centre of the bowl and moving horizontally with velocity V: prove that, if θ be the angle which the common normal makes with the vertical, and ϕ the angle through which the vertical plane containing the normal has turned at the end of a time t,

$$\sin^2\theta\frac{d\phi}{dt} = \frac{V}{b}, \quad \left(\frac{d\theta}{dt}\right)^2 = \frac{10g}{7b}\cos\theta - \frac{V^2}{b^2}\cot^2\theta;$$

and, if R, F, S be the reactions at the point of contact along the common normal, along the tangent which lies in the same vertical plane with the common normal, and at right angles to both these directions, that

$$R = m\left(\frac{V^2}{b} + \frac{17g}{7}\cos\theta\right), \quad F = \frac{2mg}{7}\sin\theta, \quad S = 0;$$

also, if ω_1, ω_2, ω_3 be the angular velocities of the sphere about these three directions,

$$\omega_1 = 0, \quad a\omega_2\sin = \theta V, \quad a\omega_3 = -b\frac{d\theta}{dt}.$$

2692. A rough sphere of radius a rolls in a spherical bowl of radius $a+b$ in a state of steady motion, the normal making an angle α with the vertical: prove that the time of small oscillations about this position is

$$2\pi\sqrt{\frac{7b\cos\alpha}{5g(1+3\cos^2\alpha)}}.$$

HYDROSTATICS.

[In the questions under this head, a fluid is supposed to be uniform, heavy, and incompressible, unless otherwise stated: and all cones, cylinders, paraboloids, &c. are supposed to be surfaces of revolution and their bases circles.]

2693. A cylinder is filled with equal volumes of n different fluids which do not mix; the density of the uppermost is ρ, of the next 2ρ, and so on, that of the lowest being $n\rho$: prove that the mean pressures on the corresponding portions of the curve surfaces are in the ratios

$$1^2 : 2^2 : 3^2 : \ldots : n^2.$$

2694. A hollow cylinder containing a weight W of fluid is held so that its axis makes an angle α with the horizon: prove that the resultant pressure on its curve surface is $W \cos \alpha$ in a direction making an angle α with the vertical.

2695. Equal volumes of three fluids are mixed and the mixture separated into three parts; to each of these parts is then added its own volume of one of the original fluids, and the densities of the mixtures so formed are in the ratios $3 : 4 : 5$: prove that the densities of the fluids are as $1 : 2 : 3$.

2696. A thin tube in the form of an equilateral triangle is filled with equal volumes of three fluids which do not mix and held with its plane vertical: prove that the straight lines joining the common surfaces of the fluids form an equilateral triangle whose sides are in fixed directions; and that, if the densities be in *A. P.*, the straight line joining the surfaces of the fluid of mean density will be always vertical.

2697. A thin tube in the form of a square is filled with equal volumes of four fluids which do not mix, whose densities are $\rho_1, \rho_2, \rho_3, \rho_4$, and held with its plane vertical; straight lines are drawn joining adjacent points where two fluids meet so as to form another square: prove that, if $\rho_1 + \rho_4 = \rho_2 + \rho_3$, the diagonals of this square will be vertical and horizontal respectively; but, if $\rho_1 = \rho_3$ and $\rho_2 = \rho_4$, every position of the fluids will be one of equilibrium.

2698. A fine tube in the form of a regular polygon of n sides is filled with equal volumes of n different fluids which do not mix and held with its plane vertical: prove that the sides of the polygon formed by joining adjacent points where two fluids meet will have its sides in fixed directions; and, if the densities of the fluids satisfy two certain conditions, every position will be one of equilibrium.

[These conditions may be written

$$\rho_1 \cos \alpha + \rho_2 \cos 2\alpha + \ldots + \rho_n \cos n\alpha = 0,$$

$$\rho_1 \sin \alpha + \rho_2 \sin 2\alpha + \ldots + \rho_n \sin n\alpha = 0,$$

where $n\alpha = 2\pi$.]

2699. A circular tube of fine uniform bore is half filled with equal volumes of four fluids which do not mix and whose densities are as $1 : 4 : 8 : 7$, and held with its plane vertical: prove that the diameter joining the free surfaces will make an angle $\tan^{-1} 2$ with the vertical.

2700. A triangular lamina ABC, right-angled at C, is attached to a string at A and rests with the side AC vertical and half its length immersed in fluid: prove that the density of the fluid is to that of the lamina as $8 : 7$.

2701. A lamina in the form of an equilateral triangle, suspended freely from an angular point, rests with one side vertical and another side bisected by the surface of a fluid: prove that the density of the lamina is to that of the fluid as $15 : 16$.

2702. A hollow cone, filled with fluid, is suspended freely from a point in the rim of the base: prove that the total pressures on the curve surface and on the base in the position of rest are in the ratio

$$1 + 11 \sin^2 \alpha : 12 \sin^2 \alpha,$$

where 2α is the vertical angle of the cone.

2703. A tube of small bore, in the form of an ellipse, is half filled with equal volumes of two given fluids which do not mix: find the inclination of its axes to the vertical in order that the free surfaces of the fluids may be at the ends of the minor axis.

2704. A hemisphere is filled with fluid and the surface is divided by horizontal planes into n portions, on each of which the whole pressure is the same: prove that the depth of the rth of these planes is to the radius as $\sqrt{r} : \sqrt{n}$.

2705. A hemisphere is just filled with fluid and the surface is divided by horizontal planes into n portions, the whole pressures on which are in a geometrical progression of ratio k: prove that the depth of the rth plane is to the radius as

$$\sqrt{k^r - 1} : \sqrt{k^n - 1}.$$

2706. A lamina $ABCD$ in the form of a trapezium with parallel sides AB, CD is immersed in fluid with the parallel sides horizontal: prove that the depth of the centre of pressure below E, the point of intersection of AB, CD is

$$\frac{c}{2(1+m)}\frac{c(3m^2-4m+3)-4h(1-m^2)}{3h(1+m)-2c(1-m)},$$

where h is the depth of E, c the distance between AB, CD, and m the ratio $CD : AB$; and that, when the centre of pressure is at E, the depths of AB, CD will be as

$$3m^2-1 : 3-m^2.$$

2707. The co-ordinates of the centre of pressure of a triangular lamina immersed in fluid are

$$\frac{x_1^2+x_2^2+x_3^2}{x_1+x_2+x_3}, \quad \frac{x_1y_1+x_2y_2+x_3y_3}{x_1+x_2+x_3},$$

where (x_1, y_1), (x_2, y_2), and (x_3, y_3) are the co-ordinates of the middle points of the sides of the lamina, the axis of y being the intersection of the plane of the lamina with the surface of the fluid and the axis of x any other straight line in the plane of the lamina.

2708. The co-ordinates of the centre of pressure of any lamina immersed in fluid are

$$\frac{x_1^2+x_2^2+x_3^2}{x_1+x_2+x_3}, \quad \frac{x_1y_1+x_2y_2+x_3y_3}{x_1+x_2+x_3},$$

where (x_1, y_1), (x_2, y_2), (x_3, y_3) are the co-ordinates of the corners of a maximum triangle inscribed in an ellipse whose equation referred to the principal axes of the lamina at its C.G. is

$$Ax^2+By^2=2AB,$$

$\sqrt{A}$, $\sqrt{B}$ being the principal radii of gyration. The axes to which the centre of pressure is referred are as in the previous question.

2709. Prove the following construction for finding the centre of pressure of a lamina always totally immersed in fluid which is capable of motion in its own plane about its C.G.: find A, B the highest and lowest positions of the centre of pressure, through A draw a straight line parallel to that straight line of the lamina which is horizontal when A is the centre of pressure, and another straight line similarly determined through B; their point of intersection is the centre of pressure.

2710. Prove that, when the C.G. is fixed below the surface of a fluid and the lamina move about the C.G. in its own plane, the centre of pressure describes a circle in space and in the lamina the ellipse whose equation referred to the principal axes is

$$\frac{x^2}{A^2}+\frac{y^2}{B^2}=\frac{1}{c^2},$$

where $\sqrt{A}$, $\sqrt{B}$ are the principal radii of gyration, and c the depth of the C.G. measured in the plane of the lamina below the surface of the fluid: also that, of the four points in which the circle and ellipse intersect, the centre of pressure is the lowest and the other three are corners of a triangle whose sides touch a fixed circle with its centre at the C.G.

2711. A lamina totally immersed in fluid moves in its own plane so that the centre of pressure is a point fixed in space: prove that the path of the C.G. is the curve whose equation is, referred to the centre of pressure as origin,

$$x^2\{(x-a)^2+y^2\}+x(x-a)(A+B)+AB=0\,;$$

where $\sqrt{A}$, $\sqrt{B}$ are the principal radii of gyration at the C.G., and a the depth (measured in the plane of the lamina) of the centre of pressure below the surface of the fluid.

2712. A rectangular lamina $ABCD$ is immersed in fluid with the side AB in the surface of the fluid; a point P is taken in CD and the lamina divided into two parts by the straight line AP: determine for what position of P the distance between the centres of pressure of the two parts is a maximum.

[If the sides AB, BC be denoted by a, b, and DP by x, the distance will be a maximum when $27ax=4(9a^2-2b^2)$, and since x must be positive and less than a, there will be no maximum unless $b:a$ lie between $3:2\sqrt{2}$ and $3:\sqrt{2}$.]

2713. A lamina in the form of the sector of a circle is immersed in fluid with the centre of the circle in the surface: prove that the co-ordinates of its centre of pressure are

$$\frac{3a}{8}\,\frac{\alpha-\sin\alpha\cos\alpha\cos 2\theta}{\sin\alpha\sin\theta},\quad \frac{3a}{4}\cos\alpha\cos\theta,$$

where the axis of y is in the surface of the fluid, and $\theta-\alpha$, $\theta+\alpha$ are the angles which the bounding radii make with the axis of y.

2714. A lamina, bounded by the epicycloid generated by a circle of radius a rolling on a circle of radius $2a$, is placed in fluid with the cusp line in the surface: prove that the co-ordinates of the centre of pressure of half the part immersed are $\frac{1575}{2048}\pi a$, $\frac{75}{64}a$; and those of the centre of pressure of the part lying outside the fixed circle are $\frac{1455}{1328}\pi a$, $\frac{5}{4}a$; the axis of y lying in the surface.

2715. An isosceles triangle is immersed with its axis vertical and its base in the surface of a fluid: prove that the resultant pressure on the area intercepted between any two horizontal planes acts through the C.G. of that portion of the volume of a sphere, described with the axis for diameter, which is intercepted between the planes.

2716. A conical shell is placed with its vertex upwards on a horizontal table and fluid is poured in through a small hole in the vertex; the cone begins to rise when the weight of the fluid poured in is equal to its own weight: prove that this weight bears to the weight of fluid which would fill the cone the ratio $9-3\sqrt{3}:4$.

2717. A parabolic lamina bounded by a double ordinate perpendicular to the axis floats in fluid with its focus in the surface and its axis inclined at an angle $\tan^{-1} 2$ to the vertical: prove that the density of the fluid is eight times the density of the lamina, and that the length of the axis bears to the latus rectum the ratio 15 : 4.

2718. A lamina in the form of an isosceles triangle floats in a fluid with its plane vertical and (1) its base totally out of the fluid, (2) its base totally immersed, and its axis in these two positions makes angles θ, ϕ with the vertical: prove that

$$\sin^2 \alpha (\sec^2 \theta + \sec^2 \phi) = 4 \cos \alpha,$$

and that both positions will not be possible unless $\cos \alpha > \sqrt{2} - 1$; where α is the vertical angle.

2719. A cone with its axis vertical and vertex downwards is filled with two fluids which do not mix and their common surface cuts off one-fourth of the axis from the vertex: prove that, if the whole pressures of the fluids on the curve surfaces be equal, their densities will be as 45 : 1.

2720. A right cone just filled with fluid is attached to a fixed point by a fine extensible string attached to the vertex, and initially the string is of its natural length and the cone at rest: prove that the pressure of the fluid on the base of the cone in the lowest position is six times the weight of the fluid.

2721. A barometer stands at 29·88 inches and the thermometer is at the dew-point; a barometer and a cup of water are placed under a receiver from which the air is removed and the barometer then stands at ·36 of an inch: find the space that would be occupied by a given volume of the atmosphere if it were deprived of its vapour without changing its pressure or temperature.

2722. In Hawksbee's air-pump, the machine is kept at rest when the nth stroke is half completed; find the difference of the tensions of the two piston rods.

2723. In Smeaton's air-pump, during the nth stroke, find the position of the piston at that instant of time when the upper valve begins to open.

2724. The volumes of the receiver and barrel of an air-pump are A, B; ρ, σ are the densities of atmospheric air and of the air in the receiver respectively, and Π the atmospheric pressure: prove that the work done in slowly raising the piston through one stroke is

$$\Pi \left\{ B - A \frac{\sigma}{\rho} \log \left(1 + \frac{B}{A} \right) \right\},$$

gravity being neglected.

2725. A portion of a cone cut off by a plane through the axis and two planes perpendicular to the axis is immersed in fluid in such a manner that the axis of the cone is vertical and the vertex in the surface: prove that the resultant horizontal pressure on the curve surface passes through the C. G. of the body immersed.

2726. Assuming that the temperature of the atmosphere in ascending from the earth's surface decreases slowly by an amount proportional to the height ascended, prove that the equation connecting the pressure p and the density ρ at any height will be of the form $p = k\rho^{1+m}$, where m is a small fraction.

2727. A cylinder floats in fluid with its axis inclined at an angle $\tan^{-1}\frac{3}{5}$ to the vertical, its upper circular boundary just out of the fluid and the lower one completely immersed: prove that the length of the axis is nine-eighths of the diameter of the generating circle.

2728. Two equal and similar rods AB, BC, fixed at an angle α at B, rest in a fluid of twice the specific gravity with the angle B out of the fluid, and the axis of the system makes an angle θ with the horizon: prove that

$$\cos 2\theta = 2 - \sec \alpha.$$

2729. A uniform solid tetrahedron has each edge equal to the opposite edge: prove that it can float partly immersed in fluid with any two opposite edges horizontal.

2730. A lamina in the form of a parabola bounded by a double ordinate rests in liquid with its plane vertical, its focus in the surface of the fluid, and its base just out of the fluid: prove that the ratio of the densities of the solid and liquid is $1 : (1 + \cos \alpha)^3$, where α is the angle given by the equation $2 \cos 2\alpha = 3 (1 - \cos \alpha)$.

2731. A cone of density ρ floats with a generator vertical in a fluid of density σ, the base being just out of the fluid: prove that, if 2α be the vertical angle,

$$\frac{\rho}{\sigma} = (\cos 2\alpha)^{\frac{3}{2}},$$

and that the length of the vertical side immersed is to the length of the axis as $\cos 2\alpha : \cos \alpha$.

2732. A cone is moveable about its vertex, which is fixed at a given distance c below the surface of a liquid, and rests with its axis, h in length, inclined at an angle θ to the vertical and its base completely out of the fluid: prove that

$$\frac{\cos\theta \cos^8\alpha}{(\cos^2\theta - \sin^2\alpha)^{\frac{5}{2}}} = \frac{\sigma h^4}{\rho c^4};$$

2α being the vertical angle and ρ, σ the densities of the liquid and cone. Also prove that this position will be stable, but that it cannot exist unless $\sigma h^4 \cos^2\alpha > \rho c^4$.

2733. A homogeneous solid in the form of a cone rests with its axis vertical and its vertex at a depth c below the surface of a liquid whose density varies as the depth: prove that the condition for stable equilibrium is that $\cos\alpha < \sqrt{\frac{4c}{5h}}$, where h is the length of the axis and 2α the vertical angle. Prove also that this is the condition that no positions of equilibrium in which the axis is not vertical can exist.

2734. An elliptic tube half full of liquid revolves about a fixed vertical axis in its plane with angular velocity ω: prove that the angle which the straight line joining the free surfaces of the fluid makes with the vertical will be $\tan^{-1}\left(\frac{g}{p\omega^2}\right)$, where p is the distance of the axis from the centre of the ellipse.

2735. A hollow cone very nearly filled with liquid revolves uniformly about a vertical generator: prove that the pressure on the base is

$$\tfrac{3}{8} W \left\{\frac{a\omega^2}{g}(1 + 5\cos^2\alpha)\tan\alpha + 8\sin\alpha\right\};$$

where W is the weight of the fluid, 2α the vertical angle, a the radius of the base, and ω the angular velocity.

2736. A hollow cone very nearly filled with liquid revolves about a horizontal generator with uniform angular velocity ω: prove that the whole pressure on the base in its highest or lowest position is

$$\frac{\pi}{8}\rho a^4\omega^2\left(1 + \frac{8g}{a\omega^2}\cos\alpha + 5\cos^2\alpha\right);$$

where a is the radius of the base and 2α the vertical angle.

2737. A cone the length of whose axis is h and the radius of the base a floats in liquid with $\frac{27}{128}$ of its volume below the surface: prove that, when the liquid revolves about the axis of the cone with angular velocity $\sqrt{[\text{illegible}]\,\frac{gh}{a^2}}$, the cone will float with the length h or $\frac{3}{4}h$ of its axis immersed; and investigate which of the two positions is stable.

2738. A sphere of radius a floats in liquid, which is revolving with uniform angular velocity ω about a vertical axis, with its centre at the vertex of the free surface of the liquid: prove that

$$4(p^2 + 4a^2)(a - pq) = a(p + 4aq)^2;$$

where $p\omega^2 = 2g$ and $1 + q : 2$ is the ratio of the densities of the sphere and liquid.

2739. A hollow paraboloid whose axis is equal to the latus rectum is placed with its axis vertical and vertex upwards and contains seven-eighths of its volume of liquid: find the angular velocity with which this liquid must revolve about the axis in order that its free surface may be confocal with the paraboloid; and prove that in this case the pressure on the base is greater than when the liquid was at rest in the ratio

$$2\sqrt{2} : 2\sqrt{2} - 1.$$

2740. A liquid is acted on by two central forces, each varying as the distance from a fixed point and equal at equal distances from those points, one attractive and one repulsive: prove that the surfaces of equal pressure are planes.

2741. A liquid is at rest under the action of two forces tending to two fixed points and each varying inversely as the square of the distance, one attractive and one repulsive: prove that one surface of equal pressure is a sphere.

2742. A mass of elastic fluid is confined within a hollow sphere and repelled from the centre of the sphere by a force $\mu \div$ distance: prove that the whole pressure on the sphere bears to the whole pressure which would be exerted if no such force acted the ratio $3k + \mu : 3k$; where $p = k\rho$ is the relation between the pressure and density.

2743. A quantity of liquid not acted on by gravity just fills a hollow sphere and is repelled from a point on the surface of the sphere by a force equal to μ (distance); the liquid revolves about the diameter through the centre of force with uniform angular velocity ω: find the whole pressure on the sphere, and prove that, if when the angular velocity is diminished one half the pressure is also diminished one half,

$$\omega^2 = 6\mu.$$

2744. All space being supposed filled with an elastic fluid the total mass of which is known, which is attracted to a given point by a force varying as the distance; find the pressure at any point.

2745. Water is contained in a vessel having a horizontal base and a cone is supported partly by the water and partly by the base on which the vertex rests: prove that, for stable equilibrium, the depth of the fluid must be greater than $h\sqrt{m}\cos a$, m^2 being the specific gravity of the cone, h the length of its axis, and $2a$ the vertical angle.

2746. A solid paraboloid is divided into two parts by a plane through the axis and the parts united by a hinge at the vertex; the system is placed in liquid with its axis vertical and vertex downwards and floats without separation of the parts prove that the ratio of the density of the solid to that of the liquid must be greater than x^4, where x is given by the equation

$$3hx^3 = 7l(1-x),$$

and l, h are the lengths of the latus rectum and axis respectively.

2747. A cone is floating with its axis vertical in a fluid whose density varies as the depth: prove that, for stable equilibrium,

$$\cos^2 a < \tfrac{4}{5}\sqrt[4]{\frac{4\rho}{\sigma}},$$

where $2a$ is the vertical angle, ρ the density of the cone, and σ the density of the fluid at a depth equal to the height of the cone.

2748. A uniform rod rests in an oblique position with half its length immersed in liquid and can turn freely about a point in its length whose distance from the lower end is one-sixth of the length: compare the densities of the rod and liquid, and prove that the equilibrium is stable.

2749. A uniform rod is moveable about one end, which is fixed below the surface of a liquid and, when slightly displaced from its highest position, it sinks until just immersed: prove that, when at rest in the highest position, the pressure on the point of support was zero.

2750. Two equal uniform rods AB, BC, freely jointed at B, are capable of motion about A, which is fixed at a given depth below the surface of a liquid: find the position in which both rods rest partly immersed, and prove that for such a position to be possible the density of the rods must not exceed one-third the density of the liquid.

2751. A hemisphere, a point in the rim of whose base is attached to a fixed point by a fine string, rests with the centre of the sphere in the surface of the liquid and the base inclined at an angle a to the horizon: prove that

$$\frac{\rho}{\sigma}=\frac{16(\pi-a)\cos a-3\pi\sin a}{2\pi(8\cos a-3\sin a)};$$

where ρ, σ are the densities of the solid and liquid.

2752. A cone is floating with its axis vertical and vertex downwards in fluid and $\frac{1}{n}$th of its axis is immersed; a weight equal to the weight of the cone is placed upon the base and the cone then sinks until just totally immersed before rising: prove that

$$n^3+n^2+n=7.$$

2753. A hollow cylinder with its axis vertical contains liquid, and a solid in the form of an ellipsoid of revolution is allowed to sink freely in the liquid with axis also vertical: the solid just fits into the cylinder and sinks until just immersed before rising: prove that its density is one-half that of the liquid.

2754. A hollow cylinder with its axis vertical contains liquid and a solid cylinder is allowed to sink freely in it with axis also vertical: prove that, if it sink until just immersed before rising, the densities of the solid and liquid must be in the ratio 1 : 2. Also, if the density of the liquid initially vary as the depth, prove that the density of the solid must be the initial density of the liquid at a depth of one-sixth of the whole distance sunk by the solid.

2755. A hollow cylinder with vertical axis contains a quantity of liquid and a solid of revolution (of the curve $y^2 \propto x^n$ about the vertical axis of x) is allowed to sink in the liquid, starting when its vertex is in the surface and coming to instantaneous rest when just immersed: prove that the density of the solid must bear to the density of the liquid the ratio $1 : 2(n+2)$; and that, if a similar solid be allowed to sink in an unlimited mass of liquid of half the density of the former, this solid will also come to rest when just immersed.

2756. A cylinder whose axis is vertical contains a quantity of fluid whose density varies as the depth and into this is allowed to sink a solid of revolution whose base is equal to that of the cylinder, which sinks until just immersed before rising; in the lowest position of this solid the density of the surrounding fluid varies as the nth power of the depth: prove that the weight of the solid is to the weight of the displaced fluid as $n-1 : 3(2n+1)$, whereas if the solid can rest in this position the ratio must be $n-1 : n+1$. Also prove that the generating curve of the solid will be

$$1-\frac{y^2}{a^2}=\left(1-\frac{x}{h}\right)^{n-1},$$

where a is the radius of the base and h the height.

[If $n=2$ the solid is a paraboloid, if $n=3$, an ellipsoid.]

2757. A hollow cylinder with vertical axis contains a quantity of fluid whose density varies as the depth and into it is allowed to sink slowly, with vertex downwards, a solid cone the radius of whose base is equal to the radius of the cylinder; the cone rests when just immersed: prove that the density of the cone is equal to the initial density of the fluid at a depth equal to one-twelfth of the length of the axis of the cone. If the cone be allowed to sink freely into the fluid, starting with its vertex at the surface and just sinking until totally immersed, the density of the cone will be to the density of the fluid at the vertex of the cone in its lowest position as $1 : 30$.

2758. A tube of fine bore whose plane is vertical contains a quantity of fluid which occupies a given length of the tube; a given heavy particle just fitting the tube is let fall through a given vertical height: find the impulsive pressure at any point of the fluid; and prove that the whole kinetic energy after the impact bears to the kinetic energy dissipated the ratio of the mass of the particle to the mass of the fluid.

[If m, m' be the masses of the particle and fluid, V the velocity of the particle just before impact, the impulsive pressure at a point whose distance along the arc from the free end is s will be $\frac{mm'V}{m+m'}\cdot\frac{s}{l}$, where l is the whole length of arc occupied by the fluid.]

2759. A flexible inextensible envelope when filled with fluid has the form of a paraboloid whose axis is vertical and vertex downwards and whose altitude is five-eighths of the latus rectum: prove that the tension of the envelope along the meridian will be greatest at points where the tangent makes an angle $\frac{\pi}{4}$ with the vertical.

[In general, if $4a$ be the latus rectum and h the altitude, the tension per unit of length at a point where the tangent makes an angle θ with the vertical will be $\frac{\sigma a}{2}\left(\frac{2h+a}{\sin\theta}-\frac{a}{\sin^3\theta}\right)$, where σ is the specific gravity of the fluid.]

2760. Fluid without weight is contained in a thin flexible envelope in the form of a surface of revolution and the tensions of the envelope at any point along and perpendicular to the meridian are equal: prove that the surface is a sphere.

2761. A quantity of homogeneous fluid is contained between two parallel planes and is in equilibrium in the form of a cylinder of radius b under a pressure ϖ; that portion of the fluid which lies within a distance a of the axis being suddenly annihilated, prove that the initial pressure, at a point whose distance from the axis is r, is

$$\varpi \log\left(\frac{r}{a}\right) \div \log\left(\frac{b}{a}\right).$$

2762. A thin hollow cylinder of length h, closed at one end and fitted with an air-tight piston, is placed mouth downwards in fluid, the weight of the piston is equal to that of the cylinder, the height of a cylinder of equal weight and radius formed of the fluid is a, the height of fluid which measures the atmospheric pressure is c, and the air enclosed in the cylinder would just fill it at atmospheric density: prove that, for small vertical oscillations, the distances of the piston and of the top of the cylinder from their respective positions of equilibrium are of the form $A \sin(\lambda t + \alpha) + B \sin(\mu t + \beta)$, λ, μ being the positive roots of the equation

$$x^4 - \frac{g}{a}(2m+1)\,x^2 + m\frac{g^2}{a^2} = 0,$$

and $m \equiv (a+c)^2 \div ch$.

2763. A filament of liquid PQR is in motion in a fixed tube of small uniform bore which lies in a vertical plane with its concavity always upwards; on the horizontal ordinates to P, Q, R at any instant are taken points p, q, r, whose distances from the vertical axis of abscissæ are equal to the arcs measured to P, Q, R from a fixed point of the tube: prove that the fluid pressure at Q is always proportional to the area of the triangle pqr.

2764. A centre of force, attracting inversely as the square of the distance, is at the centre of a spherical cavity within an infinite mass of liquid, the pressure in which at an infinite distance is ϖ, and is such that the work done by this pressure on a unit of area through a unit of length is one half the work done by the attractive force on a particle whose mass is that of a unit of volume of the liquid as it moves from infinity to the initial boundary of the cavity: prove that the time of filling up the cavity will be $\pi a \sqrt{\frac{\rho}{\varpi}}\{2 - (\frac{3}{2})^{\frac{1}{2}}\}$; where a is the initial radius of the cavity and ρ the density of the fluid.

GEOMETRICAL OPTICS.

2765. Three plane mirrors are placed so that their intersections are parallel to each other and the section made by a plane perpendicular to their intersections is an acute-angled triangle; a ray proceeding from a certain point of this plane after one reflexion at each mirror proceeds on its original course: prove that the point must lie on the perimeter of a certain triangle.

2766. In the last question a ray starting from any point after one reflexion at each mirror proceeds in a direction parallel to its original direction: prove that after another reflexion at each mirror it will proceed on its original path, and that the whole length of its path between the first and third reflexions at any mirror is constant and equal to twice the perimeter of the triangle formed by joining the feet of the perpendiculars.

2767. A ray of light whose direction touches a conicoid is reflected at any confocal conicoid: prove that the reflected ray also touches the first conicoid.

2768. In a hollow ellipsoidal shell small polished grooves are made coinciding with one series of circular sections and a bright point placed at one of the umbilics in which the series terminates: prove that the locus of the bright points seen by an eye in the opposite umbilic is a central section of the ellipsoid, and that the whole length of the path of any ray by which a bright point is seen is constant.

2769. A ray proceeding from a point on the circumference of a circle is reflected n times at the circle: prove that its point of intersection with the consecutive ray similarly reflected is at a distance from the centre equal to $\frac{a}{2n+1}\sqrt{1+4n(n+1)\sin^2\theta}$, where a is the radius and θ the angle of incidence of the ray: also prove that the caustic surface generated by such rays is the surface of revolution generated by an epicycloid in which the fixed circle has the radius $\frac{a}{2n+1}$ and the moving circle the radius $\frac{na}{2n+1}$.

2770. A ray of light is reflected at two plane mirrors, its direction before incidence being parallel to the plane bisecting the angle between the mirrors and making an angle θ with their line of intersection: prove that the deviation is $2\sin^{-1}(\sin\theta\sin 2\alpha)$, where 2α is the angle between the mirrors. More generally, if D_r be the deviation after r successive reflexions,

$$\cos \tfrac{1}{2} D_{2n-1} = \sin\theta \sin(\overline{2n-1}\, a - \phi), \quad \sin \tfrac{1}{2} D_{2n} = \sin\theta \sin 2na,$$

where ϕ is the angle which a plane through the intersection of the mirrors parallel to the incident ray makes with the plane bisecting the angle between the mirrors.

2771. Two prisms of equal refracting angles are placed with one face of each in contact and their other faces parallel and a ray passes through the combination in a principal plane: prove that the deviation will be from the edge of the denser prism.

2772. The radii of the bounding surfaces of a lens are r, s, and its thickness is $\left(1 + \frac{1}{\mu}\right)(s - r)$: prove that all rays incident on the lens from a certain point will pass through without aberration but also without deviation.

2773. Prove that a concave lens can be constructed such that the path of every ray of a pencil proceeding from a certain point after refraction at the first surface shall pass through the centre of the lens; that in this case there will be no aberration at the second refraction, and that the only effect of the lens is to throw back the origin of light a distance $(\mu - 1)t$, where t is the thickness of the lens.

2774. What will be the centre of a lens whose bounding surfaces are confocal paraboloids on a common axis? Prove that the distance between the focal centres of such a lens is $\frac{\mu - 1}{\mu + 1}(a \pm b)$, $4a$, $4b$ being the latera recta.

2775. The path of a ray through a medium of variable density is an arc of a circle in the plane of xy: prove that the refractive index at any point (x, y) must be $\frac{1}{x - a} f\left(\frac{x - a}{y - b}\right)$, where f is an arbitrary function and (a, b) the centre of the circle.

2776. A ray of light is propagated through a medium of variable density in a plane which divides the medium symmetrically: prove that the path is such that when described by a point with velocity always proportional to μ, the index of refraction, the accelerations of the point parallel to the (rectangular) axes of x and y will be proportional to $\frac{d(\mu^2)}{dx}$, $\frac{d(\mu^2)}{dy}$ respectively.

2777. A ray is propagated through a medium of variable density in a plane (xy) which divides the medium symmetrically: prove that the projection of the radius of curvature at any point of the path of the ray on the normal to the surface of equal density through the point is equal to $\mu : \sqrt{\left(\frac{d\mu}{dx}\right)^2 + \left(\frac{d\mu}{dy}\right)^2}$.

2778. A small pencil of parallel rays of white light, after transmission in a principal plane through a prism, is received on a screen whose plane is perpendicular to the direction of the pencil: prove that the length of the spectrum will be proportional to

$$(\mu_v - \mu_r) \sin i \div \cos^2 D \cos (D + i - \phi) \cos \phi';$$

where i is the refracting angle, ϕ, ϕ' the angles of incidence and refraction at the first surface, and D the deviation, of the mean ray.

2779. Prove that, when a ray of white light is refracted through a prism in a principal plane so that the dispersion of two given colours is a minimum,

$$\frac{\sin (3\phi' - 2i)}{\sin \phi'} = 1 - \frac{2}{\mu^2};$$

where ϕ' is the angle of refraction at the first surface and i the refracting angle. Hence prove that minimum dispersion cannot co-exist with minimum deviation.

2780. A transparent sphere is silvered at the back: prove that the distance between the images of a speck within it formed (1) by one direct refraction, (2) by one direct reflexion and one direct refraction, is

$$2\mu a c (a - c) \div (a + c - \mu c)(\mu c + a - 3c),$$

where a is the radius of the sphere and c the distance of the speck from the centre towards the silvered side.

2781. The focal length of the object-glass of an Astronomical Telescope is 40 inches, and the focal lengths of four convex lenses forming an erecting eye-piece are respectively $\frac{3}{2}$, $\frac{1}{8}$, $\frac{3}{8}$, $\frac{3}{2}$ inches, the intervals being the first and second, and the second and third being 1 inch and $\frac{1}{2}$ inch respectively: find the position of the eye-lens and the magnifying power when the instrument is in adjustment; and trace the course of a pencil from a distant object through the instrument.

[The eye-lens must be at a distance of 41·5 inches from the object-glass.]

2782. Two thin lenses of focal lengths f_1, f_2 are on a common axis and separated by an interval a; the axis of an excentric pencil, before incidence, cuts the axis of the lenses at a distance d from the first lens: prove that, if F be the focal length of the equivalent single lens,

$$\frac{1}{F} = \frac{1}{f_1} + \frac{1}{f_2} + \frac{a}{f_2}\left(\frac{1}{f_1} + \frac{1}{d}\right).$$

2783. The focal length F of a single lens equivalent to a system of three lenses of focal lengths f_1, f_2, f_3 separated by intervals a, b, for an excentrical pencil parallel to the axis, is given by the equation

$$\frac{1}{F} = \frac{1}{f_1} + \frac{1}{f_2} + \frac{1}{f_3} + \frac{a}{f_2}\left(\frac{1}{f_1} + \frac{1}{f_3}\right) + \frac{b}{f_3}\left(\frac{1}{f_1} + \frac{1}{f_2}\right) + \frac{ab}{f_1 f_2 f_3}.$$

2784. Prove that the magnifying power of a combination of three lenses of focal lengths f_1, f_2, f_3 on a common axis at intervals a, b will be independent of the position of the object, if

$$(f_1 - a)(f_3 - b) + f_2(f_1 + f_3 - a - b) = 0.$$

SPHERICAL TRIGONOMETRY AND ASTRONOMY.

2785. In a spherical triangle ABC, $a=b=\frac{\pi}{3}$, $c=\frac{\pi}{2}$: prove that the spherical excess is $\cos^{-1}\frac{7}{9}$.

2786. In an equilateral spherical triangle ABC, a, b, c are the middle points of the sides: prove that $2\sin\frac{bc}{2}=\tan\frac{BC}{2}$.

2787. In an equilateral spherical triangle whose sides are each a and angles A, prove that $2\cos\frac{a}{2}\sin\frac{A}{2}=1$.

2788. Each of the sides of a spherical triangle ABC is a quadrant, and P is any point on the sphere: prove that

$$\cos^2 AP+\cos^2 BP+\cos^2 CP=1,$$

$$\cos AP\cos BP\cos CP+\cot BPC\cot CPA\cot APB=0,$$

and that
$$\tan BCP\tan CAP\tan ABP=1;$$
the angles BPC, CPA, APB being measured so that their sum is always four right angles, and sign regarded in the third equation.

2789. Each of the sides of a spherical triangle ABC is a quadrant and P is any other point on the sphere within the triangle; another spherical triangle is described with sides equal to $2AP$, $2BP$, $2CP$ respectively: prove that the area of the latter triangle is twice that of the former.

2790. A spherical triangle ABC is equal and similar to its polar triangle: prove that

$$\sec^2 A+\sec^2 B+\sec^2 C+2\sec A\sec B\sec C=1.$$

2791. Solve a spherical triangle in which the side a, the sum or the difference of the other two sides b, c, and the spherical excess E, are given.

[Either of the equations

$$1 + \cos\frac{E}{2} = \frac{\left(\cos\frac{a}{2} + \cos\frac{b+c}{2}\right)\left(\cos\frac{a}{2} + \cos\frac{b-c}{2}\right)}{\cos\frac{a}{2}\left(\cos\frac{b-c}{2} + \cos\frac{b+c}{2}\right)},$$

$$1 - \cos\frac{E}{2} = \frac{\left(\cos\frac{b-c}{2} - \cos\frac{a}{2}\right)\left(\cos\frac{a}{2} - \cos\frac{b+c}{2}\right)}{\cos\frac{a}{2}\left(\cos\frac{b-c}{2} + \cos\frac{b+c}{2}\right)},$$

suffice to determine $b \pm c$ when a, E and $b \mp c$ are given.]

2792. The sum of the sides of a spherical triangle being given, prove that the area is greatest when the triangle is equilateral.

2793. In a spherical triangle ABC, $a + b + c = \pi$, prove that

$$\cos A + \cos B + \cos C = 1, \quad \cos a = \tan\frac{B}{2}\tan\frac{C}{2},$$

and that $$\sin\frac{A}{2} = \cos\frac{B}{2}\cos\frac{C}{2}\sin a.$$

2794. In a spherical triangle $A + B + C = 2\pi$: prove that

$$\cos A + \cot\frac{b}{2}\cot\frac{c}{2} = 0, \text{ \&c.}$$

2795. In a spherical triangle ABC, $A = B + C$: prove that

$$\sin^2\frac{a}{2} = \sin^2\frac{b}{2} + \sin^2\frac{c}{2}.$$

2796. The pole of the small circle circumscribing a spherical triangle ABC is O: prove that

$$\sin^2\frac{b}{2} + \sin^2\frac{c}{2} - \sin^2\frac{a}{2} = 2\sin\frac{b}{2}\sin\frac{c}{2}\cos\frac{BOC}{2};$$

and that, if P be any point on this circle,

$$\sin\frac{a}{2}\sin\frac{PA}{2} + \sin\frac{b}{2}\sin\frac{PB}{2} + \sin\frac{c}{2}\sin\frac{PC}{2} = 0,$$

that arc of the three PA, PB, PC being reckoned negative which crosses one of the sides.

2797. Prove that

$$\sin s > \cos a \sin(s - a) + \cos b \sin(s - b) + \cos c \sin(s - c),$$

and $$\cos S < \cos A \cos(S - A) + \cos B \cos(S - B) + \cos C \cos(S - C),$$

where a, b, c are the sides and A, B, C the angles of a spherical triangle, and

$$2s \equiv a + b + c, \quad 2S \equiv A + B + C.$$

2798. The centre of the sphere on which lies a spherical triangle ABC is O, and forces act along OA, OB, OC proportional to $\sin a$, $\sin b$, $\sin c$ respectively: prove that their resultant acts through the pole of the circle ABC.

2799. The great circle drawn through a corner of a spherical triangle perpendicular to the opposite side divides the angle into parts whose cosines are as the cotangents of the adjacent sides, and divides the opposite side into parts whose sines are as the cotangents of the adjacent angles.

2800. Prove that a spherical triangle can be equal and similar to its polar triangle only when coincident with it, each side being a quadrant.

2801. In a spherical triangle $A + a = \pi$: prove that

$$\tan\left(\frac{\pi}{4} - \frac{b}{2}\right)\tan\left(\frac{\pi}{4} - \frac{c}{2}\right) = \pm \tan\left(\frac{\pi}{4} - \frac{a}{2}\right) = \pm \tan\left(\frac{\pi}{4} - \frac{B}{2}\right)\tan\left(\frac{\pi}{4} - \frac{C}{2}\right).$$

2802. Prove the formula

$$\cos\frac{a+b+c}{2} = \frac{\cos A + \cos B + \cos C - 1}{4\sin\frac{A}{2}\sin\frac{B}{2}\sin\frac{C}{2}}.$$

2803. Two sides of a spherical triangle are given in position, and the included angle is equal to the spherical excess: prove that the middle point of the third side is fixed.

2804. Two sides of a spherical triangle are given in position, including an angle 2α, and the spherical excess is 2β; on the great circle bisecting the given angle are taken two points S, S', such that

$$\cos SA = -\cos S'A = \tan(\alpha - \beta)\cot\alpha:$$

prove that, if P be the middle point of the base,

$$\sin\alpha\sin\tfrac{1}{2}(SP + S'P) = \sin(\alpha - \beta).$$

2805. Two fixed points A, B are taken on a sphere, and P is any point on a fixed small circle of which A is pole; the great circle PB meets the great circle of which A is pole in Q: prove that the ratio $\cos PQ : \cos BQ$ will be constant.

2806. Prove that, when the Sun rises in the N.E. at a place in latitude l, the hour angle at sunrise is $\cot^{-1}(-\sin l)$.

2807. In latitude 45° the observed time of transit of a star in the equator is unaffected by the combined errors of level and of deviation in a transit: prove that these errors must be very nearly equal to each other.

2808. The ratio of the radius of the Earth's orbit to that of an inferior planet is $m : 1$, and the ratio of their motions in longitude (considered uniform) is $n : 1$: prove that the elongation of the planet as seen from the Earth when the planet is stationary is

$$\tan^{-1}\sqrt{\frac{1 - m^2n^2}{m^2 - 1}}.$$

2809. The mean motions in longitude of the Earth and of an inferior planet are m, m', and the difference of their longitudes is ϕ: prove that the planet's geocentric longitude is increasing at the rate

$$(mm')^{\frac{1}{3}}\left(m^{\frac{1}{3}}+m'^{\frac{1}{3}}\right)\frac{(mm')^{\frac{1}{3}}-\left\{m^{\frac{2}{3}}-(mm')^{\frac{1}{3}}+m'^{\frac{2}{3}}\right\}\cos\phi}{m^{\frac{2}{3}}-2(mm')^{\frac{1}{3}}\cos\phi+m'^{\frac{2}{3}}};$$

and verify that the mean value of this during a synodic period is m.

2810. The maximum value of the aberration in declination of a given star is

$$20''\cdot 5\sqrt{1-(\cos\delta\cos\omega+\sin\delta\sin\omega\cos a)^2};$$

where a, δ are the right ascension and declination of the star, and ω the obliquity of the ecliptic.

2811. Prove that all stars whose aberration in right ascension is a maximum at the same time that the aberration in declination vanishes lie either on a quadric cone whose circular sections are parallel to the ecliptic and equator, or on the solstitial colure.

2812. The right ascensions and declinations of two stars are a, a'; δ, δ' respectively, and A is the Sun's right ascension at a time when the aberrations in declination of both stars vanish: prove that

$$\tan A = \frac{\tan\delta\sin a - \tan\delta'\sin a'}{\tan\delta\cos a - \tan\delta'\cos a'}.$$

2813. In the Heliostat, if the diurnal change of the Sun's declination be neglected, the normal to the mirror, and the intersection of the plane of the mirror with the plane of reflexion will each trace out a quadric cone whose circular sections are perpendicular to the axis of the Earth and to the reflected ray.

2814. The latitude of a place has been determined by observation of two zenith distances of the Sun and the time between them and each observed distance was too great by the same small quantity Δz: prove that the consequent error in the latitude is

$$\Delta z\cos(a+a')\div\cos(a-a');$$

where $2a$, $2a'$ are the azimuths at the times of observation.

2815. The hour angle is determined by observation of two zenith distances of a known star and the time between; each observed zenith distance is too great by Δz: prove that the consequent error in hour angle is

$$\Delta z\sin(a+a')\div\cos l\cos(a-a');$$

where l is the latitude of the place and $2a$, $2a'$ the azimuths of the star at the two observations.

[See a paper by Mr Walton, *Quarterly Journal*, Vol. v., page 289.]

THE END.

CAMBRIDGE: PRINTED BY C. J. CLAY, M.A. AT THE UNIVERSITY PRESS.

www.ingramcontent.com/pod-product-compliance
Lightning Source LLC
Chambersburg PA
CBHW031933220326
41598CB00062BA/1719
* 9 7 8 3 3 3 7 1 4 1 6 1 5 *